Series in BioEngineering

The Series in Bioengineering serves as an information source for a professional audience in science and technology as well as for advanced students. It covers all applications of the physical sciences and technology to medicine and the life sciences. Its scope ranges from bioengineering, biomedical and clinical engineering to biophysics, biomechanics, biomaterials, and bioinformatics.

Indexed by WTI Frankfurt eG, zbMATH.

Seyed Morteza Naghib

Nano-Biosensors, Drug Delivery and Tissue Engineering

Advances in Nanobiotechnology

Springer

Seyed Morteza Naghib
Division of Nano-Bioengineering Research
and Education (DNBRE)
Department of Nanotechnology
School of Advanced Technologies
Iran University of Science and Technology
Tehran, Iran

ISSN 2196-8861 ISSN 2196-887X (electronic)
Series in BioEngineering
ISBN 978-981-95-1887-6 ISBN 978-981-95-1888-3 (eBook)
https://doi.org/10.1007/978-981-95-1888-3

This Springer imprint is published by the registered company Springer Nature Singapore Pte Ltd.
The registered company address is: 152 Beach Road, #21-01/04 Gateway East, Singapore 189721, Singapore

Competing Interests The author has no competing interests to declare that are relevant to the content of this manuscript.

Contents

Chapter 1
Conductive Polymers-Based on Polyaniline for Biosensing in Cancer Diagnosis: A Practical Approach for Diagnosis in Early Stages

Abstract Developments in biosensor and bioelectronics technology have helped the biomedical profession with diagnosis and treatment. With an emphasis on cancer detection, this approach offers the potential to identify molecular targets as a new generation of diagnostics. Given that cancer is the second primary cause of death from all illnesses, more research is being done in this area with an emphasis on early detection. Traditional cancer screening techniques have low specificity and sensitivity and are expensive, unpleasant, and time-consuming. Recently, scientists have developed a range of biosensors to detect specific molecular markers of cancer. One kind of point-of-care device that patients may use at home is a biosensor. By incorporating nanocomposite (NC) into the electrochemical sensing interface, signal amplification is achieved, increasing detection accuracy and limits. PANI (polyaniline) and its composites found widespread usage in electrochemical (EC) sensors and bio-sensors due to their attractive properties and their ability to be modified by carefully choosing the suitable components. The concepts and detection limit that may aid in the early detection of cancer are described in this work along with a detailed examination of a number of biosensors based on PANI and tumor biological markers that have been developed to discover biomarkers.

Keywords Biosensor · PANI · Cancer · Tumor biological markers

1.1 Introduction

One of the most persistent diseases, cancer has shown the second-largest decline in death rates within the last several decades [1–7]. Accurately predicting the prognosis of the patient is challenging because of the shortcomings of the existing analysis of cancer techniques. Successful cancer therapy implementation depends on early cancer identification [8]. Reliable diagnostic methods must be created since a higher life expectancy and better patient outcomes are associated with early cancer detection [9–16]. In order to do this, cancer is now identified utilizing a variety of methods, including biopsy, CT (computed tomography), MRI (magnetic resonance imaging),

S. M. Naghib, *Nano-Biosensors, Drug Delivery and Tissue Engineering*, Series in BioEngineering, https://doi.org/10.1007/978-981-95-1888-3_1

PET (positron emission tomography), X-ray, and ultrasound [17–19]. However, a number of these approaches are expensive, labor-intensive, insensitive, and inconvenient [20–26]. Moreover, genetic detection methods include the PCR (polymerase chain reaction) and DNA sequencing for locating cancer biomarkers (BM) [27, 28]. Researchers have recently identified cancer-related BM, which can serve as markers for the presence of cancer, using immunoassay methods such as ELISA [29–31]. Despite their high selectivity and sensitivity, these methods are expensive, time-consuming, and often ineffective in detecting BM in the small amounts present early in the illness's progression, which is why Clark developed biosensor technology in the 1960s [32]. Cancer biomarker (BM) is a crucial method that aids in the early detection of cancer [33]. BM are biological substances found in blood, urine, cancer cells, and other biofluids that are used as markers of the health and status of several diseases [34]. These molecules may include DNA, RNA, proteins, and enzymes [35–39]. Biofluid-detected BM are particularly essential for keeping an eye on the efficacy of cancer therapies like radiation therapy and chemotherapy because they might indicate alterations in the expression of a particular protein [40]. There are several cancer BM that can detect different kinds of cancer. The two main groups into which BM may be separated are DNA and protein BM. Protein BM are often used to detect cancer since they are prognostic markers. Although they are unable to detect malignancies in their early stages, DNA markers may also be used to diagnose tumors. Rather, they provide insights about how tumor development has changed throughout time. The costly chemicals needed for every experiment and the sluggish detection rate are two technological drawbacks of the conventional PCR and ELISA-based methods for BM identification. Recent decades have seen the rise in popularity of the biosensing technique as a viable diagnostic tool because to its numerous benefits, which include being highly sensitive, fast, and high-throughput. In the past, identification and detection of BM have often relied on EC biosensors. Because of its ease of use, affordability, quick reaction time, and high sensitivity, biosensors for EC are valuable analytical tools in several medical specialties, encompassing environmental surveillance, food evaluation, and disease diagnosis, including cancer [41].

Analytical tools called biosensors connect biological components, including monosaccharides, nucleic acids, and cells, to a detector or transducer [42, 43]. The detector or transducer of the biomolecules interprets the reaction, transforming the biochemical process that leads to a biological reaction into a signal suitable for computer analysis [18, 44]. A bio-recognition element, an immobilization surface like nanoparticles (NPs) and conducting polymers (CP), and a detector/transducer unit are the three primary parts of a biosensor [45, 46]. The market for biosensors is growing rapidly due to the demand for further reliable, correct, and reasonably priced solutions for diagnostics. The growing frequency of long-term illnesses like cancer and diabetes is one of several factors affecting the sector. Moreover, thanks to technological advancements in this field, biosensor vendors may now offer more sophisticated and efficient solutions, which have further increased as a result of demand and market expansion. The biosensors market was worth 28.80 billion $ in 2022 and is anticipated to rise at a 7.9% CAGR (compound annual growth rate) from 2023 to 2030, reaching 52.93 billion $ in 2030. Due to their distinct polymeric characteristics, stability, and beneficial electrical and mechanical properties, CP, a family of polymers, has inherent electrical conductivity, which makes them among the most promising substances for a range of uses, including energy storage [47], flexible electronics [48], and bioelectronics [49, 50]. CP's market, which was valued at $6.30 billion in 2022, is expected to rise at a CAGR of 6.82% from 2023 to 2030, reaching $10.68 billion. The most intriguing compounds for CP are principally polypyrrole (PPY), PANI, and polythiophene (PTH), and their byproducts [51]. Conductive polymers, and PANI in particular, have some unique advantages, but they also have certain disadvantages. For example, PANI's biological applications are limited because of its poor processing capacity, lack of biodegradability, and rigidity [52, 53]. The worldwide market for PANI is growing due to the increasing demand for PANI in various applications such as electronics, biosensors, and energy storage devices [54].

CV (cyclic voltammetry), SWV (square wave voltammetry), EIS (electrochemical impedance spectroscopy), CA (chromatoamperometry), and DPV (differential pulse voltammetry) are examples of EC techniques. The necessity to increase the sensitivity of the EC biosensors has led to the adoption of novel approaches like the use of NPs (oxide of metal, metal, and based on carbon), nanomaterials, or polymers (membrane of cellulose, chitosan, and CPs) is being discussed. The development of CP, with its high stability, tunable electrical conductivity, and simple and reversible undoping and doping, heralded the emergence of a new class of biosensors [55, 56]. An enzyme further facilitates the transfer of charge between a polymer and itself; the PANI exhibits two redox couples. Additionally, it is not necessary to mix any other diffusers for the biosensing apparatus's electron transport because the PANI works on its own. Because the mediator is localized to the electrode surface, it cannot

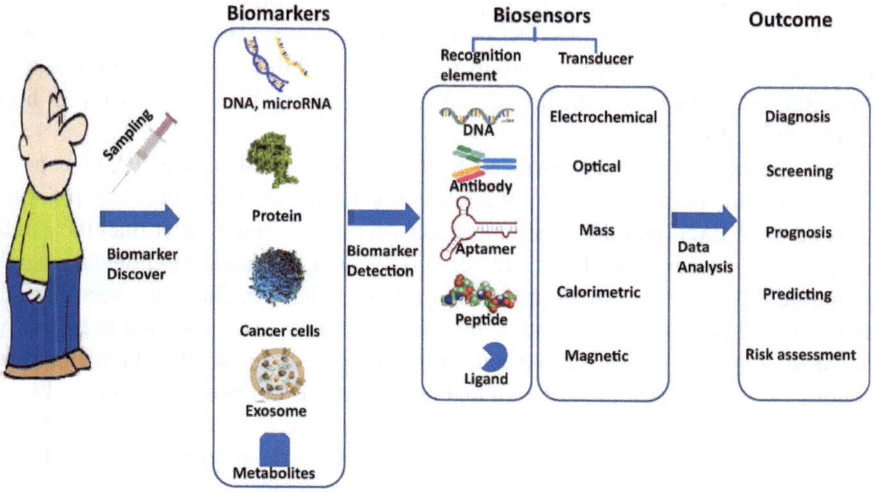

Fig. 1.1 Illustration of the working process of biosensors for cancer diagnosis [58]

leak into the medium, ensuring the biosensor's exceptional long-term stability. The PANI-based NC, with its superior EC properties, can identify biomolecules through optical detection and in vivo biocompatibility, demonstrating extraordinarily high sensitivity and fast reactions. There are many different types of biosensors available today, including piezoelectric, heat, tissue, enzyme, immunosensors, and DNA biosensors [57]. Figure 1.1 illustrates how the biosensor works for cancer detection; identification of BM and biosensor-based BM detection analysis of data are the three primary steps in the procedure [58].

1.2 PANI-Based Materials and Nanostructures

PANI may oxidize into five different forms: nigraniline, pernigraniline, emeraldine, leucoemeraldine, and protoemeraldine. However, MacDiarmid proposed three plausible models for the PANI structure based on the oxidation states [59]. The emeraldine bases (EB) are blue, semi-reduced and semi-oxidized, the LB (leucoemeraldine base) is entirely reduced (achromatic), and the PB (pernigraniline base) is fully oxidized (violet/blue). Benzenoid and quinonoid rings are present in the EB type; only benzenoid rings are present in the LB type, and only quinoid rings are present in the PB type, where the EB is doped (green) to become a conducting emeraldine salt (ES) [46, 60]. The transformation of PANI's EB type, an insulating ES type, into a conductive state depends on the creation of protonated species in the ES, also known as polaronic carriers. Since imines are more fundamental than amines, the imine N is protonated rather than the amine N. A species that is bipolar (both imine Ns are protonated) is therefore produced when proton doping is applied to the ES form, and it undergoes the extremely conductive polarons that are produced by an internal relaxation. When polarons split apart, charge flows, and after acid doping, the electrical conductivity of PANI in its ES form is high. While EB is considered the most stable kind of PANI due to its remarkable stability at ambient temperature. Nevertheless, even after being doped with an acid, LB and PB continue to show high conductivity [61, 62].

Sensing performance is enhanced by the production of several PANI nanoarchitectures, including spherical particles, wires, tubes, and rods (Fig. 1.2) [65–68]. For example, PANI nanofibers are more sensitive and respond faster than their bulk counterparts because they have a larger surface area and some biomolecules can't penetrate as deeply into them [63]. Due to its simplicity in changing from base to salt forms when acid (H^+) or basic (OH^-) is added, PANI is considered to be one of the most appealing CPs from an electrical perspective. Because of its high electrical conductivity, resistance to oxidation and reduction, ease of synthesis and modification, and environmental durability, PANI was one of the greatest numbers of CPs materials that the researchers used. In the past, PANI was produced by oxidatively polymerizing aniline monomers in a medium that is acidic. But as PANI's development and research progressed, a variety of processes and approaches, such as vapor-phase polymerization (VPP), chemical, electrochemical, and enzyme-catalyzed, and

Fig. 1.2 SEM pictures of PANI and its NCs in various morphologies like: **a** spheres [65], **b** cubes. Adapted with permission from [66], Copyright 2009 American Chemical Society. **c** flower-type. Reprinted with permission from [67], Copyright 2014 American Chemical Society. **d** rods [68]

others, were used to manufacture PANI. Numerous methods, including electrochemical polymerizations, heterophase interfacial, self-assembling, and solution, have been used to create and manufacture PANI nanostructures, like nanotubes, nanofibers, nanoflowers, nanorods, nanospheres, and nanogranules. Furthermore, a number of variables and processes, including pH, radiochemistry, template (hard or soft), oxidant or initiator, temperature, electrochemistry, the process of interfacial reaction (chemical oxidation), sonochemistry, and additives made of chemicals, like π-bonding chemicals and oligoaniline, should be considered when designing unique PANI nanostructures [61, 64].

The most conventional and widely used method for large-scale PANI production is chemical oxidative polymerization. However, this method isn't very good at precisely controlling the finished product's chemical and physical characteristics because it has a fast radical reaction that could lead to over-oxidation and/or side-chain branching into the backbone. Although attempts to produce PANI in solvents that are organic have been documented in the literature of science, the three reactants needed for PANI are typically an acidic medium, an oxidant, and an aniline monomer. potassium dichromate ($K_2Cr_2O_7$), potassium iodate (KIO_3), and hydrogen peroxide (H_2O_2), sodium vanadate ($NaVO_3$), ammonium sulfate ($(NH_4)_2S_2O_8$), potassium-free cyanide (K_3 (iron $(CN)_6$), and cerium sulfate ($Ce(SO_4)_2$) are potential oxidizers, while H_2SO_4 and HCl are common acids for PANI fabrication [69]. Oxidative polymerization is the most widely used technique for producing PANI, which involves reacting an aniline with anything acidic to form a polymer and applying a neutralizing substance at various temperatures in the presence of an oxidizing agent like APS or ammonium persulfate. Filtration was the polymerization technique that separated the aqueous solution after realization (3 h), which was washed six times using DI water to obtain pure PANI. To verify that everything is non-reactive and to make the filter colorless, components are eliminated, and acetone and alcohol are then introduced. Because to the presence of weapons, the material, called PANI polymeraldine salt, is notoriously unsteady and slime green. Consequently, at room temperature, the salt of PANI changes into the EB form that is inherently stable, which allows the way to strike a balance with the appropriate concentration of hydroxide of ammonium [70].

Since many applications need the production of polymers with a wide surface area, the EC approach of the synthesis of conductive polymers is essential to the procedure. Since the study of conductive polymers' electrical and optical properties primarily relies on the formation of thin polymer coatings on the electrode, monitoring the electrical conductivity (EC) process of chemically created polymers that conduct electricity presents a challenge. This method of producing conductive polymers is similar to the electrodeposition method used to manufacture metals. Both galvanic and electrodynamic methods were used in the EC operations to manufacture PANI. A simple cell chamber comprising an electrode, a power supply, and a solution of electrolytes is used for the procedure. Compared to the chemical method, the EC approach is more economical, easier to use, and produces a highly uniform and pure polymer that is electrodeposited on the electrode surface. The electro-polymerization process used to create PANI includes the following steps: (1) An ionic positive free radical is made when aniline monomers are oxidized at the anode; (2) the aromatic rings' electrons reorganize, and protons are taken away from the buildings that were made in the earlier phase to make dimers; (3) as these dimers grow, new, stronger structures are made; (4) the polymeric chain that is made in the acidic solution has to activate on its own to make the denatured PANI that results; this is the recommended way to electro-polymerize PANI [71] (Fig. 1.3).

By using enzymes (laccase, glucose oxidase, and horseradish peroxidase (HRP)) that catalyze the production of matching free radicals with aniline, PANI synthesis

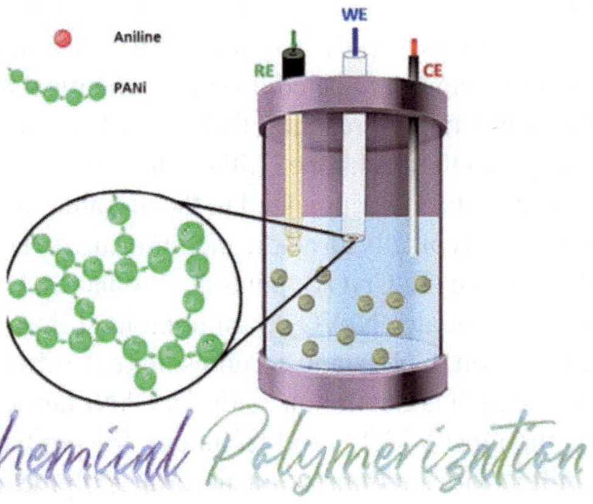

Fig. 1.3 Diagrammatic representation of the PANI EC synthesis system

has shown enzymatic polymerization [72–76]. Heme cofactors, also known as redox-active cysteine or selenium cysteine residues, make up the active sites of peroxidases. In order to convert peroxides to alcohols or water, peroxidases mainly use (i) a phase of dual electron transfer or (ii) two successive, clearly distinct steps of solitary electron transport. The optimum substrate Certain cell oxidation processes produce hydrogen peroxide, a common peroxidase. Hydrogen peroxide acts as an oxidant in oxidative polymerization processes, generating Cation radicals and CP that are prone to polymeric structure formation. Low molecular weight branching polymers have been found in many examinations, despite the fact that the peroxide is converted into water throughout this process, rendering it ecologically innocuous. To overcome these problems, Samuelson et al. used different electrolytes and templates, such as poly(styrene sulfonate) (PSS) to create The HRP synthesis mixture catalyzes the formation of regular polymeric chains inside the PANI. Three tasks are carried out by PSS in this process: (i) To overcome these problems, Samuelson et al. used different electrolytes and templates, such as poly(styrene sulfonate) (PSS), to create The HRP synthesis mixture catalyzes the formation of regular polymeric chains inside the PANI [77]. (ii) It provides important doping during PANI's activation phase, which produces ES, which conducts electricity. (iii) This method aids in making the PANI/PSS product soluble in water in an acidic solution by polymerizing the polymer chains. Since H_2O_2 is present, HRP may initiate the oxidation process of a variety of compounds, such as phenols and aromatic amines, to generate the related free radicals [78–82].

In this instance, the native enzyme (HRP) equivalents from hydrogen peroxide receive two oxidizing agents, creating a transitional HRP I. HRP I (RH) oxidizes the foundation, which is either an aromatic amine monomer (RH) or phenol in this instance, by doing two consecutive reductions of one electron operation. Aromatic amine or phenol may produce the radical species represented by the symbol R*. These free radicals pair to create the dimer after further coupling and oxidation events, which ultimately lead to the polymer's synthesis. However, since the polymer forms in liquid solutions after precipitating out, enzymatic polymerization has a major drawback in that it only produces a very small number of polymers with low molecular weights, also known as oligomers. To get around the problem and Several modified enzymatic polymerizations, such as solvent mixes, air-water interface polymerizations, changed monomers in aqueous solutions, reverse micelles, and micelles, have been studied to increase processability [83].

Using vapor phase polymerization, aniline monomer is introduced to an oxidant-coated substrate in vapor form to create PANI. The polymerization occurs at the oxidant vapor contact. Chemical and electrochemical vapor phase polymerization (CVPP and EVPP) are two types of vapor phase polymerization (VPP). The very consistent conductive PANI produced by the solvent-free CVPP technique enables the creation of PANI layers of any thickness on an insulating substrate. VPP prevents particle agglomeration by applying monomers as vapor rather than solution or liquid. VPP produces intrinsic conjugated polymers that are free of agglomeration and have good electrical conductivity. VPP offers very pure polymers with exceptional conductivity and synthesis range at the nanoscale [84].

When PANI with a desired shape is required, template synthesis is a helpful alternative. This technique may use a soft, hard, or a combination of the two templates to create PANI nanostructures with the appropriate dimensions. Although the yield is a little lower than with other conventional chemical methods, the key element of the synthesis based on templates is the authority above the necessary parameters (length, diameter, orientation, etc.). In each case, the polymer grows on the surface of the template before being eliminated or demolished at the end of the response. Most commonly used templates are made of organic polymers and metal oxides, with a wide range of hard and soft templates often used to control the size and shape of PANI. Nevertheless, harsh patterning techniques need post-processing to eliminate the design, making it challenging to keep the pattern; the structured order has been eliminated. So, to create PANI, the soft template approach has been frequently used with nanostructure through a technique that self-assembles because it is easy to use and doesn't harm the environment. Surfactants are called soft templates, whereas Polystyrene (PS) microspheres, silica spheres, and particles of polymethyl methacrylate (PMMA) are often utilized as hard templates. In an acidic solution, methyl orange (MO) may efficiently form supramolecular aggregates by self-assembling. It is possible to create supramolecular aggregates like flakes and dendrites by varying hydrochloric acid to MO molar ratio. Cao's straightforward, single-step synthesis method achieves large PANI structures' aspect ratios (up to 20) and may be used as soft templates in polymerization. Compared to spherical PANI nanoparticles, MO improves the form of the microtubules [85].

Interfacial polymerization created the bio-inspired PANI hierarchical structure on the pore surface of the polymer film. The benzoyl peroxide (BPO)-containing porous polystyrene (PS) honeycomb patterned (HCP) film was made in a humid atmosphere using the BF or (breath figure) technique [86]. Due to its great pH sensitivity, the film may be used in many biosensing applications. Miao produced hollow nanofibers of PANI by using the polymerization of aniline in situ using a membrane made with electrospun poly(amic acid) as a template. Nanofibers that have tunable thickness of the wall demonstrated 601 F/g of specific capacitance at 1 A/g for supercapacitors [87]. Pang uses a direct oxidative polymerization method for chemicals in situ that uses urea as a soft model. The PANI-3 nanotubes that were made have square capillaries that are stable and normal, with a 3:1 molar ratio for aniline and urea. The results of facile ion transport, significant electrode-electrolyte contact, and increased electroactive areas throughout the electrochemical process ultimately contribute to improved electrochemical performance and low internal resistance [68]. Kaneda produced using electrochemical techniques for polymerization to produce unique microporous PANI structures. This method creates a foam template in a solution of electrolytes with surfactant by ultrasonography using a constant-potential technique. By electrochemically polymerizing aniline, this foam template subsequently facilitates the development of a microporous PANI nanostructure (Fig. 1.4) [88]. There were a lot of interesting results and PANI shapes when different types of surfactants were added to the previous polymerization system, as expected. Specifically, transparent PANI rods, spheres, cubes, flower-like shapes, cylinders, and honeycombs were each individually prepared in a 0.010 M HCl solution containing PVP and

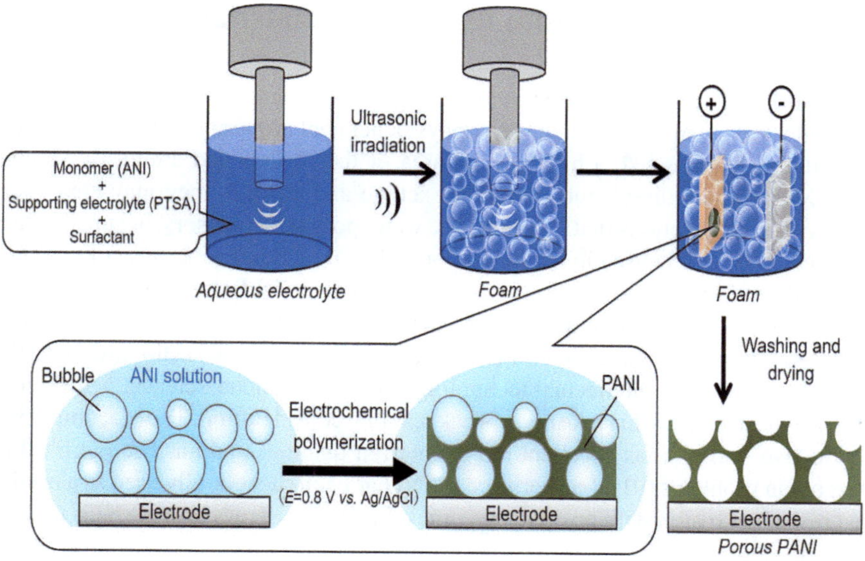

Fig. 1.4 Ultrasonic foam-templated electrochemical polymerization technique for producing porous PANI nanostructures [88]

PEG and different amounts of SDS, SLS, CTAC, CTAB, and SDBS; the first two are non-ionic, the next three are anionic, and the last two are cationic [89].

Because of its numerous beneficial properties, including its cost, stability, and ease of synthesis; a broad spectrum of conductivity upon doping; and a distinct technique of protonic doping, PANI is the most extensively studied electrical area CP. However, PANI's processability is limited in two ways: When doped, the polymer's powdered form is insoluble in any common organic solvents and breaks down at high temperatures, resulting in poor mechanical power. A number of solutions have been investigated to get around the process-ability PANI restrictions, and the two most popular ones are blending and compositing. Over the years, a lot of research has been done to find out whether conducting PANI fillers can take the place of conventional conductive inorganic additives, like transmitting carbon black and metal particles. According to tests of the PANI's ionic liquid's electrical and thermal conductivity, the polymer is classified as a semiconductor of the p-type. The PANI's conductivity is often greater than 10 S/cm in undoped EB to doped ES and lower than 10−10 S/cm. PANI has been used in a number of studies and shows a keen interest in magnetic behavior due to its high spin density. During the oxidation process, PANI that conducts electricity is doped with hydrochloric acid, giving rise to the nature of paramagnetics that is directly and closely related to the PANI's electrical and magnetic properties [90].

1.3 Cancer Detection with PANI-Based Materials

Because PANI has many surface morphologies (such as nanoflowers, nanofibers, and nano-leaves), it exhibits a range of physiochemical properties. Even if they have attractive properties like a high special area of the surface and are easily soluble in PANI's organic dissolvents, including ketone and alcohol, they may also boost reactivity in applications that don't include water pollutants, sensors, or super capacitors. PANI is widely used in the field of optoelectronic devices because of its semiconducting characteristics, like light-emitting diodes (OLEDs) and organic solar cells, excellent electrical and optical properties, and increased heat stability. Since medicine's primary focus right now is engineering, developments in this area need the development of new intellectual technology. To advance neuroscience, neuroscientists need instruments that take nerve weakness into consideration. The biocompatible conductive scaffold has excellent bio-counterfeit qualities and has been utilized to treat organ problems [91]. Furthermore, PANI has been successfully used as a barrier that resists corrosion with positive outcomes. Figure 1.5 shows how PANI is used in a variety of areas [92].

Compared to its more widely used counterpart, rechargeable batteries, supercapacitors, one of the most anticipated energy storage technologies, offer a number of advantages. These sophisticated intermediate energy units comprise the line that separates PANI electrically powered supercapacitors and carbon-based supercapacitors. PANI is appropriate for capacitive applications due to its rapid redox process in the material's bulk. Because of its porous and ionic surface, PANI works nicely for usage as a support catalyst with regard to various metals and metal complexes.

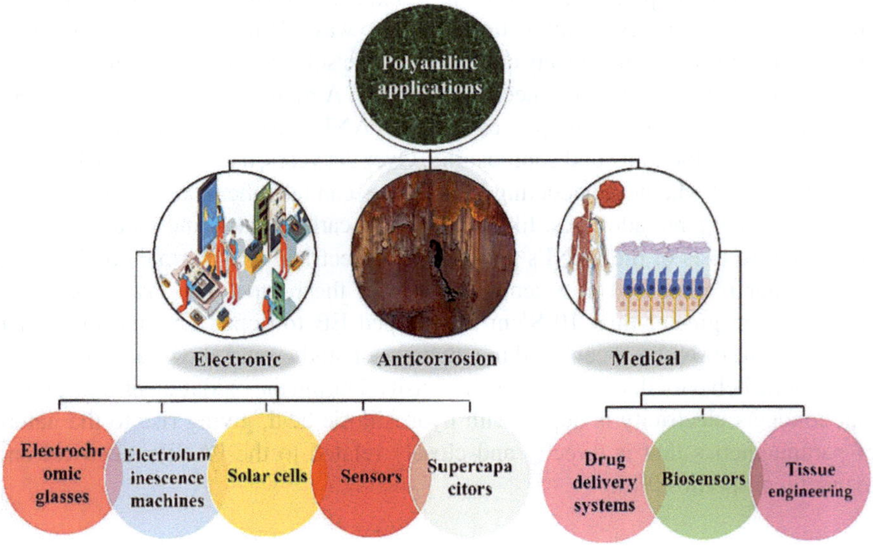

Fig. 1.5 Applications of PANI [92]

TiO_2, Pd, Pt, Zn, Ni, and Fe are among the metals and metal oxides that are used [93, 94]. Because of the elevated demand and restricted access to fossil fuels, energy is one of the most important global issues. In this regard, PANI is a very promising contender as a cost-effective, environmentally friendly substitute for conventional materials in the construction of competitive, diversified energy devices such as fuel cells, light-emitting diodes, solar cells, and supercapacitors. Because of their biological characteristics that can be tuned, medical professionals have looked into the therapeutic and diagnostic applications of PANI-NCs. The primary applications are organ transplantation, tissue engineering, and photodynamic therapy [95].

Unchecked cell division ultimately led to the development of a mass known as a tumor. Tumors may be categorized as either malignant or benign, with the latter eventually developing into cancer. Lung cancer is a serious disease that is challenging to treat because of its natural propensity to metastasize in its early stages. Although tumors of lung cancer may affect any bodily organ, they particularly prefer to target the liver, bones, brain, and adrenal glands. However, it has been shown that 90–95% of lung cancers originate from the cells of the epithelium that line the airways, both big and little (bronchioles and bronchi); consequently, lung cancers are often called cancers or bronchogenic carcinomas. Over the years, 5-year lung cancer patient survival rates have been continuously low around 15% for men and 22% for women. CT and traditional chest X-rays are the most often used imaging modalities for tumor diagnosis; however, they have limitations. Lung cancer comes in two primary forms: NSCLC (nonsmall cell lung cancer) and SCLC (small cell lung cancer) [96].

Approximately 80% of all instances of lung cancer are classified as squamous cell carcinoma, large cell carcinoma, adenocarcinoma, and mixtures. Furthermore, SCLC, which is divided into three categories combined carcinoma with tiny cells, mixed large/small cell carcinoma, and small cancer of the cell represents almost 20% of instances of lung cancer, with SCLC having a high likelihood of metastasis [97]. Chemotherapy/radiation therapy is often used to treat two primary forms of lung cancer, NSCLC and SCLC, respectively. CEA, CYFRA-21, NSE, and VEGF165 are important and typical lung cancer BM. Scientists first identified carcinoembryonic antigen (CEA), about 200 kDa in molecular weight; this glycoprotein is transmembrane, which is intricate and heavily glycosylated, and is a crucial cell adhesion protein that is a member of the immunoglobulin superfamily [98]. CEA serum started to act as a cancer biomarker when its levels were above the usual maximum of the typical values employed in studies, which is around 5.0 ng/ml; structural proteins called cytokeratins resemble keratin intermediate filaments, which are 10–12 nm long. Among other biological functions, these filaments sustain the membrane of plasma, endure mechanical strain, and regulate migration and signaling as the components of epithelial intermediates. A part of cytokeratin 19, which is typically seen, is referred to as cytokeratin 19 fragment 21-1 (CYFRA 21-1) in cells of epithelial origin. The glycolytic process, which is mostly mediated by the enzyme neuron-specific enolase (NSE), breaks down glucose to create NADH (nicotinamide adenine dinucleotide) and ATP (adenosine triphosphate). Generally speaking, cut-off values for the NSE for serum, or the highest level of serum NSE needed to distinguish distinct lung malignancies [99].

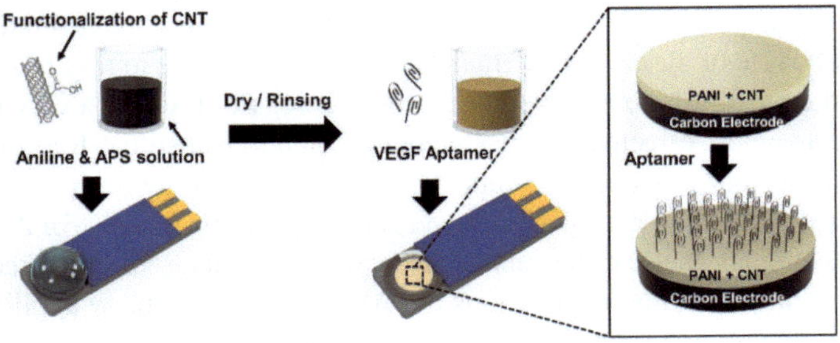

Fig. 1.6 Diagrammatic representation of the assembly of the VEGF165 aptamer and the fabrication of the PANI/carbon nanotube (PANI/CNT) nanocomposite on the sensor surfaces for the VEGF sensors [102]

The sensitive CYFRA 21-1 was recently attached to a combination of Au/PANI/CdS that resembles a raspberry and established detection of the Ab2-Au/PANI/CdS bioconjugate in blood, making use of a microfluidic sensing platform, with the NC bioconjugate being a signal probe [100]. A genosensor was developed using PANI with reduced graphene oxide in three dimensions (3D-rGO). SsDNA covalent immobilization on a GC electrode with composite modification provided surface-sensitive identification of the CYFRA 21-1 sensor, and then a dsDNA probe was generated by adding cDNA. The rGO-PANI composite provided the DNA immobilization with a stable microenvironment and surface, enabling it to maintain its biological processes and activities [101]. In related research, the tumor indicator VEGF165 was found utilizing an electrode based on CNT/PANI, where aniline was polymerized at the CNT edge sites using APS to fabricate the device. Aptamers' selectivity for the target biomolecule and the complementing PANI/CNT surface area impacts were credited with the electrode's high sensitivity (Fig. 1.6) [102].

PANI and poly (1, 2-diaminobenzene) were used to build a two-layer surface treated with polymers for the aim of detecting NSE. The initial phase of electropolymerization included the usage of poly (1, 2-diaminobenzene), and in the subsequent stage of electropolymerization, PANI was conducted and deposited. According to the AC impedance approach, the range of lines of the sensor was 0–50 pg/Ml. The successful synthesis of a novel pCBMA-PANI, NC poly (methacrylate of carboxybetaine), led to its use as a surface for modification in the development of the CEA biosensor. Because of the exceptional antifouling ability of pCBMA and the better PANI electrical conductivity, the sensor demonstrated high sensitivity and minimal fouling characteristics. To directly detect miRNAs, RuO_2-NC-initiated aniline polymerization constituted the basis for the EC electrode. It was discovered that the sensor system's use of polymerization for amplification was a crucial component that greatly raised the sensor sensitivity [103].

The American Cancer Society has released projections in 2023 that about 41,210 new instances of intrahepatic bile duct cancer and 13,230 women and 27,980 men will be detected in the US. These illnesses will take the lives of around 29,380 persons (10,380 women and 19,000 men) [104]. An increased risk of liver cancer is associated with cirrhosis of the liver, hepatitis C or B virus infection (HCV or HBV), smoking, alcohol misuse, and a fatty liver condition that is not alcoholic, and other elements that harm the parenchyma of the liver [105]. In 2018, around 826,000 cases of hepatocellular carcinoma (HCC) accounted for 80% among all instances of primary liver cancer, with liver cancer being recorded worldwide. According to estimates, 905,677 people (4.7% incidence) will receive a liver cancer diagnosis in 2020, making it the sixth most common disease globally. According to predictions, liver cancer will rank as the third most common cause of cancer-related fatalities worldwide in 2020, accounting for 830,180 deaths (8.3%) compared to other cancers [106]. Several tactics that have been proposed to solve the dilemma of liver cancer include the discovery of novel and targeted diagnostic BMs, the use of sensitive progress technologies for analysis that are essential for the early identification of HCC, and compliance with monitoring initiatives to identify vulnerable people. To corroborate the diagnosis of HCC, alpha-fetoprotein (AFP) is often used as an additional BM. The most studied biomarkers in HCC are AFP, agglutinin-reactive lens culinaris alpha-fetoprotein (AFP-L3) [107], and DCP, or des-gamma-carboxy prothrombin [108].

AFP, a highly specific and sensitive BM, detects HCC; the molecular weight of a glycoprotein is around 67 kDa, produced from tissue cells of the embryonic endoderm. The content in the blood of healthy people is 5–10 ng mL^{-1}, but it significantly rises in the serum of individuals with carcinoma of the liver [109]. By isolating total AFP, three different types of glycoforms AFP-L3, AFP-L1, and AFP-L2 can be produced. People having chronic hepatitis B (CHB) or cirrhosis make AFP-L1, which is the main glycoform made by noncancerous hepatocellular. Since HCC cells that express higher levels of this protein have a tendency for intrahepatic metastases and early vascular invasion, this glycoform does not bind to LCA, which currently has a limit of 10% for AFP-L3. Yolk sac tumors make AFP-L2, which binds strongly to LCA. This protein can be found in the mother's blood throughout the course of a typical pregnancy. In DCP, prothrombin, a coagulation factor produced by the liver, is abnormally expressed, and there is no interaction between DCP and other coagulation factors. Some of the N-terminal domain's glutamic acid residues of prothrombin are preserved by its predecessor, DCP [110].

The researchers developed an EC immunosensor using PANI in 3D with macropores for the measurement of AFP. Several AFP Abs could be immobilized using this modified electrode because it has 3.4 times the electroactive surface area compared to a flat electrode modified with PANI. Without the need for additional redox reactions, the natural redox signal of PANI generated the EC signal of this sensor. The porous PANI-based immunosensor showed a wide linear range under ideal conditions [111]. Hui et al. made use of AFP immunosensors in a PANI nanowire-based hybrid construction of EC, AuNPs, and PEG (polyethylene glycol) in a different study. The AuNPs/PANI electrodes that were changed were then used to immerse the 4-armed PEG-amine. PEG-NH2 and AuNPs groups interacted to form a very non-fouling network of cross-linked hydrated polymers by drop-casting PEG into the gold-PANI. This device has a broad linear range and an ultra-low threshold for detection [112]. The AFP device was developed by Feng et al. HFG/PANI/rGO/PB is PANI, rGO, prussian blue, and a hierarchically flower-like gold microstructure composite substrate. PANI/rGO was able to effectively halt the PB leak, whereas HFG permitted the large quantity immobilization of the Abs [113].

Ag NPs are also used to detect AFP. Amarnath created an EC biosensor for AFP detection by altering the surface with Ag NPs, bovine serum albumin, and PANI. The signal probe was made by conjugating silver nanoparticles with HRP-secondary Ab (HRP-Ab2) included. Then, HRP catalyzed the reaction of hydroquinone (HQ) and H_2O_2 and produced an EC signal; for AFP detection, the sensor's detection limit was 4.7 pg/mL. The thiols in the PANI backbone covalently functionalize carboxylic acid groups; Anbalagan also created carboxylic acid-tethered PANI (PCOOH), which increased the effectiveness of Ab immobilization. The AFP antigen, immobilized on the GCE modified by PCOOH, was captured with a low limit of detection of 2 pg/mL. An additional HRP-labeled Ab catalyzes the oxidation-reduction reaction of H_2O_2/HQ; Zhao developed a different immunosensor for accurate AFP detection. High carbonized nanospheres (HCS)-PANI are used as an EC sensing platform to enhance AFP conductivity and are adsorbed on the surface of the electrode. PANI mediated by protonic acid is used as an electron transfer to further amplify EC signals since it has strong EC activity. Under optimal conditions, the EC immunosensor's linear range is 0.001–80 ng/mL with a limit of detection of 0.33 pg/mL (Fig. 1.7) [114].

There are several factors of risk that might raise the possibility of breast cancer, like age, estrogen, sex, family history, poor living habits, and gene alterations [115, 116]. Important and conventional indicators for breast cancer include UBE2C, BRCA1, MUC1, and HER2. The EGFR is the first stop in the avalanche of signals that regulates cell division and proliferation [117, 118]. HER2 receptors are overexpressed in breast

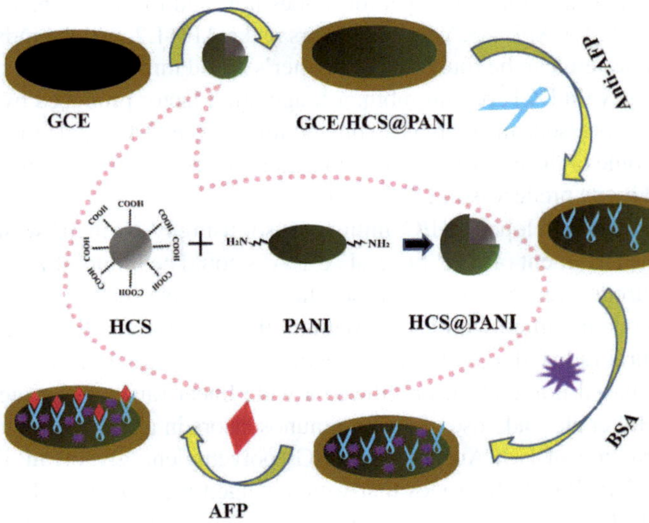

Fig. 1.7 The schematic illustration of the label-free electrochemical immunosensor: preparation process of HCS@PANI the assembly [114]

cancer in 15–30% of cases. HER2 blood concentrations in patients with breast cancer vary from 15 to 75 ng/mL, which is greater than in healthy individuals (2–15 ng/mL) [119]. The gene linked to breast cancer susceptibility, also known as BRCA1, is a human tumor-suppressor gene that aids in the repair of DNA damage [13]. Researchers have shown that a significant chance of BRCA1 mutation is associated with 40–50% of cases of inherited breast cancer. A crucial component of inherited breast cancer that regulates the cell cycle is the oncogene ubiquitin-conjugating enzyme 2C (UBE2C) [120]. UBE2C primarily helps regulate the checkpoint for the cell cycle and the mitotic spindle. The mucin family of glycoproteins with a high molecular weight expresses on the apical cell surface of epithelia. Compared to healthy cells, tumor cells exhibit overexpression and a marked absence of cell polarity [2]. There are several members of the human mucin (MUC) family, including the transmembrane and secretory forms of MUC1 through MUC21. MUC5B, MUC6, MUC5AC, and MUC2 are the secreted mucins. In healthy human serum, MUC1 expression is typically low [115, 121].

Researchers constructed an EC biosensor for breast cancer by modifying and using chitosan-co-PANI copolymer for indium tin oxide (ITO) and attaching complementary DNA (cDNA) probes that have 42 bases associated with the BRCA1 finding [15]. The present amperometric sensor for the modification of electrode dropped according to the cDNA's linear-chitosan-co-PANI modification appeared when the single-stranded target DNA is existed. Furthermore, the cancer-related BM was assayed by PANI NFs; sadly, problems like biofouling may result from the adsorption of non-specific proteins on electrodes. By decreasing the target analyte's adsorption, this might impair the biosensor's function, which is dependent on the anti-fouling surface [11]. The researchers aimed to develop an electrochemical sensor for the detection of UBE2C, a biomarker associated with breast cancer progression. In this sensor, anti-UBE2C antibodies were immobilized using glutaraldehyde on a PANI (PANI) layer modified on a glassy carbon electrode, which has high biocompatibility and conductivity. The sensor was able to provide a wide detection range with a very low detection limit. Also, experiments with MCF-7 breast cancer cell extracts confirmed the high selectivity, reproducibility, and stability of the sensor [122].

Another study uses the unique design of AgNPs-PANI-functionalized graphene (NFG) doped with nitrogen, which significantly affects signal gain, to produce an extremely sensitive sensor without a label for the identification of miRNA-21, a recognized biomarker for breast cancer. By applying PEG to the PANI NFs surface, Hui et al. were able to resolve this problem. The BRCA1 test specifically adsorbed the single-stranded capture DNA, avoiding fouling and enabling subsequent interaction with MB (methylene blue). After that, hybridization of the target DNA, releasing MB and creating double-stranded DNA, both of which showed decreased signals in the EC assembly [123]. Xia et al. successfully developed a DNA biosensor without labels that uses PANI NC@3D-reduced GCE-GO for the very sensitive BRCA1 detection. Biosensor nanotechnology has the potential to directly identify BRCA1 in real blood samples [124]. Kaur made an impedimetric immunosensor based on silver-loaded copper oxide-linked PANI (Ag-CuO@PANI NC) that can very accurately find MUC1 (Fig. 1.8) [125]. MUC1 is predominantly overexpressed in multiple malignancies, especially breast cancer, making its reliable, cost-effective, and convenient detection essential in healthcare. Biosensors, such as the developed impedimetric immunosensor based on silver-doped copper oxide conjugated polyaniline (Ag-CuO@PANI) nanocomposite, offer promising advantages including point-of-care diagnosis while overcoming limitations of conventional methods. The Ag-CuO@PANI nanocomposite has been synthesized hydrothermally, deposited on an FTO platform, and linked covalently with MUC1-specific antibodies via a cysteamine linker for sensitive detection using electrochemical impedance spectroscopy (EIS). The immunosensor demonstrated excellent analytical performance with a broad linear dynamic range (1–100 pg/mL and 1–100 ng/mL), a low detection limit of 3.23 pg/mL, high sensitivity (0.68 kΩ pg mL^{-1} cm^{-2}), reproducibility, selectivity, and stability for 25 days at 4 °C, along with satisfactory results in spike-in recovery tests, highlighting its potential for clinical applications in MUC1 detection.

The mean Patients with PCa are 65 years of age or older, indicating that age is a contributing factor to the illness. Since PCa patients sometimes do not show symptoms until much later in the illness, early PCa identification may enhance the prognosis and hopes of curative therapy. PCa metastasis, advanced prostate tumorigenesis with local invasion, localized prostate carcinogenesis, and prostate intraepithelial neoplasia which is very difficult to treat are the four phases in which prostate tissue in normal condition is transformed into a tumor [25]. Since 1991, the primary BM for PCa has been prostate-specific antigen, or PSA. PSA measurements have enhanced PCa prediction when paired with digital rectal examination (DRE) [12]. However, certain protein BM candidates seem potential for predictive or diagnostic patient screening. These include human glandular kallikrein 2 (hK2), serum amyloid

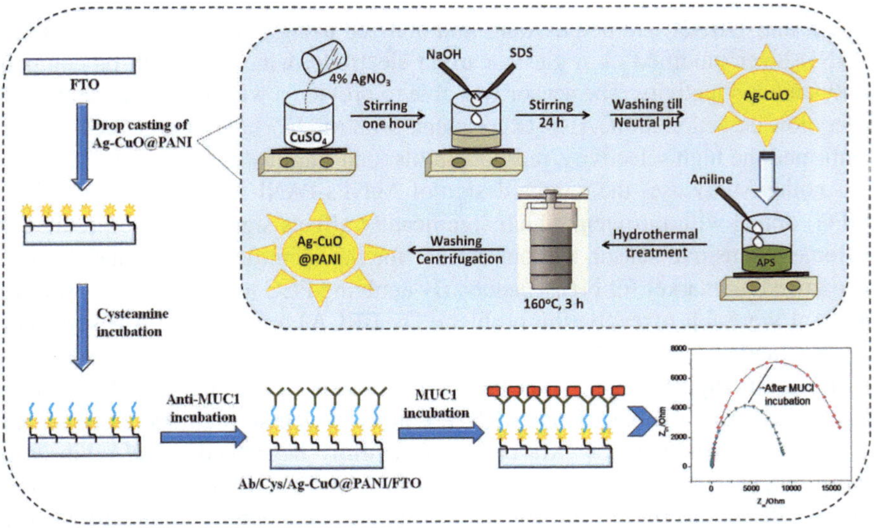

Fig. 1.8 Schematic diagram of MUC1 biosensor fabrication based on Ag-CuO@PANI [125]

A (SAA), annexins 3 (Annex 3), PSA, microsomalinoprotein-beta (MSMB), and Engrailed-2 (EN2).

To measure PSA in raw samples, Vural et al. developed an immunosensor that is a disposable amperometric sandwich immunosensor, this work uses a PGE modified with PANI (pencil graphite electrode), AuNP and PNT (peptide nanotubes), similar to the use of HRP which gave it the labels anti-PSA and anti-PSA. With an RSD of 5.6%, the immunoassay demonstrated excellent repeatability, robust stability, and a range of lines of 1–100 ng/mL^{-1} with a LOD of 0.68 ng/mL^{-1} [126]. It was found that the highest RSD was 14.7% and the largest relative error was 10.9%. This immunoassay is also less costly than the ELISA method and takes 65 min to finish. Additionally, Tezerjani et al. have created a modified GO and PANI-based PCa detection method using an amperometric DNA biosensor. CV and EIS were used to examine the combination of complimentary threads at the electrode surface altered by PANI/GO. A limit of detection of 3.3×10^{-17} M and a wide range of 1.0×10^{-16} M to 1.0×10^{-8} M. For this NC platform, there have been reports [127]. Dey et al. also used PANI nanowires and an AuNP-altered Au electrode to create an immunosensor that can detect PSA without using any mediators. After immobilizing the anti-PSA introduced PSA as the analyte on the altered gold electrode and measured different PSA concentrations using the DPV mode. In fact, the results showed that PANI-AuNP improves the transfer of electrons and provides a valuable anti-PSA surface area immobilization [128].

CEA is a 200 kDa (MW) protein that is heavily glycosylated and is among the most popular BCs for the assessment of prognosis and monitoring of malignant growths, encompassing liver, pancreas, breast, colorectal, and stomach malignancies. CEA is quite low in healthy human blood, but it rises to more than 5 ng mL^{-1} when cancer cells proliferate. Numerous methods, such as colorimetric, luminous methods, EC, fluorescent, and surface-enhanced Raman scattering, have been used to detect CEA [129]. An EC sensor is among the most effective techniques for detecting Due of its extreme sensitivity, CEA, fast response, portability, and simplicity of use. Sun and associates used a composite hybrid reticular, composed of PANI nanowires filled with an ionic liquid (IL) and grown on PEDOT (poly (3, 4-ethylenedioxythiophene). The redox current of PANI, which is typically measured at 0.16 V, provides the analytical information [130]. In a different work, Amarasiri developed the Fe_3O_4, PANI-immobilized screen-printed carbon electrode, or SPCE, and gold nanoparticles, or AuNPs, to rapidly evaluate the CEA BM. PANI was electropolymerized onto the electrode substrate while Fe_3O_4 NPs were drop-cast onto it [131]. Wang recently the few-layered MXene (f-MXene)-PANI was decorated using the EPD, or electrophoretic deposition technique, in combination with fluorine-doped glass made of indium tin oxide (FTO) [132].

Ganganboina et al. developed the N, S-GQDs@Au-PANI graphene quantum dots-Au-PANI nanowire immunosensor filled with nitrogen and sulfur, with the goal of CEA detection. The AuNP particles are between 11 and 4 nm in size and are spread out evenly on the PANI nanowire surface, which has a diameter of 30–50 nm. This is because the S-GQD attachment needs a large active area to work. The uniform distribution of AuNP throughout the PANI surface may enhance the quantity of Au-PANI NC's active sites [133].

1.4 Conclusion

With a focus on the advantages associated with structure, synthesis, properties, and potential uses in a variety of cancer biosensor applications, this article offers a comprehensive summary of NCs and their PANI. NCs based on conductive PANIs are a new and intriguing subject of research that has not yet been thoroughly investigated due to unresolved technical challenges and encouraging advancements within the biomaterial science sector. Advanced study and more thorough examination are needed in this field. In fact, PANI is gaining more and more attention from the scientific community. It's important to remember from a business standpoint that a number of these challenges also have a major detrimental impact on quality assurance and regulatory issues, which contributes to the paucity of clinical data about PANI usage. One of the most encouraging discovery methods for the purpose of diagnosing cancer clinically is EC biosensors. Outstanding cost, portability, robustness, selectivity, ease, sensitivity, and speed of detection are only a few of their advantages; nevertheless, their stability is sometimes limited to a narrow temperature range. To tackle this issue, it is essential to maintain the EC samples and biosensors that are as stable as possible at the right temperatures. In the production of EC biosensors, PANI composites and PANI have been extensively used as substrates for electrodes. Electron delocalization across the polymer backbones is probably the reason for PANI's excellent charge transfer and EC redox efficiency properties. These occurrences point to methods for improving the EC biosensors' stability, sensitivity, and selectivity. However, most measurements were conducted in traditional settings, and just a few papers employed in-vitro research. Biocompatibility, interferences, nonspecific adsorption, and the use of EC biosensors are some of the problems that continue to impede the development of POC diagnostic devices. The creation of ultraselective and sensitive interface materials is one method to address these problems and make it possible to transfer in vitro research to in vivo and clinical research. Future advancements for these biosensors made of EC will focus on making wearable, non-invasive, or minimally invasive, and cost-effective diagnostic tools smaller and shorter sample processing times. Additionally, the development of ultra-small arrays and multiplexed biosensors and label-free EC biosensors may help with tumor diagnosis, disease prevention, clinical treatment, and patient health data.

References

1. H.A. Saputra, K.A. Jannath, K.B. Kim, D.-S. Park, Y.-B. Shim, Conducting polymer composite-based biosensing materials for the diagnosis of lung cancer: a review. Int. J. Biol. Macromol. **126149** (2023)
2. O. Abazari, P. Dayati, M. Shahidi, J. ZavarReza, M. Rahmanian, S.M. Naghib, Exploring a desirable quadr-mRNAs panel for non-invasive and ultrasensitive bladder cancer diagnosis: in-silico and clinical studies. Curr. Med. Chem. (2024). https://doi.org/10.2174/010929867 3302140240627053904
3. N.A. Abtahi, S.M. Naghib, S.J. Ghalekohneh, Z. Mohammadpour, H. Nazari, S.M. Mosavi, S.M. Gheibihayat, F. Haghiralsadat, J.Z. Reza, B.Z. Doulabi, Multifunctional stimuli-responsive niosomal nanoparticles for co-delivery and co-administration of gene and bioactive compound: in vitro and in vivo studies. Chem. Eng. J. **429**, 132090 (2022). https://doi.org/10.1016/j.cej.2021.132090.
4. N.A. Abtahi, S.M. Naghib, F. Haghiralsadat, J.Z. Reza, F. Hakimian, F. Yazdian, D. Tofighi, Smart stimuli-responsive biofunctionalized niosomal nanocarriers for programmed release of bioactive compounds into cancer cells in vitro and in vivo. Nanotechnol. Rev. **10**(1), 1895–1911 (2021)
5. S. Afereydoon, F. Haghiralsadat, N. Hamzian, A. Shams, M. Hemati, S.M. Naghib, M. Shabani, B. Zandieh-Doulabi, D. Tofighi, Multifunctional PEGylated niosomal nanoparticle-loaded herbal drugs as a novel nano-radiosensitizer and stimuli-sensitive nanocarrier for synergistic cancer therapy. Front. Bioeng. Biotechnol. **10**, 917368 (2022)
6. M. Akhlaghi, M. Taebpour, N.N. Lotfabadi, S.M. Naghib, N. Jalili, L. Farahmand, B.F. Haghiralsadat, M. Rahmanian, D. Tofighi, Synthesis and characterization of smart stimuli-responsive herbal drug-encapsulated nanoniosome particles for efficient treatment of breast cancer. Nanotechnol. Rev. **11**(1), 1364–1385 (2022)
7. M.A. Farzin, S.M. Naghib, Laccase-mimicking Cu@CDs-enabled bioinspired immunosensor using CuO nanoparticles for voltametric detection of serum thyroxine level in patients with hyperthyroidism or hypothyroidism. Sens. Actuators Rep. **9**, 100291 (2025). https://doi.org/10.1016/j.snr.2025.100291
8. J. Kappen, M. Skorupa, K. Krukiewicz, Conducting polymers as versatile tools for the electrochemical detection of cancer biomarkers. Biosensors **13**(1), 31 (2022)
9. M.A. Farzin, S.M. Naghib, N. Rabiee, Bio-inspired and biomimetic composites based on biodegradable polymers for sensing applications with emphasis on early diagnosis of cancer. Chem. Eng. J. **493**, 152445 (2024). https://doi.org/10.1016/j.cej.2024.152445
10. H.R. Garshasbi, S.M. Naghib, Smart stimuli-responsive alginate nanogels for drug delivery systems and cancer therapy: a review. Curr. Pharm. Des. **29**(44), 3546–3562 (2023). https://doi.org/10.2174/0113816128283806231211073031
11. M. Hosseine, A. Bakhshi, S.M. Naghib, N. Rabiee, Recent advancements in polyaniline-based biosensors for diagnosis of cancers: a comprehensive review. TrAC Trends Anal. Chem. **181**, 118040 (2024). https://doi.org/10.1016/j.trac.2024.118040
12. B.M. Kangarshahi, S.M. Naghib, Nanogenosensors based on aptamers and peptides for bioelectrochemical cancer detection: an overview of recent advances in emerging materials and technologies. Discover Appl. Sci. **6**(2), 47 (2024). https://doi.org/10.1007/s42452-024-05681-z
13. A. Matini, S.M. Naghib, M.R. Mozafari, Quantum dots in cancer theranostics: a thorough review of recent advancements in bioimaging, tracking, and therapy across various cancer types. Curr. Pharm. Biotechnol. (2024). https://doi.org/10.2174/0113892010294163240407153842
14. Z. Rahimzadeh, S.M. Naghib, E. Askari, F. Molaabasi, A. Sadr, Y. Zare, M. Afsharpad, K.Y. Rhee, A rapid nanobiosensing platform based on herceptin-conjugated graphene for ultrasensitive detection of circulating tumor cells in early breast cancer. Nanotechnol. Rev. **10**(1), 744–753 (2021)

15. R. Salahandish, A. Ghaffarinejad, E. Omidinia, H. Zargartalebi, K. Majidzadeh-A, S.M. Naghib, A. Sanati-Nezhad, Label-free ultrasensitive detection of breast cancer miRNA-21 biomarker employing electrochemical nano-genosensor based on sandwiched AgNPs in PANI and N-doped graphene. Biosens. Bioelectron. **120**, 129–136 (2018). https://doi.org/10.1016/j.bios.2018.08.025
16. A.-A. Zare, H. Naderi-Manesh, S.M. Naghib, M. Shamsipur, F. Molaabasi, Label-free electrochemical cancer cell detection leveraging hemoglobin-encapsulated silver nanoclusters and Cu-MOF nanohybrids on a graphene-assisted dual-modal probe. Sci. Rep. **13**(1), 21980 (2023). https://doi.org/10.1038/s41598-023-49418-1
17. F. Kazemi, S.M. Naghib, Y. Zare, K.Y. Rhee, Biosensing applications of polyaniline (PANI)-based nanocomposites: a review. Polym. Rev. **61**(3), 553–597 (2021). https://doi.org/10.1080/15583724.2020.1858871
18. T. Karasu, C. Armutcu, K. Elkhoury, E. Özgür, A. Maziz, L. Uzun, Conducting polymers as a functional recognition interface to design sensors for pathogen and cancer diagnosis. TrAC Trends Anal. Chem. **117705** (2024)
19. M. Yazdan, S.M. Naghib, M.R. Mozafari, (2024) Liposomal nano-based drug delivery systems for breast cancer therapy: recent advances and progresses, anti-cancer agents in medicinal chemistry (formerly current medicinal chemistry—Anti-Cancer Agents) **24**(12):896–915. https://doi.org/10.2174/0118715206293653240322041047
20. F. Castro-Giner, S. Gkountela, C. Donato, I. Alborelli, L. Quagliata, C.K. Ng, S. Piscuoglio, N. Aceto, Cancer diagnosis using a liquid biopsy: challenges and expectations. Diagnostics **8**(2), 31 (2018)
21. M. Sadeghi, S. Sadeghi, S.M. Naghib, H.R. Garshasbi, A comprehensive review on electrochemical nano biosensors for precise detection of blood-based oncomarkers in breast cancer. Biosensors (2023)
22. M. Hosseine, S.M. Naghib, N. Rabiee, Metal-organic frameworks: biosensor applications for diagnosis of cancers. Microchem. J. **208**, 112341 (2025). https://doi.org/10.1016/j.microc.2024.112341
23. B. Mikaeeli Kangarshahi, S. Sojdeh, H. Daneshgar, M. Bagherzadeh, S.M. Naghib, N. Rabiee, Electroconductive polymer-based biosensors for early cancer detection via liquid biopsy: Advances, challenges, and future prospects. TrAC Trends Anal. Chem. **183**, 118062 (2025). https://doi.org/10.1016/j.trac.2024.118062
24. M.A. Farzin, S.M. Naghib, Paper-based immunosensor integrated with bioinspired Cu-polydopamine nanozyme for voltammetric detection of CA-15-3 tumor marker. Sens. Actuators Rep. **8**, 100209 (2024). https://doi.org/10.1016/j.snr.2024.100209
25. B. Mikaeeli Kangarshahi, S.M. Naghib, N. Rabiee, DNA/RNA-based electrochemical nanobiosensors for early detection of cancers. Crit. Rev. Clin. Lab. Sci. **61**(6):473–495 (2024)
26. M. Hosseine, S.M. Naghib, A. Khodadadi, Label-free electrochemical biosensor based on green-synthesized reduced graphene oxide/Fe_3O_4/nafion/polyaniline for ultrasensitive detection of SKBR3 cell line of HER2 breast cancer biomarker. Sci. Rep. **14**(1), 11928 (2024)
27. J.H. Luong, T. Narayan, S. Solanki, B.D. Malhotra, Recent advances of conducting polymers and their composites for electrochemical biosensing applications. J. Funct. Biomater. **11**(4), 71 (2020)
28. Y. Hao, H. Wang, C. Wang, B. Zhang, H. Ma, D. Wu, X. Ren, Y. Li, Q. Wei, Polyaniline-derived lotus-pod nano-hybrid incorporated luminescent N-CDs combined with novel co-reaction accelerator for ultra-sensitive electrochemiluminescence detection of trenbolone. Microchem. J. **112761** (2025)
29. E. Ito, K. Iha, T. Yoshimura, K. Nakaishi, S. Watabe, Early diagnosis with ultrasensitive ELISA. Adv. Clin. Chem. **101**, 121–133 (2021)
30. N. Shoaie, M. Daneshpour, M. Azimzadeh, S. Mahshid, S.M. Khoshfetrat, F. Jahanpeyma, A. Gholaminejad, K. Omidfar, M. Foruzandeh, Electrochemical sensors and biosensors based on the use of polyaniline and its nanocomposites: a review on recent advances. Microchim. Acta **186**, 1–29 (2019)

31. H.Z. Yamani, N. Safwat, A.M. Mahmoud, M.F. Ayad, M.F. Abdel-Ghany, M.M. Gomaa, Point-of-care diagnostics for rapid determination of prostate cancer biomarker sarcosine: application of disposable potentiometric sensor based on oxide-conductive polymer nanocomposite. Anal. Bioanal. Chem. **415**(22), 5451–5462 (2023)
32. L.C. Clark Jr., C. Lyons, Electrode systems for continuous monitoring in cardiovascular surgery. Ann. N. Y. Acad. Sci. **102**(1), 29–45 (1962)
33. A. Mehra, M. Chaudhary, F. De Souza, R.K. Gupta, Recent advancements in conducting polymers for biomedical sensors, in *NanoCarbon: A Wonder Material for Energy Applications: Volume 1: Basics to Advanced Applications for Energy Production* (2024) 325-349.
34. X. Zhang, X. Tan, P. Wang, J. Qin, Application of polypyrrole-based electrochemical biosensor for the early diagnosis of colorectal cancer. Nano **13**(4), 674 (2023)
35. N.A. Abtahi, S. Salehi, S.M. Naghib, F. Haghiralsadat, M.A. Edgahi, S. Ghorbanzadeh, W. Zhang, Multi-sensitive functionalized niosomal nanocarriers for controllable gene delivery in vitro and in vivo. Cancer Nanotechnol. **14**(1), 22 (2023). https://doi.org/10.1186/s12645-023-00175-w
36. A. Beigi, S.M. Naghib, A. Matini, M. Tajabadi, M.R. Mozafari, Lipid-based nanocarriers for targeted gene delivery in lung cancer therapy: exploring a novel therapeutic paradigm. Curr. Gene Ther. **25**(2), 92–112 (2025). https://doi.org/10.2174/0115665232292768240503050508
37. G.K. Kashani, S.M. Naghib, S. Soleymani, M.R. Mozafari, A review of DNA nanoparticles-encapsulated drug/gene/protein for advanced controlled drug release: Current status and future perspective over emerging therapy approaches. Int. J. Biol. Macromol. **268**, 131694 (2024). https://doi.org/10.1016/j.ijbiomac.2024.131694
38. M.A. Khorasani, S.M. Naghib, A review of chitosan-based multifunctional nanocomposites for drug/gene/protein delivery and gene therapy in cancer treatments: promises, challenges and outlooks. Int. J. Biol. Macromol. **306**, 141394 (2025). https://doi.org/10.1016/j.ijbiomac.2025.141394
39. S.M. Naghib, B. Ahmadi, B. Mikaeeli Kangarshahi, M.R. Mozafari, Chitosan-based smart stimuli-responsive nanoparticles for gene delivery and gene therapy: Recent progresses on cancer therapy, International Journal of Biological Macromolecules 278 (2024) 134542. https://doi.org/10.1016/j.ijbiomac.2024.134542.
40. V.M. Vijayan, L. Jothi, R.S. Arunagirinathan, G. Nageswaran, Recent advances in the electrochemical sensing of lung cancer biomarkers. Biosens. Bioelectron.: X **12**, 100235 (2022)
41. M. Freitas, H.P. Nouws, E. Keating, C. Delerue-Matos, High-performance electrochemical immunomagnetic assay for breast cancer analysis. Sens. Actuators B Chem. **308**, 127667 (2020)
42. T. Acar, K. Nur Çimen, E. Özalp, Ö. Ilıca, E. Ahlatcıoğlu Özerol, Recent advances in biosensors based on conducting polymers for biomedical applications. ChemistrySelect **8**(30), e202300819 (2023)
43. E.N. Zare, P. Makvandi, B. Ashtari, F. Rossi, A. Motahari, G. Perale, Progress in conductive polyaniline-based nanocomposites for biomedical applications: a review. J. Med. Chem. **63**(1), 1–22 (2019)
44. S. Shahrokhian, R. Salimian, Ultrasensitive detection of cancer biomarkers using conducting polymer/electrochemically reduced graphene oxide-based biosensor: Application toward BRCA1 sensing. Sens. Actuators B Chem. **266**, 160–169 (2018)
45. A. Mohammadpour-Haratbar, Y. Zare, K.Y. Rhee, Electrochemical biosensors based on polymer nanocomposites for detecting breast cancer: recent progress and future prospects. Adv. Colloid Interf. Sci. **309**, 102795 (2022)
46. Y. Pan, J. Zhang, X. Guo, Y. Li, L. Li, L. Pan, Recent advances in conductive polymers-based electrochemical sensors for biomedical and environmental applications. Polymers **16**(11), 1597 (2024)
47. A.M. Bryan, L.M. Santino, Y. Lu, S. Acharya, J.M. D'Arcy, Conducting polymers for pseudocapacitive energy storage. Chem. Mater. **28**(17), 5989–5998 (2016)

48. S. Peng, Y. Yu, S. Wu, C.-H. Wang, Conductive polymer nanocomposites for stretchable electronics: Material selection, design, and applications. ACS Appl. Mater. Interfaces **13**(37), 43831–43854 (2021)
49. O. Bettucci, G.M. Matrone, F. Santoro, Conductive polymer-based bioelectronic platforms toward sustainable and biointegrated devices: a journey from skin to brain across human body interfaces. Adv. Mater. Technol. **7**(2), 2100293 (2022)
50. X. Gao, Y. Bao, Z. Chen, J. Lu, T. Su, L. Zhang, J. Ouyang, Bioelectronic applications of intrinsically conductive polymers. Adv. Electron. Mater. **9**(10), 2300082 (2023)
51. M.G. Sumdani, M.R. Islam, A.N.A. Yahaya, S.I. Safie, Recent advancements in synthesis, properties, and applications of conductive polymers for electrochemical energy storage devices: a review. Polym. Eng. Sci. **62**(2), 269–303 (2022)
52. K. Namsheer, C.S. Rout, Conducting polymers: a comprehensive review on recent advances in synthesis, properties and applications. RSC Adv. **11**(10), 5659–5697 (2021)
53. C.I. Idumah, Novel trends in conductive polymeric nanocomposites, and bionanocomposites. Synth. Met. **273**, 116674 (2021)
54. J. Ouyang, Application of intrinsically conducting polymers in flexible electronics. SmartMat **2**(3), 263–285 (2021)
55. L.P. Lin, M.T.T. Tan, Biosensors for the detection of lung cancer biomarkers: a review on biomarkers, transducing techniques and recent graphene-based implementations. Biosens. Bioelectron. **115492** (2023)
56. H. Khan, M.R. Shah, J. Barek, M.I. Malik, Cancer biomarkers and their biosensors: a comprehensive review. TrAC Trends Anal. Chem. **158**, 116813 (2023)
57. S. Sarkar, S. Hazra, S. Patra, M. Gogoi, Biosensors for cancer detection: a review. TrAC Trends Anal. Chem. **117978** (2024)
58. L. Qian, Q. Li, K. Baryeh, W. Qiu, K. Li, J. Zhang, Q. Yu, D. Xu, W. Liu, R.E. Brand, Biosensors for early diagnosis of pancreatic cancer: a review. Transl. Res. **213**, 67–89 (2019)
59. A.G. MacDiarmid, Polyaniline and polypyrrole: where are we headed? Synth. Met. **84**(1-3), 27–34 (1997)
60. S. Sun, J. Chen, Recent advances in hydrogel-based biosensors for cancer detection. ACS Appl. Mater. Interfaces **16**(36), 46988–47002 (2024)
61. C.D. Pina, E. Falletta, Advances in polyaniline for biomedical applications. Curr. Med. Chem. **29**(2), 329–357 (2022)
62. C. Dhand, N. Dwivedi, S. Mishra, P.R. Solanki, V. Mayandi, R.W. Beuerman, S. Ramakrishna, R. Lakshminarayanan, B.D. Malhotra, Polyaniline-based biosensors. Nanobiosensors Dis. Diagn., 25–46 (2015)
63. A. Akbar, M. Das, D. Sarkar, Room temperature ammonia sensing by CdS nanoparticle decorated polyaniline (PANI) nanorods. Sens. Actuators A Phys. **310**, 112071 (2020)
64. A.H. Majeed, L.A. Mohammed, O.G. Hammoodi, S. Sehgal, M.A. Alheety, K.K. Saxena, S.A. Dadoosh, I.K. Mohammed, M.M. Jasim, N.U. Salmaan, A review on polyaniline: synthesis, properties, nanocomposites, and electrochemical applications. Int. J. Polym. Sci. **2022**(1), 9047554 (2022)
65. M. Matsuguchi, T. Nakamae, R. Fujisada, S. Shiba, A highly sensitive ammonia gas sensor using micrometer-sized core—shell-type spherical polyaniline particles. Sensors **21**(22), 7522 (2021)
66. J. Fei, Y. Cui, X. Yan, Y. Yang, K. Wang, J. Li, Controlled fabrication of polyaniline spherical and cubic shells with hierarchical nanostructures. ACS Nano **3**(11), 3714–3718 (2009)
67. L. Hu, Y. Ren, H. Yang, Q. Xu, Fabrication of 3D hierarchical MoS_2/polyaniline and MoS_2/C architectures for lithium-ion battery applications. ACS Appl. Mater. Interfaces **6**(16), 14644–14652 (2014)
68. S. Pang, W. Chen, Z. Yang, Z. Liu, X. Fan, D. Fang, Facile synthesis of polyaniline nanotubes with square capillary using urea as template. Polymers **9**(10), 510 (2017)
69. M. Beygisangchin, S. Abdul Rashid, S. Shafie, A.R. Sadrolhosseini, H.N. Lim, Preparations, properties, and applications of polyaniline and polyaniline thin films—a review. Polymers **13**(12), 2003 (2021)

70. H.N. Heme, M.S.N. Alif, S.S.M. Rahat, S.B. Shuchi, Recent progress in polyaniline composites for high capacity energy storage: a review. J Energy Storage **42**, 103018 (2021)
71. C.-W. Lee, J.-K. Wu, C.-H. Chang, C.-W. Cheng, H.-Y. Chang, P.-C. Wang, F.-G. Tseng, Sulfonated polyaniline as zwitterionic and conductive interfaces for anti-biofouling on open electrode surfaces in electrodynamic systems. ACS Appl. Mater. Interfaces **12**(16), 19102–19109 (2020)
72. P. Walde, K. Kashima, G. Ćirić-Marjanović, Synthesizing polyaniline with laccase/O_2 as catalyst. Front. Bioeng. Biotechnol. **7**, 165 (2019)
73. S.M. Naghib, M. Rabiee, E. Omidinia, P. Khoshkenar, Investigation of a biosensor based on phenylalanine dehydrogenase immobilized on a polymer-blend film for phenylketonuria diagnosis. Electroanalysis **24**(2), 407–417 (2012). https://doi.org/10.1002/elan.201100391
74. S.M. Naghib, M. Rabiee, E. Omidinia, P. Khoshkenar, D. Zeini, Biofunctionalization of dextran-based polymeric film surface through enzyme immobilization for phenylalanine determination. Int. J. Electrochem. Sci. **7**(1), 120–135 (2012) https://doi.org/10.1016/S1452-3981(23)13325-6
75. S.M. Naghib, M. Rabiee, E. Omidinia, Electroanalytical validation of a novel nanobiosensing strategy and direct electrochemistry of phenylalanine dehydrogenase for clinical diagnostic applications. Int. J. Electrochem. Sci. **9**(5), 2301–2315 (2014) https://doi.org/10.1016/S1452-3981(23)07928-2
76. S.M. Naghib, M. Rabiee, E. Omidinia, Electrochemical biosensor for L-phenylalanine based on a gold electrode modified with graphene oxide nanosheets and chitosan. Int. J. Electrochem. Sci. **9**(5), 2341–2353 (2014). https://doi.org/10.1016/S1452-3981(23)07931-2
77. L.A. Samuelson, A. Anagnostopoulos, K.S. Alva, J. Kumar, S.K. Tripathy, Biologically derived conducting and water soluble polyaniline. Macromolecules **31**(13), 4376–4378 (1998)
78. V. Tsakova, Conductive polymer-based materials for medical electroanalytic applications, in *Applications of Electrochemistry in Medicine* (Springer, 2013), pp. 283–342
79. E. Omidinia, S.M. Naghib, A. Boughdachi, P. Khoshkenar, D.K. Mills, Hybridization of silver nanoparticles and reduced graphene nanosheets into a nanocomposite for highly sensitive L-phenylalanine biosensing. Int. J. Electrochem. Sci. **10**(8), 6833–6843 (2015). https://doi.org/10.1016/S1452-3981(23)06765-2
80. S.M. Naghib, Fabrication of nafion/silver nanoparticles/reduced graphene nanosheets/glucose oxidase nanobiocomposite for electrochemical glucose biosensing. Anal. Bioanal. Electrochem. **8**(4), 453–465 (2016)
81. S.M. Naghib, E. Parnian, H. Keshvari, E. Omidinia, M. Eshghan-Malek, Synthesis, characterization and electrochemical evaluation of polyvinylalchol/graphene oxide/silver nanocomposites for glucose biosensing application. Int. J. Electrochem. Sci. **13**(1), 1013–1026 (2018). https://doi.org/10.20964/2018.01.74
82. E. Askari, S.M. Naghib, A novel approach to facile synthesis and biosensing of the protein-regulated graphene. Int. J. Electrochem. Sci. **13**(1), 886–897 (2018). https://doi.org/10.20964/2018.01.73
83. L. Duan, Y. Zhao, F. Guo, W. Liu, C. Hou, Z. Ni, Enzymatic-catalyzed polymerization of water-soluble electrically conductive polymer PEDOT: PSS. Polym. Adv. Technol. **25**(8), 896–899 (2014)
84. G. Mandal, J. Bauri, D. Nayak, S. Kumar, S. Ansari, R.B. Choudhary, Synthesis, Structural Study and Various Applications of Polyaniline and its Nanocomposites (2023)
85. G. Cao, J. Xu, S. Cai, Y. Chen, D. Zhou, H. Zhang, C. Jiang, G. Zhang, Y. Tian, Highly conductive and dispersible polyaniline microtubes controlled by methyl orange. ACS Appl. Polym. Mater. **5**(1), 593–601 (2022)
86. P. Kulshrestha, B.K. Shin, S. Falak, Bio-inspired hierarchical structure of polyaniline in the pore surface of polymer film through interfacial polymerization as a smart material sensitive to pH. Eur. Polym. J. **187**, 111893 (2023)
87. Y.-E. Miao, W. Fan, D. Chen, T. Liu, High-performance supercapacitors based on hollow polyaniline nanofibers by electrospinning. ACS Appl. Mater. Interfaces **5**(10), 4423–4428 (2013)

88. C. Kaneda, Y. Sueyasu, E. Tanaka, M. Atobe, Electrochemical synthesis of microporous polyaniline films using foam templates prepared by ultrasonication. Ultrason. Sonochem. **64**, 104991 (2020). https://doi.org/10.1016/j.ultsonch.2020.104991
89. M. Hosseine, A. Bakhshi, S.M. Naghib, N. Rabiee, Recent advancements in polyaniline-based biosensors for diagnosis of cancers: a comprehensive review. TrAC Trends Anal. Chem. **118040** (2024)
90. A.K. Poddar, S.S. Patel, H.D. Patel, Synthesis, characterization and applications of conductive polymers: a brief review. Polym. Adv. Technol. **32**(12), 4616–4641 (2021)
91. F.F. Garrudo, P.E. Mikael, C.A. Rodrigues, R.W. Udangawa, P. Paradiso, C.A. Chapman, P. Hoffman, R. Colaço, J.M. Cabral, J. Morgado, Polyaniline-polycaprolactone fibers for neural applications: electroconductivity enhanced by pseudo-doping. Mater. Sci. Eng. C **120**, 111680 (2021)
92. M. Beygisangchin, A.H. Baghdadi, S.K. Kamarudin, S.A. Rashid, J. Jakmunee, N. Shaari, Recent progress in polyaniline and its composites; synthesis, properties, and applications. Eur. Polym. J. **112948** (2024)
93. J.A. Oyetade, R.L. Machunda, A. Hilonga, Photocatalytic degradation of azo dyes in textile wastewater by polyaniline composite catalyst-a review. Sci. Afr. **17**, e01305 (2022)
94. A. Yadav, H. Kumar, R. Sharma, R. Kumari, Influence of polyaniline on the photocatalytic properties of metal nanocomposites: a review. Colloid Interface Sci. Commun. **40**, 100339 (2021)
95. R. Rai, J.A. Roether, A.R. Boccaccini, Polyaniline based polymers in tissue engineering applications: a review. Prog. Biomed. Eng. **4**(4), 042004 (2022)
96. R. Zheng, A. Wu, J. Li, Z. Tang, J. Zhang, M. Zhang, Z. Wei, Progress and outlook on electrochemical sensing of lung cancer biomarkers. Molecules **29**(13), 3156 (2024)
97. G. Bolat, O.A. Vural, Y.T. Yaman, S. Abaci, Polydopamine nanoparticles-assisted impedimetric sensor towards label-free lung cancer cell detection. Mater. Sci. Eng. C **119**, 111549 (2021)
98. S. Lantuejoul, L. Fernandez-Cuesta, F. Damiola, N. Girard, A. McLeer, New molecular classification of large cell neuroendocrine carcinoma and small cell lung carcinoma with potential therapeutic impacts. Transl. Lung Cancer Res. **9**(5), 2233 (2020)
99. G.P. Caviglia, M. Ciruolo, A. Olivero, P. Carucci, E. Rolle, C. Rosso, M.L. Abate, A. Risso, D.G. Ribaldone, F. Tandoi, Prognostic role of serum cytokeratin-19 fragment (CYFRA 21-1) in patients with hepatocellular carcinoma. Cancer **12**(10), 2776 (2020)
100. L. Dai, R. Xu, M. Cui, X. Ren, X. Wang, J. Feng, R. Wu, H. Ma, Q. Wei, A high throughput dual-signal ultra-sensitive electrochemical and photoelectrochemical microfluidic immunoassay platform with raspberry au/PANI@ CdS for Cyfra 21-1 detection. Biosens. Bioelectron. : X **11**, 100207 (2022)
101. Y.-M. Xia, M. Xia, Y. Zhao, M.-Y. Li, X. Ou, W.-W. Gao, Photocatalytic electrochemical sensor based on three-dimensional graphene nanocomposites for the ultrasensitive detection of CYFRA21-1 gene. Microchem. J. **166**, 106245 (2021)
102. Y. Park, M.-S. Hong, W.-H. Lee, J.-G. Kim, K. Kim, Highly sensitive electrochemical aptasensor for detecting the VEGF165 tumor marker with PANI/CNT nanocomposites. Biosensors **11**(4), 114 (2021)
103. A. Khanmohammadi, A. Aghaie, E. Vahedi, A. Qazvini, M. Ghanei, A. Afkhami, A. Hajian, H. Bagheri, Electrochemical biosensors for the detection of lung cancer biomarkers: a review. Talanta **206**, 120251 (2020)
104. D. Anwanwan, S.K. Singh, S. Singh, V. Saikam, R. Singh, Challenges in liver cancer and possible treatment approaches. Biochim. Biophys. Acta (BBA) Rev. Cancer **1873**(1), 188314 (2020)
105. D.Q. Huang, A.G. Singal, Y. Kono, D.J. Tan, H.B. El-Serag, R. Loomba, Changing global epidemiology of liver cancer from 2010 to 2019: NASH is the fastest growing cause of liver cancer. Cell Metab. **34**(7), 969–977 (2022) e2
106. S. Enitan, G. Avwioro, E. Adejumo, A review of complementary and alternative medicine used in cancer care: challenges and prospects. TMR Integr. Med. **7**, e23021 (2023)

107. J.-M. Zhou, T. Wang, K.-H. Zhang, AFP-L3 for the diagnosis of early hepatocellular carcinoma: a meta-analysis. Medicine **100**(43), e27673 (2021)
108. Y.-S. Zhang, J.-H. Chu, S.-X. Cui, Z.-Y. Song, X.-J. Qu, Des-γ-carboxy prothrombin (DCP) as a potential autologous growth factor for the development of hepatocellular carcinoma. Cell. Physiol. Biochem. **34**(3), 903–915 (2014)
109. P.R. Galle, F. Foerster, M. Kudo, S.L. Chan, J.M. Llovet, S. Qin, W.R. Schelman, S. Chintharlapalli, P.B. Abada, M. Sherman, Biology and significance of alpha-fetoprotein in hepatocellular carcinoma. Liver Int. **39**(12), 2214–2229 (2019)
110. C.-W. Lee, H.-I. Tsai, W.-C. Lee, S.-W. Huang, C.-Y. Lin, Y.-C. Hsieh, T. Kuo, C.-W. Chen, M.-C. Yu, Normal alpha-fetoprotein hepatocellular carcinoma: are they really normal? J. Clin. Med. **8**(10), 1736 (2019)
111. S. Falahi, H.-A. Rafiee-Pour, M. Zarejousheghani, P. Rahimi, Y. Joseph, Non-coding RNA-based biosensors for early detection of liver cancer. Biomedicine **9**(8), 964 (2021)
112. N. Hui, X. Sun, Z. Song, S. Niu, X. Luo, Gold nanoparticles and polyethylene glycols functionalized conducting polyaniline nanowires for ultrasensitive and low fouling immunosensing of alpha-fetoprotein. Biosens. Bioelectron. **86**, 143–149 (2016)
113. D. Feng, K. Zhang, Y. Lu, J. Chen, Y. Wei, Gold microstructures/polyaniline/reduced graphene oxide/prussian blue composite as stable redox matrix for label-free electrochemical immunoassay of α-fetoprotein. Anal. Sci. **36**(12), 1501–1505 (2020)
114. C. Zhao, K. Gao, C. Ma, M. Wu, K. Cao, Y. Yang, C. Hong, X. Qiao, Functionalized polyaniline based on protonic acid doping as a direct electron mediator to amplify sensor signals. J. Mol. Struct. **1209**, 127924 (2020). https://doi.org/10.1016/j.molstruc.2020.127924
115. N. Shahbazi, R. Zare-Dorabei, S.M. Naghib, Multifunctional nanoparticles as optical biosensing probe for breast cancer detection: a review. Mater. Sci. Eng. C **127**, 112249 (2021)
116. S.M. Naghib, B. Ahmadi, M. Mozafari, Stimuli-sensitive chitosan-based nanosystems-immobilized nucleic acids for gene therapy in breast cancer and hepatocellular carcinoma. Curr. Top. Med. Chem. **24**(17), 1464–1489 (2024)
117. R. Salahandish, A. Ghaffarinejad, S.M. Naghib, K. Majidzadeh-A, H. Zargartalebi, A. Sanati-Nezhad, Nano-biosensor for highly sensitive detection of HER2 positive breast cancer. Biosens. Bioelectron. **117**, 104–111 (2018). https://doi.org/10.1016/j.bios.2018.05.043
118. M. Shamsipur, F. Molaabasi, A.-A. Zare, H. Naderi-Manesh, S.M. Naghib, Scientific Reports 13 21980 (2023)
119. M. Sadeghi, S. Kashanian, S.M. Naghib, F. Haghiralsadat, D. Tofighi, An efficient electrochemical biosensor based on pencil graphite electrode mediated by 2D functionalized graphene oxide to detect HER2 breast cancer biomarker. Int. J. Electrochem. Sci. **17**(4), 220459 (2022). https://doi.org/10.20964/2022.04.62
120. F. Dang, L. Nie, W. Wei, Ubiquitin signaling in cell cycle control and tumorigenesis. Cell Death Differ. **28**(2), 427–438 (2021)
121. B.K. Kim, J.W. Lee, P.J. Park, Y.S. Shin, W.Y. Lee, K.A. Lee, S. Ye, H. Hyun, K.N. Kang, D. Yeo, The multiplex bead array approach to identifying serum biomarkers associated with breast cancer. Breast Cancer Res. **11**, 1–12 (2009)
122. V.S.A. Jayanthi, A.B. Das, U. Saxena, Fabrication of an immunosensor for quantitative detection of breast cancer biomarker UBE2C. RSC Adv. **9**(29), 16738–16745 (2019)
123. N. Hui, X. Sun, S. Niu, X. Luo, PEGylated polyaniline nanofibers: Antifouling and conducting biomaterial for electrochemical DNA sensing. ACS Appl. Mater. Interfaces **9**(3), 2914–2923 (2017)
124. Y.-M. Xia, M.-Y. Li, C.-L. Chen, M. Xia, W. Zhang, W.-W. Gao, Employing label-free electrochemical biosensor based on 3D-reduced graphene oxide and polyaniline nanofibers for ultrasensitive detection of breast cancer BRCA1 biomarker. Electroanalysis **32**(9), 2045–2055 (2020). https://doi.org/10.1002/elan.202060039
125. A. Kaur, S. Sharma, S. Rana, R. Rohilla, N. Prabhakar, Ag-CuO integrated polyaniline based impedimetric detection of cancer biomarker MUC1. Microchem. J. **194**, 109266 (2023). https://doi.org/10.1016/j.microc.2023.109266

126. T. Vural, Y.T. Yaman, S. Ozturk, S. Abaci, E.B. Denkbas, Electrochemical immunoassay for detection of prostate specific antigen based on peptide nanotube-gold nanoparticle-polyaniline immobilized pencil graphite electrode. J. Colloid Interface Sci. **510**, 318–326 (2018). https://doi.org/10.1016/j.jcis.2017.09.079
127. M.D. Tezerjani, A. Benvidi, M. Rezaeinasab, S. Jahanbani, S.M. Moshtaghioun, M. Youssefi, K. Zarrini, An impedimeric biosensor based on a composite of graphene nanosheets and polyaniline as a suitable platform for prostate cancer sensing. Anal. Methods **8**(41), 7507–7515 (2016). https://doi.org/10.1039/C6AY01524G
128. A. Dey, A. Kaushik, S.K. Arya, S. Bhansali, Mediator free highly sensitive polyaniline–gold hybrid nanocomposite based immunosensor for prostate-specific antigen (PSA) detection. J. Mater. Chem. **22**(29), 14763–14772 (2012). https://doi.org/10.1039/C2JM31663C
129. N. Shoaie, M. Daneshpour, M. Azimzadeh, S. Mahshid, S.M. Khoshfetrat, F. Jahanpeyma, A. Gholaminejad, K. Omidfar, M. Foruzandeh, Electrochemical sensors and biosensors based on the use of polyaniline and its nanocomposites: a review on recent advances. Microchim. Acta **186**(7), 465 (2019). https://doi.org/10.1007/s00604-019-3588-1
130. X. Sun, N. Hui, X. Luo, Reagentless and label-free voltammetric immunosensor for carcinoembryonic antigen based on polyaniline nanowires grown on porous conducting polymer composite. Microchim. Acta **184**(3), 889–896 (2017). https://doi.org/10.1007/s00604-016-2068-0
131. C. Amarasiri, T.B. Nguyen, L.T. Nguyen, V.T. Thu, N.T.M. Thuy, T. Dai Lam, Electrochemical Immunosensor based on Fe3O4/PANI/AuNP detecting Interface for carcinoembryonic antigen biomarker. J. Electron. Mater. **46**(10), 5755–5763 (2017). https://doi.org/10.1007/s11664-017-5610-x
132. Q. Wang, H. Xin, Z. Wang, Label-free Immunosensor based on polyaniline-loaded MXene and gold-decorated β-Cyclodextrin for efficient detection of carcinoembryonic antigen. Biosensors **12**(8), 657 (2022)
133. A.B. Ganganboina, R.-A. Doong, Graphene quantum dots decorated gold-polyaniline nanowire for impedimetric detection of carcinoembryonic antigen. Sci. Rep. **9**(1), 7214 (2019). https://doi.org/10.1038/s41598-019-43740-3

Chapter 2
Bio-inspired Nanocomposites Based on Biological Macromolecules for Biosensing and Early Detection of Cancers

Abstract The word biomimetics is comparatively recent in the area of biosensors, while being used for many years by biochemists. Biomimetics often aims to intelligently employ natural or synthetic materials without using live systems. Drug administration and sensing techniques are only two of the many uses for macromolecules that mimic biodegradation or their mixtures, a significant category of biomimetics. Biomimetic macromolecule-based detection methods are the interesting ways to get around the problems with biosensors and make them better at finding biological targets, especially signs of tumours. The area of biosensors based on biomimetic macromolecules has seen a number of noteworthy advancements in recent years. With an emphasis on sensing techniques based on mimetic biodegradable macromolecules and their composites, the present chapter provides an overview of these advancements. The merit indices of biosensors, including biocompatibility, biodegradability, antifouling, sensitivity, selectivity, etc., may be enhanced by using the unique properties of biodegradable macromolecule-based composites that can replicate biological systems. Furthermore, these composites provide a chance for in-situ continuous monitoring biomarkers and real-time biomarkers. In this chapter, I have attempted to demonstrate the great potential of these biomacromolecules (natural or synthetic) and their composites in implantable or wearable sensing systems and other customized sensors. Lastly, a summary of the field's existing issues and prospects for growth was provided.

Keywords Bio-inspired materials · Biomimetic nanocomposites · Biological macromolecules · Synthetic macromolecules · Biosensors · Early detection of cancers

2.1 Introduction and Basic Concepts

"Biomimetic" refers to the idea of enhancing material performance by imitating the superior characteristics of biological and natural systems [1]. Implantable biosensors may have their biotoxicity decreased via biodegradability without having to be

S. M. Naghib, *Nano-Biosensors, Drug Delivery and Tissue Engineering*, Series in BioEngineering, https://doi.org/10.1007/978-981-95-1888-3_2

taken out. In vivo biosafety and tissue compatibility are improved using biomimetic macromolecule composite-based sensors [2]. Biodegradability, biocompatibility, and non-toxicity are essential qualities regarding the biosensors' material components, particularly wearable or implanted biosensors [3]. Furthermore, these biomacromolecules may co-deposit or mix with antifouling substances to prevent serum proteins from adhering to implanted electrode surfaces. Biofouling significantly shortens the device's lifespan and interferes with the sensor's functionality. More crucially, the creation of sophisticated systems that replicate biological processes has been greatly influenced by the study of biomimetic macromolecules. Macromolecule nanozymes, for instance, have been used as elements that imitate enzymes in the creation of biosensors, increasing sensor effectiveness and allowing for customization [4]. Nanomaterials and nanocomposites having comparable enzymatic characteristics are referred to as nanozymes. Gentle biosensors created using composites based on biomimetic macromolecules may be used as implanted sensors that imitate the structure of tissue, in contrast to rigid metal nanoparticle-based biosensors [5]. Amphiphilic antimicrobial macromolecules have been used to create a vast array of biofunctional and structural imitators of host defense peptides in recent years. The fact that these biomimetic macromolecules do not cause bacterial resistance to medications is one of their benefits over antimicrobial medications [6]. Consequently, these macromolecules provide special chances for the development of novel biosensors. Composites of biomimetic macromolecules have the ability to completely transform the sensing industry, from design to market [7]. Biomimetic composites made of biodegradable macromolecules have many uses in different areas of biomedicine because of their special qualities, which include high mechanical adaptability, inherent bioactivity, excellent biocompatibility, and a high rate of microbial degradation (satisfactory degradability). Because of their great capacity to regulate certain biological processes, biomimetic scaffolds based on biodegradable macromolecules may be the best option for developing novel tissues [8]. Furthermore, these macromolecules may be extensively used as drug delivery vehicles, particularly for cancer tumors, due to their enhanced biocompatibility and ability to withstand rapid immune cell clearance [9–14].

Screening is one of the major recommendations for the general population given the rising risk of cancer worldwide [15]. ELISA is a time-consuming and demanding test method in spite of its numerous benefits. As a universal technique for oncomarker detection, several researchers think that biosensors might be a good substitute for enzyme ELISA. The removal of tagged antibodies, the lack of sophisticated equipment and highly skilled workers, and above all the possibility of real-time and in situ measurement are just a few of the numerous benefits that biosensors provide [16]. Wearable, implantable, digestible, and handheld/portable point-of-care (POC) biosensors are just a few of the customizable diagnostic biosensors that may be created. Particularly when combined with smartphones, the customized biosensors are the perfect tool for ongoing biomarker monitoring in bodily fluids. Mobile devices that use biosensors, the future's health instruments, enhance personalised medicine [7].

Combining biomimetic macromolecules with novel therapeutic and diagnostic techniques might enhance tissue engineering performance, pharmacokinetic profile, and biosensor merit indices while also getting beyond biological obstacles. In this context, a number of review papers and chapters have been released with the intention of compiling the most recent advancements and difficulties in the domains of tissue engineering [17] and delivery of drugs [18]. I am urged to provide a summary of the state, constraints, and prospects of biosensors made from biomimetic macromolecules for biological target detection, particularly tumor indicators. Certain biodegradable macromolecules or their hydrogels may be inserted into the intended tissue with little harm because of their structural resemblance to bodily tissues. Many others may mimic the actions of certain biological substances, such enzymes, and are thus crucial for catalytic sensors that do not need enzymes. Since enzymes have caused issues in the enzymatic biosensor design because of their expensive price, macromolecule composite sensors with enzyme-like activity are essential for detecting biomarkers in case of low stability and purity. For chronic patients, POC biosensors that are single-use are often the best option for quick diagnosis and monitoring without the need for a hospital. Their primary flaw, nevertheless, is their low dependability, which results from the biorecognition components' weak stability. As a result, nanozyme-based biosensors have shown great promise for real-time and in situ biomarker monitoring. The scientific discipline of biomimetics, also known as biomimicry, employs nature to address difficult human problems. Researchers have successfully created contemporary artificial intelligence that mimics the operation of synapses and neurons, drawing inspiration from the architecture of the brain (Fig. 2.1a). The schematic shows neuromorphic computing, focusing on soft biomaterials like hydrogels and organic polymers that mimic biological synapses' properties. It highlights synaptic plasticity, energy efficiency, and fabrication techniques such as inkjet printing and electrochemical deposition for creating neural networks. The study emphasizes applications in robotics, prosthetics and cognitive computing, advancing brain-like and efficient computing systems. The customized processors speed up neural network computations by simulating the brain's parallel, low-energy processing [19]. Actually, data processing may be altered via neuromorphic computing, which aims to resemble the human brain. Artificial intelligence systems based on neuromorphic circuits are still distant from their biological counterparts, despite notable advancements. These chips struggle with flexibility, energy consumption, and adaptation when compared to the human brain [20]. Advanced artificial olfactory sensors, or e-noses, have emerged as a result of advancements in artificial neural networks. These sensors have highly differentiated detecting capabilities and function as olfactory receptors, simulating the human olfactory system (Fig. 2.1b) [21]. Figure 2.1b represents artificial olfactory sensors, which detect patterns from olfactory receptors, emerging for monitoring volatile organic compounds, aided by advanced statistical and data processing methods. Technologies like deep learning and artificial neural networks enhance pattern recognition in sensor arrays, exemplified by electronic noses composed of diverse sensors tailored to specific environments. Despite current limitations in reliability and standardization, these sensors show great potential for applications in food quality, environmental monitoring and diagnostics, with promising future integration into advanced pattern recognition and learning systems. Future food quality and environmental pollution detection is anticipated to make extensive use of artificial olfactory sensors. Our everyday life will also be completely transformed by the combination of this technique with the IoT (Internet of Things). With about 1000 nerve endings, the human skin establishes the most extensive sensory link inside the brain and the external environment via pressure, temperature, and touch. New developments in human-machine interaction are made possible by the combination of artificial intelligence with soft, stretchy biosensors that replicate the sensing capabilities of skin (also known as electronic skins or e-skin). In actuality, human-robot connection is enhanced by simulating human skin and human brain data processing [22].

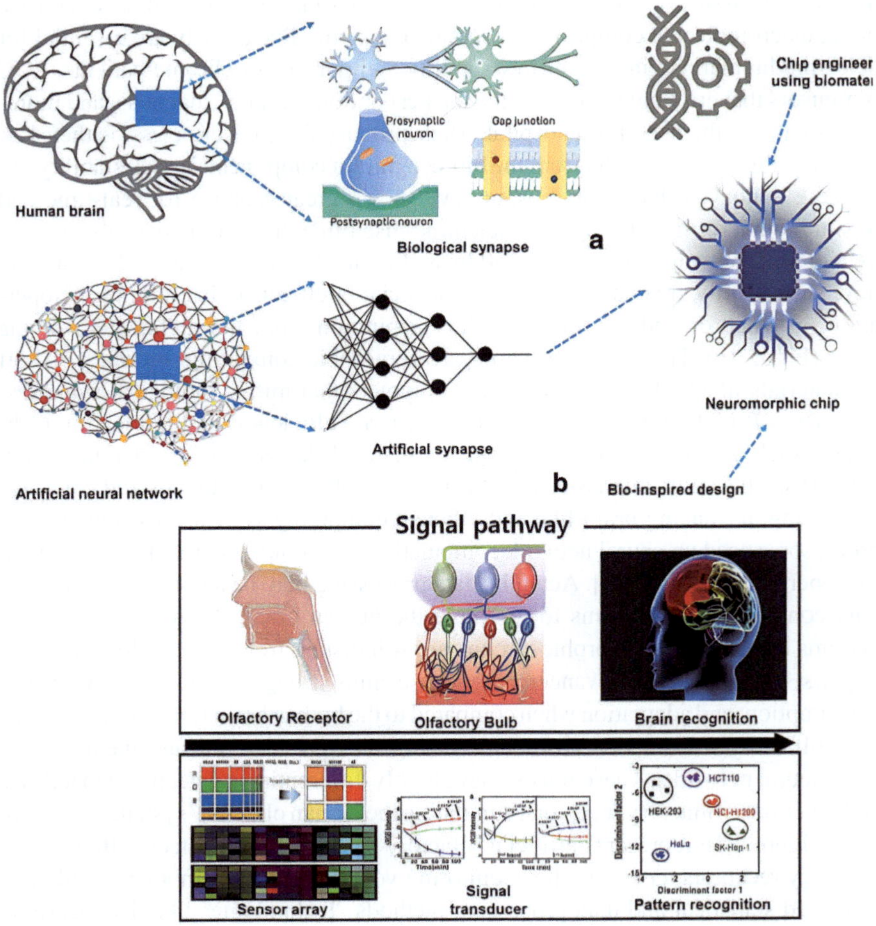

Fig. 2.1 Schematic representation of **a** a neuromorphic artificial intelligence system that mimic the function of neurons and synapses [19], **b** an artificial olfactory sensor that mimic the mammalian olfactory system [21]

Nature-inspired algorithms are very helpful for computers and intricate computations with lots of variables that don't have clear "normal" answers. Genetic algorithms are among the most crucial approaches to resolving the blind deconvolution issue in computationally demanding image processing techniques because of their nature. Using the idea of the genome, this technique is often used to separate blind digital signals. In 2012, Duman et al. [23] presented a novel optimization technique for the quadratic assignment problem that was influenced by bird migration. To handle discrete issues, Tongur and Ülker suggested an optimization technique inspired by multi-catchment migrating birds [24]. Biomimetic has also been beneficial to other engineering science disciplines. Biomimicry, for instance, has had a big impact on architecture and urban planning. Nature-inspired architecture first appeared throughout the Art Nouveau era and has peaked recently. Many architects now use a different approach when using plants and trees as structural inspiration [25]. In addition to factors like lightness, strength, earthquake resistance, and energy consumption, biomimetics-based designs are used in the building sector because they may be formed of lightweight materials and support and support heavy loads [26].

Bioinspiration and biomimetics have significantly influenced the morphology, structure, form, synthesis method, and enhancement of the characteristics and functionality of nanomaterials, transforming the field of nanotechnology. For example, Li and colleagues used biomimetic mineralization to create β-galactosidase-encapsulated nanoflowers based on $Cu_3(PO_4)_2$. These scientists drew inspiration from nature for the design and process of creating hybrid capsules. Remarkably, nanoflowers encapsulated with β-galactosidase replicated the activity of lactase, an enzyme the body produces to aid in the breakdown (digestion) of lactose [27]. Huang and colleagues created a flower-like bovine serum albumin (BSA)–$Cu_3(PO_4)_2 \cdot 3H_2O$ nanocomposite in another investigation. In the presence of H_2O_2, these nanofloweres exhibited catalytic activity similar to that of peroxidase for the 3-ethylbenzothiozoline-2-azinobis-6-sulfonic acid oxidation (ABTS) [28]. Zhang et al. created self-assembled nanoparticles based on porphyrin and chitosan, altered using HIV lipids and Tat peptide. Nanoparticles exhibited viral-like components because this peptide was generated from the HIV-1 virus's TAT protein. Additionally, because of the nano spikes that formed on their surface as a consequence of pH-responsive decomposition-induced self-assembly, the nanoparticles mimicked the shape of viruses. Compared to non-viral-mimicking nanoparticles, these virus-mimicking nanoparticles significantly increased cell internalization because of their virus-like appearance and constituents (Fig. 2.2) [29].

As will be briefly covered later in this chapter, biomimetics has also significantly advanced tissue engineering and medication delivery and other areas of biomedicine. Specifically, biomimetics has had a significant impact on biosensors/sensors that are evolving into advanced sensing technology, including energy harvester-based sensors, cellphones, IOTs, and microelectromechanical systems (MEMS). Because biomimetics is so widely used in sensors, this chapter will solely discuss the usage of natural and synthetic biomimetic macromolecules in sensor design.

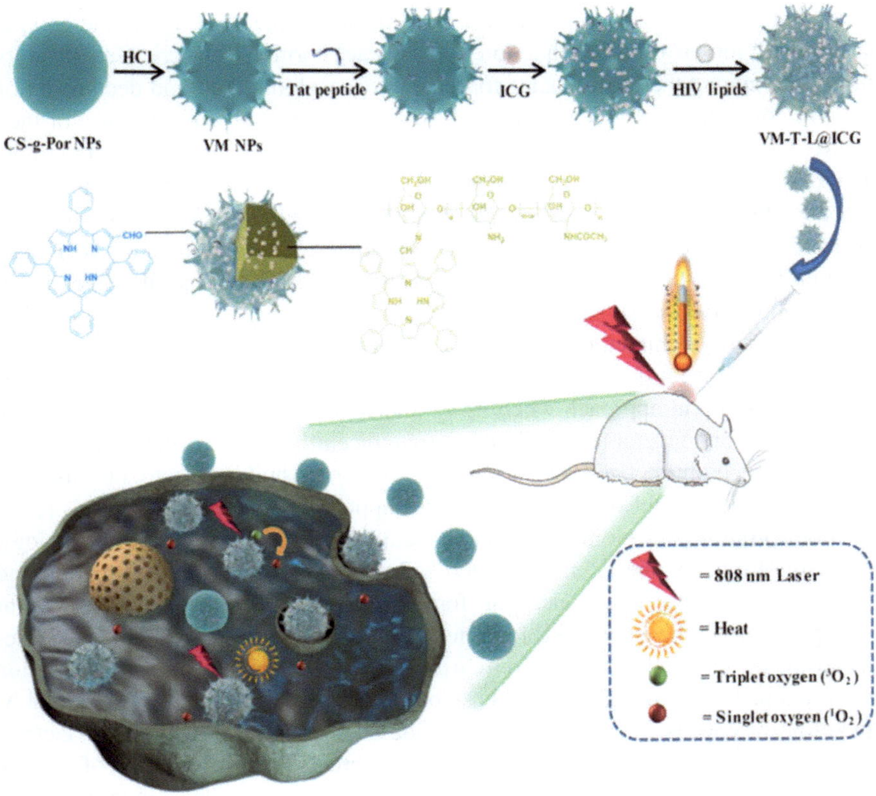

Fig. 2.2 Morphological and constituent viral-mimicking self-assembled nanoparticles for enhancement of cellular uptake [29]

2.2 Bio-inspired Biological Macromolecules in Biosensors

To address the need for the manufacture of materials for sensors, researchers are attempting to employ biomimetics. A fresh batch of sensors, including biosensors based on nanozymes, biosensors that adhere to soft tissues, biosensors that prevent fouling, etc., may be created by creating novel macromolecule composites that resemble biological systems. In addition to overcoming biological obstacles, biomimetic macromolecules have the potential to transform medication delivery, sensing, and regenerative medicine techniques. The best materials for creating biosensors are biodegradable macromolecules, which are regarded as a potent instrument to achieve ecologically friendly sensor functioning. These macromolecules have tolerable degradability and high biocompatibility, with the subsequent evolution of implanted biophysical or biochemical sensors being significantly influenced by biomimetics, especially those based on biodegradable macromolecules. This section presents bio-inspired or mimetic composites, which are based on natural or manufactured biodegradable macromolecules, and their uses in biophysical or biochemical sensing systems. The authors attempt to illustrate a number of designing biomimetic techniques that may be used in the production of biosensors. A summary is provided of the primary uses of biodegradable macromolecule-based biomimetics in the various platforms, interfaces, and recognition components of biosensors. The most promising prospects for biomimetic sensing applications techniques are presented, together with a discussion of the present drawbacks of each design approach [30].

As the most prevalent organic macromolecules, the majority of biomacromolecules exhibit exceptional biodegradability and biocompatiblity. However, the broad paradigm does not imply that every biomacromolecule degrades quickly or at a high rate. For example, only a few number of specialized microbes break down lignin, the most prevalent natural macromolecule, at a slow pace. In contrast to this macromolecule, cellulose biodegrades at a rapid pace and the biodegradation phase starts almost instantly. The macromolecule's level of acetylation determines the rate of breakdown in chitosan; molecular weight has no bearing on this [31]. Baranwal et al. [32] categorized biomacromolecules into three primary classes: macromolecules based on repeating units, synthetic macromolecules, and natural macromolecules. The most unique group of naturally occurring biomacromolecules is represented by polysaccharides, including cellulose, chitosan, and alginate, as shown in Fig. 2.3. Shell of a crustacean debris, like crab or shrimp shells, may be processed to produce chitosan. Although plants are the primary source of cellulose, numerous microorganisms are also capable of producing it. Bacteria and brown seaweed may also create alginate. On the other hand, polydopamine is produced chemically by two methods: physical self-assembly of dopamine and dihydroxyindole and oxidative macromoleculeization with covalent bonding. From a different perspective, all of the macromolecules that have been discussed fall within the category of biodegradable macromolecules [33]. Wu et al.'s categorization divides biodegradable macromolecules into two categories: remacromoleculeization macromolecules and regeneration macromolecules. For example, the fermentation remacromoleculeization of maize yields poly(lactic acid) (PLA). Processing is required to turn another class of biodegradable macromolecules into useful products. These macromolecules must dissolve and regenerate in order to be processed. In cellulose regeneration, for instance, cellulose I (native cellulose) is structurally changed into cellulose II crystals [34]. Islam and colleagues suggested regenerating and dissolving chitosan in an ionic liquid solvent in a different investigation. The mentioned biomacromolecules will be the main focus of this research since they are the most widely recognized biodegradable macromolecules, both synthetic and natural [35].

In shrimp, crabs, a few insects, fungus, microbes, and algae, chitin is deacetylated to generate chitosan, a N-acetyl glucosamine comacromolecule. Numerous researchers from a variety of biological science fields have taken notice of this linear polysaccharide because of its special qualities, which include perfect non-toxicity, biodegradability, biocompatibility, and ease of processing [36]. It has the ability to mimic the composition and operations of biological and natural systems, tissues, and microbes. One well-known and effective method for creating sticky hydrogels, which are the closest to the sticky proteins found in mussels, is grafting catechol derivatives onto the chitosan backbone's amino groups (chitosan-catechol). Mussels are able

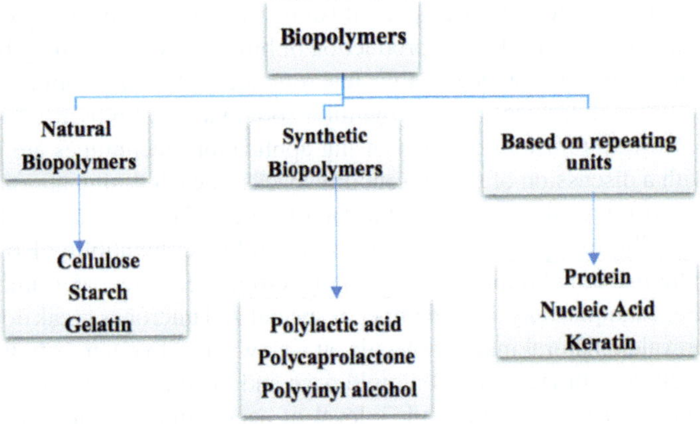

Fig. 2.3 Classification of biomacromolecules based upon their origin [32]

to adhere to and depend on submerged rocks due to the existence of proteins with multiple catechol groupings. Strong adherence to surfaces that are metal, biological, or inorganic is produced by the adhesive proteins found in mussels via the formation of hydrogen bonds, covalent bonds, π–π bonds, and electrostatic (charge-charge) interactions [37]. Another example is the development of catechol-functionalized chitosan-pluronic hydrogels by Ryu and colleagues, which have a high degree of adhesion to soft tissues and mucous layers [38]. Hydrogels derived from catechol-grafted chitosan exhibit great efficiency; nevertheless, their adherence is diminished by catechol toxicity and quick oxidation. In actuality, traditional sticky hydrogels that are bioinspired by mussels' adhesion behavior often have temporary adhesiveness. Guyot et al. proposed the use of dihydroxybenzoic acid to graft chitosan, a tiny molecule with a high resistance to oxidation, in order to overcome these issues. This hydrogel is anticipated to have applications in tissue engineering and injectable medication delivery. Because of their high tissue resemblance, varied and adaptable physicochemical characteristics, and high functionalization or doping ability, chitosan-catechol-based adhesive hydrogels are a type of three-dimensional macromolecule network that contains a significant amount of water and generally hold great promise in a variety of biomedical fields [39]. Recently, soft biosensors with implantable hydrogel adhesives have garnered significant attention due to their robust resistance to brittleness and tensile deformation. After serving their purpose, biosensors that adhere to tissue may be readily extracted from it without causing any harm, in contrast to commercially implanted biosensors [40]. By implanting biosensors, tissue degradation may be avoided thanks to the development of these flexible and soft biosensors. Conversely, there is a great need for skin-attachable soft bioelectronics or wearable biosensors to monitor human health problems in real time. Long-term use of certain wearable biosensors made of artificially produced biocompatible materials applied directly to the skin's surface might result in skin issues and allergies. Kim et al. proposed an autonomous tremor-detecting system based on diatom hydrogel, chitosan, and catechol to track Parkinson's disease patients' low-frequency vibrational movements in real time (Fig. 2.4) [41]. The authors claim that an adhesive hydrogel of chitosan, catechol, and diatoms was created for applications using wearable sensors because of the diatoms' porous structure and the chitosan's catechol and amine groups' strong adhesion ability, which caused the matrix of macromolecules and diatoms to become chemically and physically entangled. In addition to imitating mussel adhesive proteins, chitosan composites may also be used to imitate different biological systems, including the matrix outside of cells, skin, tissue of bones, growth hormones, etc. Because of its advantageous properties for wound healing, chitosan and its derivatives have garnered a lot of interest. Chitosan-based hydrogels may mimic the structure of skin and hasten the development of epithelial tissue on wounds. The extracellular matrix is a crucial component in controlling cell phenotypic, adhesion, migration, and proliferation, among other aspects of wound healing [42, 43]. Yang et al. fabricated chitosan nanofibrous microspheres that resembled extracellular matrix, a scaffold used in tissue engineering, by heating an alkaline urea solution containing chitosan molecules [44]. Additionally, chitosan fibers were modified with pieces of human collagen I to create the extracellular matrix. Furthermore, chitosan may be combined with nanoparticles and macromoleculeic materials to replicate the structure and functionality of bone [45, 46]. It should be mentioned that the materials utilized in bone tissue healing must have osteoconductivity, osteoinductivity, and osteogenicity. In an intriguing investigation, Rao and colleagues mimicked growth factors for vascular endothelium and neurotrophic factors generated from the brain by grafting chitosan nanofibers with a combination of peptides (RGI and KLT). Rats with sciatic nerve abnormalities were treated using peptide-grafted chitosan nanofibers as nerve canal fillers. More significantly, chitosan-based composites may make it possible for external tactile force/strain sensing to resemble skin, which is an essential feature of robotics, electronic skins, and personal health monitoring devices. Even though there are many different types of pressure/strain sensors, some, like skin, have very high sensitivity and a low detection limit, which allows them to detect even the smallest stresses [47]. Jin et al. used a skin-attachable hydrogel powered by poly (2-acrylamide/acrylamido-2-methylpropane sulfonic acid) and chitosan to suggest an epidermal strain biosensor. This hydrogel sensor, which resembles skin, was able to accurately identify even the smallest strains [48]. Another example is the introduction of an ultrasensitive carbon black-based strain sensor/chitosan composite sponge by Liu and colleagues for the detection of tiny stresses like vital signs [49]. One of the most important problems for scientific study has always been the creation of novel chemicals that may molecularly replicate recognition processes in biological systems. Their function as nanozymes that replicate the action of enzymes has grown in importance among the alluring uses of chitosan-based biomimetic macromolecule nanocomposites in recent years. Nanozymes are very effective nanocomposites with characteristics similar to those of enzymes. These synthetic or artificial enzymes have been set apart from natural enzymes by characteristics including excellent stability, affordability, and adjustable catalytic capabilities. With around 5000 D-glucosamine units, chitosan is a polycationic macromolecule that may mimic the amino acid environment of horseradish peroxidase (HRP). Because of these similarities, chitosan-based nanozymes have been developed [50]. For example, a nanozyme derived from Fe-doped carbon dots grafted onto chitosan was created by Pan and colleagues. This nanozyme was employed to break down hard bacterial biofilms because of its catalytic properties similar to those of HRP and chitosan's combined antibacterial action. Bacterial DNA was effectively cleaved by hydroxyl ($^{\bullet}OH$) radicals generated by the synthesized nanocomposite's peroxidase-like catalytic activity. Furthermore, the positive charge of chitosan allowed it to electrostatically adhere to the negatively charged cell membrane of the microorganism, altering its permeability and ultimately leading to bacterial cell death [51]. Drawing from the Fe, N-doped carbon dot chelated by chitosan, created a super peroxidase-active antibacterial nanozyme ($V_{max}/K_m = 1.77 \times 10^{-6}\ s^{-1}$), which was calculated to be 8.8 times greater than HRP. The designed nanozyme demonstrated several benefits, including great biocompatibility, good stability, and outstanding antibacterial activity, in contrast to conventional antimicrobial nanomaterials and natural enzymes, which have drawbacks like poor performance, instability, and poor safety. In addition to being effective in eliminating germs, nanozymes also significantly improve the detection of chemical or biological substances [52]. In sensing or biosensing applications, nanozymes have shown encouraging promise. The HRP enzyme-like characteristics of a nanozyme produced from magnetite/chitosan hydrogel were reported by Ariti et al. Using a smartphone-based readout, the reported nanozyme was used to determine thiabendazole colorimetrically. The colorless substrate for 3,3′,5,5′-tetramethylbenzidine (TMB) could be oxidized by the hydrogen peroxide/magnetite-chitosan nanozyme system to produce oxidized TMB (oxTMB) in blue, which had a 652 nm absorption peak. At acidic pH, the hydrogel nanoenzyme based on magnetite chitosan exhibited 24% higher activity of peroxidase than Fe_3O_4, confirming chitosan's ability to improve electron transport. When thiabendazole was added to the reaction solution, the nanozyme's catalytic activity was inhibited in a concentration-dependent manner. The thiabendazole concentrations were determined to be as low as the readouts from the smartphone and the UV-Vis spectrometer: 0.73 and 1.84 μM, respectively, due to the linear inhibition rate connection with the level between 0.1 and 100 μM [53]. In another study, He et al. described a colorimetric sensor for iodine measurement based on nanoparticles of palladium stabilized by carboxylated chitosan. The H_2O_2-mediated oxidation of TMB was accelerated by carboxylated chitosan-stabilized palladium nanoparticles. The produced nanoenzyme's peroxidase-like activity was validated by the color shift becoming colorless from blue. By forming iodine-palladium bonds and blocking the active websites of the palladium nanoparticles, the carboxylated chitosan-stabilized palladium nanoparticles demonstrated peroxidase-like activity, which was suppressed in the presence of iodine anions. A simple colorimetric test was used to find the anion using a detection limit of 0.19 nM, according to the inhibitory iodine's action ions [54]. Zou and his colleagues made a Cu/Fe/chitosan composite that works better than peroxidase and speeds up the oxidation of TMB. In this work, we developed a colorimetric sensor to detect mercury cations based on the fact that cysteine inhibits TMB oxidation by catalysis and that the formation of the cysteine-mercury ion complex results in the recovery of chitosan/Fe/Cu peroxidase-like activity [55]. In order to detect H_2O_2, Ragavan et al. used chitosan as a peroxidase mimic to create a sensor that uses colorimetry based on the study. Chitosan promoted the TMB's oxidation to a colorful complex in the presence of H_2O_2. In the range of 10 μM to 10 mM, the concentration of the colored complex increased linearly with increasing H_2O_2 concentration, with 1.55 μM being determined as the LOD. This paper-based sensor shown significant promise for significant uses in the measurement of H_2O_2 in a range of analytes [56]. The researchers proposed chitosan-gold nanocomposites as a peroxidase-mimicking nanozyme for measuring human blood samples containing glucose. The GO enzyme, which uses glucose as a substrate to oxidize, is often used to determine glucose. Therefore, a lot of investigations are being conducted to enhance the way this enzyme functions since its stability and reactivity have a significant impact on the effectiveness of the GO-based biosensor. A large portion of this study focuses on nanomaterials that imitate oxidase activity. A color shift proportional to the H_2O_2 concentration followed the oxidation process of TMB, which was accelerated by the chitosan–gold biomimetic. Therefore, a LOD of 0.6 μM was used to measure the H_2O_2 concentration. Additionally, glucose oxidase (GO) and chitosan-gold nanocomposite were able to detect glucose as low as 3 μM. In particular, GOx used a chitosan-gold nanocomposite to oxidize glucose, which produced H_2O_2. The glucose concentration was inferred indirectly from the chromogenic reaction's color shift [57]. Jiang and colleagues presented silver nanoparticles stabilized by chitosan that exhibit peroxidase-like properties. This nanozyme substrate, H_2O_2, oxidized OPD (O-phenylenediamine) and TMB, changing the substrates' colors from inert to blue for TMB and red for OPD. The fact that chitosan-stabilized silver nanoparticle-based nanozyme exhibits greater stability than the native HRP enzyme at high pH and temperature is one of its key benefits. Indeed, even under challenging conditions, this nanocomposite mimics HRP and exhibits consistent catalytic action. H_2O_2 was efficiently catalyzed to break down into radicals of hydroxyl ($^{\bullet}OH$) by the result of the nanocomposite. Through an interaction of electrons with silver nanoparticles stabilized by chitosan, the generated OH was stabilized, giving the nanocomposite peroxidase-like activity. Chitosan-stabilized silver nanoparticles stabilize the OH and oxidize TMB to produce a blue product. TMB and a cationic free radical form a charge transfer complex, which is responsible for the apparent blue hue. The experimental findings showed that the TMB–H_2O_2 interaction could not be catalyzed by nanoparticles of silver coated using additional macromolecules, like glycol polyethylene, to create a blue solution. Glucose was oxidized in the presence of GOx, and the main product of the GOx-catalyzed reaction was H_2O_2; therefore, the proposed method may be used for glucose detection. Based on this method, an improved colorimetric and visual biosensor for glucose measurement was developed, which has a lower limit of detection (LOD) of 100 nm and an upper limit of detection (DLR) of 5.0×10^{-6} 2.0×10^{-4} M. Although nanozymes mimicking peroxidase that use H_2O_2 as an acceptor of electrons have outstanding stability, great biocompatibility, and excellent catalytic activity, they have a number of disadvantages. In addition to oxidizing the analytes, the presence of H_2O_2 as a potent and unstable oxidant may also result in restrictions because of its breakdown. Thus, ascorbate

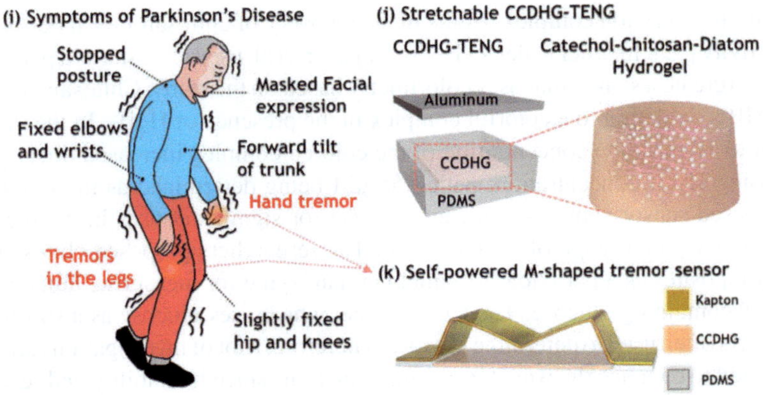

Fig. 2.4 Illustration of a skin-attachable self-powered tremor biosensor based on the catechol/chitosan/diatom hydrogel for real-time monitoring low-frequency vibrational motions of Parkinson's disease patients [41]

oxidase-like nanozymes, laccase-mimicking nanozymes, etc., are promising options for sensor design [58].

β-1,4 ($C_5H_8O_4$) m connected glucose units make up the long linear polysaccharide macromolecule known as cellulose. The composition or role of certain tissues is mimicked by this complex carbohydrate, which has a large number of hydroxyl groups. For instance, heparin-mimicking materials made from cellulose nanocrystals that have been functionalized were described by Gallagher et al. A kind of cellulose called cellulose nanocrystals is made up of rod-shaped, highly organized crystalline domains. Effective conversion of sodium carboxylate groups to primary hydroxy groups was made possible by the oxidation of cellulose nanocrystals; tetramethylpiperidine-1-oxyl radical 2,2,6,6 (TEMPO) is the mediator. Next, a sulfur trioxide-pyridine complex was used to sulfonate cellulose nanocrystals oxidized by TEMPO. Cellulose nanoparticles were also subjected to a sulfonation process comparable to the one described above (TEMPO-oxidized cellulose nanocrystals sulfonated). The strongest anticlotting effects were shown by sulfonated cellulose nanocrystals, which were followed by cellulose nanocrystals oxidized by sulfonated TEMPO [59]. Researchers created electrospun acetate-cellulose mats to simulate bone extracellular matrix in a different study. To bioactivate the surface of the mats and enhance osteoconduction and osteoinduction, they added sticky peptides such as RGD, KRSR, and the growth component BMP-2. Hydrogels based on cellulose mimicked the structure of skin and hastened the development of epithelial tissue on wounds [60, 61]. In order to make a printed gelatin gel from cellulose that could be used to make scaffolds for bone regeneration, Wang et al. treated bacterial cellulose (BC) with maleic acid (MA) and 2,2,6,6-tetramethylpiperidin-1-yl)oxidanyl (TEMPO) (Fig. 2.5). New techniques for the creation of composites that replicate skin characteristics are required for developing touch sensors, notwithstanding the attempts to replicate the composition and functionality of tissues [62].

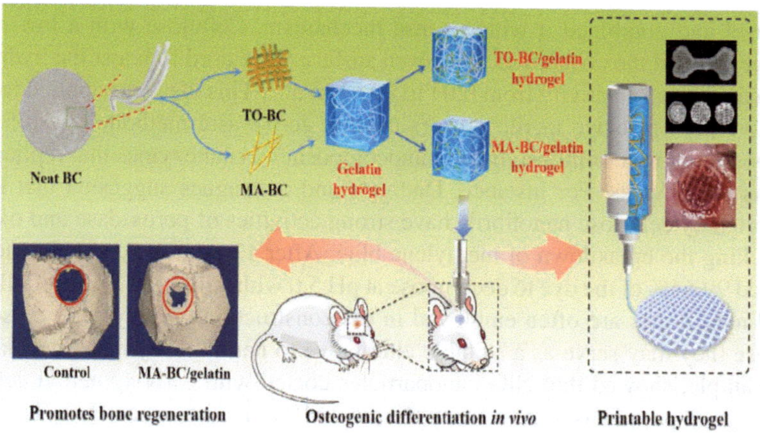

Fig. 2.5 Preparation of printable bacterial cellulose-gelatin gel for in vivo osteogenic regeneration [62]

The biggest and oldest sensory organ in the human body, the skin has many touch receptors that allow it to perceive a wide range of external stimuli, such as pressure, strain, vibration, slippage, etc. Therefore, sensors that are based on materials that imitate the skin allow for the detection of various stimuli, including pressure, strain, and other things. Their promise for healthcare is shown by their ability to detect the delicate stresses of the human body via blood pressure, heart rate, voice activity, and breathing patterns. Notably, human–machine interaction and health monitoring both use skin-inspired touch sensors. Exciting features that were previously unattainable for conventional sensors have been made feasible by these skin-like sensors. Meanwhile, viscoelastic and biocompatible cellulose fibers replicate the skin's layers of dermis and epidermis. A cellulose-based biomimetic hydrogel that mimics the structure of skin with skin-like characteristics was suggested by Zhang et al. A network of porous cellulose fibers with adjustable crystallinity and an elastic polyacrylamide that had been chemically cross-linked made up the hydrogel. This chemical demonstrated exceptional elasticity, great stretchability (~ 914.3%), high strength (0.8 MPa), and strong biocompatibility. Based on the biomimetic hydrogel generated from cellulose, a wearable strain sensor was created by using these exceptional advantages [63]. Shu and colleagues used cellulose-based skin-mimicking hydrogel to create a multipurpose sensor for temperature and mobility monitoring of humans, which was inspired by skin conductivity. Polyvinyl alcohol with cotton linter cellulose was dissolved in a focused zinc chloride solution to create the ionic conductive hydrogel. It demonstrated a broad tolerance for temperature, 2.05 MPa of compressive strength, 0.30 MPa of tensile strength, and 8.16 S m^{-1} of ionic conductivity. Apart from temperature, this sensor could precisely identify various pressure signals brought on by people writing different numbers. It's intriguing to note that this sensor could track both the intricate motions of the muscles and the fingers [64]. Similar to this, Chen et al. created a skin-like, ion-conducting cellulose hydrogel and combined it with a sense mechanism. Cellulose with a low molecular weight and the concept of an uneven surface were used to boost the hydrogel's piezocapacitive sensitivity from 0.04 to 89.81 kPa^{-1}. This idea was applied to ultrasensitive piezoresistive tactile sensors using a zero-waste methodology [65]. It is also possible to use cellulose-based nanocomposites as nanozymes that replicate the actions of enzymes. For instance, Dadigala and colleagues suggested that PdNPs supported by cellulose nanofibrils have strong activities of peroxidase and oxidase, mimicking the breakdown of methylene blue. After 12 min, the created nanozyme enabled 99.64% of the dye to decompose at pH 5.0 with 0.4 M H_2O_2 [66]. Cellulose-based nanozymes are often employed in the construction of enzymatic biosensors because they may serve as a suitable alternative to natural enzymes. Shehala et al. for example, showed that NiO nanoparticles coated with carboxymethyl cellulose exhibit good peroxidase-like activity toward TMB oxidation when H_2O_2 is present. After only one and a half minutes, TMB oxidation caused a quick visual reaction (change in color from colorless to blue). With a LOD of 0.078 μM, the presented colorimetric device showed exceptional sensitivity and may also measure H_2O_2 levels in human serum and milk samples [67]. Baretta and colleagues combined PB (Prussian blue nanoparticles) with GOx in a hydrogel based on carboxymethyl cellulose. Outstanding peroxidase-like catalytic ability is shown by Prussian blue nanoparticles with iron atoms coordinated by cyanide bridges. These nanoparticles are used to create colorimetric glucose biosensors by combining with GOx and chromogenic chemicals to detect H_2O_2. Consequently, glucose may be determined using cellulose hydrogel with carboxymethyl combined with GOx and PB. With a LOD of 1.0 μM, a strong linear relationship between the absorbance intensity at 420 nm and the glucose concentration was found in the 5.0–100.0 μM range. Serum glucose levels were determined using the colorimetric biosensor that was developed. A very sensitive chromogenic peroxidase substrate, 2,2′-Azino-bis-(3-ethylbenzothiazoline-6-sulfonic acid) (ABTS), was added to the carboxymethyl cellulose hydrogel-PB to detect H_2O_2 (Fig. 2.6a) [68]. According to Alle et al., a peroxidase-mimicking nanozyme is produced when gold nanoparticles are grown in-situ on cationic cellulose nanofibers. Considering the TMB substrate, the colorimetric measurement of glucose and H_2O_2 is produced using nanozyme. This biosensor's stability, which allows for up to five reuses without lowering activity, is one of its key benefits. More significantly, the suggested biosensor demonstrated excellent performance in measuring blood glucose levels. The results were similar to those of the conventional GOD-POD, or glucose oxidase-peroxidase technique (Fig. 2.6b) [69]. Bandi and colleagues suggested using a nanopaper made of carbon dots mixed with iron (which acts like peroxidase) to measure glucose and H_2O_2 using a smartphone and a cellulose-supported colorimetric apparatus. As in earlier research, the oxidation of TMB was used to examine the nanopaper's peroxidase-like activity [70]. With polypyrrole nanocomposite, carboxymethyl cellulose, and PB nanoparticles, Uzunçar et al. made an amperometric detector that can measure glucose and H_2O_2 [71]. Researchers developed gold nanoparticles adsorbed on a cellulose membrane for colorimetric measurement of uric acid (UA) and H_2O_2. Nanoclusters that resembled peroxidase catalyzed the H_2O_2 breakdown on the nanozyme surface, which oxidizes colorless TMB to blue oxTMB [72]. Uzuncar and colleagues proposed a catalytic nanocomposite consisting of nickel-hexacyanoferrate, PEDOT (poly (3,4-ethylene dioxythiophene)), carboxymethyl cellulose, and PB nanoparticles for the enzyme-free measurement of H_2O_2 in tap water. Nickel-hexacyanoferrate helped make the structure more stable, and the peroxidase-like action was shown by the PEDOT layer, PB nanoparticles, and carboxymethyl cellulose [73].

Because of its low toxicity, ease of gelation, cheap cost, and biocompatibility, alginate, a naturally occurring anionic biomacromolecule, is extensively used in a range of biomedical applications, like biosensors. Like cellulose and chitosan, alginate may resemble biological or tissue substances [74]. Yan et al. presented an additional example for wound healing: a sodium alginate-enriched reduced graphene oxide

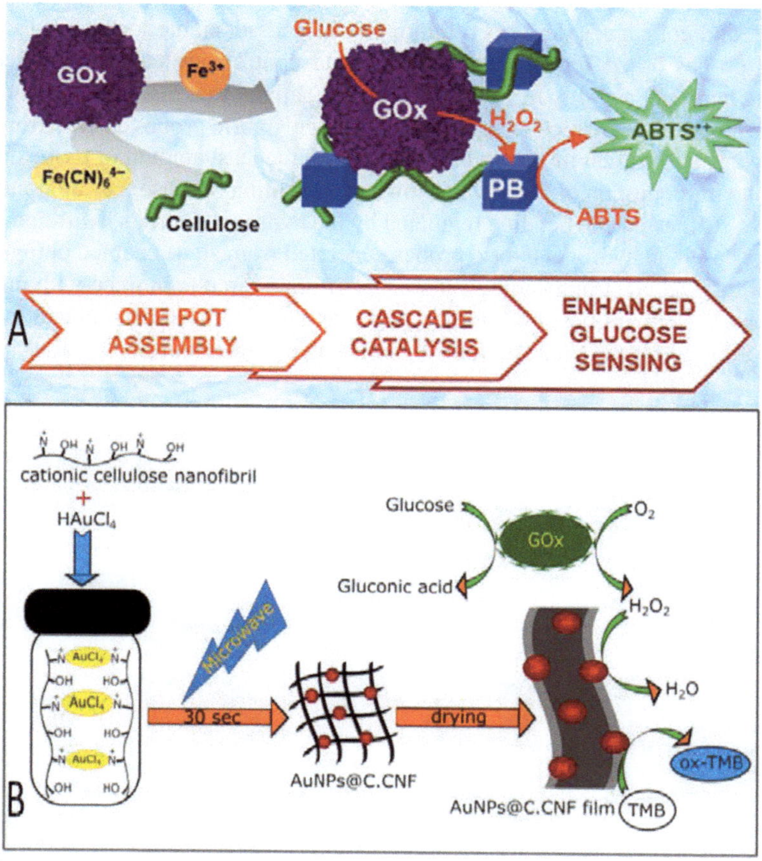

Fig. 2.6 **a** Schematic illustration of the in-situ synthesis of AuNPs@C.CNF and their application in glucose detection via the peroxidase-like activity with TMB substrate [68], and **b** schematic illustration of the glucose sensing mechanism using a carboxymethyl cellulose hydrogel-based biosensor with Prussian blue nanoparticles and GOx for enhanced H_2O_2 detection via ABTS chromogenic reaction [69]

(Ag@rGO) nanocomposite enhanced by bioinspired silver nanoparticles. Alginate molecules of sodium were adsorbed on Ag@rGO sheets to create the layer of the nacre-mimic nanocomposite, which was then sprayed onto the substrate (Fig. 2.7a). It was examined if this nanocomposite might shield the wound from infections caused by harmful microorganisms. This nanoparticle promoted closure of wounds and decreased inflammation in rats, based on in vivo experiments. The Ag@rGO-sodium alginate film was applied to the right wound, which showed substantial shrinkage, as shown in Fig. 2.7b, but the left wound, which was wrapped following disinfection using a sterile towel, did not alter much [75]. Alginate-gelatin hydrogels have been shown by Distler et al. to mimic the intricate mechanical properties of brain tissue. This material can replicate brain activity because of its combinatorial tunability. Gelatin behaves almost linearly, but alginate shows a clear nonlinear reaction and asymmetry in compressive strain [76].

Park and colleagues developed multipurpose and adaptable long-term hydrogel hybrid probes based on alginate detection and control due to fiber-coupled Ca-alginate hydrogel's ability to stimulate neuronal activity that resembles tissue of the brain. These sensors, which were based on hydrogel hybrid probes derived from alginate, were sent straight into deep brain areas with the least amount of tissue damage related to micromotion of the brain following implanting because they demonstrated adaptive stiffness while bending regulated by the hydrogel matrix's hydration states. Sensors based on hydrogel hybrid probes generated from algae enabled optogenetic, behavioral, and electrophysiological studies of brain circuits in mice with minimal reactions to alien entities. Furthermore, for over six months after implantation, this sensor allowed for the stable monitoring of isolated single-neuron potentials in mice

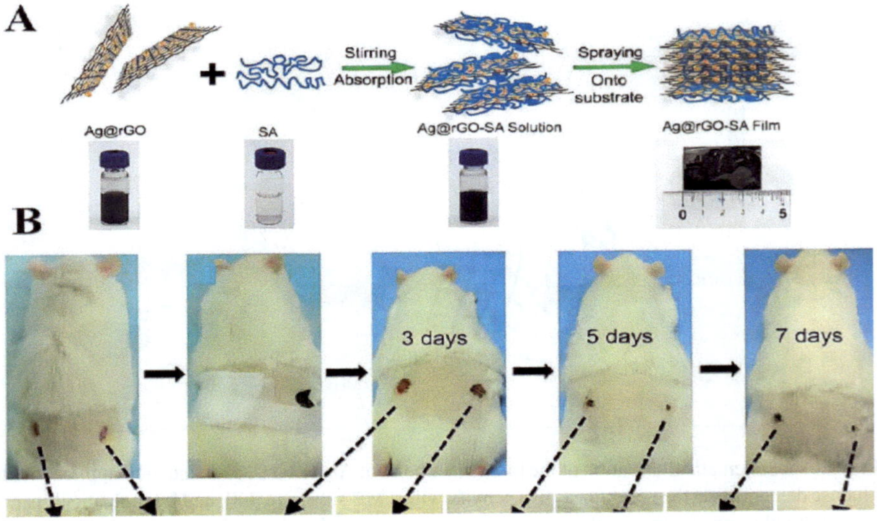

Fig. 2.7 Illustration of **a** a bioinspired Ag@reduced graphene-sodium alginate nanocomposite for the wound healing, **b** wound healing after 3, 5 and 7 days [75]

that were free to move around. Because of its mechanical strength (fracture toughness 500 Jm^{-2}), softness (shear modulus 5.5 kPa), strong surface integration with functional fiber-forming macromolecules, long-term chemical stability in physiological settings, and biocompatibility, poly(acrylamide)-alginate hydrogel was chosen as the matrix for hybrid probes in this study. The drawbacks of large, rigid sensors may be addressed by soft materials that resemble tissue or skin [77]. Strong oxidase-like activity was shown by very effective platinum nanozymes treated with sodium alginate. According to He et al., color change resulted from the catalysis of the TMB oxidation process by nanoparticles of alginate and platinum with the inherent activity of oxidase. However, the addition of oligomeric proanthocyanidins (OPC) to the TMB system, which alginate-platinum nanoparticles catalyzed, led to a competition between OPC and TMB for binding to the active website of the nanozyme. A colorless reaction was the outcome of oligomeric proanthocyanidins' substantial inhibition of the catalytic activity. Considering the exceptional activity similar to oxidase of the alginate-platinum nanozyme and the antioxidant potential of OPC, a colorimetric sensor was created for the OPC determination [78]. A colorimetric sensor was created by Zou et al. utilizing Cu^{2+}-crosslinked alginate hydrogel loaded with CoOOH nanozyme. Cu^{2+}'s electrostatic interaction with pyrophosphate ions caused the hydrogel to separate, releasing CoOOH nanoflakes to catalyze TMB's oxidation to oxTMB (a chromogenic reaction). The liberated CoOOH nanozyme's catalytic action was amplified by pyrophosphate ions and Cu^{2+}-pyrophosphate. The addition of alkaline phosphatase to the mixture catalyzed the pyrophosphate ion hydrolysis to produce phosphate ions. By activating ions of pyrophosphate and boosting the activity of the nanozyme, a regulated release of CoOOH nanozyme was used to create an alkaline phosphatase colorimetric sensor with superior performance detection. Alkaline phosphatase may be detected by this stimuli-responsive sensor with a 0.07 mU mL^{-1} LOD and 0.1–250 mU mL^{-1} DLR [79]. In a different study, Qian and his colleagues made PtS2 nanosheets inside alginate/poly-L-lysine microspheres that work like peroxidase to measure dopamine. Blue oxTMB was produced via the oxidation of TMB by H_2O_2, which was catalyzed by PtS2 nanosheets. The blue hue faded when oxTMB was decreased by dopamine, a strong reducing agent. A colorimetric sensor with a LOD of 0.22 μM and a DLR of 2.0–200 μM was created using this approach. Serum samples were used to test the viability of a nanozyme-based sensor utilizing the spiked/recovery technique (Fig. 2.8) [80].

2.3 Bio-inspired Other Biological Macromolecules

As a macromolecule containing amine and catechol functional groups, polydopamine has excellent chemical activity, adhesion, hydrophilicity, and biocompatibility. In fact, because of its sticky functional groups (amine and catechol) that replicate the inherent bioadhesive qualities of mussels, polydopamine has become one of the many adaptable bio-inspired components in the last ten years [81]; it is extensively used

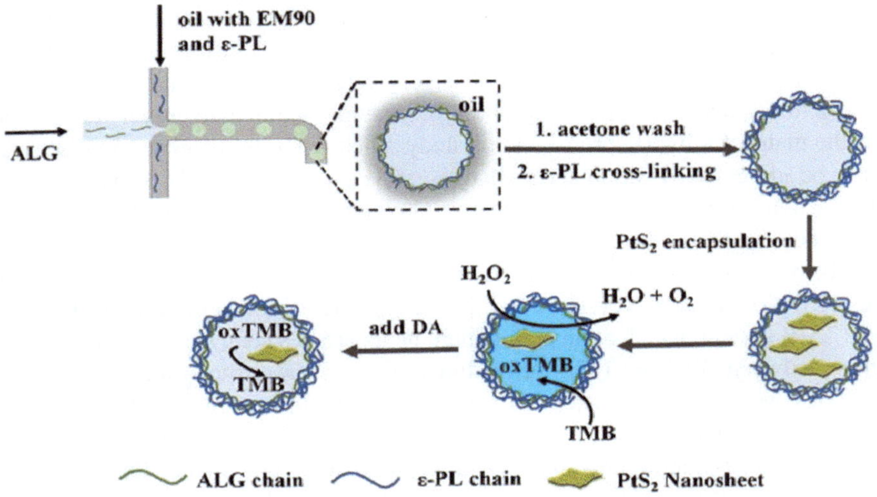

Fig. 2.8 The image illustrates the creation of PtS_2 nanosheets inside alginate/poly-L-lysine microspheres, used in a colorimetric sensor for measuring dopamine levels [80]

for skin [82] and peripheral nerve regeneration. The scaffold that resulted from decorating polydopamine with calcium-cross-linked alginate dialdehyde-gelatin demonstrated adequate biomineralization for bone regeneration [83, 84]. Because of its many catechol groups, polydopamine is known to be a self-adhesive macromolecule, much like mussel proteins. The primary cause of mussels' adherence to damp surfaces is catechol. The melanin pigment in human skin, which shields the skin's underlying layers by absorbing dangerous light wavelengths, is structurally very similar to polydopamine. Because mussels bioinspire into human skin, polydopamine is crucial [85]. In order to create polydopamine/carboxymethyl cellulose/polyacrylic acid adhesive hydrogels that were modeled after the sticky proteins found in mussels, Tang et al. used the functional polydopamine. This biocompatible hydrogel, which was inspired by mussels, was able to stick securely to a variety of surfaces, including pig skin, glass, aluminum, and stainless steel. More significantly, hydrogel demonstrated the perfect UV-protective ability to stop collagen fibers from breaking down under UV light. Therefore, skin care products may use this UV-shielding hydrogel [86]. In another intriguing study, Buscemi and colleagues created whole bacterium-based hybrid photoanodes that combine enhanced photo-induced harvesting of electrons employing redox driven by bioinspired-adhesive polydopamine [87]. Scientists used purple bacteria as examples of microorganisms because they can make photo-bioelectrochemistry systems that can make NH_3 and H_2 and also do bioassays. The polydopamine matrix fabrication method enabled the trapping of the photoanode's living purple bacterial cells. The results show that electrons move between biology and abiotics, and cell viability goes up by five times when bacterial cells are put on the electrode surface at the same time as polydopamine macromoleculeization. Implantable/wearable tissue-adhesive biosensors or polydopamine-based membranes with potentiometry may be created by imitating the adhesion process of mussels. Polydopamine is a strong adhesive primer that sticks to a variety of surfaces, and polydopamine's ability to act as a smart free radical scavenger is another advantage that makes it attractive for the development of implantable sensors. This ability would be helpful for very stable in vivo analysis in intricate matrices, even when reactive oxygen species (ROS) are present [88]. Jiang et al. made a calcium ion-selective electrode using a poly(vinyl chloride) membrane covered in polydopamine, which was based on the shells of marine mussels. The membrane made of polyvinyl chloride covered with polydopamine demonstrated an important decrease in the absorption of blood platelets when compared to the uncoated Ca^{2+}-selective electrode; nevertheless, the reactions of the two potentiometric electrodes made of membranes were identical. Biofouling, or the adsorption of lipids and proteins on the sensor surface, is a problem in sensing techniques, which often occurs when the sensor is in constant contact with living materials. The findings showed that the PVC-based hydrophilic polydopamine coating has a strong antifouling performance (Fig. 2.9a) [89]. In another example, Du and colleagues created a sticky, washable electrode for ambulatory electrocardiography sensing by combining stretchy elastomer polystyrene, poly(ethylene-ran-butylene)-block, polystyrene, polydopamine, and silk fiber with stretchy Au nanofilms. This electrode's ability to be readily removed from the skin during prolonged ECG monitoring is one of its key benefits; this prevented injury to the skin's epidermis upon detachment. More intriguingly, after stretching to 300% pressure, the developed electrode adhered to the pig's heart's wet surface. It also adhered to the pig's dry skin when the wrist was bent. In contrast, the silk fibroin-based electrode failed to adhere to dry skin when the commercial electrode was in motion easily on moist skin [90]. Another example that draws inspiration from nature is the integration of in vivo sensing using a tissue-implantable electrode with good capacity to prevent fouling, along with a biointerface based on SBMA (sulfobetaine methacrylate) and biocompatible polydopamine by Feng et al. Changing the chemical makeup of the interface with different hydrophilic coatings, such as zwitterionic or PEG (polyethylene glycol) macromolecules, is very important because biofouling of the electrode, which happens when proteins bind to the sensing interface without specificity, often makes implanted electrochemical biosensors less effective. This investigation used methacrylate of sulfobetaine as a zwitterionic compound. The polydopamine/SBMA-based microsensor accurately monitored the dynamics of ascorbic acid in the brains of both rats with Parkinson's disease and normal ones, which also lessened the initiation of the acute neuroinflammatory response locally. This sensor demonstrated 92.0% of its original sensitivity two hours after implantation in the brain tissue, indicating high stability in the tissue of the rat brain. Furthermore, nonspecific protein securing, including fibrinogen, was prevented by the SBMA-based polydopamine biointerface in the complex biological fluids. To reduce nonspecific adsorption on their surface, the biointerface that has been described may really resemble the proteins and phospholipids that make up cell membranes. More significantly, on almost every kind of electrode surface, this polydopamine-based biointerface demonstrated a high level of adhesion (Fig. 2.9b) [91]. In addition to replicating human skin and mussel adhesive proteins, polydopamine also has a role in mimicking other biomolecules, including enzymes. For instance, Mohammad et al. combined copper phosphate and mussel-inspired polydopamine with SiO_2-NH_2-coated Fe_3O_4 nanoparticles to create magnetic Cu-polydopamine nanoflowers. With H_2O_2 present, this magnetic nanocomposite is shown to have catalytic activity similar to that of peroxidase, converting from colorless TMB to oxidized blue TMB. They are very promising for organic dye reduction via catalysis and biosensing (bioassays) [92]. Gao and colleagues produced cu-doped polydopamine, which resembles laccase, on cellulose paper. A large variety of phenolic substrates and amines may be one-electron oxidized with molecular oxygen by laccases, a family of multi-copper oxidases, without the need for or generation of hydrogen peroxide. Due to the inherent fragility of this enzyme, researchers have created biosensors that effectively replace natural laccases by modeling nanomaterials after the structure of the laccase's electron transport mechanism and active site. When compared to the native laccase, the nanocomposites of Cu-doped polydopamine showed increased chemical production of benzoquinone from hydroquinone through catalytic activity. A UV-vis spectrophotometer is used to measure the absorbance of yellow 2-nitro-5 thiobenzoic acid, which was used to determine the catalytic activity of laccase-like Cu-doped nanocomposites of polydopamine. In this investigation, 2-nitro-5-thiobenzoic acid and benzoquinone, which was produced by catalytically oxidizing hydroquinone, interacted via Michael coupling to provide a colorless material [93].

Biosensors based on bio-inspired biological macromolecules for early cancer detection

The use of natural macromolecule-based biomimetic sensing techniques is an intriguing way to get beyond the limitations of biosensors and improve their ability to identify targets in biology, particularly markers of tumors. It is not hyperbole to state that the capacity of biosensors based on biomimetic macromolecules to identify cancer early on holds the key to transforming disease management [94–98]. The special qualities of scaffolds mimicking biology made of biomacromolecules, such as their excellent biocompatibility, low toxicity, ability to produce stable hydrogels, inherent catalytic capabilities, and ability to stick to skin or tissues, increase the potential for in situ and real-time tumor marker detection. The function of biomimetic biological macromolecules in cancer detection using biosensors will be the main topic of this section. Soft, wearable, and implantable biosensors have shown remarkable promise thanks to hydrogels using chitosan as an adhesive that are inspired by biological mussels' behavior of adhesion [99–104]. Chitosan-based bioinspired adhesive macromolecules possess unique properties that render them ideal for application in implanted or wearable biosensors, specifically triboelectric or piezoelectric biosensors [105–108]. These characteristics include quick flexibility, strain sensitivity, self-adhesion to skin or tissue, cheap cost, biocompatibility, and reaction to external stimuli. Bioinspired adhesive, the use of hydrogels, has been

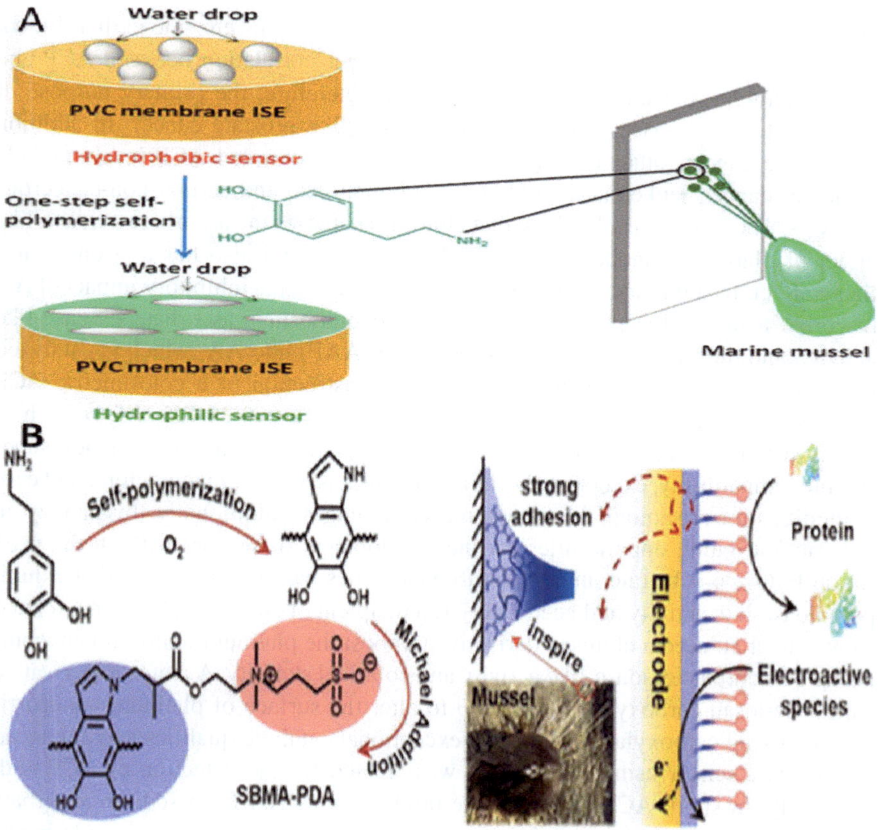

Fig. 2.9 **a** The image depicts the development of a calcium ion-selective electrode with a poly(vinyl chloride) membrane coated in polydopamine, inspired by marine mussels, to reduce biofouling in sensing applications. Adapted with permission from [89]. Copyright 2019 American Chemical Society. **b** The image illustrates the development of a polydopamine/SBMA-based biointerface for in vivo sensing, inspired by mussel adhesion, which prevents biofouling and enhances electrode stability in tissue-implantable sensors [91]

developed, despite several investigations on the construction of biosensors for pressure or tension connected to tissue based on these hydrogels [109, 110]. The development of hydrogel-based biosensors that can stick to the skin to monitor biomarkers particularly tumor markers may be sparked by this review article [111]. Composites based on chitosan have the ability to replicate both the catalytic activity of certain enzymes and the behavior of certain tissues [112–116]. Thankfully, a great deal of research has been conducted in this area, and a variety of sensors that use biology based on nanozymes have been developed to assess biomarkers for cancer in order to diagnose the incurable illness early. For instance, Deng et al. used chitosan-platinum nanocomposites to provide a sensitive method for colorimetrically determining acid phosphatase (ACP). As an enzyme that contains copper, this nanocomposite can replicate the actions of ascorbate oxidase. The prostate produces the enzyme ACP, and men with prostate cancer may have higher levels of this enzyme in their blood. Research has demonstrated a significantly stronger correlation between ACP and the morphological features of prostate cancer. Therefore, the primary purpose of the ACP test is to assess and identify diseases like prostate cancer. In addition to prostate cancer, other conditions that may cause elevated ACP levels include testicular cancer, leukemia, non-Hodgkin's lymphoma, and certain non-cancerous illnesses. The study's results show that chitosan increases the ascorbate oxidase-like effect of platinum nanoparticles on TMB oxidation and also has a strong stabilizing effect. In summary, TMB's accelerated oxidation, the inhibitory impact of AA (ascorbic acid) on the platinum-chitosan nanocomposite's ascorbate oxidase-like activity, and the hydrolysis of AA 2-phosphate (AAP) into AA, which is catalyzed by acid phosphatase, were the basis for the construction of a colorimetric ACP biosensor. The LOD was 0.016 U L^{-1}, and the DLR was 0.25–2.5 U L^{-1}, both of which were attained by this sensing technique. Outstanding catalytic efficiency, superior biocompatibility, exceptional stability of colloids, ease of manufacture, and ease of modification are among this nanozyme's advantages. Additionally, this nanozyme presents intriguing opportunities for the development of sensors without the need for an H_2O_2 co-substrate, in contrast to nanozymes based on platinum that exhibit peroxidase-like activity and need H_2O_2 as a recipient of electrons [117]. In a related study, He et al. used a chitosan derivative to boost the platinum-chitosan nanocomposite's Ascorbate oxidase-like activity and colloidal stability. A chitosan derivative called Chitosan carboxylated was used to alter the surface of platinum nanoparticles. Utilizing carboxylated chitosan's exceptional catalytic qualities and enhanced efficiency, a sensor using fluorometry was created for the detection of ACP with sensitivity. In short, ACP catalyzed the production of AA from sodium ascorbate phosphate (AAP), which was subsequently produced by oxidation using a platinum-chitosan nanocomposite that produces DHAA (dehydroascorbic acid) by mimicking the activities of ascorbate oxidase. A DFQ (derivative of fluorescent quinoxaline) with a bright 425 nm fluorescence emission was produced by the reaction of DHAA with OPDA (o-phenylenediamine). This nanozyme-based sensor demonstrated a broad A LOD of 1.31×10^{-3} U L^{-1} and a DLR of 0.25–18 U L^{-1}, indicating high selectivity for acid phosphatase. This sensor's ability to detect phosphatase acid in samples of human semen is its most significant benefit [118]. In a different work, Wang et al. proposed an electrochemical biosensor based on chitosan nanozyme, reduced graphene oxide, and Hemin@AuNPs for real-time H_2O_2 detection from live cancer cells, with promising results. One of the well-known signaling chemicals in cancer cells is hydrogen peroxide; for example, ROS (reactive oxygen species) and excessive inflammatory response are often associated with cervical cancer, the most common disease in women. HeLa cells generated from cells of cervical carcinoma were therefore used as an example for this investigation. The electrode transformed by nanozyme demonstrated outstanding electrochemical performance in relation to H_2O_2 with quick response, excellent stability, and great selectivity using the previously published nanohybrid peroxidase-like activity. This electrochemical sensor can detect H_2O_2 with a DLR between 1.0 and 0.05 M and a LOD of 9.3 nM [119]. Apart from chitosan-based composites that replicate the composition or function of biological systems, chitosan itself may also enhance the effectiveness of components used in cancer detection biosensors, where, based on hydrogen bond networks, chitosan may spontaneously form into fibers. In order to create high-performance biosensors, signal probes or bioreceptors may be included in the hydrogel of chitosan. For example, reduced graphene oxide/chitosan was suggested by Nasrollahpour et al. as a nanobiocompatible platform for the electrochemiluminescence (ECL) technique of ultrasensitive HER-2 bioassaying. As an ECL emitter, the $[Ru(bpy)_3]^{2+}$ signal probe was incorporated into biocompatible chitosan for this investigation. DLR and LOQ were determined to be 1 fM–1 nM and 1 fM, in that order, based on this sensing method [120]. Zhang et al. described a chitosan-based immunosensor and graphene oxide reduction for the electrochemical identification of malignant melanoma biomarkers. They used a cross-linking technique with glutaraldehyde to create an immunorecognition interface by grafting antibodies against S-100B, which act on the chitosan molecules as biological receptors [121]. A genosensor based on chitosan and butein grafting for the electrochemical detection of exosomal nucleic acid specific to CD24 for the diagnosis of ovarian cancer was created by Krishnan et al. When Butein contains many electroactive hydroxyl groups, it becomes a species that has a redox (electroactive) function on the electrode surface. In contrast to butein, the chitosan-modified electrode with butein grafting displayed a highly reversible redox process activity under varying pH values between 4 and 9 [122]. Cellulose has garnered a lot of interest lately as a desirable substrate for the advancement of disposable, affordable, and POC (point-of-care) target detection devices. However, cellulose-based hydrogels, which are three-dimensional porous materials made by chemically or physically linking macromolecules together, offer a lot of options for making sensing materials. Most significantly, it is simple to alter the surface, Crotch, and interlink via physical or chemical methods because of the surface's abundance of hydroxyl groups [123]. Piloto and his colleagues made CdTe QDs that were covered with molecularly imprinted macromolecules (MIPs) and mixed them into cellulose hydrogel to find out how bright the tumor marker CA19-9 was. Between 2.76×10^{-2} and 5.23×10^{-2} U mL^{-1}, the fluorescence quenching of MIP-treated CdTe QDs rose linearly with increasing CA19-9 concentration. The delicate nature of the Sensing fluorescence technique was enhanced in the direction of ranges of therapeutic importance by the addition of QDs coated with MIP with hydrogel made of cellulose [124]. As previously stated, mussel-inspired polydopamine adheres strongly to almost every kind of electrode surface, including cells. This is why Hong et al. created an immunosensor based on polydopamines-gold nanowires in three-dimensional electrodes for the tumor marker AFP's electrochemical detection. The oxidation current increased by 2.6 times due to the four-fold active area of gold nanowires. The polydopamine nanofilm, which adhered well to the gold nanowires, strongly bound antibodies and was highly compatible with living organisms, greatly improved the electrochemical signal and caused the capacitance to increase sixfold. A broad LOD of 0.01 pg mL^{-1} and a DLR of 0.1 pg mL^{-1} 50 ng mL^{-1} were attained using this approach [125]. Fan et al. created a biomimetic polydopamine-coated TiO_2-based photoelectrochemical sensor platform with nanotube arrays and the recognition element folic acid for the crucial oncomarker serum-soluble FBP (folate binding protein) by utilizing zwitterionic molecules and adhesive polydopamine. In order to prevent the electrode's biofouling caused by the complicated biofluids' non-specific protein binding, polydopamine-TiO_2 nanotubes were coupled with PEG, or poly(ethylene glycol) with an amino group at the end, which is an antifouling agent. The detection limit is 2×10^{-4} ng mL^{-1}; the biosensor presented was capable of measuring FBP falls between 1×10^{-3} and 5×10^{2} ng mL^{-1} [126]. In a separate study, Xu et al. developed a biosensor that detects fouling electrochemically for the detection of CEA (carcinoembryonic antigen). It has a DLR of 0.01–10 pg mL^{-1} and a low limit of detection (LOD) of 3.3 fg mL^{-1}. This method used the zwitterionic macromolecule PSBMA and anti-CEA aptamer-functionalized polydopamine to detect CEA in the complicated biological medium [127]. Polydopamine has a significant quenching of fluorescence capacity in the direction of the variety of colors that glow, in addition to its ability to imitate mussel proteins and the laccase enzyme. Thus, fluorescence detection of biomolecules, including tumor markers, is possible with this superquencher. To determine the CEA, for example, Deng and colleagues created an immunosensor that is photoelectrochemically based on the repeated polydopamine nanosphere quenching in the direction of CdS-modified TiO_2 nanoparticles doped with Mn^{2+}. After the antigen and antibody were incubated, nanoparticles of polydopamine approached TiO_2/CdS:Mn. As a result, the nanospheres' consumption of the photoelectrons generated in TiO/CdS:Mn and their blocking of donor electrons' access significantly reduced the photocurrent. A broad DLR of 0.1–100 ng mL^{-1} and a detection limit of 0.02 pg mL^{-1} were achieved with this immunosensor (Fig. 2.10) [128]. In a different study, Jiang and colleagues reported a fluorescence aptasensor that uses silver nanoparticles as a luminous dye and polydopamine nanospheres as a quencher for fluorescence to detect AFP and CEA. This study used a variety of ssDNA (single-stranded DNA) templates, such as sequences of anti-CEA and anti-AFP aptamers, to create silver nanoclusters. Remarkably, ssDNA strands were adsorbed via the π–π interaction on the polydopamine surface. The aptamer-target complex was formed when the tumor indicators were added to the resolution, and this complex then caused the silver nanoclusters that are ssDNA-templated to be released from the surface of polydopamine. In the 10–100 nM range, the restored fluorescence intensity rose linearly with the AFP and CEA concentrations; AFP and CEA had respective LODs of 5.6 and 2.4 nM [129]. Generally speaking, polydopamines have a lot of potential for creating antifouling sensors with exceptional stability because of their ability to adhere to various surfaces, particularly electrodes in some cancer cells and electrochemical sensors, in addition to their capacity to mix or co-deposit using antifouling agents and above all quench fluorescence in sensors that rely on fluorescence. Antifouling materials like zwitterionic macromolecules, ethylene glycol, etc., are necessary for reducing biofouling because biofouling, or serum proteins adhering to the surface of electrodes and biosensors that have been implanted, significantly impairs sensor performance and shortens the device's lifespan. Conversely, zwitterionic compounds reduce the alien entity's reaction to biosensors that are implanted. The sensors stand out due to their exceptional and consistent performance in the intricate biological medium.

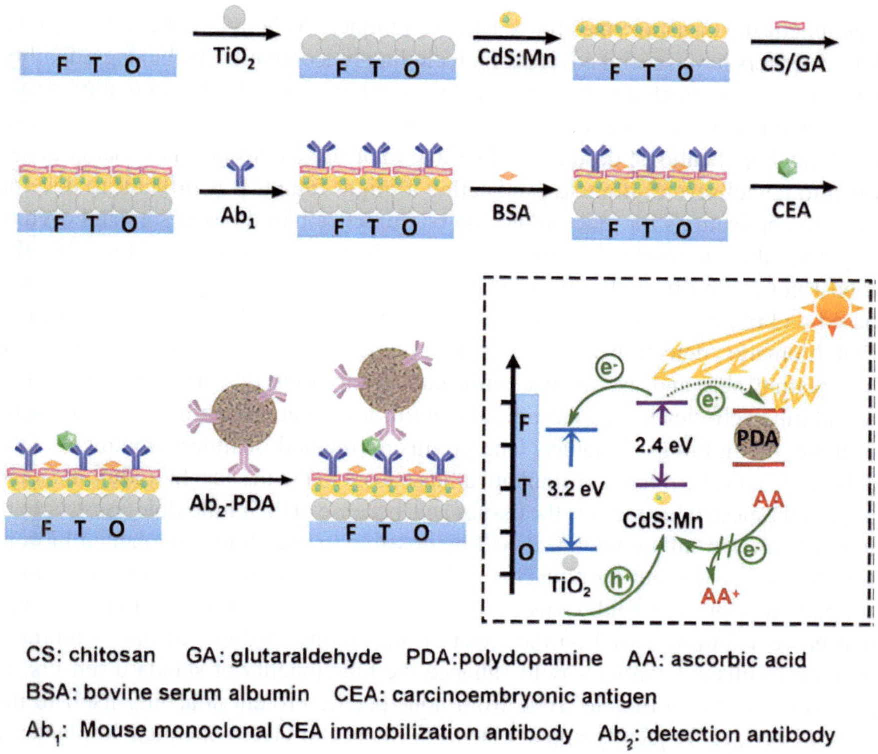

Fig. 2.10 Photoelectrochemical immunosensor for CEA detection using polydopamine nanospheres as quenchers on CdS:Mn-modified TiO_2 [128]

2.4 Conclusion

Many novel sensing technologies and biosensors have been developed as a result of biomimetics, a multidisciplinary scientific area that simulates biochemical processes in nature artificially. This research examined the uses of a number of biodegradable macromolecules in sensing systems, including polydopamine, chitosan, alginate, and cellulose, as well as their composites that may imitate systems of biology. Given the rising incidence of cancer worldwide, particularly in poorer nations, an entire field has dedicated itself to the application of biosensors for early cancer detection based on biological macromolecules that mimic one another or their composites. The function of enzyme-based sensors using biomacromolecules was emphasized, along with the benefits of nanozymes over natural enzymes. Enzymes will eventually be replaced by nanozymes in the construction of biosensors due to their limited stability and lifespan. Although nanozymes have advanced remarkably, creating multifunctional nanozymes with potent catalytic activity remains difficult. Using protein reconstruction technology and nanotechnology, researchers are now working to shed new light on biomimetic enzymology and provide new avenues for the stable and affordable manufacturing of nanozymes for sensing techniques. Although mimetic bioplymer-based sensors have several advantages, including biocompatibility, biodegradability, sensitivity, and selectivity, they also have a number of drawbacks. In reality, only a small number of macromolecule-based biomimetics have fulfilled the requirements of clinical applications, despite the fact that their use as biosensors is being developed to detect several illnesses, indicating a bright future. It is difficult to develop clinical applications for macromolecule biomimetics-based biosensors and eventually bring them to market because of their fragile macromolecule structure. To offer robust compatibility with advanced biosensors, it is crucial to create composites based on biomimetic macromolecules with outstanding performance. In the future, POC biomarker detection using wearable or implanted biosensors could be improved by creating new and useful macromolecule composites that have properties like reusability, self-cleaning, adhesion, and antifouling that are similar to those found in many living things in nature. One promising method of illness treatment is the adherence and attachment of implanted biosensors to the tissues. In traditional techniques, the electrodes encircle the tissues and, because of their inflexible construction, inflict harm and inflammation. It will be possible to glue future implanted biosensors to tissues without harming them. Scientists also attempt to replicate natural surfaces that prevent fouling using natural methods of action. Creating coatings that prevent fouling based on the superior antifouling qualities of the membranes of natural cells is a viable way to enhance the functionality of standard sensors, as bacterial growth on the electrode from non-specific protein attaching itself to the sensor interface frequently impairs the performance of electrochemical implantable biosensors. Antifouling coatings are used to create clinical biosensors that closely match clinical outcomes while maintaining low absorption in biological matrices, including serum, urine, and whole blood. As a result, the creation of these conductive biocoatings has enormous practical utility and removes the last obstacle preventing biosensors from being used in real-world settings.

References

1. S. Antony, T. Anju, B. Thomas, Nature-inspired biomimetic polymeric materials and their applications, in *Handbook of Biopolymers* (Springer, 2022), pp. 1–31
2. A.S. Teymourlouei, S.M. Naghib, M. Mozafari, Stimuli-responsive graphene-polysaccharide nanocomposites for drug delivery and tissue engineering. Curr. Org. Synth. **22**(2), 211–233 (2025)
3. G. Liu, Z. Lv, S. Batool, M.Z. Li, P. Zhao, L. Guo, Y. Wang, Y. Zhou, S.T. Han, Biocompatible material-based flexible biosensors: from materials design to wearable/implantable devices and integrated sensing systems. Small **19**(27), 2207879 (2023)
4. S.M. Naghib, M. Zarrineh, M.R. Mozafari, 3D printing chitosan-based nanobiomaterials for biomedicine and drug delivery: recent advances on the promising bioactive agents and technologies. Curr. Org. Chem. **28**(7), 510–525 (2024a)
5. X. Mei, D. Ye, F. Zhang, Implantable application of polymer-based biosensors. J. Polym. Sci. **60**(3), 328–347 (2022)

6. H. Etayash, R.E. Hancock, Host defense peptide-mimicking polymers and polymeric-brush-tethered host defense peptides: recent developments, limitations, and potential success. Pharmaceutics **13**(11), 1820 (2021)
7. M.A. Farzin, S.M. Naghib, N. Rabiee, Bio-inspired and biomimetic composites based on biodegradable polymers for sensing applications with emphasis on early diagnosis of cancer. Chem. Eng. J. **493**, 152445 (2024a). https://doi.org/10.1016/j.cej.2024.152445
8. R. Gholami, S.M. Naghib, Composites for BTE, in *Nanobioceramics for Bone Tissue Engineering and Regenerative Biomedicine: Tunable Biological Characteristics*, ed. by R. Gholami, S.M. Naghib, (Springer Nature Singapore, Singapore, 2025), pp. 251–282. https://doi.org/10.1007/978-981-96-0041-0_9
9. S.M. Naghib, B. Ahmadi, M.R. Mozafari, Stimuli-sensitive chitosan-based nanosystems-immobilized nucleic acids for gene therapy in breast cancer and hepatocellular carcinoma. Curr. Top. Med. Chem. **24**(17), 1464–1489 (2024b). https://doi.org/10.2174/0115680266293173240506054439
10. S. Soleymani, S.M. Naghib, M.R. Mozafari, pH/temperature/redox and light-responsive polymersome structure and application in cancer therapy: smart drug delivery and targeted drug release. Curr. Med. Chem. (2024). https://doi.org/10.2174/0109298673303351240507113603
11. S.M. Naghib, K. Mohammad-Jafari, Microfluidics-mediated liposomal nanoparticles for cancer therapy: recent developments on advanced devices and technologies. Curr. Top. Med. Chem. **24**(14), 1185–1211 (2024). https://doi.org/10.2174/01156802662864602402200073334
12. N. Shanehband, S.M. Naghib, Recent advances in nano/microfluidics-based cell isolation techniques for cancer diagnosis and treatments. Biochimie **220**, 122–143 (2024). https://doi.org/10.1016/j.biochi.2024.01.001
13. O. Abazari, P. Dayati, M. Shahidi, J. ZavarReza, M. Rahmanian, S.M. Naghib, Exploring a desirable quadr-mRNAs panel for non-invasive and ultrasensitive bladder cancer diagnosis: In-silico and clinical studies. Curr. Med. Chem. (2024). https://doi.org/10.2174/0109298673302140240627053904
14. A. Shamaeizadeh, A. Beigi, S.M. Naghib, M. Tajabadi, M. Rahmanian, M.R. Mozafari, Smart nanobiomaterials for gene delivery in localized cancer therapy: an overview from emerging materials and devices to clinical applications. Curr. Cancer Drug Targets (2024). https://doi.org/10.2174/0115680096288917240404060506
15. N. Nath, S. Chakroborty, D.P. Vishwakarma, G. Goga, A.S. Yadav, R. Mohan, Recent advances in sustainable nature-based functional materials for biomedical sensor technologies. Environ. Sci. Pollut. Res. **31**(46), 57289–57313 (2024)
16. L. Farzin, S. Sadjadi, M. Shamsipur, S. Sheibani, An immunosensing device based on inhibition of mediator's faradaic process for early diagnosis of prostate cancer using bifunctional nanoplatform reinforced by carbon nanotube. J. Pharm. Biomed. Anal. **172**, 259–267 (2019)
17. T. Jafari, S.M. Naghib, M. Mozafari, Bio-printing of materials for bone tissue engineering. Curr. Org. Chem. **29**(1), 19–40 (2025)
18. T. Jafari, S.M. Naghib, M. Rahmanian, M. Mozafari, 3D printing of self-healing materials for drug delivery applications: promises, advances and outlooks. Bioprinting, e00353 (2024)
19. T. Ahmed, Bio-inspired artificial synapses: neuromorphic computing chip engineering with soft biomaterials. Memories-Mater. Devices Circuits Syst **6**, 100088 (2023)
20. D. Ivanov, A. Chezhegov, D. Larionov, Neuromorphic artificial intelligence systems. Front. Neurosci. **16**, 959626 (2022)
21. C. Kim, K.K. Lee, M.S. Kang, D.-M. Shin, J.-W. Oh, C.-S. Lee, D.-W. Han, Artificial olfactory sensor technology that mimics the olfactory mechanism: a comprehensive review. Biomater. Res. **26**(1), 40 (2022)
22. Y. Liu, C. Yiu, Z. Song, Y. Huang, K. Yao, T. Wong, J. Zhou, L. Zhao, X. Huang, S.K. Nejad, Electronic skin as wireless human-machine interfaces for robotic VR. Sci. Adv. **8**(2), eabl6700 (2022)
23. E. Duman, M. Uysal, A.F. Alkaya, Migrating birds optimization: a new metaheuristic approach and its performance on quadratic assignment problem. Inf. Sci. **217**, 65–77 (2012)

24. V. Tongur, E. Ülker, PSO-based improved multi-flocks migrating birds optimization (IMFMBO) algorithm for solution of discrete problems. Soft. Comput. **23**(14), 5469–5484 (2019)
25. M.S. Aziz, Biomimicry as an approach for bio-inspired structure with the aid of computation. Alex. Eng. J. **55**(1), 707–714 (2016)
26. M. Pedersen Zari, Biomimetic urban and architectural design: illustrating and leveraging relationships between ecosystem services. Biomimetics **6**(1), 2 (2020)
27. S. Li, S. Zhang, Y. Tao, Y. Chen, Y. Yang, X. Liang, Q. Li, Construction of β-galactosidase-inorganic hybrid nanoflowers through biomimetic mineralization for lactose degradation. Biochem. Eng. J. **197**, 108980 (2023)
28. Y. Huang, X. Ran, Y. Lin, J. Ren, X. Qu, Self-assembly of an organic–inorganic hybrid nanoflower as an efficient biomimetic catalyst for self-activated tandem reactions. Chem. Commun. **51**(21), 4386–4389 (2015)
29. W. Zhang, Z. Zhai, S. Huang, Z. Mao, Y. Zhang, C. Gao, Morphological and constituent viral-mimicking self-assembled nanoparticles promote cellular uptake and improve cancer therapeutic efficiency in vivo. Giant **3**, 100026 (2020)
30. M.A. Farzin, S.M. Naghib, N. Rabiee, Bio-inspired and biomimetic composites based on biodegradable polymers for sensing applications with emphasis on early diagnosis of cancer. Chem. Eng. J. **152445** (2024b)
31. I. Aranaz, A.R. Alcántara, M.C. Civera, C. Arias, B. Elorza, A. Heras Caballero, N. Acosta, Chitosan: an overview of its properties and applications. Polymers **13**(19), 3256 (2021)
32. J. Baranwal, B. Barse, A. Fais, G.L. Delogu, A. Kumar, Biopolymer: a sustainable material for food and medical applications. Polymers **14**(5), 983 (2022)
33. S.M. Naghib, S. Amiri, M. Mozafari, Stimuli-responsive chitosan-based nanocarriers for drug delivery in wound dressing applications: a review. Carbohydr. Polym. Technolo. Appl., 100497 (2024c)
34. L. Wu, X. Shi, Z.S. Wu, Recent advancements and perspectives of biodegradable polymers for supercapacitors. Adv. Funct. Mater. **33**(16), 2211454 (2023)
35. S. Islam, L. Arnold, R. Padhye, Comparison and characterisation of regenerated chitosan from 1-Butyl-3-methylimidazolium chloride and chitosan from crab shells. Biomed. Res. Int. **2015**(1), 874316 (2015)
36. H.S. Ghazali, E. Askari, Z.S. Ghazali, S.M. Naghib, T. Braschler, Lithography-based 3D printed hydrogels: From bioresin designing to biomedical application. Colloid Interface Sci. Commun. **50**, 100667 (2022). https://doi.org/10.1016/j.colcom.2022.100667
37. C. Wang, Y. Hu, Y. Liu, Y. Shan, X. Qu, J. Xue, T. He, S. Cheng, H. Zhou, W. Liu, Tissue-adhesive piezoelectric soft sensor for in vivo blood pressure monitoring during surgical operation. Adv. Funct. Mater. **33**(38), 2303696 (2023)
38. J.H. Ryu, Y. Lee, W.H. Kong, T.G. Kim, T.G. Park, H. Lee, Catechol-functionalized chitosan/pluronic hydrogels for tissue adhesives and hemostatic materials. Biomacromolecules **12**(7), 2653–2659 (2011)
39. C. Guyot, A. Adoungotchodo, W. Taillades, M. Cerruti, S. Lerouge, A catechol-chitosan-based adhesive and injectable hydrogel resistant to oxidation and compatible with cell therapy. J. Mater. Chem. B **9**(40), 8406–8416 (2021)
40. D. Hu, Q. Ren, Z. Li, L. Zhang, Chitosan-based biomimetically mineralized composite materials in human hard tissue repair. Molecules **25**(20), 4785 (2020)
41. J.-N. Kim, J. Lee, H. Lee, I.-K. Oh, Stretchable and self-healable catechol-chitosan-diatom hydrogel for triboelectric generator and self-powered tremor sensor targeting at Parkinson disease. Nano Energy **82**, 105705 (2021)
42. H. Guo, W. Ran, X. Jin, Y. Huang, F. Long, Y. Xiao, R.-Y. Gan, Y. Wu, H. Gao, Development of pectin/chitosan-based electrospun biomimetic nanofiber membranes loaded with dihydromyricetin inclusion complexes for wound healing application. Int. J. Biol. Macromol. **278**, 134526 (2024)
43. N. Zandi, B. Dolatyar, R. Lotfi, Y. Shallageh, M.A. Shokrgozar, E. Tamjid, N. Annabi, A. Simchi, Biomimetic nanoengineered scaffold for enhanced full-thickness cutaneous wound healing. Acta Biomater. **124**, 191–204 (2021)

44. H. Yang, S. Wang, H. Bian, X. Xing, J. Yu, X. Wu, L. Zhang, X. Liang, A. Lu, C. Huang, Extracellular matrix-mimicking nanofibrous chitosan microspheres as cell micro-ark for tissue engineering. Carbohydr. Polym. **292**, 119693 (2022)
45. S. Korpayev, G. Kaygusuz, M. Şen, K. Orhan, Ç. Oto, A. Karakeçili, Chitosan/collagen based biomimetic osteochondral tissue constructs: a growth factor-free approach. Int. J. Biol. Macromol. **156**, 681–690 (2020)
46. S. Soleymani, S.M. Naghib, 3D and 4D printing hydroxyapatite-based scaffolds for bone tissue engineering and regeneration. Heliyon (2023)
47. C.M. Oddo, L. Beccai, G.G. Muscolo, M.C. Carrozza, A biomimetic MEMS-based tactile sensor array with fingerprints integrated in a robotic fingertip for artificial roughness encoding, in *2009 IEEE International Conference on Robotics and Biomimetics (ROBIO)*, (IEEE, 2009), pp. 894–900
48. R. Jin, J. Xu, L. Duan, G. Gao, Chitosan-driven skin-attachable hydrogel sensors toward human motion and physiological signal monitoring. Carbohydr. Polym. **268**, 118240 (2021)
49. Y. Liu, H. Zheng, M. Liu, High performance strain sensors based on chitosan/carbon black composite sponges. Mater. Des. **141**, 276–285 (2018)
50. S. Salehi, S.M. Naghib, H.R. Garshasbi, S. Ghorbanzadeh, W. Zhang, Smart stimuli-responsive injectable gels and hydrogels for drug delivery and tissue engineering applications: a review. Front. Bioeng. Biotechnol. **11**, 1104126 (2023)
51. T. Pan, H. Chen, X. Gao, Z. Wu, Y. Ye, Y. Shen, Engineering efficient artificial nanozyme based on chitosan grafted Fe-doped-carbon dots for bacteria biofilm eradication. J. Hazard. Mater. **435**, 128996 (2022)
52. Y. Wang, C. Chen, C. He, W. Dong, X. Yang, Q. Kong, B. Yan, J. He, Quaternized chitosan-based biomimetic nanozyme hydrogels with ROS scavenging, oxygen generating, and antibacterial capabilities for diabetic wound repair. Carbohydr. Polym. **348**, 122865 (2025)
53. A.M. Ariti, S.A. Geleto, B.T. Gutema, E.G. Mekonnen, Y.A. Workie, E.M. Abda, M.L. Mekonnen, Magnetite chitosan hydrogel nanozyme with intrinsic peroxidase activity for smartphone-assisted colorimetric sensing of thiabendazole. Sens. Bio-Sens. Res. **42**, 100595 (2023)
54. S.-B. He, F.-Q. Chen, L.-F. Xiu, H.-P. Peng, H.-H. Deng, A.-L. Liu, W. Chen, G.-L. Hong, Highly sensitive colorimetric sensor for detection of iodine ions using carboxylated chitosan–coated palladium nanozyme. Anal. Bioanal. Chem. **412**, 499–506 (2020a)
55. Y. Zou, Q. Chai, T. Zhu, X. Yu, G. Mao, N. Li, J. Chen, G. Lai, Simultaneously colorimetric detection and effective removal of mercury ion based on facile preparation of novel and green enzyme mimic. Spectrochim. Acta A Mol. Biomol. Spectrosc. **266**, 120410 (2022a)
56. K. Ragavan, S.R. Ahmed, X. Weng, S. Neethirajan, Chitosan as a peroxidase mimic: Paper based sensor for the detection of hydrogen peroxide. Sens. Actuators B Chem. **272**, 8–13 (2018)
57. R. Batool, A. Rhouati, M.H. Nawaz, A. Hayat, J.L. Marty, A review of the construction of nano-hybrids for electrochemical biosensing of glucose. Biosensors **9**(1), 46 (2019)
58. H. Jiang, Z. Chen, H. Cao, Y. Huang, Peroxidase-like activity of chitosan stabilized silver nanoparticles for visual and colorimetric detection of glucose. Analyst **137**(23), 5560–5564 (2012)
59. Z.J. Gallagher, S. Fleetwood, T.L. Kirley, M.A. Shaw, E.S. Mullins, N. Ayres, E.J. Foster, Heparin mimic material derived from cellulose nanocrystals. Biomacromolecules **21**(3), 1103–1111 (2020)
60. A.O. Ijaola, *Multifunctional Biomimetic Scaffold with Hydrogel for Skin Tissue Engineering* (Wichita State University, 2021)
61. S. Jana, in *Biomaterial-Based Hydrogels: Therapeutics Carrier and Tissue Regeneration.* (Springer Nature, 2024)
62. X. Wang, S. Tang, S. Chai, P. Wang, J. Qin, W. Pei, H. Bian, Q. Jiang, C. Huang, Preparing printable bacterial cellulose based gelatin gel to promote in vivo bone regeneration. Carbohydr. Polym. **270**, 118342 (2021)

63. D. Zhang, J. Jian, Y. Xie, S. Gao, Z. Ling, C. Lai, J. Wang, C. Wang, F. Chu, M.-J. Dumont, Mimicking skin cellulose hydrogels for sensor applications. Chem. Eng. J. **427**, 130921 (2022a)
64. L. Shu, X.-F. Zhang, Y. Wu, Z. Wang, J. Yao, Facile fabrication of strong and conductive cellulose hydrogels with wide temperature tolerance for flexible sensors. Int. J. Biol. Macromol. **240**, 124438 (2023)
65. M. Chen, H. Wan, Y. Hu, F. Zhao, X. An, A. Lu, Rationally designed cellulose hydrogel for an ultrasensitive pressure sensor. Mater. Horiz. **10**(10), 4510–4520 (2023)
66. R. Dadigala, R. Bandi, M. Alle, C.-W. Park, S.-Y. Han, G.-J. Kwon, S.-H. Lee, Effective fabrication of cellulose nanofibrils supported Pd nanoparticles as a novel nanozyme with peroxidase and oxidase-like activities for efficient dye degradation. J. Hazard. Mater. **436**, 129165 (2022)
67. K. Shehala, M. Baranwal, T. Prabha, A. Malviya, V. Gaurav Singh, Carboxymethyl cellulose-NiO nanoparticles as peroxidase mimic for sensitive colorimetric detection of hydrogen peroxide. Chem. Pap. **77**(3), 1299–1316 (2023)
68. R. Baretta, V. Gabrielli, M. Frasconi, Nanozyme–cellulose hydrogel composites enabling cascade catalysis for the colorimetric detection of glucose. ACS Appl. Nano Mater. **5**(10), 13845–13853 (2022)
69. M. Alle, S.C. Park, R. Bandi, S.-H. Lee, J.-C. Kim, Rapid in-situ growth of gold nanoparticles on cationic cellulose nanofibrils: recyclable nanozyme for the colorimetric glucose detection. Carbohydr. Polym. **253**, 117239 (2021)
70. R. Bandi, M. Alle, C.-W. Park, S.-Y. Han, G.-J. Kwon, N.-H. Kim, J.-C. Kim, S.-H. Lee, Cellulose nanofibrils/carbon dots composite nanopapers for the smartphone-based colorimetric detection of hydrogen peroxide and glucose. Sens. Actuators B Chem. **330**, 129330 (2021)
71. S. Uzunçar, N. Özdoğan, M. Ak, Amperometric detection of glucose and H_2O_2 using peroxide selective electrode based on carboxymethylcellulose/polypyrrole and Prussian blue nanocomposite. Mater. Today Commun. **26**, 101839 (2021)
72. S.N. Prasad, *Nanozyme-Based Colorimetric Sensors for the Detection of Urinary Biomarkers of Early Onset Diabetes and Renal Deterioration* (RMIT University Australia, 2024)
73. S. Uzuncar, N. Ozdogan, M. Ak, Enzyme-free detection of hydrogen peroxide with a hybrid transducing system based on sodium carboxymethyl cellulose, poly (3, 4-ethylenedioxythiophene) and Prussian blue nanoparticles. Anal. Chim. Acta **1172**, 338664 (2021)
74. R.A. Raus, W.M.F.W. Nawawi, R.R. Nasaruddin, Alginate and alginate composites for biomedical applications. Asian J. Pharm. Sci. **16**(3), 280–306 (2021)
75. X. Yan, F. Li, K.-D. Hu, J. Xue, X.-F. Pan, T. He, L. Dong, X.-Y. Wang, Y.-D. Wu, Y.-H. Song, Nacre-mimic reinforced ag@ reduced graphene oxide-sodium alginate composite film for wound healing. Sci. Rep. **7**(1), 13851 (2017)
76. T. Distler, E. Schaller, P. Steinmann, A. Boccaccini, S. Budday, Alginate-based hydrogels show the same complex mechanical behavior as brain tissue. J. Mech. Behav. Biomed. Mater. **111**, 103979 (2020)
77. S. Park, H. Yuk, R. Zhao, Y.S. Yim, E.W. Woldeghebriel, J. Kang, A. Canales, Y. Fink, G.B. Choi, X. Zhao, Adaptive and multifunctional hydrogel hybrid probes for long-term sensing and modulation of neural activity. Nat. Commun. **12**(1), 3435 (2021)
78. S.-B. He, L. Yang, X.-L. Lin, H.-P. Peng, Z. Lin, H.-H. Deng, W. Chen, G.-L. Hong, Sodium alginate modified platinum nanozymes with highly efficient and robust oxidase-like activity for antioxidant capacity and analysis of proanthocyanidins. Front. Chem. **8**, 654 (2020b)
79. Y. Zou, J. Chen, Q. Chai, T. Zhu, S. Zeng, Y. Liu, Stimuli-responsive colorimetric sensor based on bifunctional pyrophosphate-triggered controlled release and enhancing activity of CoOOH nanozyme. Sens. Actuators B Chem. **369**, 132215 (2022b). https://doi.org/10.1016/j.snb.2022.132215
80. L. Qian, Q. Peng, N. Jiang, C. Qiao, W. Yue, Peroxidase-mimicking poly-L-lysine/alginate microspheres with PtS2 nanoparticles for image-based colorimetric assays. Microchim. Acta **190**(8), 300 (2023)

81. K.S. Schanze, H. Lee, P.B. Messersmith, Ten years of polydopamine: Current status and future directions. ACS Appl. Mater. Interfaces **10**(9), 7521–7522 (2018)
82. A. Bakhshi, S.M. Naghib, N. Rabiee, Antibacterial and antiviral nanofibrous membranes, antibacterial and antiviral functional materials, volume 2. (ACS Publications, 2024), pp. 47–88
83. M. Saraf, R. Prateek, B. Ranjan, V.K. Balasubramaniam, R.K.G. Thakur, Polydopamine-enabled biomimetic surface engineering of materials: new insights and promising applications, advanced materials. Interfaces **11**(6), 2300670 (2024)
84. H. Omidian, R.L. Wilson, Polydopamine applications in biomedicine and environmental science. Materials **17**(16), 3916 (2024)
85. C. Xie, Bio-inspired nanofunctionalisation of biomaterial surfaces: a review. Biosurface Biotribology **5**(3), 83–92 (2019)
86. Z. Tang, Y. Miao, J. Zhao, H. Xiao, M. Zhang, K. Liu, X. Zhang, L. Huang, L. Chen, H. Wu, Mussel-inspired biocompatible polydopamine/carboxymethyl cellulose/polyacrylic acid adhesive hydrogels with UV-shielding capacity. Cellulose **28**, 1527–1540 (2021)
87. G. Buscemi, D. Vona, P. Stufano, R. Labarile, P. Cosma, A. Agostiano, M. Trotta, G.M. Farinola, M. Grattieri, Bio-inspired redox-adhesive polydopamine matrix for intact bacteria biohybrid photoanodes. ACS Appl. Mater. Interfaces **14**(23), 26631–26641 (2022)
88. M.L. Alfieri, T. Weil, D.Y.W. Ng, V. Ball, Polydopamine at biological interfaces. Adv. Colloid Interf. Sci. **305**, 102689 (2022)
89. X. Jiang, P. Wang, R. Liang, W. Qin, Improving the biocompatibility of polymeric membrane potentiometric ion sensors by using a mussel-inspired polydopamine coating. Anal. Chem. **91**(10), 6424–6429 (2019a)
90. X. Du, Z. Niu, R. Li, H. Yang, W. Hu, Highly adhesive, washable and stretchable on-skin electrodes based on polydopamine and silk fibroin for ambulatory electrocardiography sensing. J. Mater. Chem. C **8**(35), 12257–12264 (2020)
91. T. Feng, W. Ji, Y. Zhang, F. Wu, Q. Tang, H. Wei, L. Mao, M. Zhang, Zwitterionic polydopamine engineered interface for in vivo sensing with high biocompatibility. Angew. Chem. Int. Ed. **59**(52), 23445–23449 (2020)
92. M. Mohammad, F. Ahmadpoor, S.A. Shojaosadati, Mussel-inspired magnetic nanoflowers as an effective nanozyme and antimicrobial agent for biosensing and catalytic reduction of organic dyes. ACS Omega **5**(30), 18766–18777 (2020)
93. F. Gao, Z. Xiong, Y. Jia, H. Li, J. Li, Cu-boosted one-pot nanoarchitectonics for synthesis of polydopamine membranes as reusable laccase mimic. J. Colloid Interface Sci. **628**, 935–942 (2022)
94. A. Matini, S.M. Naghib, M.R. Mozafari, Quantum dots in cancer theranostics: a thorough review of recent advancements in bioimaging, tracking, and therapy across various cancer types. Curr. Pharm. Biotechnol. (2024). https://doi.org/10.2174/011389201029416324040 7153842
95. M. Yazdan, S.M. Naghib, M.R. Mozafari, Liposomal nano-based drug delivery systems for breast cancer therapy: recent advances and progresses. Anti-Cancer Agents Med. Chem. (Formerly Current Medicinal Chemistry Anti-Cancer Agents) **24**(12), 896–915 (2024). https://doi.org/10.2174/0118715206293653240322041047
96. B.M. Kangarshahi, S.M. Naghib, Nanogenosensors based on aptamers and peptides for bioelectrochemical cancer detection: an overview of recent advances in emerging materials and technologies. Discov. Appl. Sci. **6**(2), 47 (2024). https://doi.org/10.1007/s42452-024-056 81-z
97. H.R. Garshasbi, S.M. Naghib, Smart stimuli-responsive alginate nanogels for drug delivery systems and cancer therapy: a review. Curr. Pharm. Des. **29**(44), 3546–3562 (2023). https://doi.org/10.2174/0113816128283806231211073031
98. A.-A. Zare, H. Naderi-Manesh, S.M. Naghib, M. Shamsipur, F. Molaabasi, Label-free electrochemical cancer cell detection leveraging hemoglobin-encapsulated silver nanoclusters and cu-MOF nanohybrids on a graphene-assisted dual-modal probe. Sci. Rep. **13**(1), 21980 (2023). https://doi.org/10.1038/s41598-023-49418-1

99. S.N. Hosseini, S.M. Naghib, G.K. Kashani, M.R. Mozafari, Chitosan-grafted graphene materials for drug delivery in wound healing. Curr. Pharm. Des. **31**(9), 691–715 (2025). https://doi.org/10.2174/0113816128333493241014134711
100. S.M. Naghib, S.N. Hosseini, A. Beigi, 3D 4D printing of chitosan-based scaffolds for wound dressing applications. Carbohydr. Polym. Technol. Appl. **8**, 100594 (2024d). https://doi.org/10.1016/j.carpta.2024.100594
101. K. Khodaverdi, S.M. Naghib, M.R. Mozafari, Chitosan/bioglass nanocomposites for bone tissue engineering and regenerative medicine: an overview of promising biomaterials. Curr. Org. Chem. **28**(18), 1437–1451 (2024). https://doi.org/10.2174/0113852728314706240529052535
102. P. Vosoughi, S.M. Naghib, M.R. Mozafari, Carboxymethyl chitosan-based nanomaterials for biomedical applications: An overview of recent advances and prospects. Mini-Rev. Org. Chem. (2024)
103. S.M. Naghib, B. Ahmadi, B. Mikaeeli Kangarshahi, M.R. Mozafari, Chitosan-based smart stimuli-responsive nanoparticles for gene delivery and gene therapy: recent progresses on cancer therapy. Int. J. Biol. Macromol. **278**, 134542 (2024e). https://doi.org/10.1016/j.ijbiomac.2024.134542
104. S.M. Naghib, S. Amiri, M.R. Mozafari, Stimuli-responsive chitosan-based nanocarriers for drug delivery in wound dressing applications: A review. Carbohydr. Polym. Technol. Appl. **7**, 100497 (2024f). https://doi.org/10.1016/j.carpta.2024.100497
105. H. Garshasbi, S. Salehi, S.M. Naghib, S. Ghorbanzadeh, W. Zhang, Stimuli-responsive injectable chitosan-based hydrogels for controlled drug delivery systems, Front. Bioeng. Biotechnol. Volume 10–2022 (2023). https://doi.org/10.3389/fbioe.2022.1126774
106. O. Sartipzadeh, S.M. Naghib, F. Haghiralsadat, F. Shokati, M. Rahmanian, Microfluidic-assisted synthesis and modeling of stimuli-responsive monodispersed chitosan microgels for drug delivery applications. Sci. Rep. **12**(1), 8382 (2022). https://doi.org/10.1038/s41598-022-12031-9
107. S. Gooneh-Farahani, S.M. Naghib, M.R. Naimi-Jamal, A. Seyfoori, A pH-sensitive nanocarrier based on BSA-stabilized graphene-chitosan nanocomposite for sustained and prolonged release of anticancer agents. Sci. Rep. **11**(1), 17404 (2021). https://doi.org/10.1038/s41598-021-97081-1
108. S. Gooneh-Farahani, S.M. Naghib, M.R. Naimi-Jamal, A novel and inexpensive method based on modified ionic gelation for pH-responsive controlled drug release of homogeneously distributed chitosan nanoparticles with a high encapsulation efficiency. Fibers and Polymers **21**(9), 1917–1926 (2020). https://doi.org/10.1007/s12221-020-1095-y
109. J. Saberi, M. Ansari, B. Ebrahimi Hoseinzadeh, S.S. Kordestani, S.M. Naghib, Chitosan-Polyacrylic acid hybrid nanoparticles as novel tissue adhesive: synthesis and characterization. Fibers Polym. **19**(12), 2458–2464 (2018). https://doi.org/10.1007/s12221-018-8762-2
110. S. Gooneh-Farahani, M.R. Naimi-Jamal, S.M. Naghib, Stimuli-responsive graphene-incorporated multifunctional chitosan for drug delivery applications: a review. Expert Opin. Drug Deliv. **16**(1), 79–99 (2019)
111. S. Xia, Q. Zhang, S. Song, L. Duan, G. Gao, Bioinspired dynamic cross-linking hydrogel sensors with skin-like strain and pressure sensing behaviors. Chem. Mater. **31**(22), 9522–9531 (2019)
112. P. Vosoughi, S.M. Naghib, T. Jafari, B.M. Kangarshahi, Chitosan-encapsulated lipid-based nanovesicles for therapeutic applications and tissue engineering: a review: Carbohydr. Polym. Technol. Appl. (2025) 100805. https://doi.org/10.1016/j.carpta.2025.100805
113. M.A. Khorasani, S.M. Naghib, A review of chitosan-based multifunctional nanocomposites for drug/gene/protein delivery and gene therapy in cancer treatments: promises, challenges and outlooks. Int. J. Biol. Macromol. **306**, 141394 (2025). https://doi.org/10.1016/j.ijbiomac.2025.141394
114. H. Hasannezhad, A. Bakhshi, M.R. Mozafari, S.M. Naghib, A review of chitosan role in milk bioactive-based drug delivery, smart packaging and biosensors: recent advances and developments. Int. J. Biol. Macromol. **294**, 139248 (2025). https://doi.org/10.1016/j.ijbiomac.2024.139248

115. K. Khodaverdi, S.M. Naghib, M.R. Mozafari, M. Rahmanian, Chitosan/hydroxyapatite hydrogels for localized drug delivery and tissue engineering: a review. Carbohydr. Polym. Technol. Appl. **9**, 100640 (2025). https://doi.org/10.1016/j.carpta.2024.100640
116. S.M. Naghib, B. Ahmadi, M.R. Mozafari, Smart physicochemical-triggered chitosan-based nanogels for siRNA delivery and gene therapy: A focus on emerging strategies and paradigms for cancer therapy. Curr. Med. Chem. (2024). https://doi.org/10.2174/010929867328605224 0426044945
117. H.-H. Deng, X.-L. Lin, Y.-H. Liu, K.-L. Li, Q.-Q. Zhuang, H.-P. Peng, A.-L. Liu, X.-H. Xia, W. Chen, Chitosan-stabilized platinum nanoparticles as effective oxidase mimics for colorimetric detection of acid phosphatase. Nanoscale **9**(29), 10292–10300 (2017)
118. S.-B. He, L. Yang, Y. Yang, H.A. Noreldeen, G.-W. Wu, H.-P. Peng, H.-H. Deng, W. Chen, Carboxylated chitosan enabled platinum nanozyme with improved stability and ascorbate oxidase-like activity for a fluorometric acid phosphatase sensor. Carbohydr. Polym. **298**, 120120 (2022)
119. W. Wang, H. Tang, Y. Wu, Y. Zhang, Z. Li, Highly electrocatalytic biosensor based on hemin@ AuNPs/reduced graphene oxide/chitosan nanohybrids for non-enzymatic ultrasensitive detection of hydrogen peroxide in living cells. Biosens. Bioelectron. **132**, 217–223 (2019)
120. H. Nasrollahpour, I. Isildak, M.-R. Rashidi, E.A. Hashemi, A. Naseri, B. Khalilzadeh, Ultrasensitive bioassaying of HER-2 protein for diagnosis of breast cancer using reduced graphene oxide/chitosan as nanobiocompatible platform. Cancer Nanotechnol. **12**(1), 10 (2021)
121. H. Zhang, H. Qu, J. Cui, L. Duan, A simple electrochemical immunosensor based on a chitosan/reduced graphene oxide nanocomposite for sensitive detection of biomarkers of malignant melanoma. RSC Adv. **12**(40), 25844–25851 (2022)
122. V. Krishnan, G.R. Pandey, K.A. Babu, S. Paramasivam, S.S. Kumar, S. Balasubramanian, V. Ravichandiran, G.P. Pazhani, M. Veerapandian, Chitosan grafted butein: a metal-free transducer for electrochemical genosensing of exosomal CD24. Carbohydr. Polym. **269**, 118333 (2021)
123. D. Oliveira, M.C. Carneiro, F.T. Moreira, SERS biosensor with plastic antibodies for detection of a cancer biomarker protein. Microchim. Acta **191**(5), 238 (2024)
124. A.M.L. Piloto, D.S. Ribeiro, S.S.M. Rodrigues, J.L. Santos, P. Sampaio, M.G.F. Sales, Cellulose-based hydrogel on quantum dots with molecularly imprinted polymers for the detection of CA19-9 protein cancer biomarker. Microchim. Acta **189**(4), 134 (2022)
125. Y. Hong, Y. Wang, Y. Zhu, Highly sensitive immunosensor based on polydopamine-nanofilm modified 3D gold nanoelectrode for α-fetoprotein detection. Electrochim. Acta **364**, 137328 (2020)
126. B. Fan, Q. Fan, L. Hu, M. Cui, X. Wang, H. Ma, Q. Wei, Polydopamine-PEG–folic acid conjugate film engineered TiO_2 nanotube arrays for photoelectrochemical sensing of folate binding protein. ACS Appl. Mater. Interfaces **12**(1), 1877–1884 (2019)
127. Z. Xu, R. Han, N. Liu, F. Gao, X. Luo, Electrochemical biosensors for the detection of carcinoembryonic antigen with low fouling and high sensitivity based on copolymerized polydopamine and zwitterionic polymer. Sens. Actuators B Chem. **319**, 128253 (2020)
128. X. Deng, X. Yang, X. Guan, J. Song, S. Wu, Polydopamine nanospheres with multiple quenching effect on TiO_2/CdS: Mn for highly sensitive photoelectrochemical assay of tumor markers. Anal. Bioanal. Chem. **413**, 2045–2054 (2021)
129. Y. Jiang, Y. Tang, P. Miao, Polydopamine nanosphere@ silver nanoclusters for fluorescence detection of multiplex tumor markers. Nanoscale **11**(17), 8119–8123 (2019b)

Chapter 3
Metal-Organic Frameworks-Based Biosensors Applications: A Focus on Early Cancer Detection

Abstract The identification of cancer biomarkers that are both sensitive and specific is necessary for early cancer analysis and treatment. Nanostructure-based biosensor technology is among the most potent tools for early development of cancer detection. Recently, porous nanoparticles particularly MOFs, or metal-organic frameworks have drawn more attention. They are a novel kind of crystalline material due to their distinctive physico-chemical characteristics, large area of the surface, adjustable scales of pores, and strong adsorption capabilities. As a result, MOFs have been suggested for a number of biosensor and biomedical uses, including imaging agents, drug delivery, and tissue engineering. Analytes may be absorbed very efficiently due to their high porosity, which qualifies them for sensing applications. Three primary techniques including solid, liquid, and gas phase approaches can be used to synthesis of MOF. This chapter distinctive feature is its comprehensive description of the three distinct processes utilized to synthesize MOFs and the tools for examining their characteristics. The research also looks at how MOF has been used recently in many cancer biosensors, like those for the prostate, breast, lung, and CEA. The chapter concludes by outlining prospective pathways for more research utilizing MOFs and discussing possible future paths. It highlights how crucial it is to use cutting-edge technology and create quick and affordable manufacturing techniques. This thorough evaluation is a useful resource that may support the creation of next studies using similar nanomaterials.

Keywords Synthesis · Biosensors · Metal-organic frameworks · Cancer biomarkers · Early cancer detection

3.1 Introduction and Basic Concepts

Cancer is caused by healthy cells transforming into tumor cells, according to the WHO. This process is multiphase and often starts as a precancerous lesion and evolves into a malignant tumor [1–3]. Early detection is very critical and may improve rates of survival, reduce death, and make additional inexpensive therapy choices possible

S. M. Naghib, *Nano-Biosensors, Drug Delivery and Tissue Engineering*, Series in BioEngineering, https://doi.org/10.1007/978-981-95-1888-3_3

for several forms of cancer, which have a fair likelihood of being treated [4–7]. Consequently, some biomarkers of disease (BMs) may be used to identify and track the ailment. The synthesis of specific bodily biomolecules or an alert in their concentration is often associated with infection and the onset of illnesses, including cancer [8–11]. Clarifying the cellular growth of cancer cells is made easier by BMs for cancer capacity to predict the path of disease [7, 12–17]. Because BM detection has no temporal constraint on sample collection for disease monitoring, it may be used to simply and less invasively sample tears [18], sweat [19], urine [20], saliva [21], and blood [22]. Therefore, there is an urgent requirement for sensitive and rapid techniques for detecting cancer BM since postponing cancer identification and management may notably impair patients' quality of life. Laboratory studies such as ELISA (enzyme-linked immunosorbent assays) [23], examining the body for bumps [24], and imaging techniques such as PET (positron emission tomography) [7], CT, MRI [5], biopsy [25], and ultrasound [7] are now considered the gold-standard diagnostic approaches. These methods do have some disadvantages, however, such as limited sensitivity throughout the illness's early phases, time-consuming, arduous processes, the need for big laboratory equipment and expert workers, and the potential for being intrusive. In order to get over these restrictions, research is being done on creating biosensors that use nanomaterials. They are more affordable, sensitive, specific, and compact, and they may take the place of conventional diagnostic techniques [12, 26–29]. Furthermore, due to technological advancements in the field, biosensor vendors may now provide more advanced and effective solutions. MOFs are a kind of material that is porous and made up of organic molecules and clusters of metals [20, 30–33]. This materials' exceptional synthetic tunability, which permits fine authority over chemistry and structure, is made possible by their modular architecture. Furthermore, by creative synthetic design, they may be customized to specific applications because of their remarkable performance, remarkable charge mobility [34], biocompatibility [35–38], excellent mechanical power [10], and stability of heat [35]. In 2023, 2024, and 2033, the global market for MOF was valued at USD 2056.09, USD 751.65, and USD 672.14 million, respectively. Compared to other nanomaterials, MOFs have exceptional promise as biosensors [39–43]. MOF is perfect for electrochemical (EC) and optical sensing because to its exceptional electrical and luminous properties. For the creation of a platform for biosensing cancer detection, MOFs are very desirable [22, 44, 45]. The number of research articles in the field of growing MOF-based biosensors and the creation of MOF-driven cancer biosensors to find out what tumors or cancer are during the past ten years is shown in Fig. 3.1. This figure illustrates various applications of MOF-based materials including signal amplification with chemical reaction diagrams and corresponding graphs (Fig. 3.1a), MOF materials functioning as nanozymes for colorimetric and electrochemical detection using mobile devices (Fig. 3.1b), MOF materials as catalysts involved in detection and treatment processes with visual cell and reaction representations (Fig. 3.1c) and MOF materials as platforms, detailing aptamer immobilization and detection of MCF-7 cells and CA 125 through labeled diagrams (Fig. 3.1d). In this work, I outline the latest developments in the synthesis of MOFs and the creation of biosensors and their nanomaterials for the detection of different types of cancer, such as living cancer

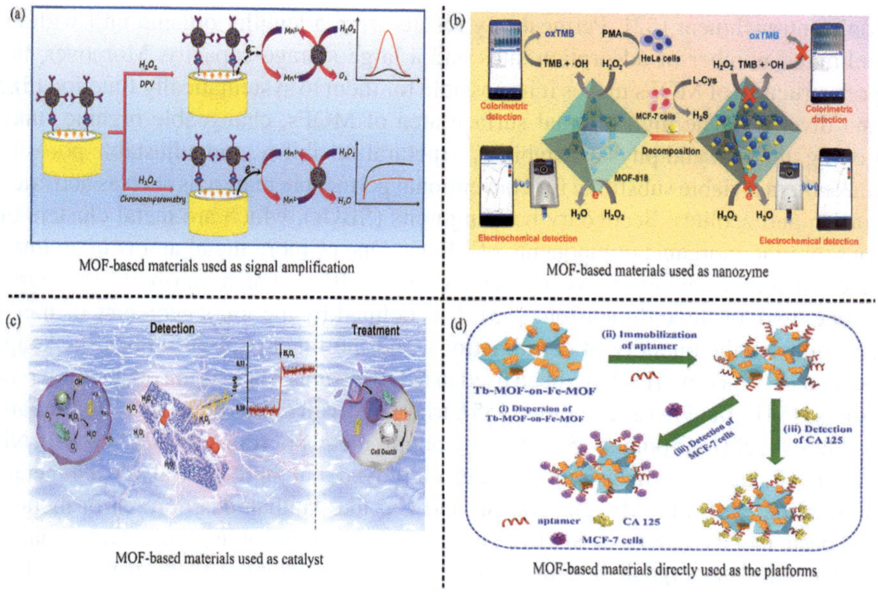

Fig. 3.1 Examples of typical EC sensors created from materials based on MOFs are **a** signal amplification [47], **b** nanozyme [48], **c** catalyst [50], **d** platform [49]

cells. Moreover, microRNA is believed to be a marker for early cancer detection, in addition to an interpretation of the biosensors' manufacturing process. Four types of biosensors based on MOFs are now on the market for detecting analytes linked to cancer. To make probes that look like hairpins using MOF materials with matching probes, changes occur in the EC signals when extra probes touch the sensor area. To detect electrochemically active substances and EC signals in compounds in watery solutions, MOFs and their nanomaterials may be used as nanozymes and catalysts. MOF-derived nanoparticles' detection signals rely on how the modified electrode and electrolyte interact. These substances are possibly used as substrates for the direct adsorption of probes (like aptamers and antibodies) [46–50].

3.2 MOF Structure, Synthesis and Physicochemical Characterization

Metal nodes in combination with natural linkers can create several million different MOFs. Yaghi and co-workers were the first to synthesize these chemicals, and currently, they have synthesized over 90,000 MOFs, with over 500,000 more predicted [51]. To produce revealed structures that, at the moment, exhibit the 5 primary features of a solid structure unbounded porosity, an enormous amount of pores, and high MOFs' surface area organic linkers are used to coordinate a metal ion's entrenchment [52]. Permeability results from a lengthy organic unit with an unlimited number of adsorption sites and a large storage capacity. Moreover, the pore structure of MOFs makes it impossible for them to systematically functionalize and diversify. The high internal surface area of MOFs, changeable organic functioning, thermal and physical stability, structural flexibility, and adjustable porosity make them a viable substitute for conventional permeable materials such as activated carbon and zeolites. Secondary building units (SBUs), which are metal clusters or ions, are the inorganic components of MOFs. One or more metal atoms in an intricate collection among atoms may create these SBUs, which can build structures in one, two, or three dimensions [53]. It is defined by the ion's geometry of metal cooperation, the quantity of coordinations, and the functional groupings' makeup. Several SBU geometries have been used in MOF designs, such as the octahedron (six points) [54], triangle (three points) [55], trigonal prism (six points) [56], and square paddle-wheel (four points) [57]. Nodes contain metals in transition like Co, Sc, Ni, Fe, Zn, Cr, and Cu, which are most often employed, but there have also been metals of alkaline earth like Mg, Ba, Ra, Sr, and Ca; like lanthanides, rare earth metals; and basic metals, like Al or Sn, from the primary periodic table group. Additionally, known as linkers, organic ligands are an essential component of MOF structures. The most typical types of linkers used are O-donors like sulfonates, carboxylates, and phosphonates, as well as N-donors such as sulfonates, triazines, and pyridines. In the lattice structure of the MOF, the linkers' and nodes' alignment and angle influence the size and form of the pores. The degree of binding between the nodes and linkers also strongly influences the chemical and mechanical stability of the generated MOFs. Moreover, these complexes' chemical activity is influenced by metal nodes of the type and the activity of the ligands that are organic (Fig. 3.2) [40, 46].

In this assessment, we will discuss 3 different techniques of categorization for obtaining MOFs: gas phase techniques, liquid phase techniques that operate in high ratios of surface area to volume environments (like 3D printing and microfluidic technology), and solid phase (mechanochemical) methods. As will be covered in the next part, all of the techniques make it possible to synthesize MOF, where the thickness and the crystal faces' orientation can be adjusted. Furthermore, since gas-phase methods eliminate the need for solvents, they are very appealing for the production of microelectronic devices. MOFs are typically created by mixing metal salt solutions with ligands in the liquid phase. When selecting a solvent, factors including its reactivity, solubility, and redox potential are considered. The solvent possesses a major influence on the activation energy and thermodynamics of any process. The majority of MOF crystal formation has been accomplished by the reaction dissolution's slow evaporation. The majority of MOFs are produced in solvo (hydro) thermal environments, which include elevated pressure and temperature. This is the "classic" way of preparing MOFs. Mechanochemical, EC, sono-chemical, and microwave approaches are some of the alternative synthetic methods that have been used recently [58].

The BM (ball mill) is a device that frequently smashes and breaks up solid substances into tiny pieces. Among the surface alterations that occur when raw materials are trapped in ammunition is bond breakage. The potential energy generated during rotation crushes the components. Thus, this process not just initiates responses

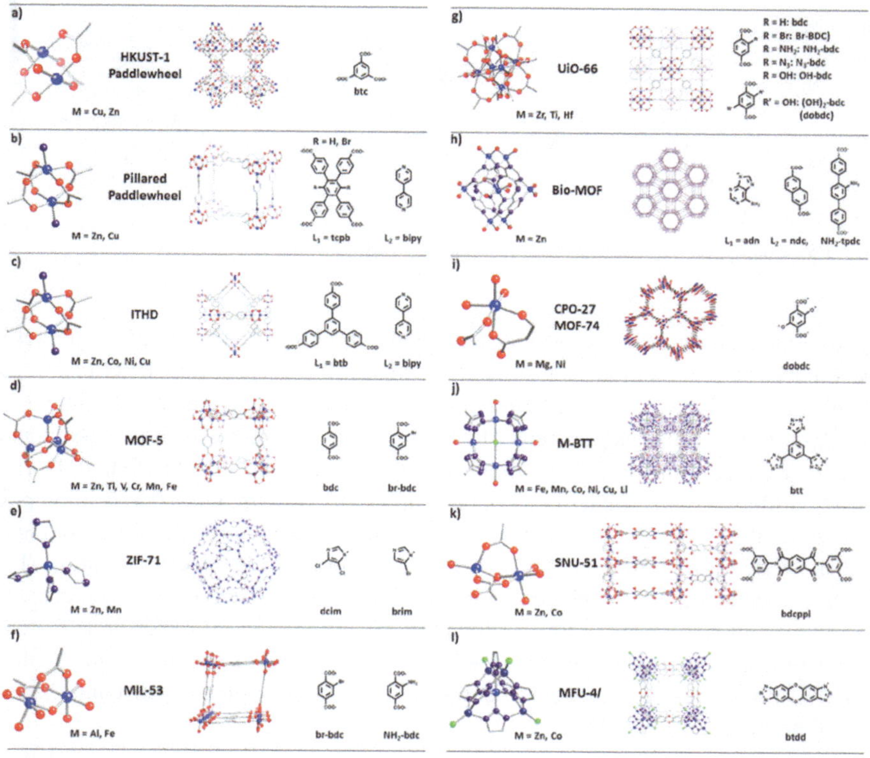

Fig. 3.2 Diverse MOF structures formed by metal nodes and organic linkers illustrate their tunable porosity, high surface area, and structural versatility [46]

but also reduces the size of the particles by generating active sites [59]. Controlling a number of variables, including the milling temperature, diameter of the grinding ball, and the weight ratio of the ball to powder, process control agent, and the relative quantity of the reactants [60]. Because it is cost-effective and ecologically safe, this method is widely utilized to produce mineral, metallurgical, and organic nanoparticles. The advantages of this approach, such as the free solvent, the ability to avoid external heating, and the quick reaction time, have led to a considerable advancement in MOF synthesis (Fig. 3.3) [58, 61, 62].

The EC approach might logically generate a vast number of MOFs. BASF was the first business to make HKUST-1 using the EC method in 2005. Their primary objective was to segregate anions, such as chloride, nitrate, and perchlorate, throughout the synthesis process in order to avoid the production of caustic anions (such as nitrate and chloride). Because of this, the EC approach may seem to be a green chemical approach to fabricating MOFs without the need for intricate manufacturing processes. The synthesis encourages an environmentally friendly deposition process because it doesn't produce harmful anions, which metal salts like nitrate help to avoid. Using an EC to introduce metal ions procedure rather than a solution of metal salt or by

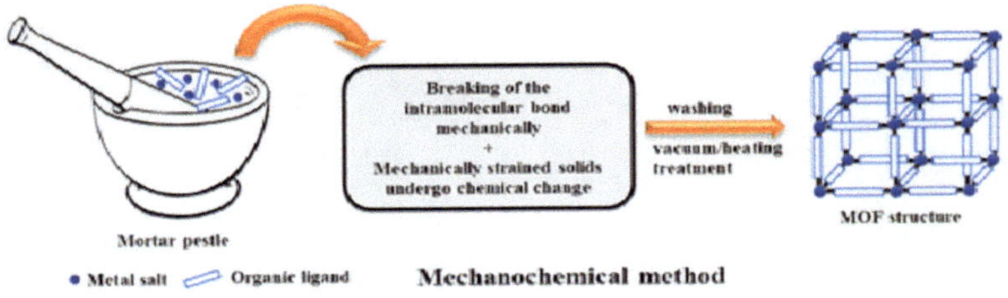

Fig. 3.3 Schematic of mechanochemical synthesis of MOF [62]

producing such ions as a consequence, showing how the metal and acid interact is necessary for this method of synthesizing MOFs [63]. The MOF's EC technique employs the dissolved organic linker's anodic dissolution and electrolyte molecules in addition to metal ions generated. Instead of using the metal as the source, metal salts cause metal ions to flow through the reaction mixture via electrons [64]. In a basic building block, which often contains an electrolyte, the electrode is immersed in water, and a voltage sufficient to dissolve the metal is applied, while the metal nodes needed for MOF development are located outside the electrode region. The MOF is located near to the electrode, and the solution's organic linker interacts with the metal ions right away. Using a solvent that is protic releases the cathode's metal deposition, although H_2 is still produced. To get more solid matter than with a typical batch response, it's too practical to carry out a methodical process utilizing the EC technique (Fig. 3.4) [62, 65].

The first published microwave MOF synthesis was Cr-MIL-100. The method, with the aid of microwave, has gained widespread use for the rapid synthesis of MOFs in thermothermal environments. The microwave approach makes it easier to produce small metal and oxide particles. Mobile solvent charges and electromagnetic waves, such as ions or polar solvent molecules, combine to produce this efficient heating

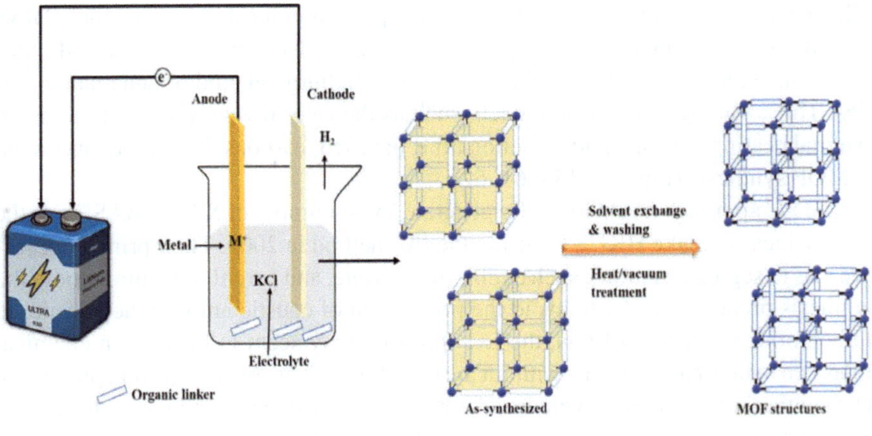

Fig. 3.4 Electrochemical synthesis of MOF structures [62]

technique. This approach, which is mostly applied to organic chemistry, is often used to make metal oxides at the nanoscale [66]. The dissolution should be increased to form metal crystals at the nanoscale. This approach involves filling a vessel made of Teflon with a foundation mixture and the right dissolving agent, sealing it, and then heating it to the proper temperature and duration in a microwave. When a field of electricity is applied, the microwave's function is to convert thermal energy from electromagnetic energy by linking the molecules' persistent dipole moment. Polar molecules attempt to align in an electromagnetic field and oscillate when a substrate is joined, causing permanent orientation changes. Molecules will collide at the right frequency, raising the method's temperature and motion energy. Because the radiation and the solution or reactant have direct interactions, it is a very heating technique that uses less energy. Taking into account should be given to the solvent selection and the particular energy input brought about by uniformly sized nanocrystals. Microwaves that operate in the frequency range of 300 MHz to 300 GHz are considered electromagnetic waves. The main goal of microwave-assisted MOF synthesis is the quick crystallization and production of goods at the nanoscale to increase the purity of the product for the specific synthesis of polymorphs (Fig. 3.5) [62, 67].

Because the sonochemical method employs ultra-radiation to shorten the crystallization time, it has been used lately to synthesis MOFs fast. This method uses a mechanical vibration that cycles (0.20–10) MHz to manufacture the MOFs. A Pyrex reactor shaped like a horn, connected to a sonicator bar, can produce electricity from a metal salt and an organic linker combination without needing external cooling (Fig. 3.6) [62, 68].

Researchers have successfully used a variety of powder fixing methods to secure MOFs onto a support. Microwaves that operate in the frequency range of 300 MHz to 300 GHz are considered electromagnetic waves. Unfortunately, the compaction procedure drastically decreases the granules' porosity, which is unsuitable for permitting macropore development in the finished product. Techniques like pressure, extrusion, granulating, spray drying, and 3D printing can further mold them to create stronger materials. 3D printing allows for straightforward creations having intricate structural networks that have far less waste than earlier techniques. Intricate 3D

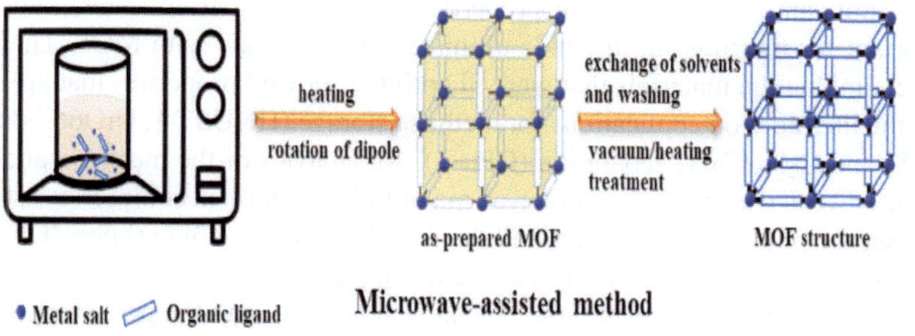

Fig. 3.5 Microwave-assisted synthesis of MOF structures [62]

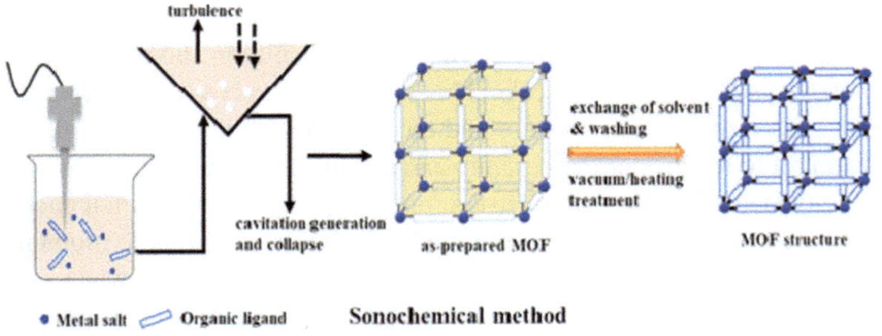

Fig. 3.6 Sonochemical synthesis of MOF structures [62]

structures may be built using this technique, which initially facilitates rapid prototyping and then establishes the foundation for future industrial expansion. With this technique, The MOF-containing ink is put straight onto the printer bed, causing it to harden. Usually, supplementary preparations Add solvents, thermoset polymers, and sticky polymers. This method was used by Thakkar et al. to manufacture MOF-74 (Ni) and UTSA-16 (Co). Generally, 80% MOFs, 5% PVA, and 15% bentonite clay dispersed in deionized water and ethanol (5:95) constitute the Ni-MOF's ink formulation. To print monoliths, the dough was put into a syringe and emanated from a 0.85 mm-diameter nozzle by utilizing 5–2 psig air pressure. A specific area decreases due to the powders and monoliths (from 737 and 568 $m^2\ g^{-1}$ to 1180 and 727 $m^2\ g^{-1}$ for MOF-74 and UTSA-16, respectively, at 5–2 psig of air pressure), showing nitrogen sorption isotherms of type I [69].

Dhainaut et al. aim to create ZIF-8 (Zn), HKUST-1 (Cu), UiO-66-NH2 (Zr), and CPL-1 (Cu) through the use of specialized extrusion-based 3D printers that utilize organic binder-based inks. A 0.65% combination consisting of a comparable proportion of water and ethanol by volume, 0.01% poly(vinyl alcohol), 0.02% hexylethylcellulose, and 0.25% MOF was typically used to make the ink for HKUST-1. The item that contains HKUST-1 shows a precise 1188 m^2/g area after shaping, which is 20% less than the main powder's specific area of 1466 m^2/g. Since MOF breaks down a bit when printed in a polar solvent, the researchers believe the process leads to a significant drop in physical properties (like thinning from shear), which are important for figuring out the ink's ultimate porosity made with MOF. The scientists claim that the materials that were 3D printed showed properties that are in line with the literature on applications for methane storage (HKUST-1, UiO66-NH_2, and ZIF-8) and C_2H_6/C_2H_4 separation (CPL-1). An analysis of the mechanical characteristics revealed a one- to two-order-of-magnitude variation in compressive strength between the pills without binders and the samples that were 3D printed. This study emphasizes how important it is to choose the appropriate wording to strike a decent equilibrium between the materials' mechanical stability and the persistent porosity of the solid 3D-printed [70].

Microfluidic devices have the potential to revolutionize material synthesis. The regime of laminar flow generated within devices that use microfluidics provides a complex artificial setting that guarantees constant bulk transmission of reagents in space and time and is very efficient. The microfluidic device has opened the door for the development of innovative methods for the MOFs of synthesis because of its small loudness, minimal reagent or sample needs, and fast response times. Since MOF breaks down a bit when printed in a polar solvent, the researchers believe the process leads to a significant drop in physical properties (like thinning from shear), which are important for figuring out the ink's ultimate porosity made with MOF [71]. Recently, many people have started using microfluidic synthesis to produce nano/micro MOFs with controlled size and form. The two most popular conventional methods for creating MOFs have a lengthy crystallization period (from hours to days) and limited control over crystal size. According to this viewpoint, continuous flow processes, such microfluidic synthesis, are more appropriate for industrial requirements than conventional laboratory-scale batch approaches. ZIF-8 nanocrystals were successfully created at room temperature using a T-type microfluidic device, as described by Yamamoto et al. The results indicate that changing the ratio of zinc ions to dimethylimidazole can control the shape and size of the nanoparticles [72]. Furthermore, using continuous silicone oil, organic solvents, and metal precursors with ligands in a phase of dispersion, various MOF-based nanostructures, including IRMOF-3, UiO-66 (Zr-BDC MOF), and Zn_4O(1,4-benzenedicarboxylate)$_3$, have been produced in a line of T-shaped microchips. The core-shell structure of ZIF-8@Fe_3O_4 nanoparticles was created in the interim using Fe_3O_4 as a precursor (Fig. 3.7) [73].

The use in the field of microfluidics production of Cu (II)-aspartate (ASP), an amino acid, was first reported by Luis et al. in 2011. Luis et al. incorporated two watery dissolutions into a microfluidic device with four input channels, using a pump as the injection mechanism. The speed of the solution in the centre of the channel changed, affecting how metal ions and organic linkers formed after these substances were added to the microfluidic device through two additional tubes. Nanofibers of Cu(II)-ASP for – 200 nm D were produced; additionally, this combination technique

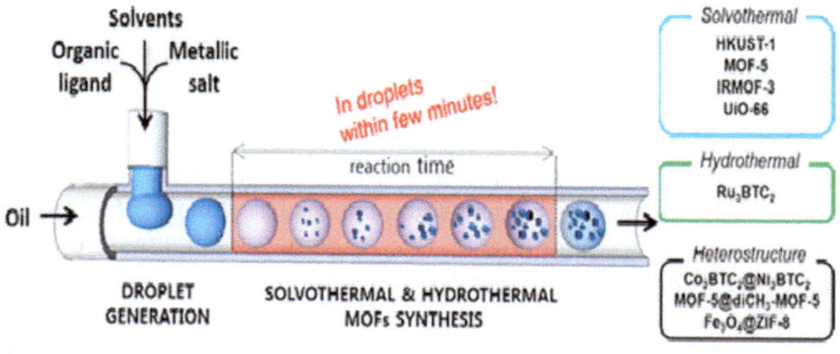

Fig. 3.7 Continuous microfluidic droplet synthesis enables the rapid formation of various MOF-based nanostructures, including IRMOF-3, UiO-66, and ZIF-8@Fe_3O_4 core-shell structures, using metal precursors and organic ligands in T-shaped microchips. Adapted with permission from [73]. Copyright 2013 American Chemical Society

has been expanded to produce other MOF nanostructures at the micro/nano level [74]. The recent successful production of 2D MOF nanosheets by Galve et al. used the microfluidic approach to regulate the dynamics. Making use of microfluidic synthesis, which further regulated the region of contact diffusion of the reactants, several phenomena of crystal formation were generated by altering the flow rate of the tube. Furthermore, synthesis using microfluidics may be accustomed to producing extreme thinness in 2D MOF nanosheets [75]. Wang et al. produced very thin two-dimensional Zr-1,3,5-tris(4-carboxyphenyl)benzene (BTB) nanosheets comprising superior spread using the microfluidic approach. The linear novel mixed compounds suppressed the development of 2D Zr-BTB reactions in manufactured microfluidic devices that were dominated by forced convection and diffusion. The length of time Zr-BTB remained in the pipeline had an effect on the stack development of layers. Zr-BTB generated 2.8 nm thick nanosheets after passing by for 8 min, according to the AFM analysis, but the crystallization was poor. After 64 min, where one resides, we observed greater symmetry; flat nanosheets have a substantial thickness of about 25 nm [76].

To show that microfluidics can also be used to make nanofibers, Yu and associates used spinning microfluidically to construct microfibers of the core-shell type, employing a saline gel or alginate as the shell element and a copper vitamin or a zinc vitamin combo as the essential element. There are several biological applications for the resultant long, thin, and flexible microfibers [77]. In another study, enzyme-containing ZIF-8 nanoparticles (about 500 nm in size) were made utilizing a microfluidic gradient mixing approach to preserve the enzymes' bioactivity and optimize the potential of microfluidic devices. It was shown that, unlike the artificial nonporous ZIF-8 produced through the conventional large approach, the precursor concentration of ZIF-8 during gradient combination may constantly alter the ZIF-8 microstructure and surface shape, increasing the loaded enzymes' activity. As a proof-of-principle, ZIF-8 with GOx (glucose oxidase) loaded was created using the microfluidic technique. The maximum observed activity of the enzymes in the combination GOx-MOF, about 98%, was sustained by this technique [78].

As a viable alternative to MOF crystal formation, gas-phase deposition techniques may help address a few of the previously mentioned difficulties with solution-based methodologies. After being extensively employed for the creation of thin films and low-dimensional materials, such as nanowires for semiconductors and dichalcogenides of transition metals (TMDs), graphene, and synthesis in the gas phase, it was only exploited for the development of materials of MOFs [79]. Researchers have since shown the process of depositing copper terephthalate (II) and UiO-66 films, in addition to exposing films of zinc oxide to 2-methylimidazole fumes to transform into ZIF-8. Despite the fact that these results show a notable improvement while using gas-phase techniques to deposit MOFs, there are still difficulties to be solved. Initially, these studies need cautious handling of leftovers, intermediates, and different circumstances of processing, and the many steps they require result in increased synthesis costs. Conversion from the bulk phase to the frameworks that are coordinated which may be either amorphous or polycrystalline produced by these methods is often required. Finally, while the objective of using these materials in devices is often the motivation behind these investigations, device transport data is seldom included [80].

Both the synthesis of MOFs and the defining of their form depend on the solvent system. Solvents may fill space or serve as coordinates with metal ions. Instead, they also act as a compass for the framework. When using solvents in the synthesis of MOFs, high polarity and boiling points are necessary. DMSO (Dimethyl Sulfoxide), DEF (diethyl Formamide), acetone, acetonitrile, DMA (dimethyl acetamide), and alcohols are examples of common solvents [81]. Because of the polarity of the solvent used, the organic linker's protolysis, and its solubility, the reaction media are crucial to the MOFs production process. Furthermore, distinct solvent solutions can produce MOFs with different morphologies under the same reaction circumstances. Differences in how organic linkers deprotonate in various solvent systems may explain this. According to Banerjee et al. MOFs with various crystal formations were created from PDC (3,5-pyridine dicarboxylic acid) and Mg (magnesium) using different solvent systems but under the same conditions. They found that the MOF network's dimensionality depends on the ability of the solvent to work with the metal. H_2O is especially drawn to magnesium when EtOH/H_2O and MeOH/DMF are used as solvents because EtOH and MeOH have minimal tendency to interact with cores made of metal [82]. Song describes a bimetallic MOF electrode made of Ni and Co, created using thiophene ligands in different shapes and a solvothermal method. The results demonstrate that the thienyl morphology-based Ni/Co MOF gradually changes from a round form to one that is flaky as the solvent protonation degree increases. Compared to earlier MOFs made using solvents with low protonation, Ni/Co MOF's sheet-like structure based on thienyl produced via EtOH exhibits distinct paths for the transfer of electrons across layers [83]. Another work by Song et al. used metal nodes: Ni and Co ions, ligands such as carboxyl and amino groups, and other substances to build a novel monomer of a conjugated small molecule termed DABA, or 3,5-diaminobenzoic acid. This process resulted in an original, highly bimetallic MOF that is electroactive (DABA-MOF). The solid structure of DABA-MOF exhibits a shape whereby amino and carboxyl groups coexist. This study investigated DABA-MOF's structure and capacitive capabilities in three organic solvents (methanol, ethanol, and N,N-dimethylformamide) with varying degrees of deprotonation and polarity [84].

It is well known that the acidity or basicity of the reaction medium has a major influence on the synthesis and organic-inorganic nanomaterial crystallization. The idea of acid-base chemistry suggests that making hydroxyl linkers in a water solution, depending on the pH level, would help the metal ion connect with the polycarboxylate linker based on how much the organic ligands lose protons. Interesting studies about the effect of pH on MOF synthesis have been reviewed by a number of research groups [85]. Lv et al., for example, verified that it was possible to use acetic acid to regulate the acidity of MOF-5 ($Zn_4O(BDC)_3$). Acetic acid changed the size of crystals but left the cubic form same because of its ability to cap them. However, by increasing the mixture's pH to 7.41, an alkaline modulator of sodium acetate altered the sheet's form. The pH-mediated alkaline link was further verified with TEA as a zinc acetate and modulator as a precursor; currently, the pH of the reaction mixture was 8.26. In both situations, specific square sheets of MOF-5 were produced [86]. Seoane et al. used as a capping agent, cetyltrimethylammonium bromide (CTAB), in another study, along with TMAOH (tetramethylammonium hydroxide), changed the pH between 2.7 and 2.1. They were able to get a pH relationship that was sometimes different. Extended hexagonal MIL-110 crystals formed in the acidic solution at pH 2.1; small MIL-100 octahedra started to develop at pH 2.2–2.5; and a mixture of indistinct crystals of MIL-96 in the shape of hexagons and MIL-100 formed at pH 2.6 (S7b Figure). The pH settings had a comparable effect on the MIL-100 particle size; as the concentration of TMAOH rose, the size of the particles decreased from 30 ± 5 nm to 330 ± 70 nm [87]. It was shown that the temperature conditions during sample preparation had a significant impact on the morphology of MOFs. Additionally, it has been shown that the morphology of the crystallites is influenced by crystal kinetics formation, temperature increase rate during synthesis, pace of cooling, and fluctuations within the values. Temperature may also have a major impact on the formation of sharp crystal edges [88]. Shen et al., for example, later examined the effects of solvothermal temperature on the form and EC characteristics of superstructures made of Ni-MOF. They suggested that MOFs of Ni made at 80 °C might build superstructures in hexagons using Ni-MOF nanofibers that cross each other. The shape showed a greater surface area, which contributed to the highest capacitance of 1 A g^{-1} and 30.89 mAh g^{-1} [89]. In another study, Jiang et al. produced Co-MOF at various temperatures using the soluble technique. The impact of synthesis temperature on the catalytic efficacy of NH_3-SCR at low temperatures and the structure of the Co-MOF-74 was thoroughly looked into. Co-MOF-74 has a flower-like shape at 100 °C, but hexagonal prism formations at different temperatures could occur. The flower-shaped Co-MOF-74 exhibited greater pore capacity and specific surface area than that of the hexagonal prisms [90]. The Deng group also investigated the effect of temperature on the generation of two kinds of Ni MOFs in high-performance supercapacitors. The creation of two new MOFs, Ni-MOF1 and Ni-MOF2, at 120 and 180 °C shows that using different reaction temperatures can lead to different processes and structures. Although Ni-MOF1 is stacked in two dimensions and Ni-MOF2 has a 3D framework, Ni-MOF1 has superior EC characteristics over Ni-MOF2, which could be attributed to the appropriate electron transport of the 2D layered structure [91].

3.3 MOF-Based Materials for Various Applications

The innovative porous material family known as MOFs has attracted attention throughout the last 20 years. With their very high porosity and surface area, MOFs are remarkable organic-inorganic composites. They are crystalline, possess outstanding thermal and chemical stability, and have frameworks that are both rigid and flexible. They have undergone extensive and ongoing analysis and evaluation for a number of reasons. MOF has recently been used in a variety of fields due to its distinct characteristics including biosensor devices (Food safety, environmental monitoring, lactate,

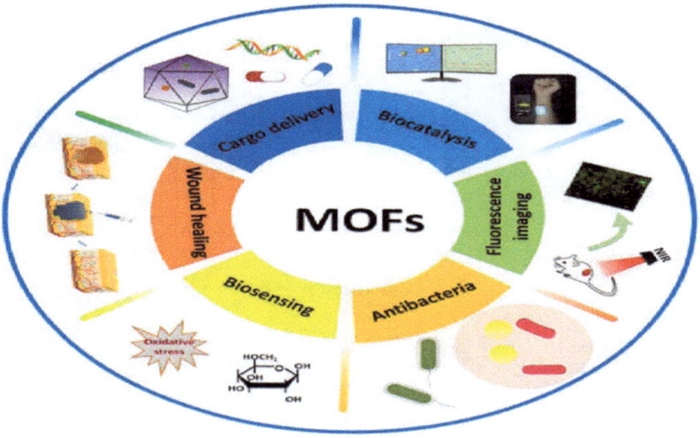

Fig. 3.8 MOF applications [101]

milk analysis, glucose, and illness detection), Additional biomedical applications include drug delivery [92–94], antimicrobial therapy [37], therapeutic approaches [95–98], including chemodynamic therapy, photodynamic therapy, photothermal therapy, and immunotherapy, as well as Wound healing, wound dressing, cardiovascular implants, tissue engineering, imaging agents, and water and wastewater treatment applications, including membrane, adsorbent, photocatalysts, and photo degradation, as well as electrical and nanoelectronic devices such as solar cells, supercapacitors, and field effect transistors (Fig. 3.8) [99–101].

3.4 MOF-Based Biosensors in Early Cancer Detection

Figure 3.9 shows a schematic representation of the components of a standard biosensor: (a) receptors that attach to the analyte with a certain affinity; (b) a design for an interface that triggers a biological occurrence that results in a signal that is retrieved by (c) The element that transduces, which transforms the signal into an electrical signal that is enhanced by a detector circuit with the relevant reference prior to being submitted for handling. (d) Computer programs that will be converted into a practical physical attribute that defines the procedure being examined. Biosensors are divided into four main classes based on signal transduction techniques: mass-sensitive, thermal, optical, and EC sensors [102].

Biosensors for EC are compatible with modern microfabrication techniques, highly sensitive, portable, and reasonably priced. SWV, EIS, and CA, CV, and DPV are examples of EC techniques [103]. The need for increase the EC biosensors' sensitivity has led to the use of novel approaches like the use of nanomaterials or polymers (with carbon-based nanoparticles, metal oxide nanoparticles, or metal nanoparticles)

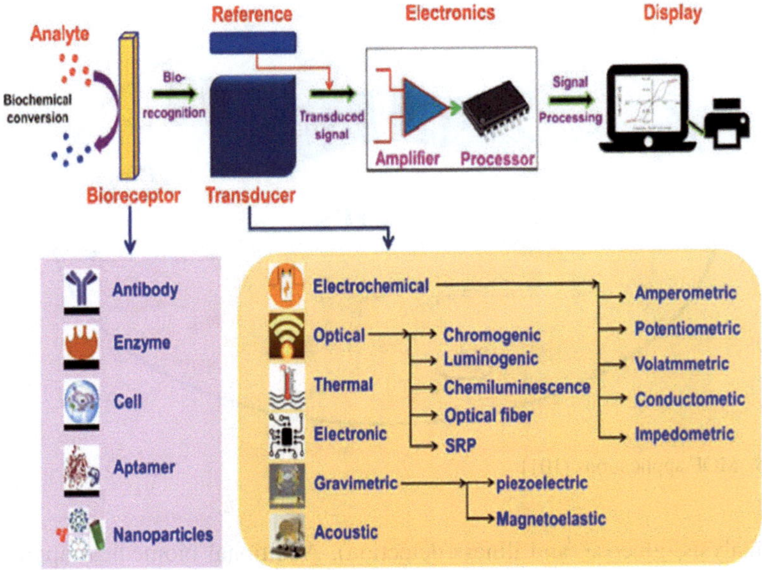

Fig. 3.9 Schematic diagram of typical biosensor consisting of bioreceptor, transducer, electronic system (amplifier and processor), and display (PC or printer) and various types of bioreceptors and transducers used in the biosensors are also shown [102]

and polymers (conductive polymers, cellulose and chitosan). EC biosensors are categorized as (a) conductometric, (b) voltammetric, (c) amperometric, (d) impedimetric, and (e) potentiometric based on their transduction mechanism [102].

(a) Conductometric biosensors: These devices monitor a change in the conductance between the two electrodes during an EC response (a change in the analyte's conductivity behavior) [104]. (b) Voltammetric biosensors: Voltammetric biosensors detect analytes by tracking the current flow during the regulated voltage change. These sensors have very high sensitivity and can detect many analytes at once [105]. (c) Amperometric biosensors: Two or three electrodes may be used in an amperometric biosensor configuration. When a steady voltage is applied to the WE compared to the RE, these sensors measure the current generated, which leads to the decrease or oxidation of electroactive substances at the WE. The analyte concentration is directly connected to the current produced on the surface of the WE in the solution. Compared to potentiometric biosensors, this method is more suited for mass production as it allows for a quick, linear response that is sensitive and accurate. However, limited interference and selectivity from other electroactive substances are disadvantages of these sensors [106]. (d) Biosensors that use impedimetry: These instruments measure the resistance to electricity that arises when a sinusoidal low-frequency signal is applied. A small alternating electric potential is applied to the sensor's electrode, and the parts of the resulting current that are in sync and out of sync are measured at different frequencies with an impedance analyzer [107]. (e)

Potentiometric biosensors: Under open-circuit conditions, these biosensors determine the voltage produced at the WE in relation to the RE, which results from the moment that the bioreceptor and analyte bind. A biological mechanism is converted into a possible reaction using ion-selective electrodes and ion-sensitive field-effect transistors [108].

Chemiluminescence (CL), fluorescence/luminescence, surface plasmon resonance and optic fiber are some of the main types of optical biosensors. The basis of optical sensing is the process of luminescence, which is the abrupt release of visible, ultraviolet, or infrared optical spectrum light devoid of heat-induced emission by a material. The emission is produced by electrons traveling from molecular orbitals with greater energy, often the ground state or the lowest vacant molecular orbitals, to lower-energy ones [109]. (a) A molecule that emits longer-wavelength light and absorbs brightness at a specific wavelength is the basic idea behind fluorescence/luminescence, which is the basis of fluorescent biosensing theory. In a fluorescence biosensor, a glowing molecule is used to label the sensing probe tag or the target molecule in reverse, and it is activated by a specific wavelength of light. As it interacts with the biosensor, the immediate surroundings of the fluorescent molecule changes, which modifies the emission properties of the molecule. Any kind of photoluminescence, such as fluorescence and phosphorescence, requires an atom, molecule, or nanostructure to have the capacity to absorb light and generate an electrical excitement. In the fluorescence process, the analyte interacts with the molecular recognition element, activates the electron linked to the molecule, and then takes in a photon. We illustrate the differences between the two using an abridged Jablonski schematic, which only shows the degree of vibrational energy changes for the ion dopant. Greater singlet energy level (S1, S2, ..., Sn) is reached when an electron is excited by incoming radiation from its ground state (S0). Ten nanoseconds will see the electron radiatively de-excite to S0 and then de-excite to the stage of lowest excitement (S1). However, in phosphorescent emission conditions, there are three different energy levels between metastases, and the electron moves from T2 to T1 after it stabilizes and loses energy from the level below. This delayed transfer of electrons can emit up to 10 s after the absorption of the radiation [110].

This delayed transfer of electrons can occur up to 10 s after the absorption of the radiation. Oxidation may directly cause this excitation, resulting in a chemiexcited intermediate that emits light as energy. An alternative method is chemiluminescence resonance energy transfer (CRET), which involves the intermediate that is enthusiastically transferring its vitality to a different substance (a fluorophore). Light is then released by the fluorophore, shows how these two pathways result in direct and indirect forms of CL. Common examples of compounds that directly generate light by CL include luciferin and luminol. Superoxide radical anion (O^{2-}) and hydrogen peroxide (H_2O_2) oxidize these compounds, changing their chemical structure and causing light to be released. Peroxides may also directly create singlet oxygen ($1O_2$) under certain circumstances. If singlet oxygen is not confined, it may oxidize other molecules and produce infrared light at 1270 nm. In indirect CL, a fluorophore or photosensitizer receives energy transfer so that it may independently produce light. Reactive oxygen species, such as $1O_2$ and superoxide radicals, also known as hydroxyl radicals (HO), may be produced by photosensitizers when oxygen is present [111]. (c) SPR: a common biosensor that reflects light from a gold-coated prism onto a detector. SPR is caused by variations in the density of free electron surface charges at the metal-dielectric contact. It may be effectively stimulated by an incident laser beam, resulting in evanescent waves. The prism-based method links the metal-dielectric incident optical signal contact by using a prism made of glass with a substantial refraction index. Measurements of the intensity of light reflected from the surface may be used to quantify the devices' sensitivity. A shift in the sensor surface, the refractive index, or the detecting area is expressed as a variation in the resonance angle for biosensing purposes [112]. (d) Optical fiber: an optical fiber biosensor consists of a sensitive layer for analyte detection, cladding, metal coating, and fiber core. Optical fiber biosensors use the total internal reflection (TIR) paradigm, where light traveling through the fiber core is encased in a cladding. A section of the cladding area is replaced by the metal covering and the element of biosensing. The components of biosensing (enzymes, antibodies, etc.) are fixed about the fiber core in order for the analyte to react [113].

In order to create MOF-based cancer biosensing approaches, MOFs have now been coupled with a variety of analytical techniques, like electrochemistry [22, 41, 45], ECL (electrochemiluminescence), PEC (photoelectrochemistry), colorimetry, and fluorescence. These methods have been useful in the identification of breast cancer [7]. It is evident from examining the many MOF types reported in the literature that the majority of them belong to the following categories: TCPP-based MOFs, lanthanide MOFs, MILs, ZIFs, NU IRMOFs, and UiO. MOFs have shown significant promise in early illness detection and diagnosis. To detect live cancer cells, exosomes, malignant tumor markers, microRNAs (miRNAs), and reactive oxygen species, many MOF-based biosensors have been developed. MOFs have attracted attention due to their tunable optical properties, which have enabled the fabrication of various MOFs in different dimensions with the inherent use of fluorescence by tunable fluorescence of specific organic ligands with metal ions. Applications such as fluorescent and colorimetric techniques have effectively used these MOFs [71, 114].

For instance, Zho et al. used aptamer-functionalized magnetic nanoparticles to create cancer-targeting medicines and bimetallic MOFs-based nanozymes to create the colorimetric signal amplification unit. Because pure Fe_3O_4 NPs also exhibit nanozyme activity with H_2O_2-like properties, $Fe_3O_4@SiO_2$ core-shell nanoparticles, which are produced by surface functionalizing Fe_3O_4 NPs that include silanes, were used in this research to raise the signal-to-noise proportion with regard to cancerous cells. Fe/Zr UIO-66 can stick directly to the cell surface because of the way Zr-O-P coordination works with the zirconium nodes in the MOF and the phosphate layers of the cell membrane. Additionally, with the help of 1 M H_2SO_4, Fe/Zr UIO-66 has a strong colorless tetramethylbenzidine (TMB) oxidation mediated by peroxide that may be catalyzed by a nanozyme with peroxidase-like activity to create oxidized ones with a yellow hue (oxTMB). For this reason, UIO-66 (Fe/Zr) is an excellent alternative to the colorimetric cancer cell detection platform. HeLa cells, with a linear range of 10×10^2–10×10^3 cells/mL and a detection limit (LOD) of 481 cells/mL, were detectable by the colorimetric aptasensor [115]. Afzalinia et al. have unveiled a novel biosensor for the quick and very accurate identification of miRNA as an indicator of cancer. This novel sensor achieves improved sensitivity for miRNA detection by combining silver nanoparticles (AgNPs) and La(III)-MOF using a hybridized oligonucleotide based on sandwich-type FRET. Ag NPs and MOFs operated as donor-acceptor energy pairs during the FRET-based mechanism for fluorescence quenching, with AgNPs acting as donors and Ln-MOFs as acceptors. To change fluorescent AgNPs and Ln-MOFs, two DNA probes were used that matched part of the target miRNA-155. The acceptor probes and energy donor link when miRNA-155 is present, which results in the FRET quenching of the Ln-MOFs' fluorescence. Not to mention that the transcriptional activities of the genome are regulated by G4s (G-quadruplexes), which are non-canonical structures made up of RNA and DNA sequences that are rich in guanine [116].

Tang and associates created a polydopamine (PDA)-coated MOF-based biosensor for the aim of detecting miRNA-122. A PDA (polydopamine), DOX (doxorubicin), and ZIF-8-based delivery system protected nanoclusters of gold coupled using AuNCs@ssDNA, or single-stranded DNA, from cellular degradation and showed an elevated level of quenching fluorescence ability. AuNCs were stuck to the ssDNA surface covered with PDA, and the nanoparticles were only released from the surface when zinc ions and the target microRNA were present, causing the target miRNA to bond with the ssDNA. The strong effect of AuNCs on aggregation was further demonstrated by the significant increase in fluorescence intensity at 498 nm as a result of Zn^{2+} emission due to acidic pH. The fluorescent miRNA-122 biosensor showed a straight line from 0.02 to 3.0 nM and a threshold for identifying 12.5 pM [117]. In another study, drDNA-BUT-88, a fluorescent nanoprobe functionalized with DNA and loaded with dye, was created by encasing the cyanine-3 (Cy3) dye molecule in the BUT-88 nanoparticle. To avoid false-positive results, the nanoprobe used in this study is a helpful instrument for ratiometric analysis identifying miRNA-21 as BM malignancy; both miRNA-21 and MUC-1 BMs may be accurately identified by it. With a 4.50 nM detection limit, the fluorescence intensity rose directly with the MUC-1 concentration. Even in the presence of other biological materials, the probe drDNA-BUT-88 has remarkable execution in live cells and is very specific for detecting miRNA-21 and MUC-1 [118].

According to estimates, lung cancer caused 238,340 fatalities in the US in 2023 more than twice as many deaths as oral cancer, breast cancer, and coupled with skin cancer. Cancer of the lung claims the lives of over 1.76 million people globally, which represents around 25% of all cancer-related deaths. Lung cancer, like other cancers, is caused by genetic and epigenetic problems, and over the years, lung cancer patients' 5-year survival rate has continuously remained poor, 18% for men and 21% for women. CT and traditional chest X-rays are the most often used imaging modalities for tumor diagnosis, however they have limitations [119]. Lung cancer comes in two primary varieties: SCLC, or small cell lung cancer, and nonsmall cell lung cancer. For instance, the BM of lung cancer is neuron-specific enolase (NSE). To find NSE, Dong et al. created an ECL biosensor that uses two signals to remove the differences caused by using just one probe or one signal. The researchers modified Ab1 and Ab2 using zinc oxide@silver NPs and $Ru(bpy)_3^{2+}$ functionalized Zn-MOF nanoflowers

(Ru-MOF-5 NFs), respectively. NSE took on the shape of a sandwich when it was attached to the immune complex. ZnO-AgNPs may increase the −1.5 V cathode ECL signal while decreasing the 1.5 V ECL signal; nonetheless, Ru-MOF-5 NFs demonstrated strong emission of ECL at both 1.5 V and −1.5 V. Employing an ECL immunosensor, a dual-signal ratio measurement technique was developed to identify NSE. It also boasts a wide linear range of 0.1 pg/mL–200 ng/mL and a detection limit of 0.041 pg/mL. In ECL, the NFs for Ru-MOF-5 and ZnO-AgNPs function as two distinct luminophores. $K_2S_2O_8$ and TEA react with $Ru(bpy)_3^{2+}$ to produce anode and cathode signals as part of the ECL process. The ECL of Ru-MOF-5 NFs is both quenched and emitted by ZnO-AgNPs. The interaction between ZnO-AgNPs and $S_2O_8^{2-}$ causes the quenching. The two luminophores interact intricately throughout the whole ECL process, with ZnO-AgNPs being essential for both emitting and modifying the ECL signal [120].

In a different study, Guo et al. investigated the application of a polyethyleneimine (PEI)-based EC device and a COFTAPB-TFPB (covalent organic framework) for the detection of ctDNA (circulating tumor DNA). Nitrogen doping of graphene (NG) was used to further functionalize the biosensor; the process by which the target and capture probes bind is verified using the DPV and CV methods. MOF-based DNA biosensors demonstrated an extensive array between 100 fM and 100 nM and a limit of detection of 7.65 fM [121]. Wang et al. developed a biosensor to detect CYFRA21-1 using an MOF and a $Ru(bpy)_3^{2+}$ combined with cyclodextrin (CD) biomaterial, which is an important application of BM for identifying lung cancer found in the epithelial cells' cytoskeleton. CD-MOF@$Ru(bpy)_3^{2+}$ is used as a substance that emits ECL; TPA (tripropylamine) is present in a Tris-HCl buffer, and it produces powerful ECL signals. As a substance that quenches, CdTe NPs absorb the light that the CD-MOF@$Ru(bpy)_3^{2+}$ emits. The combination of a quencher and an ECL emitter enables controlled ECL signal manipulation. In A549 living cells, based on a biosensor synthetic composite, it showed a high sensitivity and a threshold for identifying 0.006 ng/mL [122]. In another project, Hu et al. developed an EC biosensor that effectively identifies miRNA-126 and glial cell lines (C6) by using a solid base made of bimetallic CoNi-MOF to connect cDNA. CoNi-MOF's big cavities and profusion list anchoring locations may greatly enhance the cDNA strands' immobilization. When the presence of the analyte miRNA-126 is detected, cDNA and miRNA-126 hybridize to form the analyte miRNA-126's existence, which alters the answer from EIS. The RCT signal is raised by the CoNi-MOF coating on the electrode made of gold (AE), suggesting that its EC activity is lower. RCT is further increased by immobilizing cDNA strands because anions and probes are repelled by negative charges. When cDNA strands are hybridized utilizing miRNA-126, more oligonucleotides that are negatively charged are produced, which raises the RCT value. mRNA-126 and C6 cells were identified, and the biosensor generation process was examined using CV and EIS measurements [123].

Most often, breast cancer is a malignancy among females and the second most typical illness globally. In 2024, the American Cancer Society projects that 42,250 women will die from invasive breast cancer and that 310,720 new instances will be discovered in the US. The risk factors for breast cancer are many and include age, lifestyle, family history, sex, estrogen, and lifestyle. HER2, MUC1, BRCA1, and UBE2C are significant and conventional indicators of breast cancer [5, 40]. A bimetallic bifunctional ZrHf-MOF loaded with carbon dots (CD) was developed as an EC aptasensor to detect MCF7 and HER2 cells simultaneously and early lines; measurements of CV, DPV, and EIS were performed to examine the process of developing biosensors. Following AE modification, the Rct value utilizing CDs@ZrHf-MOF/AE would increase. The phosphate groups in the ionization of aptamer strands create negative charges, which stop electrons from moving at the surfaces of electrodes. The EIS curves show that the aptasensor can detect HER2 because of its quick response in the EIS. Additionally, the LOD shown by the sensor was 19 fg/mL and 23 cells/mL, respectively [124]. AuNPs were introduced to a PBA network enhance EC signaling and electron transport in another study that used a similar method to identify HER2. The produced nanomaterial-based assays' EC characteristics were examined using CV and EIS methods. When the aptamer strands stop moving, their negative charge interacts with the negative charge of $[Fe(CN)_6]^{3-/4-}$, which leads to an increase in the Apt/MnFePBA@AuNP-2/AE Rct value. The resulting HER2 was detected using an EC biosensor with exceptional sensing capabilities. It could detect very small amounts, as low as 36 cells per mL of MCF-7 and HER2 cells and 0.247 picograms per mL, while also being very selective, practical, and reliable [125]. On the other hand, Gu et al. proposed an independent biosensor system using multifunctional MOFs for exosome EC detection. A glucose dehydrogenase capsule was used to separate ZIF-8 to operate as an anodic biocatalyst, and UiO-66-NH_2 was used in their construction as the nanocarrier to load electroactive $K_3[Fe(CN)_6]$. The composites of ferricyanide and UiO-66 were drawn to the cathode region by the target exosomes, where they were able to absorb the electrons created by the anodic oxidation and detect their energy. Following material modification using exosomes, non-conductive CD63 antibody, and mercaptohexanol (MCH), EIS analysis revealed that the Rct values grew gradually, indicating that the alterations were well incorporated. It is significant to observe that upon immobilization of $[Fe(CN)_6]^{3-}$/UiO-66, the Rct value significantly dropped. The electroactive $[Fe(CN)_6]^3$ addition molecules are probably the cause of this; in combination with exosomes derived from the target MCF-7 cell line, the method showed an outstanding LOD of 300 particles/mL [126].

Another study used template growth to make two types of bimetallic composites of MOFs on MOFs with a shell-core shape: on Tb-MOF and Fe-MOF, respectively, at the same time. EC aptasensors were fabricated using these composites as transducer materials, and the biosensor development process was investigated using CV and EIS. The AE surface's Tb-MOF-on-Fe-MOF coating increased the electrode's Rct value, according to EIS analysis. The result suggests that there was reduced conductivity in the Tb-MOF-on-Fe-MOF layer, which prevented efficient exchange of electrons at the electrode-electrolyte contact point. After aptamer immobilization, the RCT amount increased, most likely as a result of the anchored blockage of the electrode surface. The Rct value continuously increased, indicating that the CA125 aptamer had specific binding to MCF-7 cells, and the aptasensor CV measurement showed a comparable result. Additionally, the analytical performance of the Tb-MOF was superior to that of the Tb-MOF-on-Fe-MOF. Additionally, it displayed an extremely low LOD of 58 IU/mL for the MCF-7 cells' carbohydrate antigens from 125 and 19 cells/mL [49]. Chang et al. also introduced a biosensor that uses a zirconium-MOFs method to simultaneously detect multiple cancer biomarkers, using let-7a and miR-21 as examples. Figure 3.10 shows that two types of Zr-MOFs (TMB@UiO and MB@UiO) were made by adding the electroactive molecules methylene blue (MB) and tetramethylbenzidine (TMB) to the porous UiO-66-NH_2. DNA that is double-stranded was used to enclose the porous Zr-MOFs and inhibit the EC reaction using DNA hybridization and electrostatic binding. When miR-21 and the let-7a cell line were present, there was a strand displacement that unlocked the Zr-MOF mechanism to liberate the electroactive chemicals. TMB and MB's DPV signals rose in direct proportion to their concentrations, suggesting that they might be used as dyes that are electroactive in a homogenous biosensor EC to simultaneously identify let-7a and miRNA-21. Among patients with breast cancer, the homogenous EC sensing device performed well, with LODs of 3.6 fM for let-7 and 0.0082 pM for miR-21 [127]. Wang et al. created a biosensor for ECL according to "sandwich-like" Zn-MOF, the BRCA1 gene, which is responsible for breast cancer. Under the guidance of cetyltrimethylammonium bromide, ruthenium(II) dichloride ($Ru(dcbpy)_3{}^{2+}$) and zinc ions were chelated using tris(4,4′-dicarboxylic bipyridyl) in the research to produce the flower-shaped Zn-MOF microsphere. The produced microsphere, called ZnMOF (Ru), had a 53-fold ECL luminescence higher than that of $Ru(dcbpy)_3{}^{2+}$. Successful use of the "sandwich-like" biosensor used to identify BRCA1 using one of the best ECL emitters with an incredibly low 0.71 fM LOD [128].

Xue et al., for instance, developed a paired amplitude model for the identification of miR-21 EC. In particular, the catalytic hairpin technique is paired with building a chain of hybridization reactions to provide a two-phase expansion of the intended signal input. As a consequence, there were many sites that were actively accessible for Cu-MOF-based signal tag loading [129]. Another study created FeP-pSC_4-AuNPs@CuMOFs, a new type of synthetic enzyme made from Cu-MOF, which

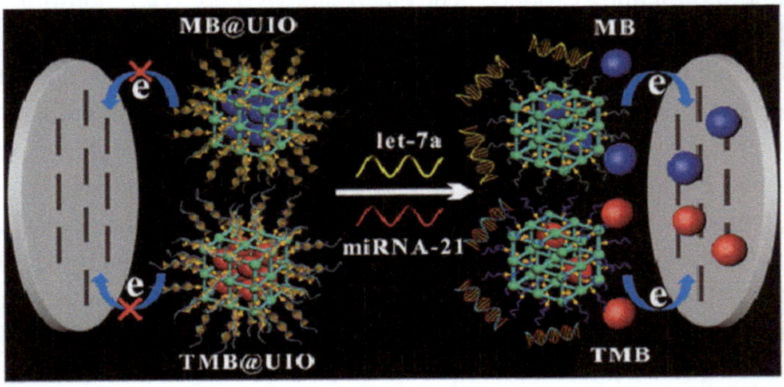

Fig. 3.10 Schematic illustration of a zirconium-MOF-based homogeneous EC biosensor for simultaneous detection of let-7a and miR-21 that releases target-induced strand displacement, electroactive MB and TMB, and generates measurable DPV signals. Adapted with permission from [127]. Copyright 2019 American Chemical Society

can self-assemble. The synthetic enzyme showed outstanding ability to speed up chemical reactions and conduct electricity, similar to peroxidase, by mimicking the Fe-N_5 and heme active site proteins. By helping to break down H_2O_2 as the electrical signal output, it could be used together with the aptamer MUC1 to identify malignant cells with elevated accuracy and sensitivity. Bionic enzymes based on Cu-MOFs were also used to find more cancer BM, encompassing nitric oxide, PSA, and CEA generated by MCF-7 cells. The popular technique of signal amplification facilitated by enzymes, which is employed to find indicators of cancer, makes Cu-MOFs very appealing candidates [130]. In one experiment, Zhang et al. used a substance created by an enzyme method based on Cu-MOF to enhance the detection of resistance when measuring CA15-3, which is a carbohydrate antigen. According to, an immunological probe comprising Cu-MOF, CA15-3, and GOx was adhered to the electrode after CA15-3 had bound. CuMOF used the H_2O_2 produced by GOx to aid in acetylacetone's (ACAC) oxidation to produce ACAC radicals. Later, ACAC It was inhibited by radical $[Fe(CN)_6]^{3-/4-}$ electron transfer while simultaneously promoting the surface area-based in-situ synthesis of N-isopropylacrylamide, an insulating polymer. The curvature of CV showed that GCE treated using AuNPs had a greater signal strength compared to bare GCE, most probably as a result of its increased conductivity and region of particular surface. After Ab1, BSA, and the CA15-3 antigen were immobilized, the current signal gradually diminished. The current signal further dropped subsequent to radical polymerization brought on by cascade catalysis, most likely as a result of the immunoprobes and polymers weak conductivity, which prevented electron transmission. The EIS findings showed that resistance values rose in response to polymer formation and immunoprobe capture, suggesting that the layer of nonconductive proteins prevents electron transmission over the user interface [131].

In another study, ascorbic acid (AA) and Acid 2,5-diaminoterephthalic were used with UiO-66-2NH_2 as ligands to create a Zr-MOF for producing multiple agents that prevent MCF-7 cells using aptasensors. The UiO-66-2NH_2/AE's modified Rct value rose in addition to EIS, showing that electrons were hindered by the nonconductive UiO-66-2NH_2 layer transport across the boundary between solid and liquid. When UiO-66-2NH_2 PO_4-Apt was added, the Rct value increased dramatically due to strong PO_4-aptamer binding. In the aqueous solution, the attached aptamer strands' PO_4^{3-} were negatively charged, which prevented electron transport at the contact and resulted in a significant distaste for the $[Fe(CN)_6]^{3-/4-}$ redox pair. The large number of HER2-positive, overexpressed MCF-7 cells that were found led to the notable rise in Rct after MCF-7 cell immobilization. The produced biosensor has various 100–100,000 cells/mL and a maximum detection of 31 cells/mL [132]. Last but not least, Sadrabadi et al. used a range of materials, like Fe@rGO, CuBTC-AIA (CuMOF), and carbon nanofibers, to raise the surface's SA:Vol and expedite the process of electron transfer to detect microRNA 155. In this research, DPV, EIS, and CV data were used to evaluate the biosensor's operating capabilities [133].

By 2023, 288,300 men in the USA are anticipated to get a prostate cancer (PCa) diagnosis. An estimated 1,414,259 people worldwide had a PCa diagnosis in 2020, of which 375,304 perished. Therefore, early and precise identification is crucial to reducing the risk of mortality from PCa. In their lives, around eleven percent of males will have cancer of the prostate, and the 5-year survival rate for localized PCa is quite high, at over 100%. The majority of widely utilized BM to treat prostate cancer is prostate-specific antigen (PSA) [134]. For instance, @UIO-66-NH_2 silver was used as a transporter to load a significant quantity of W-doped CdS luminophores into an innovative PSA ECL biosensor. Gold nanoparticles and rGO modified the GCE (glassy carbon electrode) surface to create a consistent, bustling space for the insertion of Ab1, a PSA-targeted antibody, which was then joined to PSA by way of antigen-antibody exchanges. The device, which was treated with Ab2 and Ag^+ @UIO-66-NH_2@CdWS, was used as a probe for signals. After changing the testing conditions, it was able to detect concentrations from 0.1 pg/mL to 10 ng/mL, with a detection limit of 38 fg/mL [135]. A different study developed tetracyanoquinodimethane (TCNQ)-doped Cu-MOF to increase device electrical conductivity. The composite served as an advanced platform for the development of immunoassay PSA detection. As shown by its 0.06 ng/mL LOD and strong Even PSA selectivity in the presence of other proteins, based on TCNQ-$Cu_3(BTC)_2$-conjugated PSA antibodies, the aptasensor has demonstrated use for the range of linear dynamics between 0.1 and 100 ng/mL to identify blood samples that have been tampered with [136]. To construct an ECL sensor actuated by a PSA target that functions as an enhancer and quencher all at once, Shao et al. created an Au/G-quadruplex MOF. This method allowed the ratiometric ECL sensor to achieve full amplification and signal self-calibration. MOF/Au/G-quadruplex dual potential-dependent emitters and ECL emitters (QDs and luminol) were used in its construction. The manufactured device showed an LOD of 58 pg/mL and a broad linear detection range of 0.5–500 ng/mL. Apart from $Ru(bpy)_3{}^{2+}$, several metal QDs and noble MNPs may also function as luminophores of ECL [137].

CEA (carcinoembryonic antigen), a protein that is heavily glycosylated and has an estimated molecular weight of 200 kDa, is one of the most popular BCs for the analysis, forecast evaluation, and observation of malignant growths, like cancers of the stomach, pancreas, liver, breast, and colon. CEA is very low in healthy human blood, but it rises to above 5 ng/mL when cancer cells proliferate. Numerous methods, such as colorimetric, fluorescence, surface-enhanced Raman scattering, and EC methods, have been used for CEA detection [52, 71]. Zhang and his team created a CEA EC biosensor by modifying the GCE with polyaniline nanofibers. In comparison to bare GCE, PANI/GCE exhibits comparable peak potential separation and higher currents, suggesting good conductivity and a larger EC active area. A nonconductive use of the MSA layer and a slower EC response rate are shown by the larger difference in peak potential and lower currents observed in mercaptosuccinic acid (MSA)/PANI/GCE. Investigations of the modified GCE were conducted for the concurrent detection of cancer BM, AFP, and CEA using MOF pieces made from Cd (II) or Pb (II). The nanoparticles of MOFs for both BMs were given labels with secondary antibody production to generate a sandwich test; the reference states that the comparable LODs are 30 and 100 fg/mL [138].

In a different work, Li's team created a biosensor that uses electricity that resembled a sandwich. In their investigation, they synthesized Ce MOF and loaded it using peroxidase from horseradish with silver nanoparticles to catalyze H_2O_2, which increased the signal current. The MOF-based structure was subsequently covered with hyaluronic acid, and Au NPs were added to the immunosensor to improve the capacity of Ab1, the main antibody, to bind. The immunological sensor design also used Au NPs to enhance antibody attachment due to their high electrical conductivity. After Au NPs were added to the GCE, the high-frequency radius shrank. Since the alteration of Au NPs accelerates electron transport, the interface resistance dropped when compared to the bare electrode GCE. CV and EIS curves showed a sequential drop in peak current following MoF-Ce surface area and CEA modification, suggesting a decline in conductivity of charge transport and a consequent rise in Rct in comparison to the previous stage. The immune assay assembly was analyzed for CEA detection, and the findings showed a 0.2 pg/mL LOD. The recommended immunological sensor was also shown to have outstanding stability, reproducibility, and selectivity [139]. When used as an electrode decoration, Liu and his associates showed that an MOF based on Ag may produce enduring EC signals using just a voltammetric technique. DPV curves show that the AuNP/Ag-MOFs peak current signal with the labels anti-CEA/CEA/anti-CEA/MWCNT/GCE grows in amplitude in proportion to the concentration of CEA. Adding MWCNTs, AuNP/Ag-MOFs marked with CEA, anti-CEA, and antibodies, along with other changes, raises Impedance indicates a stronger immune complex barrier for electron transport [140]. In another study, Zhou et al. used an EC biosensor using a Cu MOF and used EIS metrics for identification. After preparation, Pt@CuMOFs-hGq-GOx is added to enhance the functionality of the Cu MOF nanostructures. This biosensor assembly was found to replicate the peroxidase activity. The main blockage of the biomolecules Pt@CuMOFs-hGq-GOx, BSA, CEA, hemin, and Apt2 led to a decrease in the CV signal and a significant increase in the Rct after Apt1 was attached. When 3,3-diaminobenzidine (DAB) with glucose was added to the resulting interface between electrodes in the last stage, the production of insoluble precipitates resulted in an important increase inside the EIS signal and a decrease in the peak current of CV [141]. Additionally, to detect CEA, researchers used UiO-66 in combination with Ag NCs (silver nanoclusters) (AgNCs-Apt-UiO-66) and the CEA aptamer. Both EC and SPR methods were discovered to be used by the synthesized AgNCs-Apt-UiO-66 sensor. EIS curves demonstrated the ability of the UiO-66 and AgNCs@Apt@UiO-66 nanocomposites to increase the Rct at the electrode-analyte solution electrode contact. This was clarified by the repulsive relationship between the negatively charged surface of PO_4^{3-} within the aptamer strands and the redox pair of $[Fe(CN)_6]^{4-}$/$[Fe(CN)_6]^{3-}$. DPV curves with AgNCs@Apt@UiO-66 or CEA binding and UiO-66 showed less electrochemical activity after repeatedly changing the AE electrode. AgNCs@Apt@UiO-66, or CEA binding and UiO-66, demonstrated outstanding sensitivity with LODs of 4.93 and 8.88 pg/mL, respectively, according to impedance spectroscopy for DPV and EC. Meanwhile, The SPR biosensor's developed limit of detection was 0.3 ng/mL [142]. Zhang et al. recently developed an EC biosensor by combining gold nanoparticles with ZIF-8 and OMC. As a nanocarrier for support, ZIF-8 increased the load of the antibody on its massive surface area, rendering it immobile. To improve EC signals, the OMC was mounted atop a GCE because of its inherent electrical conductivity. DPV was employed to document the EC responses because Au NPs provide a significant quantity of antibody OMC, and binding sites enhance electron transport, so EC responses are enhanced. The findings show that Au NPs@ZIF-8 exhibits a much larger shift in the highest current of DPV than ZIF-8, and the Au NPs@ZIF-8 modified electrode has a lower ΔIDPV value than Au NPs@ZIF-8/OMC, with a 1.3 pg/mL LOD; the monitoring demonstrated excellent performance [143]. Liu et al. demonstrated the hydrothermal production of Cu MOFs to produce a sandwich-type immunosensor. Toluidine blue (TB)-loaded, PDA-coated mesoporous Cu MOFs served as a probe for signal transducers. The recommended sensor was developed employing Cu MOFs-TB/PDA, MWCNTs, and chitosan on a GCE with a big area of the surface to immobilize primary antibodies (Ab1). Because MWCNTs have better EC performance, CV curves show that the modified electrode has a greater current responsiveness. When the surface is changed using CEA, Ab1, and Ab2 labels, the electron transfer on Cu-MOFs-TB/PDA is inhibited by the selective adsorption of antigens and antibodies, causing the current signal to gradually diminish. Additionally, since the protein layer that isn't conductive prevents transmission of electrons at the contact, the resistance gradually increased in EIS curves after the immobilization phases [144].

3.5 Conclusion

The integration of MOF-based biosensors with digital health technologies, such as wireless communication systems and smartphone apps, is key to their future. Continuous monitoring and immediate response to healthcare practitioners are made possible by the real-time transmission of biosensor data via mobile devices that are coupled with MOF-based biosensors. By monitoring a patient's unique BMs at home, this integration may result in tailored cancer diagnosis and less frequent hospital stays. Wearable biosensors and further developments Technology in the Internet of Things (IoT) will make it possible for smooth health monitoring mechanisms that provide real-time data analytics and a previously unheard-of degree of patient interaction. The sensitivity and specificity of MOF-based biosensors for cancer diagnosis may still be improved, even with recent developments. Enhancing the MOFs' surface chemistry with particular antibodies, aptamers, or polymers with molecular imprints that attach to malignant cells should be the next line of investigation. The EC characteristics of MOFs may be further improved by including innovative nanomaterials like MIP, or molecularly imprinted polymer, and graphene, which will lead to lower LOD and quicker reaction times. The creation of nanostructures that are hybrid, including composites of metal nanoparticles and MOFs, may also have influences that work together to enhance sensor performance. One intriguing multidisciplinary method for printing biosensors on demand for the diagnosis of cancer is the use of MOF-based additives in bioink-based sensors for 3D printing. The creation of multiplexed MOFs-based biosensors that can identify many cancer BMs at once is one of the most important future approaches. Due to the fact that cancer is often defined via the manifestation of many BMs instead of just one signal, this capacity is essential. In addition to increasing diagnostic precision, multiplexing would provide thorough details on the stage and course of disease. To allow the simultaneous study of many BMs in one device, future designs might use microfluidic channels or microarray platforms, which would reduce the time needed for tests and the amount of sample required. The development of MOFs-based biosensors as dependable point-of-care diagnostic instruments is necessary to attain broad clinical use. This calls for creating portable, user-friendly gadgets that don't need a lot of training to use. Future research should primarily focus on reducing the size and creating reliable, disposable sensor systems that allow quick, on-site examination of BMs for cancer. The emergence of customized healthcare is highlighted because point-of-care tools that can adjust to the medical requirements of each patient allow for more individualized treatment regimens and improved cancer therapy management. The stability and reusability of MOF-based biosensors are critical to their economic viability. Future studies must deal with the problems of biofouling, sensor deterioration with time and sensitivity loss brought on by environmental variables such as temperature and pH changes. It is essential to provide coatings for protection or encapsulating methods that improve the robustness and functionality of MOF sensors. These issues may be resolved by advances in materials science, like the use of cross-linking agents or strong polymeric matrices, which would guarantee steady performance throughout time. Careful checking and following of rules are needed for biosensors using MOFs to go from the lab to real medical use. To assess the effectiveness, safety, and dependability of these biosensors in practical settings, future research must include comprehensive clinical trials. To provide uniform procedures for sensor testing and approval, cooperation between researchers, physicians, and regulatory agencies will be essential. Establishing standards for the mass manufacturing of MOF-based biosensors that satisfy quality assurance and control requirements should also be a priority. An intriguing path for improving MOF-based biosensors is the investigation of innovative technologies like artificial intelligence and machine learning. Complex datasets from biosensor readings might be analyzed by AI algorithms, increasing the precision of diagnosis and forecasting the course of cancer. The material selection, design, transduction, storage, data transmission, and analysis of MOF-based biosensors may all be aided by quantum artificial intelligence (QAI) algorithms. Furthermore, combining MOF-based biosensors with methods such as fluorescence-based detection or surface-enhanced Raman spectroscopy may provide further sensitivity and specificity. A potential area for quantum sensors, single electron/photon sensors, and NOEMS (nano-opto-electro-mechanical systems) is the integration of MOFs in optical fibers and transistor structures. Future advancements in MOF-based biosensors must take manufacturing cost-effectiveness and sustainability into account. The environmental effect and production costs of sensors may be decreased by employing biodegradable materials and green synthesis techniques. The development of low-cost, scalable manufacturing methods will be crucial to democratizing cancer diagnoses worldwide by enabling these biosensors to be used in environments with limited resources. We described the advancements and improvements in applications of MOF-based EC cancer biosensors and their production in this chapter. Even after decades of research, we still don't have enough tools to detect and treat cancer, which is thought to be the world's most fatal illness, and it's spreading fast. These treatments don't always satisfy the criteria for an accurate diagnosis of cancer, have poor results with a high chance of recurrence, and are exceedingly damaging to the body; because of their special qualities, MOFs have been the subject of our discussion. MOFs are inherently biodegradable, hybrid architectures that are biocompatible and exhibit minimal cytotoxicity because of their extremely labile metal-ligand interactions. This implies that after serving their original function, they may be swiftly removed from the body. Due to their distinct characteristics, MOFs have shown themselves to be beneficial as nanotechnology for a range of biological purposes. Among these applications are therapeutic medication administration as well as diagnostic sensing and imaging.

In its early stages, cancer is difficult to diagnose; the majority of cancer fatalities occur from late diagnosis due to individuals not obtaining consistent cancer screenings because they are time-consuming, costly, and intrusive. Future research will need to focus on MOFs' solvability and stability in water, the regularity of the size and shape of pores, and the advancement of the synthesis process. Priority should also be given to a detailed understanding of how MOFs and biomolecules interact. Because of their strong collaboration with materials science and analytical chemistry, biology, and other fields, we expect biosensors using MOFs to have promising futures and provide additional practical technological tools for detecting cancer. We are excited to see more of these in the future, and our chapter provides a platform to learn about the earlier research that has been conducted to create effective MOFs for the upcoming activities.

References

1. F. Bray, A. Jemal, N. Grey, J. Ferlay, D. Forman, Global cancer transitions according to the Human Development Index (2008–2030): a population-based study. Lancet Oncol. **13**(8), 790–801 (2012)
2. R. Salahandish, A. Ghaffarinejad, S.M. Naghib, K. Majidzadeh-A, H. Zargartalebi, A. Sanati-Nezhad, Nano-biosensor for highly sensitive detection of HER2 positive breast cancer. Biosens. Bioelectron. **117**, 104–111 (2018). https://doi.org/10.1016/j.bios.2018.05.043
3. R. Salahandish, A. Ghaffarinejad, E. Omidinia, H. Zargartalebi, K. Majidzadeh-A, S.M. Naghib, A. Sanati-Nezhad, Label-free ultrasensitive detection of breast cancer miRNA-21 biomarker employing electrochemical nano-genosensor based on sandwiched AgNPs in PANI and N-doped graphene. Biosens. Bioelectron. **120**, 129–136 (2018). https://doi.org/10.1016/j.bios.2018.08.025
4. M. Hosseine, S.M. Naghib, N. Rabiee, Metal-organic frameworks: biosensor applications for diagnosis of cancers. Microchem. J. 112341 (2024)
5. M. Sadeghi, S. Kashanian, S.M. Naghib, E. Askari, F. Haghiralsadat, D. Tofighi, A highly sensitive nanobiosensor based on aptamer-conjugated graphene-decorated rhodium nanoparticles for detection of HER2-positive circulating tumor cells. Nanotechnol. Rev. **11**(1), 793–810 (2022)

6. M. Sadeghi, S. Kashanian, S.M. Naghib, F. Haghiralsadat, D. Tofighi, An efficient electrochemical biosensor based on pencil graphite electrode mediated by 2D functionalized graphene oxide to detect HER2 breast cancer biomarker. Int. J. Electrochem. Sci. **17**(4), 220459 (2022). https://doi.org/10.20964/2022.04.62
7. M. Sadeghi, S. Sadeghi, S.M. Naghib, H.R. Garshasbi, A comprehensive review on electrochemical nano biosensors for precise detection of blood-based on comarkers in breast cancer. Biosensors **13**(4), 481 (2023)
8. M.A. Farzin, S.M. Naghib, Paper-based immunosensor integrated with bioinspired Cu-polydopamine nanozyme for voltammetric detection of CA-15-3 tumor marker. Sens. Actuators Rep. **8**, 100209 (2024). https://doi.org/10.1016/j.snr.2024.100209
9. M.A. Farzin, S.M. Naghib, Laccase-mimicking Cu@CDs-enabled bioinspired immunosensor using CuO nanoparticles for voltametric detection of serum thyroxine level in patients with hyperthyroidism or hypothyroidism. Sens. Actuators Rep. **9**, 100291 (2025). https://doi.org/10.1016/j.snr.2025.100291
10. M.A. Farzin, S.M. Naghib, N. Rabiee, Bio-inspired and biomimetic composites based on biodegradable polymers for sensing applications with emphasis on early diagnosis of cancer. Chem. Eng. J. **493**, 152445 (2024). https://doi.org/10.1016/j.cej.2024.152445
11. M.A. Farzin, S.M. Naghib, N. Rabiee, Advancements in bio-inspired self-powered wireless sensors: materials, mechanisms, and biomedical applications. ACS Biomater. Sci. Eng. **10**(3), 1262–1301 (2024). https://doi.org/10.1021/acsbiomaterials.3c01633
12. O. Abazari, P. Dayati, M. Shahidi, J. ZavarReza, M. Rahmanian, S.M. Naghib, Exploring a desirable quadr-mRNAs panel for non-invasive and ultrasensitive bladder cancer diagnosis: in-silico and clinical studies. Curr. Med. Chem. (2024). https://doi.org/10.2174/0109298673302140240627053904
13. N.A. Abtahi, S.M. Naghib, S.J. Ghalekohneh, Z. Mohammadpour, H. Nazari, S.M. Mosavi, S.M. Gheibihayat, F. Haghiralsadat, J.Z. Reza, B.Z. Doulabi, Multifunctional stimuli-responsive niosomal nanoparticles for co-delivery and co-administration of gene and bioactive compound: in vitro and in vivo studies. Chem. Eng. J. **429**, 132090 (2022). https://doi.org/10.1016/j.cej.2021.132090
14. N.A. Abtahi, S. Salehi, S.M. Naghib, F. Haghiralsadat, M.A. Edgahi, S. Ghorbanzadeh, W. Zhang, Multi-sensitive functionalized niosomal nanocarriers for controllable gene delivery in vitro and in vivo. Cancer Nanotechnol. **14**(1), 22 (2023). https://doi.org/10.1186/s12645-023-00175-w
15. N.A. Abtahi, S.M. Naghib, F. Haghiralsadat, M. Akbari Edgahi, Development of highly efficient niosomal systems for co-delivery of drugs and genes to treat breast cancer in vitro and in vivo. Cancer Nanotechnol. **13**(1), 28 (2022). https://doi.org/10.1186/s12645-022-00135-w
16. N.A. Abtahi, S.M. Naghib, F. Haghiralsadat, M. Akbari Edgahi, E. Askari, A comparative study on biopharmaceutical function of curcumin and miR-34a by multistimuli-responsive nanoniosome carrier: in-vitro and in-vivo. Front. Mol. Biosci. **9** (2022). https://doi.org/10.3389/fmolb.2022.1043277
17. N.A. Abtahi, S.M. Naghib, F. Haghiralsadat, J.Z. Reza, F. Hakimian, F. Yazdian, D. Tofighi, Smart stimuli-responsive biofunctionalized niosomal nanocarriers for programmed release of bioactive compounds into cancer cells in vitro and in vivo. Nanotechnol. Rev. **10**(1), 1895–1911 (2021)
18. A. Daily, P. Ravishankar, S. Harms, V.S. Klimberg, Using tears as a non-invasive source for early detection of breast cancer. PLoS ONE **17**(4), e0267676 (2022)
19. M. Nooshnab, S.M. Naghib, R. Rahmanifard, E. Khakpour, A review of minimally and non-invasive glucose monitoring techniques, devices and sensors. Curr. Anal. Chem. **20**(4), 217–241 (2024)
20. B. Mikaeeli Kangarshahi, S.M. Naghib, N. Rabiee, DNA/RNA-based electrochemical nanobiosensors for early detection of cancers. Crit. Rev. Clin. Lab. Sci. **61**(6), 473–495 (2024)
21. Z. Khurshid, M.S. Zafar, R.S. Khan, S. Najeeb, P.D. Slowey, I.U. Rehman, Role of salivary biomarkers in oral cancer detection. Adv. Clin. Chem. **86**, 23–70 (2018)

22. A.-A. Zare, H. Naderi-Manesh, S.M. Naghib, M. Shamsipur, F. Molaabasi, Label-free electrochemical cancer cell detection leveraging hemoglobin-encapsulated silver nanoclusters and Cu-MOF nanohybrids on a graphene-assisted dual-modal probe. Sci. Rep. **13**(1), 21980 (2023). https://doi.org/10.1038/s41598-023-49418-1
23. N. Shahbazi, R. Zare-Dorabei, S.M. Naghib, Design of a ratiometric plasmonic biosensor for herceptin detection in HER2-positive breast cancer. ACS Biomater. Sci. Eng. **8**(2), 871–879 (2022). https://doi.org/10.1021/acsbiomaterials.1c01369
24. M. Mustafa, A. Nornazirah, F. Salih, E. Illzam, M. Suleiman, A. Sharifa, Breast cancer: detection markers, prognosis, and prevention. IOSR J. Dent. Med. Sci. **15**(08), 73–80 (2016)
25. M. Shamsipur, F. Molaabasi, A.-A. Zare, H. Naderi-Manesh, S.M. Naghib, Label-free electrochemical cancer cell detection leveraging hemoglobin-encapsulated silver nanoclusters and Cu-MOF nanohybrids on a graphene-assisted dual-modal probe. Sci. Rep. **13**, 21980 (2023)
26. S.M. Naghib, Fabrication of nafion/silver nanoparticles/reduced graphene nanosheets/glucose oxidase nanobiocomposite for electrochemical glucose biosensing. Anal. Bioanal. Electrochem. **8**(4), 453–465 (2016)
27. S.M. Naghib, Two-dimensional functionalised methacrylated graphene oxide nanosheets as simple and inexpensive electrodes for biosensing applications. Micro Nano Lett. **14**(4), 462–465 (2019). https://doi.org/10.1049/mnl.2018.5320
28. S.M. Naghib, E. Parnian, H. Keshvari, E. Omidinia, M. Eshghan-Malek, Synthesis, characterization and electrochemical evaluation of polyvinylalchol/graphene oxide/silver nanocomposites for glucose biosensing application. Int. J. Electrochem. Sci. **13**(1), 1013–1026 (2018). https://doi.org/10.20964/2018.01.74
29. S.M. Naghib, M. Rabiee, E. Omidinia, Electroanalytical validation of a novel nanobiosensing strategy and direct electrochemistry of phenylalanine dehydrogenase for clinical diagnostic applications. Int. J. Electrochem. Sci. **9**(5), 2301–2315 (2014). https://doi.org/10.1016/S1452-3981(23)07928-2
30. A. Haleem, M. Javaid, R.P. Singh, R. Suman, S. Rab, Biosensors applications in medical field: a brief review. Sens. Int. **2**, 100100 (2021)
31. B.M. Kangarshahi, S.M. Naghib, Nanogenosensors based on aptamers and peptides for bioelectrochemical cancer detection: an overview of recent advances in emerging materials and technologies. Discover Appl. Sci. **6**(2), 47 (2024). https://doi.org/10.1007/s42452-024-05681-z
32. A. Matini, S.M. Naghib, M.R. Mozafari, Quantum dots in cancer theranostics: a thorough review of recent advancements in bioimaging, tracking, and therapy across various cancer types. Curr. Pharm. Biotechnol. (2024). https://doi.org/10.2174/0113892010294163240407153842
33. B. Mikaeeli Kangarshahi, S. Sojdeh, H. Daneshgar, M. Bagherzadeh, S.M. Naghib, N. Rabiee, Electroconductive polymer-based biosensors for early cancer detection via liquid biopsy: advances, challenges, and future prospects. TrAC Trends Anal. Chem. **183**, 118062 (2025). https://doi.org/10.1016/j.trac.2024.118062
34. J.J. Calvo, S.M. Angel, M.C. So, Charge transport in metal–organic frameworks for electronics applications. APL Mater. **8**(5) (2020)
35. N. Singh, S. Qutub, N.M. Khashab, Biocompatibility and biodegradability of metal organic frameworks for biomedical applications. J. Mater. Chem. B **9**(30), 5925–5934 (2021)
36. M. Aghajani, H.R. Garshasbi, S.M. Naghib, M.R. Mozafari, 3D Printing of hydrogel polysaccharides for biomedical applications: a review. Biomedicines (2025)
37. F. Behzad, S.M. Naghib, M.A.J. kouhbanani, S.N. Tabatabaei, Y. Zare, K.Y. Rhee, An overview of the plant-mediated green synthesis of noble metal nanoparticles for antibacterial applications. J. Ind. Eng. Chem. **94**, 92–104 (2021). https://doi.org/10.1016/j.jiec.2020.12.005
38. N. Dehghani, F. Haghiralsadat, F. Yazdian, F. Sadeghian-Nodoushan, N. Ghasemi, F. Mazaheri, M. Pourmadadi, S.M. Naghib, Chitosan/silk fibroin/nitrogen-doped carbon quantum dot/α-tricalcium phosphate nanocomposite electrospinned as a scaffold for wound healing application: in vitro and in vivo studies. Int. J. Biol. Macromol. **238**, 124078 (2023). https://doi.org/10.1016/j.ijbiomac.2023.124078

39. Z. Gharehdaghi, R. Rahimi, S.M. Naghib, F. Molaabasi, Cu (II)-porphyrin metal–organic framework/graphene oxide: synthesis, characterization, and application as a pH-responsive drug carrier for breast cancer treatment. J. Biol. Inorg. Chem. **26**, 689–704 (2021)
40. S. Kabiri, R. Rahimi, M.R. Mozafari, S.M. Naghib, Porphyrin-based metal–organic frameworks in drug delivery for cancer therapy: promises, advances and prospects. Discover Appl. Sci. **7**(4), 313 (2025)
41. B.M. Kangarshahi, A. Beigi Javazm, S.M. Naghib, CRISPR-integrated metal-organic frameworks for biosensing applications: recent advances and future perspective. Sens. Bio-sens. Res. **47**, 100736 (2025). https://doi.org/10.1016/j.sbsr.2025.100736
42. Z. Gharehdaghi, S.M. Naghib, R. Rahimi, A. Bakhshi, A. Kefayat, A. Shamaeizadeh, F. Molaabasi, Highly improved pH-responsive anticancer drug delivery and T2-weighted MRI imaging by magnetic MOF CuBTC-based nano/microcomposite. Front. Mol. Biosci. **10**, 1071376 (2023)
43. Z. Gharehdaghi, R. Rahimi, S.M. Naghib, F. Molaabasi, Fabrication and application of copper metal–organic frameworks as nanocarriers for pH-responsive anticancer drug delivery. J. Iran. Chem. Soc. **19**(7), 2727–2737 (2022)
44. H. Mao, L. Yu, M. Tu, S. Wang, J. Zhao, H. Zhang, Y. Cao, Recent advances on the metal-organic frameworks-based biosensing methods for cancer biomarkers detection. Crit. Rev. Anal. Chem. **54**(5), 1273–1289 (2024)
45. M. Hosseine, S.M. Naghib, N. Rabiee, Metal-organic frameworks: biosensor applications for diagnosis of cancers. Microchem. J. **208**, 112341 (2025). https://doi.org/10.1016/j.microc.2024.112341
46. P. Deria, J.E. Mondloch, O. Karagiaridi, W. Bury, J.T. Hupp, O.K. Farha, Beyond post-synthesis modification: evolution of metal–organic frameworks via building block replacement. Chem. Soc. Rev. **43**(16), 5896–5912 (2014)
47. F. Liu, L. Geng, F. Ye, S. Zhao, MOF-derivated MnO@C nanocomposite with bidirectional electrocatalytic ability as signal amplification for dual-signal electrochemical sensing of cancer biomarker. Talanta **239**, 123150 (2022)
48. K. Yu, M. Li, H. Chai, Q. Liu, X. Hai, M. Tian, L. Qu, T. Xu, G. Zhang, X. Zhang, MOF-818 nanozyme-based colorimetric and electrochemical dual-mode smartphone sensing platform for in situ detection of H_2O_2 and H_2S released from living cells. Chem. Eng. J. **451**, 138321 (2023)
49. M. Wang, M. Hu, Z. Li, L. He, Y. Song, Q. Jia, Z. Zhang, M. Du, Construction of Tb-MOF-on-Fe-MOF conjugate as a novel platform for ultrasensitive detection of carbohydrate antigen 125 and living cancer cells. Biosens. Bioelectron. **142**, 111536 (2019)
50. X. Yang, W. Qiu, R. Gao, Y. Wang, Y. Bai, Z. Xu, S.-J. Bao, MIL-47 (V) catalytic conversion of H_2O_2 for sensitive H_2O_2 detection and tumor cell inhibition. Sens. Actuators B Chem. **354**, 131201 (2022)
51. O.M. Yaghi, G. Li, H. Li, Selective binding and removal of guests in a microporous metal–organic framework. Nature **378**(6558), 703–706 (1995)
52. S. Zhang, F. Rong, C. Guo, F. Duan, L. He, M. Wang, Z. Zhang, M. Kang, M. Du, Metal–organic frameworks (MOFs) based electrochemical biosensors for early cancer diagnosis in vitro. Coord. Chem. Rev. **439**, 213948 (2021)
53. I. Sabeeh Hasan, M. Yakdhan Saleh, A.K.O. Aldulaimi, S.M. Saeed, M. Adil, A.H. Adhab, A Review: metal-organic frameworks electrochemical biosensors for bioorganic materials detection. Anal. Bioanal. Electrochem. **16**(1), 100–125 (2024)
54. K. Li, R. Qin, Y. Xiong, L. Ding, J. Wen, M. Zeng, MOF-derived octahedral-shaped Fe_3O_4@C composites for lithium storage. J. Electron. Mater. **52**(5), 3311–3320 (2023)
55. J. Wang, Q. Zhong, Y. Zeng, D. Cheng, Y. Xiong, Y. Bu, Rational construction of triangle-like nickel-cobalt bimetallic metal-organic framework nanosheets arrays as battery-type electrodes for hybrid supercapacitors. J. Colloid Interface Sci. **555**, 42–52 (2019)
56. Y. Shuku, R. Suizu, M. Tsuchiizu, K. Awaga, Ideal trigonal prismatic coordination geometry of Co(ii) in a honeycomb MOF with a triptycene-based ligand. Chem. Commun. **59**(67), 10105–10108 (2023)

57. M.J. Golomb, J. Calbo, J.K. Bristow, A. Walsh, Ligand engineering in Cu(ii) paddle wheel metal–organic frameworks for enhanced semiconductivity. J. Mater. Chem. A **8**(26), 13160–13165 (2020)
58. R. Lalawmpuia, M. Lalhruaitluangi, D. Tiwari, Metal organic framework (MOF): synthesis and fabrication for the application of electrochemical sensing. Environ. Eng. Res. **29**(5) (2024)
59. B. Saboorizadeh, R. Zare-Dorabei, M. Safavi, V. Safarifard, Applications of metal–organic frameworks (MOFs) in drug delivery. Biosensing, and therapy: a comprehensive review. Langmuir **40**(43), 22477–22503 (2024)
60. E. Binaeian, H. Nabipour, S. Ahmadi, S. Rohani, The green synthesis and applications of biological metal–organic frameworks for targeted drug delivery and tumor treatments. J. Mater. Chem. B **11**(48), 11426–11459 (2023)
61. M.T. Ulhakim, M. Rezki, K.K. Dewi, S.A. Abrori, S. Harimurti, N.L.W. Septiani, K.A. Kurnia, W. Setyaningsih, N. Darmawan, B. Yuliarto, Recent trend on two-dimensional metal-organic frameworks for electrochemical biosensor application. J. Electrochem. Soc. **167**(13), 136509 (2020)
62. V.F. Yusuf, N.I. Malek, S.K. Kailasa, Review on metal–organic framework classification, synthetic approaches, and influencing factors: applications in energy, drug delivery, and wastewater treatment. ACS Omega **7**(49), 44507–44531 (2022)
63. Z. Dourandish, S. Tajik, H. Beitollahi, P.M. Jahani, F.G. Nejad, I. Sheikhshoaie, A. Di Bartolomeo, A comprehensive review of metal–organic framework: synthesis, characterization, and investigation of their application in electrochemical biosensors for biomedical analysis. Sensors **22**(6), 2238 (2022)
64. M. Wu, Z. Zhang, F. Mo, X. Shi, D. Sun, L. Zhao, X. Zhang, Recent advances in electrochemical MOF-based aptasensors for cancer biomarker detection: a review. Microchem. J. 110924 (2024)
65. V.A. Tran, V.D. Doan, V.T. Le, T.-Q. Nguyen, T.N. Don, V. Vien, N.T. Luan, G.N. Vo, Metal–organic frameworks-derived material for electrochemical biosensors: recent applications and prospects. Ind. Eng. Chem. Res. **62**(11), 4738–4753 (2023)
66. E.C.R. Lopez, J.V.D. Perez, Current advances in the synthesis of CD-MOFs and their water stability. Eng Proc. **56**(1), 72 (2023)
67. B.J. El Taher, R. Sabouni, M. Ghommem, Luminescent metal organic framework for selective detection of mercury in aqueous media: microwave-based synthesis and evaluation. Colloids Surf. A **607**, 125477 (2020)
68. A.S. Eliwa, A.E. Ali, W.M. Hosny, G.G. Mohamed, R.G. Deghadi, Sonochemical synthesis and characterization of novel copper based metal-organic framework: its application as electrochemical sensor for determination of Cd(II) ion in real water samples. Inorg. Chem. Commun. **153**, 110733 (2023)
69. H. Thakkar, S. Eastman, Q. Al-Naddaf, A.A. Rownaghi, F. Rezaei, 3D-printed metal–organic framework monoliths for gas adsorption processes. ACS Appl. Mater. Interfaces **9**(41), 35908–35916 (2017)
70. J. Dhainaut, M. Bonneau, R. Ueoka, K. Kanamori, S. Furukawa, Formulation of metal–organic framework inks for the 3D printing of robust microporous solids toward high-pressure gas storage and separation. ACS Appl. Mater. Interfaces **12**(9), 10983–10992 (2020)
71. M.R. Ibrahim, Y.E. Greish, MOF-based biosensors for the detection of carcinoembryonic antigen: a concise review. Molecules **28**(16), 5970 (2023)
72. D. Yamamoto, T. Maki, S. Watanabe, H. Tanaka, M.T. Miyahara, K. Mae, Synthesis and adsorption properties of ZIF-8 nanoparticles using a micromixer. Chem. Eng. J. **227**, 145–150 (2013)
73. M. Faustini, J. Kim, G.-Y. Jeong, J.Y. Kim, H.R. Moon, W.-S. Ahn, D.-P. Kim, Microfluidic approach toward continuous and ultrafast synthesis of metal–organic framework crystals and hetero structures in confined microdroplets. J. Am. Chem. Soc. **135**(39), 14619–14626 (2013)
74. J. Puigmartí-Luis, M. Rubio-Martínez, U. Hartfelder, I. Imaz, D. Maspoch, P.S. Dittrich, Coordination polymer nanofibers generated by microfluidic synthesis. J. Am. Chem. Soc. **133**(12), 4216–4219 (2011)

75. Y. Du, Preparation and applications of macroscopic metal-organic framework materials (2024)
76. Y. Wang, L. Li, L. Yan, X. Gu, P. Dai, D. Liu, J.G. Bell, G. Zhao, X. Zhao, K.M. Thomas, Bottom-up fabrication of ultrathin 2D Zr metal–organic framework nanosheets through a facile continuous microdroplet flow reaction. Chem. Mater. **30**(9), 3048–3059 (2018)
77. Y. Yu, G. Chen, J. Guo, Y. Liu, J. Ren, T. Kong, Y. Zhao, Vitamin metal–organic framework-laden microfibers from microfluidics for wound healing. Mater. Horiz. **5**(6), 1137–1142 (2018)
78. C. Hu, Y. Bai, M. Hou, Y. Wang, L. Wang, X. Cao, C.-W. Chan, H. Sun, W. Li, J. Ge, Defect-induced activity enhancement of enzyme-encapsulated metal-organic frameworks revealed in microfluidic gradient mixing synthesis. Sci. Adv. **6**(5), eaax5785 (2020)
79. T. Stassin, I. Stassen, J. Marreiros, A.J. Cruz, R. Verbeke, M. Tu, H. Reinsch, M. Dickmann, W. Egger, I.F. Vankelecom, Solvent-free powder synthesis and MOF-CVD thin films of the large-pore metal–organic framework MAF-6. Chem. Mater. **32**(5), 1784–1793 (2020)
80. F.J. Claire, M.A. Solomos, J. Kim, G. Wang, M.A. Siegler, M.F. Crommie, T.J. Kempa, Structural and electronic switching of a single crystal 2D metal-organic framework prepared by chemical vapor deposition. Nat. Commun. **11**(1), 5524 (2020)
81. S. Soni, P.K. Bajpai, C. Arora, A review on metal-organic framework: synthesis, properties and application. Charact. Appl. Nanomater. **3**(2), 87–106 (2020)
82. D. Banerjee, J. Finkelstein, A. Smirnov, P.M. Forster, L.A. Borkowski, S.J. Teat, J.B. Parise, Synthesis and structural characterization of magnesium based coordination networks in different solvents. Cryst. Growth Des. **11**(6), 2572–2579 (2011)
83. S. Song, X. Ma, B. Zhang, W. Li, Y. Feng, C. Tan, Morphologies of thienyl based bimetallic metal-organic frameworks controlled by solvents for high specific capacitance supercapacitor. J. Energy Storage **47**, 103627 (2022)
84. S. Song, X. Ma, W. Li, B. Zhang, J. Sun, C. Deng, The influence of solvent controlled morphology on capacitive properties of metal-organic frameworks based on polyaminocarboxybenzene ligands. Colloids Surf. A **656**, 130350 (2023)
85. J. Łuczak, M. Kroczewska, M. Baluk, J. Sowik, P. Mazierski, A. Zaleska-Medynska, Morphology control through the synthesis of metal-organic frameworks. Adv. Coll. Interface. Sci. **314**, 102864 (2023)
86. Y. Lv, S. Wang, R. Zhang, D. Zhang, H. Yu, G. Lu, PH-modulated formation of uniform MOF-5 sheets. Inorg. Chem. Commun. **97**, 30–33 (2018)
87. B. Seoane, A. Dikhtiarenko, A. Mayoral, C. Tellez, J. Coronas, F. Kapteijn, J. Gascon, Metal organic framework synthesis in the presence of surfactants: towards hierarchical MOFs? CrystEngComm **17**(7), 1693–1700 (2015)
88. M. Yuan, R. Wang, Z. Sun, L. Lin, H. Yang, H. Li, C. Nan, G. Sun, S. Ma, Morphology-controlled synthesis of Ni-MOFs with highly enhanced electrocatalytic performance for urea oxidation. Inorg. Chem. **58**(17), 11449–11457 (2019)
89. W. Shen, X. Guo, H. Pang, Effect of solvothermal temperature on morphology and supercapacitor performance of Ni-MOF. Molecules **27**(23), 8226 (2022)
90. H. Jiang, Q. Wang, H. Wang, Y. Chen, M. Zhang, Temperature effect on the morphology and catalytic performance of Co-MOF-74 in low-temperature NH_3-SCR process. Catal. Commun. **80**, 24–27 (2016)
91. L. Deng, H. Qu, Y. Zhang, S. Jiao, X. Zhang, K. Liu, Z. Zhang, L. Wang, S. Feng, Temperature effect on the synthesis of two Ni-MOFs with distinct performance in supercapacitor. J. Solid State Chem. **281**, 121026 (2020)
92. S. Gooneh-Farahani, M.R. Naimi-Jamal, S.M. Naghib, Stimuli-responsive graphene-incorporated multifunctional chitosan for drug delivery applications: a review. Expert Opin. Drug Deliv. **16**(1), 79–99 (2019). https://doi.org/10.1080/17425247.2019.1556257
93. S.N. Hosseini, S.M. Naghib, G.K. Kashani, M.R. Mozafari, Chitosan-grafted graphene materials for drug delivery in wound healing. Curr. Pharm. Des. **31**(9), 691–715 (2025). https://doi.org/10.2174/0113816128333493241014134711
94. K. Khodaverdi, S.M. Naghib, M.R. Mozafari, M. Rahmanian, Chitosan/hydroxyapatite hydrogels for localized drug delivery and tissue engineering: a review. Carbohydr. Polym. Technol. Appl. **9**, 100640 (2025). https://doi.org/10.1016/j.carpta.2024.100640

95. E. Behjat, M. Tajabadi, S.M. Naghib, H.R. Rezaie, Multifunctional stimuli-responsive self-healing polyvinyl alcohol/tannic acid/thyme-honey/aloe vera hydrogel as promising antibacterial drug delivery platforms: physicochemical and in vitro biological assays. Polymer **317**, 127897 (2025). https://doi.org/10.1016/j.polymer.2024.127897
96. A. Ebrahimiverkiani, S.M. Naghib, G. Shafiei Sabet, P. Arghavani, M. Edrisi, N. Rabiee, Recent advances in microfluidic technologies for stimuli-sensitive alginate particles: applications in biosensing and drug delivery. TrAC Trends Anal. Chem. **184**, 118124 (2025). https://doi.org/10.1016/j.trac.2024.118124
97. H. Garshasbi, S. Salehi, S.M. Naghib, S. Ghorbanzadeh, W. Zhang, Stimuli-responsive injectable chitosan-based hydrogels for controlled drug delivery systems. Front. Bioeng. Biotechnol. **10** (2023). https://doi.org/10.3389/fbioe.2022.1126774
98. H.R. Garshasbi, S.M. Naghib, Smart stimuli-responsive alginate nanogels for drug delivery systems and cancer therapy: a review. Curr. Pharm. Des. **29**(44), 3546–3562 (2023). https://doi.org/10.2174/0113816128283806231211073031
99. C. Liu, J. Wang, J. Wan, C. Yu, MOF-on-MOF hybrids: synthesis and applications. Coord. Chem. Rev. **432**, 213743 (2021)
100. S. Chaemchuen, N.A. Kabir, K. Zhou, F. Verpoort, Metal–organic frameworks for upgrading biogas via CO_2 adsorption to biogas green energy. Chem. Soc. Rev. **42**(24), 9304–9332 (2013)
101. Q. Gao, Q. Bai, C. Zheng, N. Sun, J. Liu, W. Chen, F. Hu, T. Lu, Application of metal-organic framework in diagnosis and treatment of diabetes. Biomolecules **12**(9), 1240 (2022)
102. V. Naresh, N. Lee, A review on biosensors and recent development of nanostructured materials-enabled biosensors. Sensors **21**(4), 1109 (2021)
103. B.M. Kangarshahi, S.M. Naghib, Label-free electrochemical nanobiosensors using Au-SPE for COVID-19 detection: a comparative review of different biomarkers and recognition elements. Curr. Anal. Chem. **20**(2), 90–108 (2024)
104. R. Sahadevan, V.N. Vijayan, K. Kannan, N.A. Bovas, S. Sadhukhan, Biosensors: an overview of detection of environmental pollutants monitoring, current challenges, and future perspectives, in *Biotechnology for Environmental Sustainability* (2025), pp. 593–635
105. J. Liu, Y. Xu, S. Liu, S. Yu, Z. Yu, S.S. Low, Application and progress of chemometrics in voltammetric biosensing. Biosensors **12**(7), 494 (2022)
106. S. Borgmann, A. Schulte, S. Neugebauer, W. Schuhmann, Amperometric biosensors, in *Advances in Electrochemical Science and Engineering: Bioelectrochemistry*, vol. 13 (2011), pp. 1–83
107. Z. Štukovnik, R. Fuchs-Godec, U. Bren, Nanomaterials and their recent applications in impedimetric biosensing. Biosensors **13**(10), 899 (2023)
108. N.L. Walker, A.B. Roshkolaeva, A.I. Chapoval, J.E. Dick, Recent advances in potentiometric biosensing. Curr. Opin. Electrochem. **28**, 100735 (2021)
109. R. Vargas-Bernal, E. Rodríguez-Miranda, G. Herrera-Pérez, Evolution and expectations of enzymatic biosensors for pesticides, in *Pesticides-Advances in Chemical and Botanical Pesticides*, vol. 14 (2012), pp. 331–354
110. A.A. Oladipo, S.D. Oskouei, M. Gazi, Metal-organic framework-based nanomaterials as opto-electrochemical sensors for the detection of antibiotics and hormones: a review. Beilstein J. Nanotechnol. **14**(1), 631–673 (2023)
111. M.A. Tzani, D.K. Gioftsidou, M.G. Kallitsakis, N.V. Pliatsios, N.P. Kalogiouri, P.A. Angaridis, I.N. Lykakis, M.A. Terzidis, Direct and indirect chemiluminescence: reactions, mechanisms and challenges. Molecules **26**(24), 7664 (2021)
112. P. Singh, SPR biosensors: historical perspectives and current challenges. Sens. Actuators B Chem. **229**, 110–130 (2016)
113. Y. Zhao, R.-J. Tong, F. Xia, Y. Peng, Current status of optical fiber biosensor based on surface plasmon resonance. Biosens. Bioelectron. **142**, 111505 (2019)
114. Y. Deng, M. Guo, L. Zhou, Y. Huang, S. Srivastava, A. Kumar, J.-Q. Liu, Prospects, advances and biological applications of MOF-based platform for the treatment of lung cancer. Biomater. Sci. **12**(15), 3725–3744 (2024)

115. Y. Zhu, X. Fang, X. Lv, M. Lu, H. Xu, S. Hu, S. Zhao, F. Ye, A single aptamer-dependent sandwich-type biosensor for the colorimetric detection of cancer cells via direct coordinately binding of bare bimetallic metal-organic framework-based nanozymes. Biosensors **13**(2), 225 (2023)
116. A. Afzalinia, M. Mirzaee, Ultrasensitive fluorescent miRNA biosensor based on a "sandwich" oligonucleotide hybridization and fluorescence resonance energy transfer process using an Ln(III)-MOF and Ag nanoparticles for early cancer diagnosis: application of central composite design. ACS Appl. Mater. Interfaces **12**(14), 16076–16087 (2020)
117. K. Tang, W. Wang, Z. Song, X. Luo, Multifunctional nano-biosensor based on metal-organic framework for enhanced fluorescence imaging of intracellular miRNA-122 and synergistic chemo-photothermal therapy of tumor cells. Anal. Chim. Acta **1176**, 338779 (2021)
118. X.-J. Kong, X. Ji, T. He, L.-H. Xie, Y.-Z. Zhang, H. Lv, C. Ding, J.-R. Li, A green-emission metal–organic framework-based nanoprobe for imaging dual tumor biomarkers in living cells. ACS Appl. Mater. Interfaces **12**(31), 35375–35384 (2020)
119. Y. Zheng, Y. Zhao, M. Bai, H. Gu, X. Li, Metal–organic frameworks as a therapeutic strategy for lung diseases. J. Mater. Chem. B **10**(30), 5666–5695 (2022)
120. X. Dong, Y. Du, G. Zhao, W. Cao, D. Fan, X. Kuang, Q. Wei, H. Ju, Dual-signal electrochemiluminescence immunosensor for neuron-specific enolase detection based on "dual-potential" emitter $Ru(bpy)_3{}^{2+}$ functionalized zinc-based metal-organic frameworks. Biosens. Bioelectron. **192**, 113505 (2021)
121. L. Guo, Z. Mu, B. Yan, J. Wang, J. Zhou, L. Bai, A novel electrochemical biosensor for sensitive detection of non-small cell lung cancer ctDNA using NG-PEI-COFTAPB-TFPB as sensing platform and Fe-MOF for signal enhancement. Sens. Actuators B Chem. **350**, 130874 (2022). https://doi.org/10.1016/j.snb.2021.130874
122. Y. Wang, Y. Li, X. Zhuang, C. Tian, X. Fu, F. Luan, $Ru(bpy)_3{}^{2+}$ encapsulated cyclodextrin based metal organic framework with improved biocompatibility for sensitive electrochemiluminescence detection of CYFRA21-1 in cell. Biosens. Bioelectron. **190**, 113371 (2021)
123. M. Hu, L. Zhu, Z. Li, C. Guo, M. Wang, C. Wang, M. Du, CoNi bimetallic metal–organic framework as an efficient biosensing platform for miRNA 126 detection. Appl. Surf. Sci. **542**, 148586 (2021)
124. C. Gu, C. Guo, Z. Li, M. Wang, N. Zhou, L. He, Z. Zhang, M. Du, Bimetallic ZrHf-based metal-organic framework embedded with carbon dots: ultra-sensitive platform for early diagnosis of HER2 and HER2-overexpressed living cancer cells. Biosens. Bioelectron. **134**, 8–15 (2019)
125. Y. Zhao, J. Zhou, Z. Jia, D. Huo, Q. Liu, D. Zhong, Y. Hu, M. Yang, M. Bian, C. Hou, In-situ growth of gold nanoparticles on a 3D-network consisting of a MoS2/rGO nanocomposite for simultaneous voltammetric determination of ascorbic acid, dopamine and uric acid. Microchim. Acta **186**, 1–10 (2019)
126. C. Gu, L. Bai, L. Pu, P. Gai, F. Li, Highly sensitive and stable self-powered biosensing for exosomes based on dual metal-organic frameworks nanocarriers. Biosens. Bioelectron. **176**, 112907 (2021)
127. J. Chang, X. Wang, J. Wang, H. Li, F. Li, Nucleic acid-functionalized metal-organic framework-based homogeneous electrochemical biosensor for simultaneous detection of multiple tumor biomarkers. Anal. Chem. **91**(5), 3604–3610 (2019). https://doi.org/10.1021/acs.analchem.8b05599
128. H.-M. Wang, A.-J. Wang, P.-X. Yuan, J.-J. Feng, Flower-like metal-organic framework microsphere as a novel enhanced ECL luminophore to construct the coreactant-free biosensor for ultrasensitive detection of breast cancer 1 gene. Sens. Actuators B Chem. **320**, 128395 (2020)
129. Y. Xue, Y. Wang, S. Feng, M. Yan, J. Huang, X. Yang, A dual-amplification mode and Cu-based metal-organic frameworks mediated electrochemical biosensor for sensitive detection of microRNA. Biosens. Bioelectron. **202**, 113992 (2022)
130. R. Xiong, X. Zhu, J. Zhao, G. Ling, P. Zhang, Nanozymes-mediated cascade reaction system for tumor-specific diagnosis and targeted therapy. Small Methods **8**(10), 2301676 (2024)

131. C. Zhang, D. Zhang, Z. Ma, H. Han, Cascade catalysis-initiated radical polymerization amplified impedimetric immunosensor for ultrasensitive detection of carbohydrate antigen 15-3. Biosens. Bioelectron. **137**, 1–7 (2019)
132. G. Ashraf, T. Ahmad, M.Z. Ahmed, Murtaza, Y. Rasmi, Advances in metal-organic frameworks (MOFs) based biosensors for diagnosis: an update. Curr. Topics Med. Chem. **22**(27), 2222–2240 (2022)
133. E.A. Sadrabadi, A. Benvidi, M. Azimzadeh, L. Asgharnejad, A.S. Dezfuli, P. Khashayar, Novel electrochemical biosensor for breast cancer detection, based on a nanocomposite of carbon nanofiber, metal–organic framework, and magnetic graphene oxide. Bioelectrochemistry **155**, 108558 (2024)
134. L. Du, W. Chen, P. Zhu, Y. Tian, Y. Chen, C. Wu, Applications of functional metal-organic frameworks in biosensors. Biotechnol. J. **16**(2), 1900424 (2021)
135. X.-L. Qi, Z.-Y. Zhan, S. Jahanghiri, K. Chu, C.-Y. Zhang, Z.-F. Ding, Metal-organic frameworks for electrochemical and electrochemiluminescent immunoassay. J. Electrochem. **29**(6), 3 (2023)
136. S.K. Bhardwaj, A.L. Sharma, N. Bhardwaj, M. Kukkar, A.A. Gill, K.-H. Kim, A. Deep, TCNQ-doped Cu-metal organic framework as a novel conductometric immunosensing platform for the quantification of prostate cancer antigen. Sens. Actuators B Chem. **240**, 10–17 (2017)
137. K. Shao, B. Wang, A. Nie, S. Ye, J. Ma, Z. Li, Z. Lv, H. Han, Target-triggered signal-on ratiometric electrochemiluminescence sensing of PSA based on MOF/Au/G-quadruplex. Biosens. Bioelectron. **118**, 160–166 (2018)
138. P. Zhang, H. Huang, N. Wang, H. Li, D. Shen, H. Ma, Duplex voltammetric immunoassay for the cancer biomarkers carcinoembryonic antigen and alpha-fetoprotein by using metal-organic framework probes and a glassy carbon electrode modified with thiolated polyaniline nanofibers. Microchim. Acta **184**, 4037–4045 (2017)
139. J. Li, L. Liu, Y. Ai, Y. Liu, H. Sun, Q. Liang, Self-polymerized dopamine-decorated Au NPs and coordinated with Fe-MOF as a dual binding sites and dual signal-amplifying electrochemical aptasensor for the detection of CEA. ACS Appl. Mater. Interfaces **12**(5), 5500–5510 (2020)
140. J. Liu, Y. Shang, Q. Zhu, X. Zhang, J. Zheng, A voltammetric immunoassay for the carcinoembryonic antigen using silver(I)-terephthalate metal-organic frameworks containing gold nanoparticles as a signal probe. Microchim. Acta **186**, 1–8 (2019)
141. X. Zhou, S. Guo, J. Gao, J. Zhao, S. Xue, W. Xu, Glucose oxidase-initiated cascade catalysis for sensitive impedimetric aptasensor based on metal-organic frameworks functionalized with Pt nanoparticles and hemin/G-quadruplex as mimicking peroxidases. Biosens. Bioelectron. **98**, 83–90 (2017)
142. F. Su, Q. Jia, Z. Li, M. Wang, L. He, D. Peng, Y. Song, Z. Zhang, S. Fang, Aptamer-templated silver nanoclusters embedded in zirconium metal–organic framework for targeted antitumor drug delivery. Microporous Mesoporous Mater. **275**, 152–162 (2019)
143. Y. Zhang, Z. Zhang, S. Rong, H. Yu, H. Gao, P. Ding, D. Chang, H. Pan, Electrochemical immunoassay for the carcinoembryonic antigen based on Au NPs modified zeolitic imidazolate framework and ordered mesoporous carbon. Microchim. Acta **187**, 1–9 (2020)
144. J. Liu, Y. Shang, J. Xu, Y. Chen, Y. Jia, J. Zheng, A novel electrochemical immunosensor for carcinoembryonic antigen based on Cu-MOFs-TB/polydopamine nanocarrier. J. Electroanal. Chem. **877**, 114563 (2020)

Chapter 4
Nanobiosensing Strategies Based on Electroconductivity for Noninvasive and Early Detection of Cancers

Abstract The creation of biosensors based on electroconductive polymers (ECPs) has become a viable approach to early and noninvasive cancer detection, meeting urgent demands for accurate diagnosis and prompt treatment. ECPs, like 3,4-ethylenedioxythiophene, polypyrrole and polyaniline, are perfect for interacting with biological targets and converting biochemical signals into quantifiable outputs because of their special qualities, which include high electrical conductivity, biocompatibility and ease of functionalization. Recent developments in ECP-based biosensors are highlighted in this chapter, with particular attention paid to specificity, sensitivity and design advancements for cancer biomarkers in biofluids such as blood, urine and saliva. In order to improve sensor performance and usability, it also covers integration with wearable technologies, nanomaterials and microfluidics. The study also discusses future opportunities in customized healthcare, such as the use of artificial intelligence and wireless data transmission for real-time monitoring, and lists present issues such as signal stability, repeatability and clinical translation. The combination of digital technology, bioengineering and materials science makes ECP-based biosensors a game-changer for early and noninvasive cancer detection.

Keywords Biosensors · Immunosensors · Genosensors · Electroconductive polymers · Minimally invasive · Early cancer detection

4.1 Introduction and Conceptual Remarks

About 10 million people died from cancer in 2020, making it one of the world's top causes of death. Enhancing the results for patients, raising rates of survival, and lessening the strain on medical systems all depend on early cancer identification. Invasive biopsies, imaging procedures, and blood examinations are examples of conventional cancer diagnostic approaches that have several drawbacks, such as being time-consuming, expensive, invasive, and needing specialist staff and equipment [1–3]. An emerging device, biosensor, is a specialized analytical device that detects a specific chemical or biological substance [4–10]. Recently, biosensors have become

S. M. Naghib, *Nano-Biosensors, Drug Delivery and Tissue Engineering*, Series in BioEngineering, https://doi.org/10.1007/978-981-95-1888-3_4

important instruments for the detection of cancer [11–14]. These devices recognize and measure biological or chemical molecules using a transducer and a biorecognition element. High sensitivity, selectivity, quick reaction times, mobility, and affordability are just a few benefits that biosensors provide [15, 16]. Because they can translate biorecognition events into electrical signals that can be measured, electrochemical biosensors are chosen over other kinds of biosensors [17–20]. The efficacy of electrochemical biosensors is largely determined by the electrode materials used in these devices. Electron transport, stability, biocompatibility, and overall biosensor performance may all be significantly impacted by the electrode material choice. Conductive polymers are becoming more popular in this context as appropriate electrode materials for electrochemical biosensors [21, 22]. By allowing π-electrons to easily travel throughout their chemical structure, these organic compounds are able to transmit electricity. Adjustable conductivity, simplicity in manufacture and alteration, compatibility with various biorecognition components, and biocompatibility are only a few benefits of conductive polymers. In this chapter, the latest progress and problems in making and using biosensors that use electrochemistry that use polymers that conduct electricity to find cancer are carefully looked at. This chapter will talk about the main ideas and different kinds of conductive polymers and electrochemical biosensors. It will also talk about how to make and describe composites made of conductive polymers and other materials. The layout and construction of biosensors that use electrochemistry for the detection of biomarkers for cancer in biological materials, as well as methods to functionalize and immobilize biorecognition components on electrodes made of conductive polymers, will also be covered [23]. Additionally, using conductive polymers, we will evaluate and verify the electrochemical biosensors' performance, paying particular attention to parameters including sensitivity, dependability, reproducibility, stability, and selectivity. Furthermore, we will examine the current challenges faced by biosensors that utilize electrochemistry, which uses electrically conducting polymers to detect malignancy, and we will consider potential future developments [24]. Some of the challenges facing the development of electrochemical biosensors include improving multiplexing capabilities, detection limitations, and specificity; resolving interferences and matrix effects; and combining certain sensors with wearable and microfluidic systems to increase their usefulness and practicality in medical environments. By carefully analyzing and summarizing current information and research, this chapter seeks to be a complete resource for scientists, researchers, and clinicians [25]. It offers a starting point for future developments in the creation of electrochemical biosensors based on conductive polymers for the detection of cancer. By facilitating early identification, individualized treatment plans, and better patient outcomes, the integration of these cutting-edge biosensing technologies into clinical settings holds promise for revolutionizing cancer diagnostics [26]. Figure 4.1 illustrates a schematic representation of the biosensor categorization, highlighting the use of electrochemical biosensors in diagnostic technologies [27–30]. An electrochemical immunosensor was developed using hydrothermally synthesized reduced graphene oxide (rGO) modified with nanostructured titanium dioxide (TiO_2) electrophoretically deposited on indium tin oxide (ITO) glass. It enables covalent immobilization of monoclonal

EpCAM antibodies and detects EpCAM via differential pulse voltammetry with high sensitivity (3.24 $\mu A \cdot mL \cdot ng^{-1} \cdot cm^{-2}$), a wide linear range (0.01 to 60 ng/mL) and a low detection limit (0.0065 ng/mL). The sensor is stable, regenerable and effective for EpCAM detection in spiked human serum samples (Fig. 4.1a). Portable point-of-care biosensors were developed for rapid detection of carcinoembryonic antigen (CEA), a cancer biomarker, at room temperature using rGO and a novel nitrogen-rich polymer (PhPTz) that immobilizes antibodies and transmits charge changes. This approach promises fast, sensitive, and accurate cancer biomarker detection for early diagnosis and treatment monitoring (Fig. 4.1b). Another work demonstrates that tris(2,2′-bipyridine)ruthenium(II) (Ru(bpy)32+) generates enhanced and stable electrochemiluminescence (ECL) at a low potential on a graphene-PtPd hybrid due to the synergistic electrocatalysis of Pt and Pd. The low-potential ECL is quenched by MoS_2 nanoflowers, enabling a sensitive sandwich "signal-off" ECL immunosensor for detecting carcinoembryonic antigen (CEA) with a detection limit as low as 0.54 fg/mL. The sensor exhibits excellent specificity, stability, reproducibility and clinical applicability, providing a reliable strategy for clinical cancer biomarker detection (Fig. 4.1c). A label-free impedimetric immunosensor using N, S-graphene quantum dots on Au-polyaniline (N, S-GQDs@Au-PANI) nanowires was developed for quantitative detection of carcinoembryonic antigen (CEA). The nanowires accelerate electron transfer and anchor anti-CEA antibodies with antibody-antigen binding increasing charge transfer resistance for linear range over 0.5 to 1000 ng/mL and a detection limit of 0.01 ng/mL. This immunosensor showed high selectivity, stability and accurate CEA detection in human serum, offering a robust platform for early cancer diagnosis and monitoring (Fig. 4.1d).

4.2 Biosensors: Overview and Kinds of Conductive Polymers for Electrochemical Detection

A transducer, a biological identification element, and electrical components for signal processing and display make up a biosensor [31, 32]. Biosensors enable rapid, sensitive, and specific detection of biological or chemical substances and are widely used in medical diagnostics [33–35]. The target analyte of interest is preferentially bound by the biological recognition element. Whether it is an ROS, aptamer, DNA, RNA, enzyme, antibody, or whole cells depends on the particular application. The biological relationship between the target analyte and the recognition element is transformed into a quantifiable electrical or visual signal by the transducer. This signal is then analyzed and amplified by the electrical components, enabling quantitative analysis and result display. Because of their many benefits over traditional analytical techniques, biosensors have garnered a lot of attention and are now extensively used [36]. Its capacity to provide results quickly is one of its main advantages. Biosensors provide quick decision-making and timely action by detecting and quantifying certain compounds in an instantaneous or almost instantaneous manner. Improving patient

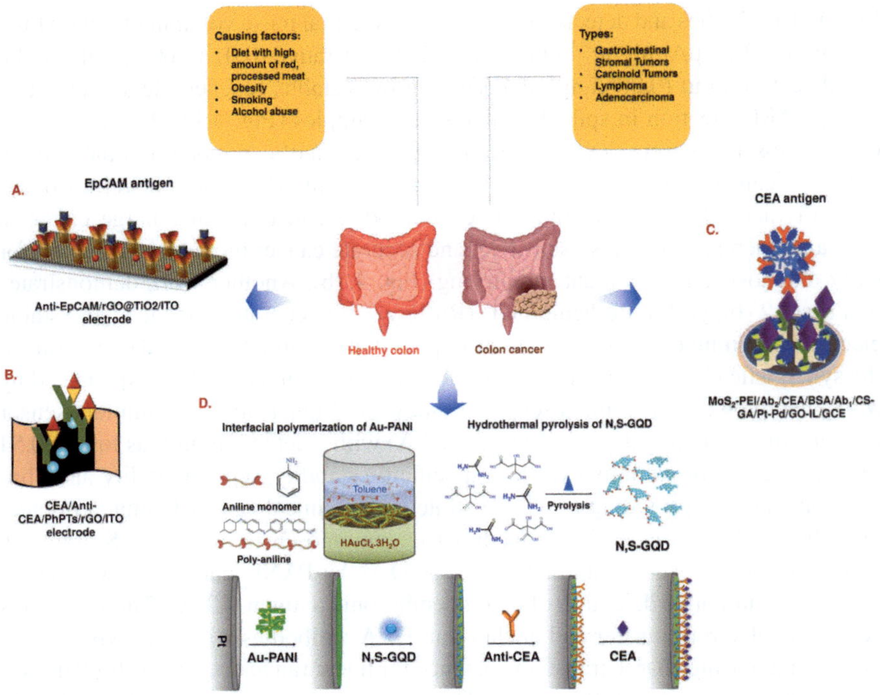

Fig. 4.1 Schematic representation of electrochemical biosensors for colon cancer detection, highlighting various antigen targets and sensor fabrication strategies [27–30]

outcomes in contexts related to medicine requires prompt diagnosis and therapy. The great sensitivity of biosensors is another important benefit. Biosensors may identify trace amounts of the analyte in complicated samples by using biological recognition components that have a high affinity and specificity for the target analytes. Early illness detection, treatment response monitoring, and environmental contaminant identification all depend on this sensitivity [37]. Biosensors provide remarkable specificity in addition to sensitivity. To reduce interference from other chemicals in the sample, the biological acknowledgment components are made to bind only to the analyte of interest. By ensuring precise and trustworthy detection, this specificity lowers the possibility of erroneous negative or positive result outcomes. For analytical testing, biosensors provide a more affordable option. They often save money by using less chemicals and sample quantities than conventional techniques [14]. Point-of-care diagnostics and on-site testing are made possible by making biosensors smaller and incorporating them into portable electronics. In the end, this mobility lowers expenses and processing time by doing away with the requirement for sample transportation and centralized laboratory facilities. There are many different kinds of biosensors, each with a unique set of uses and sensing principles. Thermal, piezoelectric, optical, and electrochemical biosensors are among the frequently utilized biosensor kinds (Fig. 4.2) [38, 39].

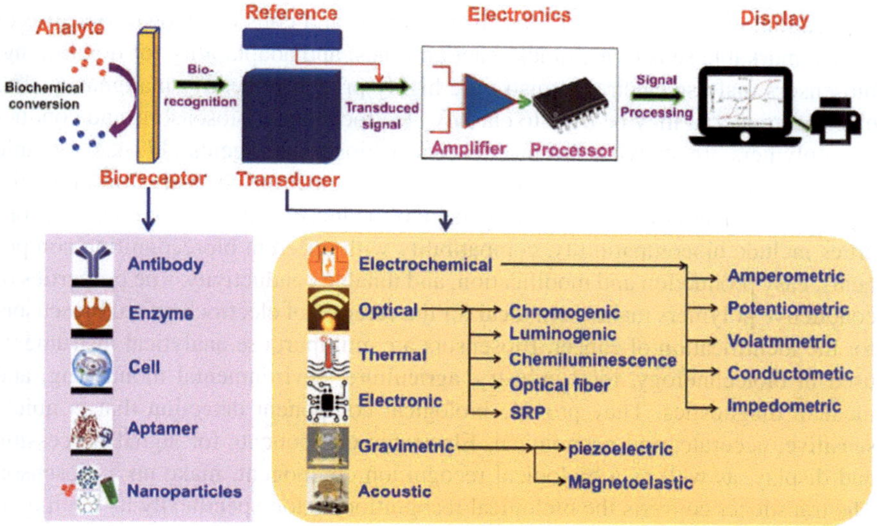

Fig. 4.2 Schematic diagram of biosensor consisting of bioreceptor, transducer, and amplifier [39]

Optical biosensors measure the relationships between matter and light to identify analytes. They often use technologies including colorimetric analysis, fluorescence, and surface plasmon resonance (SPR). SPR-based biosensors can find changes in the refractive index close to the sensor surface because the recognition element is bound to the target analyte. Variations that change the light's wavelength or angle are used to detect changes in the refractive index, which gives accurate details on the analyte's concentration. Due to their ability to translate biological exchanges into quantifiable electrical impulses, electrochemical biological sensors are beneficial. A potential or electrical current that is exactly proportionate to the analyte's focus is produced by biosensors using redox reactions on the electrode surface [40]. Among the electrochemical techniques for signal detection are impedance spectroscopy, potentiometry, and amperometry. Because of their great sensitivity, quick response times, and miniaturization potential, electrochemical biosensors are perfect for several purposes, including the detection of cancer. Based on the piezoelectric phenomenon, piezoelectric biosensors react to mechanical stress by producing an electrical signal. A crystal that is piezoelectric and covered, having a component of recognition, is the standard component of biosensors. A change in elasticity or mass caused by the target analyte's attaching itself to the acknowledgment element modifies the resonance frequency or amplitude of the crystal. Details on the analyte's concentration are obtained by detecting this change as a shift in the electrical signal's frequency or amplitude. When the recognition element and the target analyte interact, temperature changes are detected using thermal biosensors [41]. Microcalorimetry is often used in biosensors to quantify the heat generated or taken in while binding, the analyte concentration is then linked with the generated thermal signal. The biosensor selection relies on the particular needs regarding the application, and each form of biosensor has pros and cons of its own. Because of their remarkable sensitivity, quick reaction times, and adaptability for downsizing, biosensors that use electrochemistry are highly prized to identify malignancy. The electrode material may be used to classify electrochemical biosensors, and conductive polymers are showing promise as cancer biosensing agents [17, 42]. Organic substances known as conductive polymers allow π-electrons to easily flow along the backbone of their molecules, allowing them to conduct electricity. Their unique properties include biocompatibility, compatibility with different biorecognition components, easy production and modification, and tunable conductivity. The properties of conductive polymers make them ideal for the creation of electrochemical biosensors for the identification of cancer. Biosensors are multipurpose analytical instruments used in biotechnology, food industry, agriculture, environmental monitoring, and medical diagnostics. They provide biological component detection that is quick, sensitive, accurate, and economical. Electronic components for signal processing and display, as well as a biological recognition component, make up a biosensor. The transducer converts the biological recognition factor, specifically its binding to the analyte of interest, into a measurable signal. There are several kinds of biosensors, such as thermal, piezoelectric, optical, and electrochemical biosensors [43]. Because of their outstanding sensitivity, quick reaction times, and adaptability for downsizing, biosensors that use electrochemistry are especially preferred for the identification of cancer. Electrochemical biosensors have shown significant promise when based on conductive polymers. Organic compounds that may conduct electricity are called conductive polymers. Their benefits include biocompatibility, compatibility with different biorecognition components, easy synthesis and modification, and tunable conductivity. The main topics of the following sections of the chapter will include the synthesis and characterization of conductive polymers, techniques for affixing biorecognition components to electrodes made from these polymers, and the creation and manufacturing of biosensors that utilize electrochemistry with conductive polymers to identify cancer biomarkers. Performance assessment, the difficulties these biosensors face, and the field's hopes for the future will all be covered in the book [44]. Figure 4.3 shows the chemical structures of the main polymers that conduct electricity, including Polythiophene, Polypyrrole, and Polyaniline [45].

Analytical tools known as biosensors are able to identify alterations in biological mechanisms and translate them into quantifiable signals [5, 18]. They are made up of electrical parts for signal processing and display, a transducer, and a biological identification element. The food industry, biotechnology, agriculture, environmental monitoring, and medicine are just a few of the industries that employ biosensors. Proteins, nucleic acids, metabolites, hormones, toxins, and infections are just a few of the biological substances they can analyze. Biosensors provide benefits including high sensitivity, specificity, and cost-effectiveness together with quick results. Depending on the kind of transducer being utilized, biosensors are often categorized. The biological signal must be transformed into a quantifiable output by the transducer. Transducers of several kinds, including as electrochemical, optical, piezoelectric, and calorimetric transducers, are used in biosensors [36, 46]. Electrodes are used in electrochemical biosensors to monitor the electrical characteristics of the biological

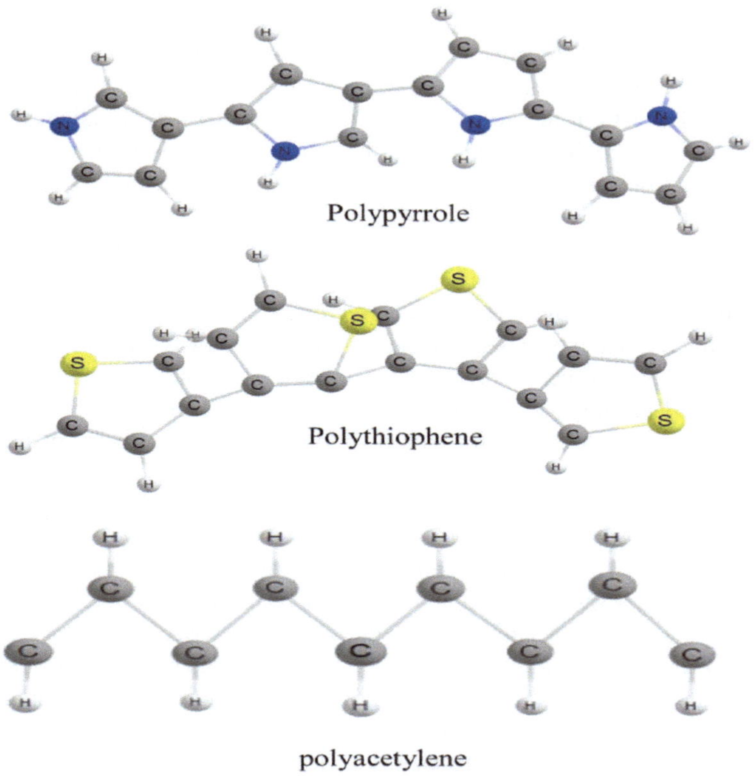

Fig. 4.3 The 3D-view of polypyrrole, polythiophene, and polyacetylene conducting polymers [45]

reaction. They are able to measure impedance, resistance, voltage, and current, and because of their ease of use, sensitivity, selectivity, and compatibility with integration and downsizing, electrochemical biosensors are extensively used. The measurement mode may be used to further categorize them; amperometric biosensors measure the current produced by reducing or oxidizing electroactive substances involved in the biological process [3, 47, 48]. The potential difference brought about by the electrochemical process is measured by potentiometric biosensors. Variations in conductivity brought on by ion binding or releasing throughout a biological process are measured by conductometric biosensors. When a possible sweep is given toward the electrode, the current produced is measured using voltammetric biosensors. Impedimetric biosensors measure changes in impedance caused by the relationship between the recognition element and the target analyte. On the other hand, optical biosensors assess the optical characteristics linked to the biological response using light [49]. Reflectance, luminescence, fluorescence, absorbance, and refractive index are among the properties they may measure. Optical biosensors provide excellent multiplexing, sensitivity, and specificity. In some applications, they also frequently provide the benefit of detection without labels. Optical and electrochemical biosensors are further divided into many categories according to the method of measurement. Like their electrochemical counterparts, visual biosensors may be categorized according to their manner of measurement. The light given off by fluorophores attached to the recognition factor is measured with fluorometric biosensors. The light produced when the recognition element connects with the target analyte is measured by chemiluminescent biosensors. When Enzymes or creatures that glow in the dark attach to the target analyte, they emit light, which is how bioluminescent biosensors work [50]. SPR biosensors monitor changes in the refractive index near the sensor surface that happen when the target binds to the analyte. Interferometric biosensors measure the modifications to the interference pattern brought about by the dialogue between the target analyte and the recognition element. Piezoelectric biosensors measure the mechanical or mass properties related to the biological response by using a piezoelectric material. They are able to measure things like phase, amplitude, and frequency. Particularly sensitive, piezoelectric biosensors may pick up on even the smallest variations in viscosity or mass [51]. They may furthermore provide light on the biological interaction's affinity and binding kinetics. Depending on how they measure, piezoelectric biosensors may be further divided into many categories. The recognition element-coated quartz crystal resonator is used in quartz crystal microbalance (QCM) biosensors to assess variations in amplitude or resonance frequency brought on by the target analyte's binding. Modifications to the way sound waves travel over the piezoelectric substrate's surface are detected by surface acoustic wave (SAW) biosensors. Variations in a bulk acoustic wave resonator's resonance frequency brought on by the relationship between the target analyte and the recognition element are measured by bulk acoustic wave (BAW) biosensors [52]. Figure 4.4 shows a graphic depiction of Conductive polymers' electrical conductivity and how it varies with different parameters [53–56].

Calorimetric biosensors detect the amount of heat generated or absorbed during the biological process, as the name implies. To measure the temperature variations brought on by the reaction, they use a thermistor or thermocouple. The thermodynamics, kinetics, and related enthalpy and entropy changes of the biological interaction may all be better understood using calorimetric biosensors. Depending on the measuring technique, they may be further divided into several sorts. At a fixed temperature during the process, biosensors that use isotherms detect the warmth produced or eaten. Differential biosensors measure the temperature difference between a reference and the response. Flow-through biosensors measure the heat the response produces or consumes while an example is flowing. Among these many biosensors, electrochemical biosensors have drawn a lot of interest for the detection of cancer. DNA, RNA, proteins, enzymes, neurotransmitters, and cells are among the cancer biomarkers that they can test with great sensitivity and specificity [57]. Microfluidic systems may also include electrochemical biosensors, which provide benefits including portability, quick analysis, and reduced sample and reagent consumption. One possible type of cancer diagnosis with electrochemical biosensors is based on polymers that conduct electricity. In electrochemical biosensors, polymers that conduct electricity with improved performance, stability, and

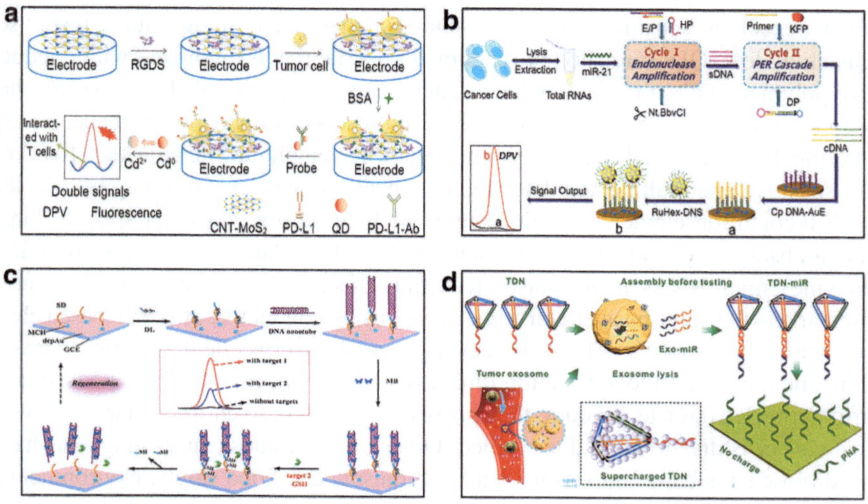

Fig. 4.4 Overview of conductive polymer-based biosensors for tumor biomarker detection: **a** dual-signal detection of PD-L1 using CNT-MoS_2 and quantum dots [53]. **b** miRNA sensing via endonuclease-assisted and PER amplification. Reprinted with permission from [54]. Copyright 2022 American Chemical Society. **c** Multiplexed DNA-based detection of miRNAs and proteins. Reprinted with permission from [55]. Copyright 2022 American Chemical Society. **d** Tetrahedral DNA nanostructures for exosomal miRNA detection. Reprinted with permission from [56]. Copyright 2021 American Chemical Society

biocompatibility including polythiophene, polypyrrole, polyaniline, and poly(3,4-ethylenedioxythiophene). These polymers make it easier for biological recognition components to become immobile and for electrons to go from them to the electrode surface. The ability to pick out specific signals, detect them accurately, and boost the signal strength of electrochemical biosensors has been improved by adding helpful substances to conductive polymers. To improve the detecting capabilities, conductive polymers may be mixed with magnetic, silver, and gold nanoparticles [58]. Through a variety of processes, including localized surface plasmon resonance and catalytic reactions, these nanoparticles may improve signal amplification and provide a greater surface area for immobilizing recognition components. Electrochemical biosensors have also extensively used CNTs (Carbon nanotubes) in conjunction with conducting polymers. CNTs have a large surface area, excellent mechanical properties, and high conductivity of electricity. By accelerating the electron transport between the electrode surface and the recognition elements, researchers may increase the sensitivity of the biosensors. Furthermore, CNTs may enhance the biomolecules' activity and stability as well as aid in their immobilization. Carbon nanomaterial in two dimensions, graphene, has shown significant promise in biosensing via electrochemistry. It is the perfect platform for rendering recognition of immobile components and improving kinetics of electron transfer because of its unique characteristics, such as high electrical conductivity, a large surface area, and exceptional mechanical strength [59]. For the detection of cancer biomarkers, biosensors based on graphene have shown excellent sensitivity, low thresholds for identification, and quick response. Quantum dots (QDs) are another functional material that is employed with conductive polymers. QDs are semiconductor nanocrystals with optical characteristics that vary with size. When excited, they produce fluorescence or phosphorescence, and by altering the nanocrystals' size, their emission wavelength may be adjusted. To allow the sensitive and multiplexed detection of cancer biomarkers, QDs may be coupled with recognition elements. Their benefits include narrow emission spectra, large molar absorptivity, and strong photostability [60]. Circulating tumor cells, proteins, enzymes, microRNAs, neurotransmitters, and DNA alterations are among the cancer biomarkers that have been effectively detected using conductive polymer-based electrochemical biosensors. These biosensors have shown excellent selectivity, broad dynamic ranges, and low detection limits. Because they provide quick, precise, and affordable detection techniques, they possess the promise to entirely change the way that cancer is diagnosed and surveilled. Other fields besides cancer diagnosis have used biosensors based on various kinds of transducers. Food safety, environmental monitoring, infectious illness detection, and glucose monitoring for diabetes treatment all make use of electrochemical biosensors. Environmental monitoring, drug screening, protein-protein interactions, and DNA sequencing all make use of optical biosensors [61]. Piezoelectric biosensors are used in pathogen detection, label-free biomolecule identification, and cell growth and behavior monitoring. Environmental analysis, enzyme activity assessment, and drug development all use calorimetric biosensors. In conclusion, transducer types such as those that are calorimetric, piezoelectric, optical, and electrochemical can be utilized to categorize biosensors. Every kind of biosensor has special benefits and is appropriate for a certain use case. Because of their excellent sensitivity, specificity, and suitability for electrochemical biosensors and microfluidic systems especially conductive polymer-based ones they have shown significant promise for cancer detection. Combining additional useful compounds with conductive polymers, such as QDs, graphene, CNTs, and nanoparticles, has further enhanced the performance of electrochemical biosensors. Applications for biosensors derived from diverse types of transducers have also been discovered in various domains, from environmental monitoring to healthcare. Further research and creation of biosensor technology have promise for improving monitoring, treatment, and diagnosis across a range of fields [62]. Figure 4.5 shows a comparative chart that shows the mechanical characteristics of several conductive polymers with flexibility and strength [63–65].

4.3 Types, Architectures, and Conductivity Processes of Conductive Polymers

A family of organic substances referred to as conductive polymers (CPs) has gained notoriety due to its unique capacity to transmit electricity. CPs are positioned as attractive prospects for use in biology, electronics, and energy storage because of

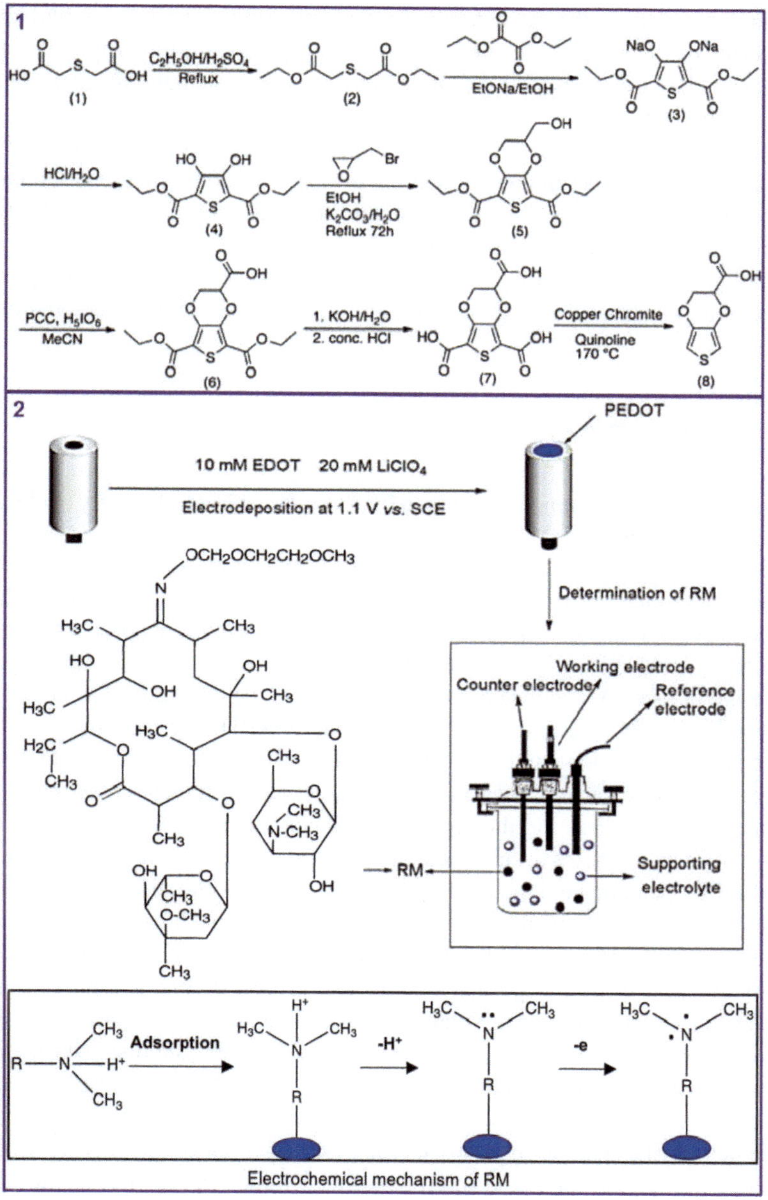

Fig. 4.5 Gives information on two important areas. The synthesis of 3,4-(ethylenedioxy)thiophene (EDOT acid) functionalized with carboxylic acid is described in Sect. 4.1. This section clarifies the synthesis of EDOT acid with carboxylic acid functionality by outlining the procedures needed to create this particular chemical structure [63]. The molecular structure of RM is emphasized in Sect. 4.2, providing a graphic depiction of the precise arrangement and makeup of RM. Both parts provide important insights into these molecular entities and their creation, laying the groundwork for future research and use across a range of scientific fields [64]. Reprinted with permission from [65] Copyright 2018 American Chemical Society

this feature as well as their adaptability and affordability. This section will provide a summary of the several kinds of conductive polymers, go into depth about their chemical makeup, and describe the processes that give these materials their electrical conductivity [66].

Many organic compounds with electrical conductivity are classified as conductive polymers. They have a number of benefits over conventional inorganic conductors like metals, including as processability, flexibility, and electrical property tunability. Among the conductive polymers that are most often researched are polythiophene (PT), polypyrrole (PPy), and polyaniline (PANi). These polymers are appropriate for a range of applications due to their distinctive architectures and characteristics [67].

4.3.1 Polypyrrole

A polymer with high conductivity known for its remarkable conductivity of electricity and durability is polypyrrole (PPy). Its connected structure, made of pyrrole rings linked by nitrogen atoms, helps charge carriers move more easily along the polymer backbone. Polymerization by oxidation, which adds positive charges, called polarons, to the chain of polymers, is the procedure used in the conventional production of PPy. Its electrical conductivity is improved by this treatment, which makes PPy appealing for a range of uses. Along the conjugated backbone, the delocalized π-electrons are responsible for PPy's excellent electrical conductivity. The pyrrole rings' switching between double and single bonds enables the intersection of π-orbitals, which makes it easier for charge carriers such as holes or electrons to travel across the polymer structure. This contributes to the high electrical conductivity of PPy by assisting the passage of charge carriers (holes or electrons) across the structure of the polymer. In addition, PPy has outstanding stability and resistance to chemicals, which allow it to endure extreme ecological factors like high temperatures, humidity, and chemical exposure. Its polymer structure's strong covalent connections are what give it its stability. Because of its exceptional stability and electrical conductivity, PPy is very versatile for a variety of uses. Notably, devices like sensors, supercapacitors, and batteries frequently use it as an electrode material in electrochemistry [68]. PPy has fascinating optical and electrochromic features in addition to its electrical ones. PPy-based electrochromic devices can change color back and forth when they respond to outside factors like changes in pH or electrical potential. These gadgets are necessary for optical sensors, displays, and smart windows. Moreover, PPy may be modified or doped to improve certain characteristics. Significant changes in the material's electrical and optical properties are possible by doping chemicals with different dopants, like cations or anions. PPy can be easily adjusted for its stability, conductivity, and color-changing properties, making it suitable for many specific uses. In conclusion, polypyrrole is a remarkable polymer that conducts electricity and is famous for its outstanding endurance and conductivity. Produced by oxidative polymerization, its structure facilitates effective charge transmission, and its resilience in harsh environments highlights its use in a range of sectors. The applicability of PPy for a variety of applications in electrochemistry, electrochromic devices, and other fields is further enhanced by its adaptability and simplicity of modification or doping [69].

4.3.2 Polyaniline

Because of its remarkable electrical characteristics and wide variety of possible uses, polyaniline (PANi), a flexible conductive polymer, has attracted a lot of attention. Its conjugated architecture, which is made up of interconnected aniline rings joined by nitrogen atoms, makes it easier for charge carriers to travel down the polymer backbone. One of polyaniline's notable characteristics is its ability to exist in several states of oxidation, each with distinct electrical characteristics. The completely diluted version (leucoemeraldine), the form that is partially oxidized (emeraldine base), and the form of pernigraniline that has completely oxidized are among the oxidation states that PANi may change between. PANi takes on the emeraldine base configuration in its conductive state, which is distinguished due to a quinoid structure that promotes π-electrons' displacing along the polymer backbone. High electrical conductivity is the outcome of effective charge transfer made possible by this electron delocalization. There are many ways to create polyaniline, like Electrochemical reactions and chemical oxidation. Oxidation of chemicals is the process of oxidizing aniline monomers using substances like hydrogen peroxide, ammonium persulfate, or ferric chloride. The emeraldine base is created by this oxidative polymerization, which results in a chain of polymers having units that alternate between being reduced and oxidized. In contrast, electrochemical synthesis uses Aniline monomer oxidation via electrochemistry to deposit the substance onto a surface of an electrode. Depending on the reaction conditions, this technique permits the production of tailored structures, such as films, nanofibers, or nanotubes, and permits exact control over polymer growth. Doping also makes it possible to adjust polyaniline's electrical characteristics. By adding dopant ions or molecules to the matrix of polymers, this method creates carriers of charges and alters PANi's behavior. Acids like hydrochloric acid or camphorsulfonic acid are common dopants that increase the stability and electrical conductivity of the polymer. Because of its flexibility, polyaniline shows a broad variety of possible applications. It is often used in electrical devices such as transistors, capacitors, and batteries because of its characteristics as a conductive polymer. Its strong electrical conductivity enables the effective transfer of electric charges. PANi is appropriate for application in electrochromic devices, actuators, and sensors due to its ability to transition between several oxidation states. This makes it possible to regulate its electrical characteristics in reaction to outside inputs. Additionally, PANi has advantageous properties in the energy storage domain. The material is appropriate for application in supercapacitors, which are energy storage devices that accomplish producing high concentrations of power due to their elevated specific capacitance and excellent cycle steadiness. In various applications, PANi-based supercapacitors may be used in addition to or rather than conventional storage of energy devices, such as batteries. A very versatile conductive polymer, polyaniline

(PANi) may exist in various states of oxidation, each of which confers distinct electrical characteristics on it. This substance's chemical composition makes it simple for charge carriers to travel along the polymer structure. Its electrical characteristics may be tailored using a variety of manufacturing processes and doping strategies. Numerous industries, such as sensors, energy storage devices, electronics, and others, use polyacrylonitrile (PAN) [3, 70, 71].

4.3.3 Polythiophene

A very conductive polymer, polythiophene, has attracted a lot of attention because of its remarkable electric characteristics and wide variety of uses in many industries. Its structure, made up of thiophene rings joined by sulfur atoms, creates a conjugated system necessary for effective charge mobility along the polymer's vertebral column. The spread-out π-electrons, which are important for the high electrical conductivity of PT, happen because of this conjugation, which is due to the alternating double bonds. The main techniques used to synthesize PT are electrochemical and oxidative polymerization. Oxidants like potassium chloride or ferric tosylate are used to oxidize thiophene monomers in oxidative polymerization, which starts the process and creates chains of polythiophene. In contrast, thiophene monomers are controlled electrochemically oxidized on an electrode surface during electrochemical polymerization. This technique offers precise control over the conductivity of electricity, dispersion, and molecular weight of the polymer, enabling customization for a range of uses. The adaptability of polythiophene is attributed to its capacity for structural changes via copolymerization or chemical functionalization. By adding various functional groups to the polymer backbone, functionalization modifies the material's electrical and physical characteristics. Other monomers may be added by copolymerization, allowing for modifications to solubility, stability, and optical properties. Because of its remarkable electrical qualities, PT is an essential component of organic electronic devices. It is extensively used in organic photovoltaics (OPVs), organic field-effect transistors (OFETs), and organic light-emitting diodes (OLEDs), which improve the performance of the gadget by enabling effective charge transfer. PT is useful in supercapacitors for energy storage because of its excellent electrical conductivity and electrochemical stability, which enable high power densities and quick charge-discharge cycles. PT has significant optical features, such as wide absorption in the visible spectrum, in addition to its electrical qualities. Because of this, it works well in optoelectronic devices like sensors and photodetectors, where charge transfer is facilitated by its capacity to absorb light. In conclusion, the structure of polythiophene (PT), a highly conductive polymer, permits effective charge transfer. Its adaptability for a variety of applications stems from its flexible synthesis techniques and structural modification possibilities. The critical role of PT in energy storage, optoelectronics, and organic electronics highlights its importance in a variety of technical domains [72]. The electrical characteristics various conductive polymers, including polythiophene (PT), polypyrrole (PPy), and polyaniline (PANi), are closely linked to their chemical structures. This allows charge carriers to travel efficiently down the polymer backbone by delocalizing π-electrons [73].

PPy: The way it is structured consists of rings of pyrrole joined by atoms of nitrogen, creating a conjugated framework in which π-orbitals may overlap due to alternating single and double bonds. High electrical conductivity is the outcome of this, since it enhances the flow of carriers of charges and encourages the propagation of π-electrons. Similar to this, polyaniline (PANi) is a compound composed of alternating double and single bonds, with nitrogen atoms connecting the rings of aniline. Effective charge transfer is made possible by the π-electrons' dispersion along the backbone of the polymer.

PANi: PANi is a similar conjugated structure made up of rings of aniline joined by nitrogen atoms. Its conductivity may be adjusted by chemical doping or electrochemical control and varies depending on the oxidation state, including leucoemeraldine, emeraldine base (which is the most conductive), and pernigraniline [74].

PT: PT is a connected system with alternating double and single bonds, consisting of thiophene rings linked by sulfur atoms. High conductivity and effective π-electron delocalization are made possible by this configuration. By altering the electronic structure of the polymer chains or introducing charge carriers, this technique increases electrical conductivity. Depending on the particular conductive polymer under consideration, dopants might be acids, anions, or cations. Moreover, doping may result in changes to optical properties, stability, and electrochromic behavior. The effectiveness of conductive polymers may be improved in a number of domains, like electronics, sensors, energy storage, and optoelectronic devices, by altering their chemical composition by doping. In addition to doping, other techniques like chemical functionalization and copolymerization may be used to change the chemical makeup of conductive polymers. Copolymerization makes it possible to add different monomers to the polymer chain, changing properties like electrical conductivity, stability, and solubility. Adding specific functional groups to the main structure of the polymer, a process called chemical functionalization, helps change the properties of conductive polymers. The substance architectures' numerous conductive polymers, such as polythiophene, polypyrrole, and polyaniline, dictate their electrical characteristics. These characteristics make it easier for π-electrons to disperse and for charges to travel efficiently throughout the main chain of the polymer. These configurations may be changed to improve the conductivity and other properties of polymers that conduct electricity for a variety of uses by using methods including doping, copolymerization, and functionalization [75]. Figure 4.6 displays electrochemical graphs that illustrate redox reactions and how conductive polymers behave with electricity [76, 77].

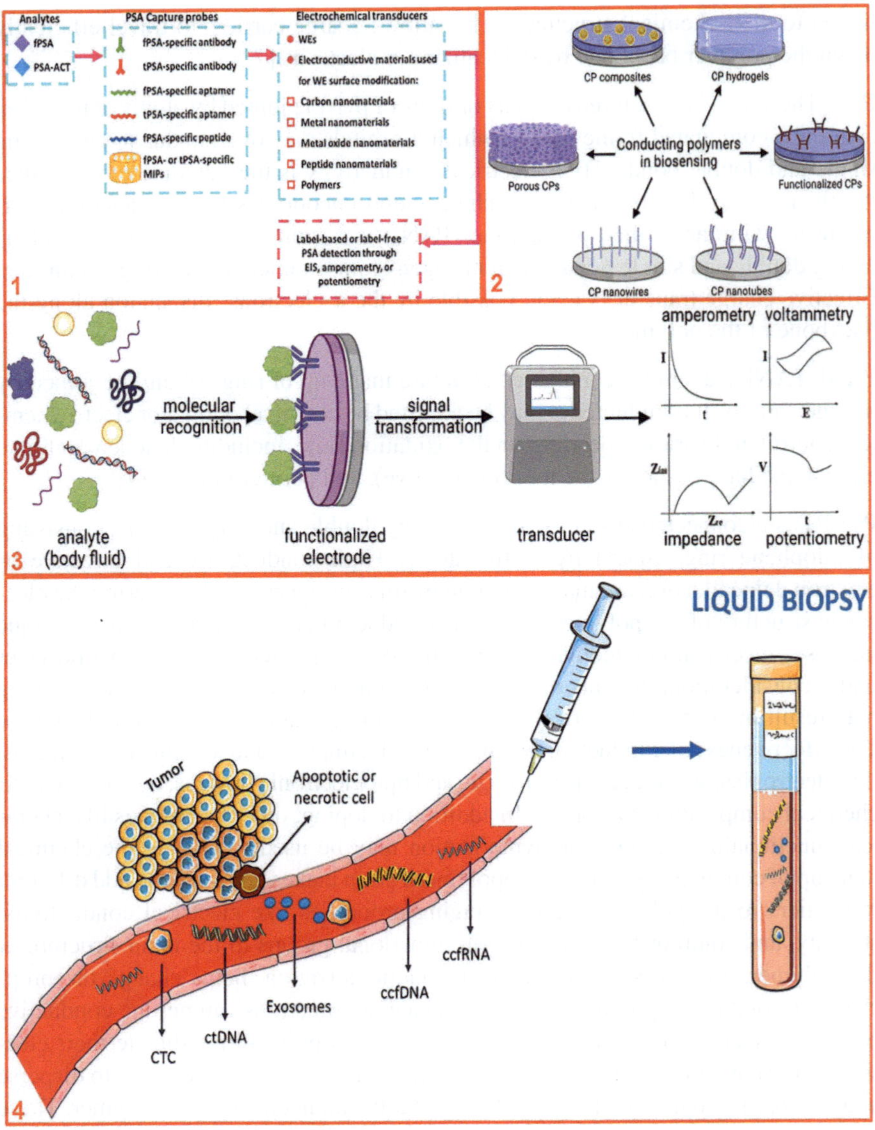

Fig. 4.6 Talks about complete electrochemical biosensors. Section 4.1 provides a description of prostate-specific antigen (PSA) electrochemical biosensors and its constituent parts [76]. In Sect. 4.2, the adaptability and usefulness of electrochemical biosensors are shown using conducting polymers (CPs). Section 4.3 provides a schematic illustration of the electrochemical biosensors' operation as well as a breakdown of its components and interactions. Section 4.4 describes the mechanism by which tumor cells release circulating tumor cells (CTC), circulating cell-free DNA (ccfDNA), circulating tumor cells (exosomes), ctDNA, and ccfRNA into the circulation. This section emphasizes the application of electrochemical biosensors to liquid biopsy blood analysis for cancer biomarker identification and study [77]

4.4 Important Elements Impacting Electrical Conductivity

One important characteristic of polymers that conduct that affects how well they work in electrical applications is their electrical conductivity. The conjugated segments' duration in the polymer chain and dopant levels, the particular polymer's shape and chemical makeup, as well as environmental conditions are some of the elements that affect it. The conjugated segments' duration in the polymer chain and dopant levels, the particular polymer's shape and chemical makeup, and environmental conditions are some of the elements that affect it. Higher conductivity is the consequence of more effective charge carrier movement made possible by longer conjugation lengths. This is due to the fact that a longer conjugated system offers more delocalized π-electrons, which are capable of easily moving charges down the backbone of the polymer. Another significant element influencing conductive polymers' electrical conductivity is doping. To alter the electrical characteristics of a polymer, doping entails adding impurities or dopants to its structure. Dopants raise the quantity to enhance conductivity and charge carriers by either donating or accepting electrons. The conductivity of the polymer may be greatly impacted by the dopant selection and concentration. For particular purposes, the electrical characteristics of conductive polymers may be customized using a variety of dopants, including inorganic ions or organic compounds. The electrical conductivity of a polymer is also influenced by its particular shape and chemical structure. The movement of charge carriers and, as a result, the polymer's conductivity can be influenced by factors like how much the polymer is cross-linked, how crystalline it is, and how the polymer chains are arranged. For instance, because of their improved charge transport channels, highly crystalline polymers with chains that are oriented tend to have better conductivities. Conductive polymers' electrical conductivity may also be affected by environmental variables such as humidity, temperature, and solvent concentration. On the other hand, moisture or solvents may modify the polymer's morphology and structure, which can result in variations in conductivity; temperature changes can also impact the mobility of charge carriers [78, 79].

Although polymers that conduct electricity are valued for their electrical qualities, they additionally need to have strong mechanical qualities to function dependably in real-world applications. Their durability, flexibility, and mechanical strength are crucial factors, especially in industries like energy storage devices, flexible sensors, and wearable electronics. The capacity of conductive polymers to tolerate tensile stresses without cracking or deforming is known as their mechanical strength. Applications needing mechanical robustness, such energy storage systems or structural parts in electronics, should pay particular attention to this. Applications requiring mechanical stability may benefit from the high tensile strength and modulus of conductive polymers, such as polyphenylene. Another essential mechanical characteristic of conductive polymers is their flexibility. These polymers' capacity to undergo deformation without losing their electrical characteristics is very desired, especially for wearable technology and flexible electronics applications. High stretchability and the ability to adapt to different forms and deformations without sacrificing electrical conductivity are characteristics of elastomeric conductive polymers, including polydimethylsiloxane (PDMS). Another crucial factor for conductive polymers is durability. Their long-term performance depends on their capacity to tolerate environmental elements including temperature, humidity, and mechanical stress. For use in challenging environments where they could be subjected to high temperatures, dampness, or mechanical stresses, stable conductive polymers with exceptional endurance are recommended (Fig. 4.7) [80, 81].

Because of their unique electrochemical characteristics, conductive polymers are very important for various cutting-edge uses, especially in vitality storage and electrochemical sensing. Two notable features that allow for their extensive use in industries including smart materials, screens, supercapacitors, and batteries are its electrochromic properties and reversible redox activity. The capacity of conductive polymers to go through reversible redox reactions is one of its noteworthy electrochemical characteristics. This indicates that the movement of electrons between the dopant ions and the polymer backbone allows them to store and release charge. Because of this characteristic, they show promise as potential components for electrochemical energy storage systems as supercapacitors and batteries. Conjugated structures that quickly collect or give electrons are usually included in conductive polymers to aid the redox processes. The capacity of some conductive polymers to

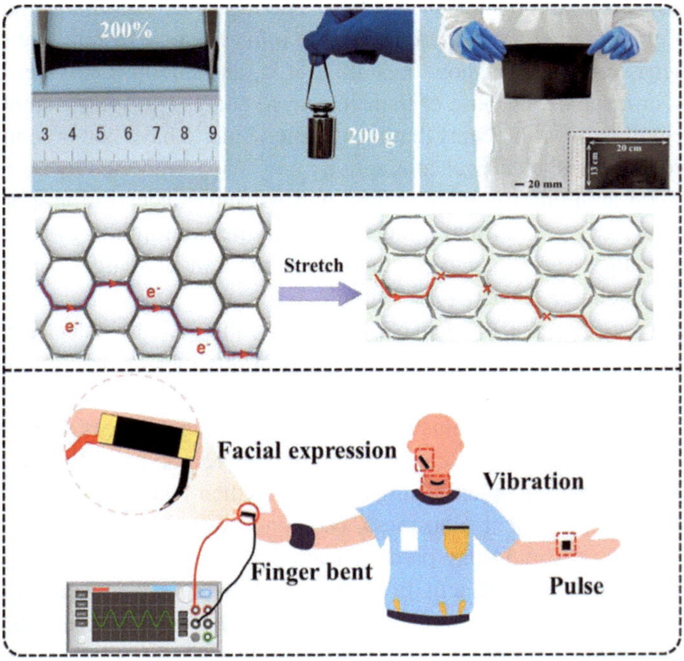

Fig. 4.7 Demonstration of a stretchable, durable conductive polymer that maintains conductivity under strain, enabling real-time physiological monitoring for wearable electronics [81]

alter their color or optical characteristics in response to the application of an electrical potential is known as electrochromic, another electrochemical trait. This effect, which results in variations in light absorption or reflection, is caused by alteration of the electrical structure of the polymer. Electrochromic conductive polymers are used in sensors, intelligent Windows, and electrochromic displays, where their changing color properties may be used for various objectives [82].

Polymers that are conductive may possess unique visual qualities, making them suitable for optoelectronic applications alongside their electrochemical and electrical properties. The capacity of conductive polymers to absorb light in certain wavelength bands is one of their optical characteristics. The existence of a bandgap, or the energy differential between the HOMO (highest occupied molecular orbital) and the LUMO (lowest unoccupied molecular orbital), is what causes this characteristic. The wavelength of light that the polymer can absorb is determined by its bandgap. Polymers that conduct may be engineered to take in light within certain electromagnetic spectrum areas by adjusting their chemical structure and conjugation length. Because the energy of light that is absorbed may be transformed into light or electrical emission, this feature finds use with optoelectronic devices like light-emitting diodes and organic solar cells. Photoconductivity is another optical characteristic of some conductive polymers. When a substance is exposed to light, its electrical conductivity increases, a phenomenon known as photoconductivity. This characteristic of conductive polymers results from the creation of extra charge carriers, like holes or electrons, when photons are absorbed. Other optoelectronic devices, such as photodetectors that need detection of light and transformation into electrical signals, might benefit from the photoconductive nature of conductive polymers [83].

For conductive polymers to be used in biomedical applications like biosensors and scaffolds for tissue engineering, they must be biocompatible. Materials that can interact with biological systems without having negative consequences are known as biocompatible materials. It is possible to change the surface and chemical structure characteristics of conductive polymers to make them show superior biocompatibility. Their biocompatibility may be improved and any possible cytotoxicity can be decreased by adding certain functional groups or coatings. Numerous biomedical technologies, including medication delivery systems and implanted sensors, use biocompatible conductive polymers. The biocompatibility of conductive polymers is particularly crucial for the use of biosensors, which are analytical tools that employ a transducer and a biological recognition element to identify and quantify specific analytes. The transducer, which converts the biological signals from the recognition element into electrical impulses, may consist of conductive polymers. The conductive polymer's biocompatibility guarantees that it may engage with the biological material without compromising the detection of analytes procedure or endangering the cells or tissues in any way. In tissue engineering, conductive polymers may also be used as scaffolds. By fusing cells with biomaterials, tissue engineering seeks to produce functioning tissues or organs. Good biocompatibility conductive polymers may provide a matrix that promotes tissue regeneration and cell proliferation. In order to encourage tissue growth and cell differentiation, they may also be utilized to provide electrical stimulation. To sum up, conductive polymers have a variety of qualities that make them appealing for a range of uses. Environmental circumstances, conjugation length, doping, and polymer structure are some of the variables that may be used to adjust their electrical conductivity. The appropriateness of conductive polymers for various applications is determined by their mechanical characteristics, such as strength, flexibility, and durability. Energy storage and visual modulation are made possible by electrochemical characteristics like electrochromic and redox processes. Conductive polymers are appropriate for optoelectronic devices due to their optical characteristics, which include light absorption and photoconductivity. Lastly, conductive polymers' biocompatibility is essential for their use in tissue engineering and biosensors as it guarantees that they are compatible with biological systems [84].

4.5 Techniques for Synthesis and Manufacturing

One flexible method for creating conductive polymers is to create them chemically; to produce polymers with the required qualities, chemical processes are used [85]. Conductive polymers are created using a number of standard chemical techniques:

4.5.1 Solution-Based Synthesis

One popular technique for creating conductive polymers is solution-based synthesis. This method uses a solvent to facilitate the polymerization process. The solvent's role as a medium for the reactants enables improved regulation of polymer synthesis and reaction conditions. The soluble nature of the monomers and the needs of the response determine which solvent is best [86]. There are many ways to do the solution-based synthesis, including:

(a) **Polymerization via chemical oxidation**: This process uses chemical oxidants such ammonium persulfate (APS), iodine (I_2), or iron (III) chloride ($FeCl_3$) to oxidize monomers. By removing electrons from the monomers, the oxidants speed up the polymerization process and create conductive polymers. Usually, the reaction is carried out at ambient temperature or with moderate heating [87].
(b) **Electrochemical polymerization**: An electrochemical cell is used in the process of electrochemical synthesis to cause monomers to polymerize. An electric current is introduced into the system while the monomers are submerged in a solution of electrolytes. Depending on the kind of monomer, the current causes oxidation or reduction processes, which result in the creation of the electrode surface's conductive polymers. Conductive polymer films with certain characteristics may be deposited via electrochemical synthesis, which provides exact control over the polymerization process [88].

(c) **Polymerization across the interface**: The interaction between monomers when two immiscible phases meet typically, an aqueous phase and an organic phase is known as interfacial polymerization. A conductive polymer is created when the monomers, which have been dissolved in distinct phases, come into touch with one another and undergo a chemical reaction at the interface. One benefit of interfacial polymerization is that it may create conductive polymers with precise thickness and shape [89].

(d) **Polymerization in situ**: The process of creating polymers that conduct electricity on a substrate or within a solid matrix is known as "in-situ polymerization." The matrix incorporates, covers, or injects the reactants typically oxidants and monomers into the substrate. An appropriate stimulation, like a chemical initiator, heat, or light, is added to start polymerization. Conductive polymers may be directly incorporated into certain materials or surfaces using in-situ polymerization [90].

4.5.2 Solid-State Synthesis

Another technique used to create conductive polymers is solid-state synthesis; it includes the direct, solvent-free solid-state polymerization of monomers [91]. There are several methods for achieving solid-state synthesis, such as:

(a) **Thermal polymerization**: In order to start the polymerization process, polymerization by heat entails heating the monomers. To facilitate the interaction between the monomers and create the conductive polymer, it heats the monomers beyond their points of melting or at extreme temperatures. To avoid the polymer's oxidation or breakdown, the process might be conducted under vacuum or in an inert environment [92].

(b) **Synthesis of mechanical**: Mechanical force induces the polymerization process in the Synthesis of Mechanicals, also known as the synthesis of mechanochemicals. When there are sturdy supports, the monomers are mechanically ground or milled along with appropriate initiators or catalysts. Conductive polymers are created when the mechanical stress disrupts the links between molecules and starts the process of polymerization. Benefits of mechanical synthesis include its simplicity, scalability, and solvent-free nature [93].

(c) **High-pressure polymerization**: In order to promote polymerization, high-pressure polymerization entails applying high pressures to the monomers. The monomers are put in a pressurized vessel or reactor with high pressure, usually with an appropriate initiator present. Conductive polymers are created as a result of the increased pressure encouraging the interaction between the monomers. High-pressure polymerization makes it possible to create polymers with special structures and characteristics that are difficult to achieve with conventional techniques [94].

4.5.3 Grafting

One method for affixing conductive polymers to non-conductive surfaces is grafting. In order to create a conductive surface, chemical linkages must be formed between the substrate and the conductive polymer [95]. Grafting may be accomplished in a number of ways, such as:

(a) **Chemical grafting**: Chemical grafting is the process of attaching the conductive polymer to the substrate by use of chemical reactions. Reactive groups like amino groups, carboxylic acids, or hydroxyl groups are introduced by functionalizing the substrate surface. Covalent bonds are created when the functionalized substrate and the conductive polymer, which is usually in solution, come into contact and undergo a chemical reaction between the reactive groups [96].
(b) **Plasma grafting**: A method called plasma grafting makes use of plasma treatment to alter the substrate's surface and make it easier for conductive polymers to adhere. Reactive species are produced on the surface of the substrate when it is subjected to a low-temperature plasma. Following their reaction with the conductive polymer, these reactive species cause chemical bonds to be formed and the polymer to graft onto the substrate. High accuracy, ability to control, and the capacity to attach conductive polymers to intricate geometries are among the benefits of plasma grafting [97].
(c) **Photochemical grafting**: Light is used to start the grafting process in photochemical grafting; a photosensitive layer containing an originator of photos is applied to the substrate surface. A photochemical reaction occurs in the photoinitiator when it is exposed to a certain wavelength of light, producing species that respond and interact with the polymer that is conductive. To create chemical connections between the polymer and the substrate, the species that reacts may either react straight from the polymer or start a secondary response [98].

4.5.4 Electrochemical Synthesis

One method for creating conductive polymers is electrochemical synthesis, which makes use of an electrochemical cell, which is made up of a monomer-containing electrolyte solution, an anode, and a cathode. When a current of electricity flows via the cell, the monomer is polymerized, and films of conductive polymers are created on the surfaces of the electrodes. The following are some benefits of synthesis via electrochemistry:

(a) **Accurate regulation**: The polymerization procedure may be precisely controlled by electrochemical synthesis. The density, shape, and characteristics of the formed films of conductive polymers may be controlled by varying parameters including the electrolyte composition, current density, and applied voltage. This degree of control is essential for modifying the conductive polymers' characteristics to satisfy certain application needs.

(b) **Versatility**: Numerous conductive polymers may be created using electrochemical synthesis. It is possible to use a variety of monomers, each with special qualities and traits. This adaptability makes it possible to create polymers that conduct electricity with various uses and functions, from electronic parts and energy storage devices to sensors and actuators [99].
(c) **Scalability**: A scalable method that is readily adaptable for large-scale manufacturing is electrochemical synthesis. The manufacturing rate may be increased by using electrochemical cells to deposit conductive polymer films on numerous electrodes at the same time. Furthermore, consistent quality across large batches is ensured by the capacity to manage the films' thickness and homogeneity [100].
(d) **Compatibility**: Metals, conductive polymers, and carbon-based materials are among the electrode types that may be used for electrochemical synthesis. This adaptability increases the potential uses of conductive polymers by enabling their integration into various substrates and device topologies [101].

4.5.5 Composite Materials

One tactic used to improve the characteristics and broaden the uses of the host materials as well as the conductive polymers is the inclusion of conductive polymers into composites. To obtain desired qualities, composites blend conductive polymers combined with metals and other materials, ceramics, or plastics. Conductive polymers may improve the composites' functional attributes, thermal stability, electrical conductivity, and mechanical strength, among other characteristics [102]. Conductive polymers may be added to composites in a number of ways:

4.5.6 Matrix Composites of Polymers

Polymers that conduct are dispersed in a polymer matrix to create a matrix of polymer composites. The polymers that conduct electricity may be blended with the polymer matrix or exist as nanoparticles or nanofibers. The dispersion of conductive materials and polymers throughout the polymer matrix enhances the electrical conductivity of the composite substance. Applications for this improved conductivity include flexible electronics, electrostatic dissipation, and electromagnetic shielding. Additionally, mechanical reinforcement is provided by conducting polymer dispersion inside the matrix of polymers, which improves the composite material's strength, toughness, and impact resistance. By serving as reinforcements or fillers, the conductive polymers may enhance the composite matrix's ability to transmit load and distribute stress. Based on the particular matrix of host polymers and conductive polymers, the addition of conductive polymers may also improve other characteristics, including heat stability, flame retardancy, and barrier qualities. In-situ polymerization, solution mixing, and melt blending are techniques for adding conductive polymers to polymer matrix composites. Blending of melts is the process of combining polymers that transmit and have a polymer matrix while they are still molten, then solidifying and molding the mixture. Solution mixing involves dissolving the electrically conducting polymers in a solvent, combining them with the solvent, and then removing the polymer matrix. Using conductive polymers that have already been polymerized may undergo additional polymerization when processing composites, or by combining monomers and initiators, polymerization in situ entails in the polymer matrix the conductive polymers are synthesized [103]. Conductive polymers may be added to metal matrix composites to enhance their electrical conductivity, corrosion resistance, wear resistance, and thermal stability. These composites are used for electromagnetic shielding, heat sinks, and electronic packaging and structural elements that require a blend of metal and conductive polymer-based useful qualities. In summary, electrochemical synthesis, chemical synthesis, and the integration of composites made of conductive polymers are the methods used in the synthesis and manufacture of polymers that conduct. These techniques enable the manufacturing of conductive polymers using specific characteristics as well as their incorporation into various materials and devices. The specific requirements for applications and the necessary characteristics of conducting polymers dictate the synthesis and manufacturing process [104].

4.6 Conductive Polymers Function in Biosensors

In biosensors, conductive polymers are necessary for enabling the detecting mechanisms. These polymers are perfect for converting biological signals into quantifiable electrical signals because of their special electrical characteristics. In biosensors, the sensing mechanism usually consists of biorecognition components, such as enzymes or antibodies, recognizing and binding certain analytes and then converting an electrical signal from this binding event. When polymers that conduct are included via the design of a biosensor, the target analyte's presence might cause changes in the biosensor's electrical resistance, capacitance, or impedance. A numerical assessment of the molecule of interest may be obtained by directly correlating this change in electrical characteristics with the analyte's concentration or activity. Conductive polymers' conjugated molecular structure, which permits electron delocalization and charge carrier generation, is responsible for their capacity to display such electrical property changes. Through functionalization or chemical changes, conductive polymers may be made to work with the analyte of interest in certain ways. For instance, functional groups that increase the polymer surface's affinity for a certain analyte may be added, enhancing the biosensor's selectivity. Furthermore, the polymer that conducts may be engineered to communicate with a certain ability to recognize them, improving the binding effectiveness and enabling the sensor surface's immobilization of them [5, 36, 105–108].

Conductive polymers provide significant benefits in this area, and surface alteration is a crucial tactic used with biosensors to improve their identification capacities. An electrically conductive and biocompatible contact is produced via chemical means by attaching conducting polymers to the surface of the sensor. A permanent and long-lasting immobilization is made possible by this alteration, which enhances the biorecognition components' adherence to the sensor surface. Thin films or coatings of conductive polymers may be applied to the sensor surface, providing a surface on which biorecognition components like enzymes, antibodies, or DNA probes can be attached. Several methods, including covalent bonding and physical adsorption, may immobilize these biorecognition components onto the conductive polymer layer. By serving as a link between the biorecognition components and the sensor surface, the conductive polymer layer makes it easier for them to communicate effectively with the analyte of interest. The conductive polymer content on the sensor surface further improves the biosensor's electrical functionality. Electrical signals may be efficiently transferred throughout the biorecognition components and the biosensor's transducing components thanks to the polymer's conductive character. This improves the biosensor's overall signal-to-noise ratio and sensitivity, which raises the accuracy and detection limits [5, 36, 106–110].

The biosensors' sensitivity and stability are greatly influenced by the characteristics of conductive polymers. For biosensors to function well over the long term, conductive polymers must have exceptional stability and resistance to fouling and degradation. The biosensor's dependability and longevity under a range of working settings are guaranteed by these polymers' resilience to variations in pH, temperature, or humidity. Rapid mass transfer of analytes is also made possible by the engineering of conductive polymers with large surface area and porous architectures. By promoting the effective target analyte diffusion to the biorecognition components, this characteristic raises the sensitivity of biosensors. Better binding effectiveness and amplification of the signal result from the higher loading capacity of biorecognition components made possible by the conductive polymer sheets' porous nature. Biosensor sensitivity is influenced by the power source characteristics of polymers that conduct electricity, like their capacity to experience variations in capacitance or electrical resistance. The target analyte may be precisely quantified by amplifying and detecting the signal from the electricity produced in reaction to the analyte binding event by the conductive polymer. To sum up, Conductive polymers are necessary to biosensors due to the fact that they streamline the sensing processes, allow surface alteration for better identification, and enhance the sensitivity and stability of these instruments. They are useful materials for the creation of sophisticated biosensing systems because of their distinct electrical characteristics, stability, and biocompatibility. Conductive polymers are anticipated to become more significant in the development of biosensor technologies for various uses, like food safety, tracking of the environment, and healthcare, as this field's research and development continue [111].

Because of their special blend of mechanical flexibility, electrical conductivity, and processability, conductive polymers have garnered a lot of scientific attention. New conductive polymers with enhanced properties have been synthesized as a result of recent advances in the sector, and methods for customizing their properties for particular uses have been developed. The latest findings in newly developed conductive polymers will be highlighted in this part, along with the methods used to modify their characteristics [112]. Figure 4.8 provides an illustration of how conductive polymers are used to change sensor surfaces for better detection [113–115].

The combination of novel polymers that conduct electricity and enhanced processability, stability, and electrical conductivity has advanced significantly recently. These new polymers often have distinctive architectures and characteristics that allow them to be used in several applications. For instance, the variety of materials accessible for electrical and optoelectronic devices has increased with the identification of novel conjugated polymers, including polypyrroles, polythiophenes, and polyanilines. The creation of conductive polymers with functionalization for particular uses, like bioelectronics, sensing, and energy storage, is one well-known field of study. To add desired features, functionalization entails adding certain side chains or chemical

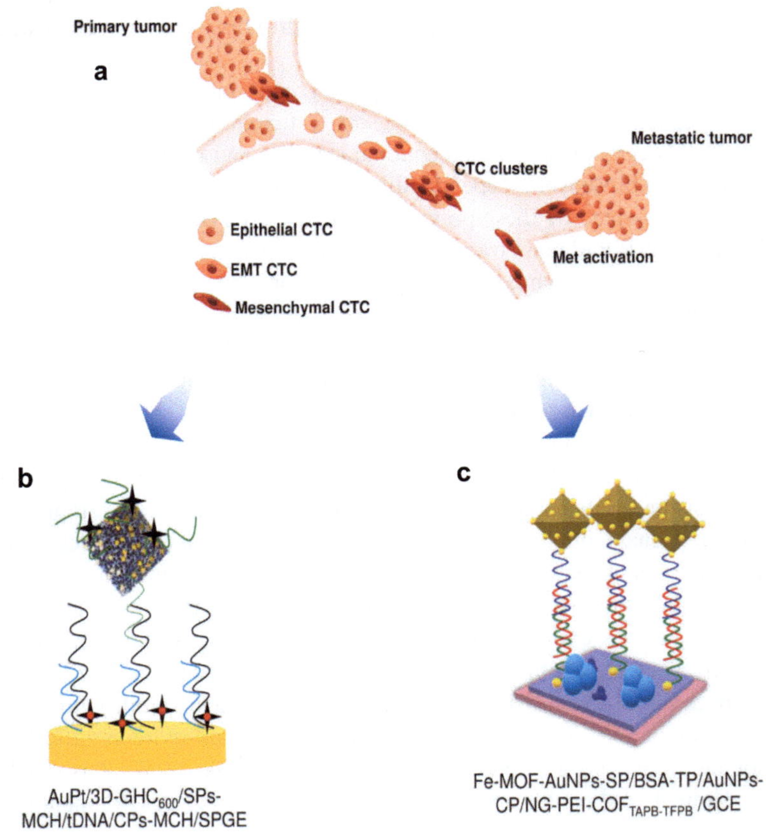

Fig. 4.8 Biosensing technique for ctDNA detection. **a** Circulating tumor cells from the original tumor to its dissemination via bodily fluids [114]; **b** electrochemical sensor fabrication for ctDNA detection [115]; and **c** ctDNA-based biosensor for non-small cell lung cancer [113]

groups to the polymer backbone. For example, adding groups that donate or remove electrons might improve the polymer's stability or electrical conductivity. These functionalized conductive polymers open up new possibilities for use in bioelectronic devices, sensors, batteries, and supercapacitors [116].

Researchers use various techniques to alter the characteristics of polymers that conduct to satisfy the unique needs of diverse applications. These tactics seek to alter the polymers' mechanical power, suppleness, electrical conductivity, ability to be biocompatible, and other essential properties. Doping using chemicals, mixing, surface alteration, and the use of nanoparticles are a few noteworthy tactics. One popular method for altering the power source characteristics of polymers that conduct electricity is chemical doping. Doping is the process of adding dopants usually tiny ions or molecules to the backbone of a polymer. By donating or accepting electrons, the dopants may change the polymer's conductivity and charge carrier concentration. This method may be utilized to maximize Conductive polymer performance for particular uses and gives exact control over electrical conductivity. Another method used to personalize the characteristics of polymers that conduct electricity is blending them. Researchers may create a cooperative impact that improves stability, conductivity, or other desirable features by combining two or more polymers with complementary qualities. The blending technique enables the combination of many functions into a single material and provides flexibility in adjusting the characteristics of conductive polymers. In order to improve the characteristics of conductive polymers, surface modification is essential. Adding functional groups to the polymer surface may enhance adhesion, biocompatibility, and other surface-related characteristics. For instance, adding hydrophilic groups may improve the polymer's water compatibility, which qualifies it for use in biomedical applications. The spectrum of uses for conductive polymers is increased by surface modification methods, which also make it possible to attach proteins or other functional entities. One intriguing method to improve the machine strength and electric conductivity of nanoparticles is incorporating conductive polymers with them. Special benefits are provided by nanocomposites that combine conductive polymers with substances like metal nanoparticles, graphene, or nanotubes of carbon. The total behavior of the polymer matrix may be greatly enhanced by the high conductivity and high aspect ratio of graphene and nanotubes of carbon. Furthermore, the inclusion of nanoparticles may improve surface area, mechanical qualities, and other capabilities. To sum up, new polymers with better qualities have been found and methods for modifying their properties have been developed as a result of recent developments in conductive polymer research. The possible uses of polymers that conduct electricity in bioelectronics, energy storage, and sensing have increased due to the combination of novel ones and the process of making preexisting ones. Furthermore, methods including surface modification, mixing, chemical doping, and the creation of nanocomposite materials provide flexible ways to alter the characteristics of polymers that conduct electricity. These developments open the door for the creation of novel materials and gadgets that make use of the special qualities of conductive polymers in a variety of industries, such as energy, electronics, medical care, and more [117, 118].

Polymers that conduct are organic substances that have the ability to conduct electricity, much like semiconductors or metals. π electron delocalization is facilitated by the molecules' conjugated structure, which is defined by switching between double and single bonds across the backbone of the polymer. To boost conductivity and increase the quantity of charge carriers, polymers that conduct electricity may be infused with electron acceptors or donors. The following unique properties of conductive polymers make them desirable for various applications:

- They are easily produced and processed, mostly by electrochemical deposition or dispersion.
- Depending type, quantity, and mode of molecular shape, doping, and their mechanical, optical, and electrical properties, properties are possibly modified.
- They are essential for biological applications due to their exceptional environmental stability and biocompatibility.
- They may be combined with other materials, like carbon nanotubes, graphene, quantum dots, and nanoparticles, or altered with various groups to improve their functionality and performance.
- Polyacetylene, polyphenylene, poly(phenylene vinylene), poly(3,4-ethylenedioxythiophene), polyaniline, polypyrrole, polythiophene, and polyfluorene are examples of conductive polymers [13].

Transmission of electrons between biological aspects of recognition and electrodes may be facilitated by conductive polymers acting as redox catalysts or mediators. To transform the organism's input into a signal of electricity that the biosensor can recognize, electron transfer is essential. By serving as a conduit for electricity, conductive polymers facilitate this process by facilitating the effective electron transport from the electrode surface to the recognition element. Quick and accurate identification of the target substance is made possible by enhanced electron transfer kinetics, which also increase the biosensor's response time and sensitivity. By boosting the biosensor's signal, conductive polymers may enhance its capacity to identify chemicals. They may increase the surface area of the electrode, generating more binding sites for components involved in recognition and raising the sensitivity of the biosensor. By using electroactive elements such as metal ions or nanoparticles, conductive polymers may enhance signal amplification. By participating in redox reactions, these electroactive species may produce electrical signals that can be detected and correlate to the target analyte's concentration. Furthermore, as soon as the analyte attaches to some conductive polymers, they undergo conformational changes that cause observable changes in their optical or electrical properties. This characteristic might be used to develop biosensors with improved sensitivity and selectivity. By improving the gadget's stability and selectivity, conductive biosensor polymers provide an essential edge. By reducing the impact of other chemicals in the sample matrix, conductive polymers may help identify the target analyte more precisely. By selectively keeping species that interfere with getting to the factors of recognition, they may act as a barrier and lower the possibility of false-negative or false-positive findings. Conductive polymers have the capacity to protect biological recognition components from deterioration, extending their useful lifetimes and ensuring the biosensor's long-term stability. One advantage of conductive polymers is that they make it possible to control the biosensor's conductivity and surface charge. By changing the polymer's composition or doping level, one can improve the biosensor's electrical conductivity. For the biosensor to have the best sensitivity and dynamic range, conductivity must be controlled. Additionally, the conductive polymer's surface charge possibly adjusted to selectively bind and identify analytes that are charged. By making it possible to distinguish between analytes with differing charges, this characteristic enhances the biosensor's accuracy and discriminating power. Utilizing electrochemical biosensors to identify several biomarkers for cancer is often developed using conductive polymers. MicroRNAs, enzymes, proteins, neurotransmitters, circulating tumor cells, and DNA mutations are some examples of the biomarkers. By incorporating conductive polymers into the biosensor's architecture, researchers have achieved notable advancements in cancer monitoring and detection. When it comes to detecting cancer biomarkers, conductive polymer biosensors have shown remarkable selectivity, rapid response times, wide detection ranges, and excellent sensitivity. By offering quick, precise, and economical biosensors, detecting techniques can revolutionize cancer diagnosis. Biosensor applications have made use of various polymers that conduct electricity, including PEDOT, polyaniline (PANI), polypyrrole (PPy), and polythiophene (PTh). Conductive polymers may be tailored to meet the particular specifications of the biosensor application and provide a number of advantages. PPy is perfect for prolonged biosensing because of its exceptional environmental stability and biocompatibility. PANI is a fantastic choice for biosensors that need quick kinetics of electron transfer because it exhibits strong redox processes and may operate as a catalyst for redox. PTh is perfect for developing biosensors with increased sensitivity and selectivity because of its exceptional stability, high Ability to process, and outstanding suitability for biological substances. PEDOT has exceptional conductivity and could be easily modified to include functional groupings, allowing for accurate control of its properties. The development of novel polymers that conduct electricity and their integration into systems of biosensors has been the focus of recent research. There has been a lot of interest in amalgamating conductive polymers with nanomaterials like metal nanoparticles, graphene, or carbon nanotubes. By combining the unique properties of the nanoparticles with the conductivity of the polymer, the hybrid materials exhibit synergistic capabilities that improve the biosensors' performance. Adding nanomaterials may increase surface area, speed up electron transmission, and provide additional properties like enhanced visuals or catalytic activity. Improvements in production processes, such as nanofabrication and microfabrication, have made the development of integrated and small-scale biosensor devices possible. Since conductive polymers are easily created and integrated into lab-on-a-chip systems or microelectrode arrays, they have played a crucial role in these innovations. Miniaturized biosensors are perfect for field application and point-of-care diagnostics because they provide advantages such as reduced reagent and sample use, faster examination times, and movement. Even though Conductive biosensors based on polymers have advanced significantly, there continue to be issues and limitations that need to be addressed. Optimizing the long-term stability and functionality of polymers that conduct electricity is one issue. Under some conditions, conductive polymers may be stable, but they may degrade when subjected to harsh environments or reactive chemicals. For conductive polymers to be successfully used in actual biosensing applications, methods that improve their stability and long-term efficacy must be developed. Adding conductive polymers to wearable and flexible biosensor devices is another challenge. For applications such as continuous health monitoring and customized therapy, flexible biosensors are attracting much attention. It is necessary to create new production processes and examine the mechanical characteristics of the resulting biosensors for the purpose of integrating conductive polymers applied on pliable surfaces such as fabrics or polymers. Cross-reactivity or non-specific binding with structurally similar analyzers may affect the specificity of biosensors based on conductive polymers. Techniques for improving the biosensors' selectivity, including developing new surface modification and recognition component techniques, require further research. When developing and manufacturing electrochemical biosensors, conductive polymers are essential. They help immobilize identification elements, improve electron transfer, amplify signals, increase stability and repeatability, improve sensitivity and selectivity, and more. Biosensors for cancer biomarkers have been successfully developed using conductive polymers, providing rapid, accurate, and economical diagnostic and monitoring tools. Biosensor technology is expected to advance because of more study and advancement in this area, improving the identification and tracking of various substances that have higher mobility, selectivity, and sensitivity [119, 120].

Since one of the main causes of mortality globally is cancer, early identification and therapy are essential to enhancing patient quality of life and survival rates. Imaging, biopsy, and histopathology are examples of traditional cancer detection methods that are often invasive, expensive, take a long time, and need specialized staff and tools. Therefore, in order to provide rapid, accurate, and economical cancer detection at the point of treatment, new techniques are needed. Signals from electricity generated when an aspect of biological recognition engages with a particular sample's analyte are detected by electrochemical biosensors. By detecting various cancer biomarkers, including cells, enzymes, proteins, microRNA, DNA, and neurotransmitters, with high selectivity and sensitivity, electrochemical biosensors provide several advantages for cancer detection. Without the need for complex sample preparation or purification, they can handle minuscule sample amounts, such as tissue, blood, serum, urine, or saliva [121].

- They help with early cancer identification and monitoring by providing quick and fast results, often in a matter of minutes.
- They may be used in conjunction with wireless, nanomaterial, and microfluidic technologies to provide portable, multipurpose, small-scale biosensors that are appropriate for applications at the point of care.
- They could be made and utilized cheaply using simple tools and materials, which can increase the accessibility and cost of cancer diagnosis.

By examining specific cancer biomarkers in various samples, electrochemical biosensors have promise for the identification of several cancer types, including colorectal, cervical, ovarian, pancreatic, lung, breast, and prostate cancers. Early detection, staging, prognosis, and treatment of cancer, response to treatment, and resistance evaluation may all benefit from the use of electrochemical biosensors [122].

Finding and quantifying cancer biomarkers chemicals associated with the presence, progression, or response to cancer treatment is a crucial part of cancer detection. According to their type and place of origin, cancer biomarkers may be classified as cellular, proteomic, metabolomic, or genomic biomarkers. Cancer biomarkers may be found in blood, serum, urine, saliva, or tissue, depending on where they come from and how long they last. In order to properly diagnose cancer, a few requirements must be met:

- The cancer biomarker must be specific, meaning it must precisely reflect the characteristics of a particular cancer type or be present or altered only in cancer cells and not in healthy ones.
- Even in the earliest phases of cancer development, it is important that the cancer biomarker is highly sensitive and observable at low concentrations in the specimen.
- The cancer biomarker has to be reliable, showing reproducible and consistent results across different samples, methods, and facilities.
- The cancer biomarker has to have clinical significance, showing a clear relationship to cancer treatment, staging, prognosis, or diagnosis, and providing useful information for clinical judgment and patient care.
- Along with the requirements for the cancer biomarker, further requirements must be met for the cancer detection method.
- In comparison to an idealized standard method, the detection of cancer technology needs to be precise, with high sensitivity and specificity.
- The method used to identify cancer should be precise, have low variability, and provide findings that are repeatable across a variety of tests, samples, and operators.
- To enable timely diagnosis and response, the cancer detection method should be rapid, yielding results in a short amount of time, preferably within minutes.
- The cancer identification method ought to be simple, involving few steps, equipment, and chemicals, and it ought to be easy for non-specialists to use and comprehend.
- The cancer detection method ought to be economical, needing little in the way of supplies, machinery, and operating costs and providing a good cost-benefit ratio for both patients and the healthcare system [123].

Because delocalized π electrons are present in their conjugated structures, conductive polymers are organic materials that have the ability to conduct electricity. Due to their increased electron transport, biological recognition elements and electrodes such as cells, enzymes, antibodies, or DNA are often used to create electrochemical biosensors, which raises the biosensor's signal-to-noise ratio and sensitivity. They may improve the biosensor's lifetime and selectivity by supplying a steady and biocompatible setting for immobilizing and maintaining the biological acknowledgment element. These resources might be easily produced, modified, and used on a variety of surfaces, such as silicon, carbon, or metal, using chemical or electrochemical processes. This makes it possible to produce biosensors with a broad variety of shapes, dimensions, and configurations. To provide them special binding sites, charges, or polarity, they could be altered with different groupings that are functional such as Sulfonic acid, hydroxyl, carboxyl, or amine. This makes it possible to detect a variety of analytes, including biomolecules, ions, and tiny molecules. To enhance their electrical, optical, or catalytic properties, biosensors may be combined using resources like nanoparticles, graphene, nanotubes, or quantum dots to form nanocomposites [124].

The signal from the electricity generated when an aspect of biological recognition engages with a particular sample's analyte is assessed using electrochemical techniques. They are often used to diagnose cancer because of their many advantages, which include:

- They provide live, quick, and instantaneous results without the need for labeling, amplification, or separation processes.
- They don't need complex sample preparation or purification to handle small sample amounts, such as tissue, blood, serum, urine, or saliva.
- They developed portable, multipurpose, small-scale Biosensors for use at the point of care using wireless, nanomaterial, and microfluidic technologies.
- The accessibility and affordability of cancer detection may be improved by their low-cost production and utilization of basic materials and technologies.

To detect circulating tumor cells (CTCs), a novel electrochemical technique based on electrochemical impedance spectroscopy (EIS) was developed. CTCs, or circulating tumor cells, were extracted from the sample of blood using a microfluidic apparatus equipped with various interdigitated electrodes covered with antibodies against EpCAM. As an indication for identifying cancer, the change in impedance brought on by CTCs, or circulating tumor cells, on the electrodes was measured. The technique successfully detected Tumor cells in circulation in cancer patients' blood samples and showed outstanding selectivity, stability, and repeatability. To identify methylation in DNA as a biomarker for many cancer types, a novel electrochemical technique using EC-SPR, or electrochemical surface plasmon resonance, was created. The technique used a gold electrode that had been modified to enable DNA probes to attach to the sample's target DNA. By contrasting the SPR signals of methylated and unmethylated DNA, the degree of DNA methylation was determined. The technique successfully detected Genomic DNA methylation in cancer patients' samples and showed excellent selectivity, stability, and repeatability. To detect exosomes, which are extracellular vesicles that carry macromolecules from parent cells, a novel electrochemical technique based on electrochemiluminescence (ECL) was developed. The technique collected and concentrated exosomes from the material using an immunoassay based on magnetic beads. ECL probes, which responded to an electrical potential by emitting light, were used to tag the exosomes. The signal utilized to detect malignancy was measured as the ECL intensity. Blood

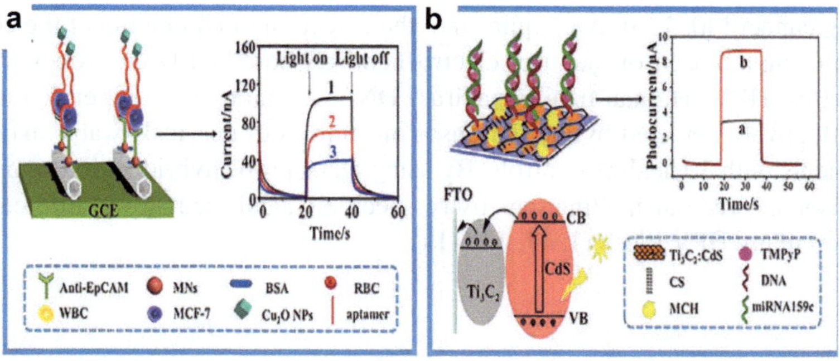

Fig. 4.9 **a** Schematic representation of the assessment of MCF 7 cells in whole blood using Cu_2O-aptamer as a signal probe and Anti-EpCAM-MNs separation of CTCs [131]. **b** A schematic representation of the procedures involved in building an electronic delivery system and photoelectrochemical biosensor [132]

samples from cancer patients that had exosomes were successfully identified using a method that demonstrated excellent selectivity, stability, and repeatability [3, 125–130]. Figure 4.9 provides a visual review of current developments in the study of conductive polymers, emphasizing new polymers and customized features [131, 132].

Through the measurement of the concentration of cancer biomarkers in bodily fluids, including Saliva, urine, serum, or blood. Electrochemical biosensors based on conductive polymers are used in several applications for cancer detection, such as early cancer diagnosis. CEA, or carcinoembryonic antigen, a colorectal cancer biomarker, was detected in people's blood samples using a biosensor that uses electrochemistry that included aptamers and PPy nanowires. By measuring the levels of cancer-fighting drugs or their byproducts in physiological tissues or fluids, they evaluate the efficacy and durability of cancer therapy. The amount of the chemotherapy medication doxorubicin in human plasma samples was measured using a biosensor that is electrochemical and used PEDOT antibodies and nanofibers [40]. By measuring epigenetic or genetic alterations in body tissues or fluids, it evaluates the risk and prognosis of cancer [133].

PANI nanofibers and antibodies are used in an electrochemical biosensor study of breast cancer detection. In blood samples taken from fifty patients with breast cancer and fifty healthy controls, the biosensor was able to detect 3 breast cancer indicators (PR, ER, and HER2). By examining the expression levels of the biomarkers, the biosensor showed outstanding accuracy, sensitivity, and selectivity, enabling it to distinguish among many subtypes of breast cancer [134]. Researchers examined the identification of lung cancer using a biosensor that employs electrochemistry and integrates aptamers and PEDOT nanotubes [135]. In a study involving blood samples from 40 lung cancer patients and 40 healthy individuals, the biosensor was able to detect two lung cancer biomarkers: Cytokeratin 19 fragment and carcinoembryonic antigen. Through the analysis of biomarker concentration levels, the biosensor's high sensitivity, selectivity, and accuracy allowed for the identification of different stages of lung cancer [40, 136]. An inquiry into the detection of carcinoma of the cervical region using a biosensor that uses electrochemistry based on DNA probes and PTh nanowires. HPV (Human papillomavirus) DNA, a major cause of cervical cancer, was effectively detected by the biosensor in samples of cervical swabs taken from 30 patients with 30 healthy controls. By using DNA probe hybridization signals, the biosensor showed outstanding sensitivity, specificity, and accuracy in differentiating between many HPV types [1, 47, 137–141].

4.7 Challenges in MOF Cancer Biosensing

Conductive polymer-based electrochemical biosensors have advanced significantly, yet there are still obstacles and restrictions. These include contamination of the electrode surface due to non-specific interfering species adsorption, loss of biological activity in recognition probes, and problems considering the stability and repeatability of the biosensors as a result of conductive polymer degradation. The poor binding and the recognition probes' specificity, ineffective transmission of electrons between biological elements and conductive polymers, and electrochemical disruptions brought on by the matrix of samples are the usual limitations on biosensor performance. Because biosensors must be compatible with electrical and microfluidic components, have fine control over polymerization settings, and optimize immobilization processes, their fabrication and integration might be challenging.

There are several possible directions for further study and advancement in the area of electrochemical biosensors based on conductive polymers for the detection of cancer to get over the present obstacles and constraints:

- Creation of novel conductive polymers: Creating novel polymers that carry electricity and have improved stability, conductivity, biocompatibility, and functionalization is one exciting avenue. To create conductive polymers with better qualities for biosensing applications, researchers might investigate novel monomers and polymerization processes. To provide accurate and repeatable measurements, those polymers ought to demonstrate long-term stability, resilience, and suitability for biological specimens.
- Nanomaterial incorporation: Adding nanoparticles to conductive polymers presents an intriguing chance to improve their electrochemical characteristics. Conductive polymer matrices may be enhanced in electrocatalytic activity, loading capacity, and surface area by including nanoparticles, nanotubes, nanowires, and nanocomposites. The special qualities of nanomaterials may increase the sensitivity, selectivity, and response time of biosensors, which allows for the identification of low amounts of cancer biomarkers.
- Investigation of new probes for recognition: The biosensors' specificity and affinity might be greatly enhanced by sophisticated recognition probes. As substitute recognition elements, MIPs (molecularly imprinted polymers), aptamers, peptides, and nanobodies are becoming more popular. By precisely binding to

cancer biomarkers, these probes may be made to be more selective and less susceptible to interference from other molecules. Highly sensitive and selective biosensors may result from investigating and refining the combination of these identification conductive polymer-based probes.

- Use of advanced electrochemical methods: The biosensors based on conductive polymers' sensitivity and selectivity may be further improved by advanced electrochemical techniques. Techniques that may be used include EIS (Electrochemical impedance spectroscopy), SWV (square wave voltammetry), and DPV (differential pulse voltammetry). EIS makes it possible to detect variations in impedance, which provides details regarding the existence of other molecules and the kinetics of binding. The identification of minute levels of cancer biomarkers is made possible by the enhanced resolution and signal-to-noise ratios provided by SWV and DPV methods.
- Integration and downsizing into wearable technology: There are intriguing prospects for non-invasive, real-time cancer biomarker observing due to the shrinking and incorporation of biosensors into flexible and wearable gadgets like tattoos, contact lenses, and smart patches. Biomarkers from physiological fluids or tissues may be continually collected and analyzed by these devices. By incorporating conductive polymers into these devices, customized solutions for healthcare for early cancer identification, observation, and treatment may be created.

In conclusion, the manufacturing of novel conductive polymers, the use of nanoparticles, the investigation of innovative identification probes, the use of sophisticated electrochemical methods, and the incorporation and reduction in size of flexible and wearable gadgets should be the main areas of upcoming studies on electrochemical biosensors based on conductive polymers for the detection of cancer. By tackling these issues, scientists may go beyond the present constraints and create biosensors for cancer detection and treatment that are more practical, sensitive, and specific.

4.8 Conclusion

Conductive biosensors based on electrochemical polymers have enormous potential for cancer diagnosis. These biosensors are useful instruments for cancer diagnostics because of their many benefits, which include high sensitivity, cost, simplicity of manufacture, and suitability for different nanomaterials and recognition probes. The most recent developments in the creation, manufacturing, and use of biosensors that use electrochemistry based on conductive polymers for cancer biomarker detection, including proteins, cells, and nucleic acids, are highlighted in this paper. Significant advancements have been achieved in the use of polymers that conduct electricity to improve the consistency, sensitivity, and selectivity of biosensors. Direct contact with target analytes has been made easier by the inclusion of acknowledgment components into matrices of conductive polymers. This has improved electron transport, signal amplification, and control over the biosensor's surface charge and conductivity. The capabilities of cancer biosensors have been greatly enhanced by these developments. But there are still difficulties. Enhancing conductive polymers' durability and long-term functionality is essential to their useful use in biosensing. To improve the precision and dependability of conductive polymer-based biosensors, problems with cross-reactivity, non-specific binding, and selectivity must be resolved. Furthermore, incorporating these polymers into wearable and flexible biosensor systems has the potential to provide continuous health monitoring and individualized treatment, which calls for further study. Future prospects for advancement in electrochemical biosensors based on conductive polymers for the detection of cancer are promising. Future studies will concentrate on creating novel conductive polymers that have enhanced characteristics and investigating how to combine them with cutting-edge nanomaterials. These combinations may result in the creation of biosensors with increased sensitivity and selectivity, which would greatly enhance diagnostic performance. Furthermore, quick on-site cancer detection might be made possible by downsizing and integration into portable and point-of-care technologies, which would completely transform medical procedures. The function of conductive polymers in cancer detection biosensors is thoroughly covered in this article. By discussing recent advancements, present difficulties, and potential futures, we want to motivate scholars and experts in the subject.

References

1. M. Sadeghi, S. Sadeghi, S.M. Naghib, H.R. Garshasbi, A comprehensive review on electrochemical nano biosensors for precise detection of blood-based oncomarkers in breast cancer. Biosensors **13**(4), 481 (2023)
2. M.A. Farzin, S.M. Naghib, Based immunosensor integrated with bioinspired Cu-polydopamine nanozyme for voltammetric detection of CA-15-3 tumor marker. Sens. Actuators Rep. **8**, 100209 (2024)
3. F. Kazemi, S.M. Naghib, Y. Zare, K.Y. Rhee, Biosensing applications of polyaniline (PANI)-based nanocomposites: a review. Polym. Rev. **61**(3), 553–597 (2021). https://doi.org/10.1080/15583724.2020.1858871
4. O. Abazari, P. Dayati, M. Shahidi, J. ZavarReza, M. Rahmanian, S.M. Naghib, Exploring a desirable quadr-mRNAs panel for non-invasive and ultrasensitive bladder cancer diagnosis: in-silico and clinical studies. Curr. Med. Chem. (2024). https://doi.org/10.2174/0109298673302140240627053904
5. B. Mikaeeli Kangarshahi, S. Sojdeh, H. Daneshgar, M. Bagherzadeh, S.M. Naghib, N. Rabiee, Electroconductive polymer-based biosensors for early cancer detection via liquid biopsy: Advances, challenges, and future prospects. TrAC Trends Anal. Chem. **183**, 118062 (2025). https://doi.org/10.1016/j.trac.2024.118062
6. S.M. Naghib, E. Parnian, H. Keshvari, E. Omidinia, M. Eshghan-Malek, Synthesis, characterization and electrochemical evaluation of polyvinylalchol/graphene oxide/silver nanocomposites for glucose biosensing application. Int. J. Electrochem. Sci. **13**(1), 1013–1026 (2018). https://doi.org/10.20964/2018.01.74
7. S.M. Naghib, M. Rabiee, E. Omidinia, Electroanalytical validation of a novel nanobiosensing strategy and direct electrochemistry of phenylalanine dehydrogenase for clinical diagnostic

applications. Int. J. Electrochem. Sci. **9**(5), 2301–2315 (2014). https://doi.org/10.1016/S1452-3981(23)07928-2
8. S.M. Naghib, M. Rabiee, E. Omidinia, Electrochemical biosensor for L-phenylalanine based on a gold electrode modified with graphene oxide nanosheets and chitosan. Int. J. Electrochem. Sci. **9**(5), 2341–2353 (2014). https://doi.org/10.1016/S1452-3981(23)07931-2
9. S.M. Naghib, M. Rabiee, E. Omidinia, P. Khoshkenar, Investigation of a biosensor based on phenylalanine dehydrogenase immobilized on a polymer-blend film for phenylketonuria diagnosis. Electroanalysis **24**(2), 407–417 (2012). https://doi.org/10.1002/elan.201100391
10. S.M. Naghib, M. Rabiee, E. Omidinia, P. Khoshkenar, D. Zeini, Biofunctionalization of dextran-based polymeric film surface through enzyme immobilization for phenylalanine determination. Int. J. Electrochem. Sci. **7**(1), 120–135 (2012). https://doi.org/10.1016/S1452-3981(23)13325-6
11. M.A. Farzin, S.M. Naghib, Paper-based immunosensor integrated with bioinspired Cu-polydopamine nanozyme for voltammetric detection of CA-15-3 tumor marker. Sens. Actuators Rep. **8**, 100209 (2024). https://doi.org/10.1016/j.snr.2024.100209
12. M.A. Farzin, S.M. Naghib, Laccase-mimicking Cu@CDs-enabled bioinspired immunosensor using CuO nanoparticles for voltammetric detection of serum thyroxine level in patients with hyperthyroidism or hypothyroidism. Sens. Actuators Rep. **9**, 100291 (2025). https://doi.org/10.1016/j.snr.2025.100291
13. M.A. Farzin, S.M. Naghib, N. Rabiee, Bio-inspired and biomimetic composites based on biodegradable polymers for sensing applications with emphasis on early diagnosis of cancer. Chem. Eng. J. **493**, 152445 (2024). https://doi.org/10.1016/j.cej.2024.152445
14. M.A. Farzin, S.M. Naghib, N. Rabiee, Advancements in bio-inspired self-powered wireless sensors: materials, mechanisms, and biomedical applications. ACS Biomater. Sci. Eng. **10**(3), 1262–1301 (2024). https://doi.org/10.1021/acsbiomaterials.3c01633
15. S.M. Naghib, M. Bakhshi, B. Ahmadi, A. Bakhshi, *Gold Nanoparticles Based Antibacterial and Antiviral Functional Materials*, Antibacterial and Antiviral Functional Materials, vol. 2 (ACS Publications, 2024), pp. 221–256
16. B.M. Kangarshahi, S. Sojdeh, H. Daneshgar, M. Bagherzadeh, S.M. Naghib, N. Rabiee, Electroconductive polymer-based biosensors for early cancer detection via liquid biopsy: advances, challenges, and future prospects. TrAC Trends Anal. Chem. 118062 (2024)
17. H. Hasannezhad, A. Bakhshi, M.R. Mozafari, S.M. Naghib, A review of chitosan role in milk bioactive-based drug delivery, smart packaging and biosensors: recent advances and developments. Int. J. Biol. Macromol. **294**, 139248 (2025). https://doi.org/10.1016/j.ijbiomac.2024.139248
18. M. Hosseine, A. Bakhshi, S.M. Naghib, N. Rabiee, Recent advancements in polyaniline-based biosensors for diagnosis of cancers: a comprehensive review. TrAC Trends Anal. Chem. **181**, 118040 (2024). https://doi.org/10.1016/j.trac.2024.118040
19. M. Hosseine, S.M. Naghib, A. Khodadadi, Label-free electrochemical biosensor based on green-synthesized reduced graphene oxide/Fe_3O_4/nafion/polyaniline for ultrasensitive detection of SKBR3 cell line of HER2 breast cancer biomarker. Sci. Rep. **14**(1), 11928 (2024)
20. M. Hosseine, S.M. Naghib, N. Rabiee, Metal-organic frameworks: biosensor applications for diagnosis of cancers. Microchem. J. **208**, 112341 (2025). https://doi.org/10.1016/j.microc.2024.112341
21. E. Omidinia, S.M. Naghib, A. Boughdachi, P. Khoshkenar, D.K. Mills, Hybridization of silver nanoparticles and reduced graphene nanosheets into a nanocomposite for highly sensitive L-phenylalanine biosensing. Int. J. Electrochem. Sci. **10**(8), 6833–6843 (2015)
22. M.M. Karim, T. Lasker, Electrochemical biosensors for cancer biomarker detection: basic concept, design strategy and cutting-edge development. Electrochem. Sci. Adv. e70007 (2025)
23. H.A. Saputra, K.A. Jannath, K.B. Kim, D.-S. Park, Y.-B. Shim, Conducting polymer composite-based biosensing materials for the diagnosis of lung cancer: a review. Int. J. Biol. Macromol. **252**, 126149 (2023)
24. E. Burcu Aydın, M. Aydın, M. Kemal Sezgintürk, Biosensors and the evaluation of food contaminant biosensors in terms of their performance criteria. Int. J. Environ. Anal. Chem. **100**(5), 602–622 (2020)

25. A. Mohammadpour-Haratbar, Y. Zare, K.Y. Rhee, Electrochemical biosensors based on polymer nanocomposites for detecting breast cancer: recent progress and future prospects. Adv. Coll. Interface. Sci. **309**, 102795 (2022)
26. V. Sanko, F. Kuralay, Label-free electrochemical biosensor platforms for cancer diagnosis: recent achievements and challenges. Biosensors **13**(3), 333 (2023)
27. S. Joshi, K.A. Raj, M.R. Rao, R. Ghosh, An electronic biosensor based on semiconducting tetrazine polymer immobilizing matrix coated on rGO for carcinoembryonic antigen. Sci. Rep. **12**(1), 3006 (2022)
28. L. Shang, X.-H. Zhao, W. Zhang, L.-P. Jia, R.-N. Ma, Q.-W. Xue, H.-S. Wang, A.-X. Guo, L. Si, Graphene-PtPd nanocomposite for low-potential-driven electrochemiluminescent determination of carcinoembryonic antigen using $Ru(bpy)_3^{2+}$. Microchim. Acta **189**, 1–8 (2022)
29. A.B. Ganganboina, R.-A. Doong, Graphene quantum dots decorated gold-polyaniline nanowire for impedimetric detection of carcinoembryonic antigen. Sci. Rep. **9**(1), 7214 (2019)
30. O. Jalil, C.M. Pandey, D. Kumar, Electrochemical biosensor for the epithelial cancer biomarker EpCAM based on reduced graphene oxide modified with nanostructured titanium dioxide. Microchim. Acta **187**, 1–9 (2020)
31. S. Morteza Naghib, S. Mahdi Katebi, S. Ghorbanzade, Introduction to electrochemical biosensors, in *Electrochemical Biosensors in Practice: Material and Methods* (Bentham Science Publishers, 2023)
32. S. Morteza Naghib, S. Mahdi Katebi, S. Ghorbanzade, Electrochemical biosensors design steps, in *Electrochemical Biosensors in Practice: Material and Methods* (Bentham Science Publishers, 2023)
33. S. Morteza Naghib, S. Mahdi Katebi, S. Ghorbanzade, Material and biomaterial for biosensing platform, in *Electrochemical Biosensors in Practice: Material and Methods* (Bentham Science Publishers, 2023)
34. S. Morteza Naghib, S. Mahdi Katebi, S. Ghorbanzade, Basic practical principles for studying electro-chemical biosensor, in *Electrochemical Biosensors in Practice: Material and Methods* (Bentham Science Publishers, 2023)
35. S. Morteza Naghib, S. Mahdi Katebi, S. Ghorbanzade, Biosensor application, in *Electrochemical Biosensors in Practice: Material and Methods* (Bentham Science Publishers, 2023)
36. S.M. Naghib, S.M. Katebi, S. Ghorbanzade, *Electrochemical Biosensors in Practice: Materials and Methods* (Bentham Science Publishers, 2023)
37. S.M. Naghib, F. Behzad, M. Rahmanian, Y. Zare, K.Y. Rhee, A highly sensitive biosensor based on methacrylated graphene oxide-grafted polyaniline for ascorbic acid determination. Nanotechnol. Rev. **9**(1), 760–767 (2020)
38. B.M. Kangarshahi, S.M. Naghib, Label-free electrochemical nanobiosensors using Au-SPE for COVID-19 detection: a comparative review of different biomarkers and recognition elements. Curr. Anal. Chem. **20**(2), 90–108 (2024)
39. V. Naresh, N. Lee, A review on biosensors and recent development of nanostructured materials-enabled biosensors. Sensors **21**(4), 1109 (2021)
40. Z. Rahimzadeh, S.M. Naghib, Y. Zare, K.Y. Rhee, An overview on the synthesis and recent applications of conducting poly(3,4-ethylenedioxythiophene) (PEDOT) in industry and biomedicine. J. Mater. Sci. **55**(18), 7575–7611 (2020). https://doi.org/10.1007/s10853-020-04561-2
41. M. Khorshid, S.B. Sichani, P. Cornelis, G. Wackers, P. Wagner, The hot-wire concept: towards a one-element thermal biosensor platform. Biosens. Bioelectron. **179**, 113043 (2021)
42. L. Zhang, W. Guo, C. Lv, X. Liu, M. Yang, M. Guo, Q. Fu, Electrochemical biosensors represent promising detection tools in medical field. Adv. Sens. Energy Mater. **2**(4), 100081 (2023)
43. T. Chalklen, Q. Jing, S. Kar-Narayan, Biosensors based on mechanical and electrical detection techniques. Sensors **20**(19), 5605 (2020)

44. J. Full, Y. Baumgarten, L. Delbrück, A. Sauer, R. Miehe, Market perspectives and future fields of application of odor detection biosensors within the biological transformation—a systematic analysis. Biosensors **11**(3), 93 (2021)
45. O. Folorunso, P. Olukanmi, S. Thokozani, Conductive polymers' electronic structure modification for multifunctional applications. Mater. Today Commun. **35**, 106308 (2023)
46. O. Adil, New electrochemical platforms for the detection of neurodegenerative disease biomarkers, Southern Illinois University at Carbondale (2024)
47. R. Salahandish, A. Ghaffarinejad, S.M. Naghib, A. Niyazi, K. Majidzadeh-A, M. Janmaleki, A. Sanati-Nezhad, Sandwich-structured nanoparticles-grafted functionalized graphene based 3D nanocomposites for high-performance biosensors to detect ascorbic acid biomolecule. Sci. Rep. **9**(1), 1226 (2019). https://doi.org/10.1038/s41598-018-37573-9
48. N.F. Vahid, M.R. Marvi, M.R. Naimi-Jamal, S.M. Naghib, A. Ghaffarinejad, X-Fe_2O_4-buckypaper-chitosan nanocomposites for nonenzymatic electrochemical glucose biosensing. Anal. Bioanal. Electrochem **11**(2019), 930–942 (2019)
49. Y. Yu, H.Y.Y. Nyein, W. Gao, A. Javey, Flexible electrochemical bioelectronics: the rise of in situ bioanalysis. Adv. Mater. **32**(15), 1902083 (2020)
50. M.M. Calabretta, A. Lopreside, L. Montali, M. Zangheri, L. Evangelisti, M. D'Elia, E. Michelini, Portable light detectors for bioluminescence biosensing applications: a comprehensive review from the analytical chemist's perspective. Anal. Chim. Acta **1200**, 339583 (2022)
51. P. Singh, K. Sharma, I. Puchades, P.B. Agarwal, A comprehensive review on MEMS-based viscometers. Sens. Actuators, A **338**, 113456 (2022)
52. N. Alanazi, M. Almutairi, A.N. Alodhayb, A review of quartz crystal microbalance for chemical and biological sensing applications. Sens. Imaging **24**(1), 10 (2023)
53. X. Du, Y. Li, Z. Zhang, C. Zhang, J. Hu, X. Wang, R. Zhang, J. Yang, L. Zhou, H. Zhang, An electrochemical biosensor for the assessment of tumor immunotherapy based on the detection of immune checkpoint protein programmed death ligand-1. Biosens. Bioelectron. **207**, 114166 (2022)
54. S. Yu, S. Chen, Y. Dang, Y. Zhou, J.-J. Zhu, An ultrasensitive electrochemical biosensor integrated by nicking endonuclease-assisted primer exchange reaction cascade amplification and DNA nanosphere-mediated electrochemical signal-enhanced system for microRNA detection. Anal. Chem. **94**(41), 14349–14357 (2022)
55. L. Kong, Z. Han, M. Zhao, X. Zhang, Y. Zhuo, Y. Chai, Z. Li, R. Yuan, Versatile electrochemical biosensor based on the target-controlled capture and release of DNA nanotubes for the ultrasensitive detection of multiplexed biomarkers. Anal. Chem. **94**(32), 11416–11424 (2022)
56. N. Liu, H. Lu, L. Liu, W. Ni, Q. Yao, G.-J. Zhang, F. Yang, Ultrasensitive exosomal microRNA detection with a supercharged DNA framework nanolabel. Anal. Chem. **93**(14), 5917–5923 (2021)
57. B. Sajeevan, M. Gopika, Sreelekshmi, R. Rejithammol, S. Antherjanam, B. Saraswathyamma, Biosensors for salivary biomarker detection of cancer and neurodegenerative diseases, in *Biosensors Nanotechnology* (2023), pp. 215–244
58. N. Abhishek, A. Verma, A. Singh, T. Kumar, Metal-conducting polymer hybrid composites: a promising platform for electrochemical sensing. Inorg. Chem. Commun. **157**, 111334 (2023)
59. Z. Štukovnik, R. Fuchs-Godec, U. Bren, Nanomaterials and their recent applications in impedimetric biosensing. Biosensors **13**(10), 899 (2023)
60. T. Aubert, A.A. Golovatenko, M. Samoli, L. Lermusiaux, T. Zinn, B. Abécassis, A.V. Rodina, Z. Hens, General expression for the size-dependent optical properties of quantum dots. Nano Lett. **22**(4), 1778–1785 (2022)
61. Y. Tao, L. Chen, M. Pan, F. Zhu, D. Zhu, Tailored biosensors for drug screening, efficacy assessment, and toxicity evaluation. ACS Sens. **6**(9), 3146–3162 (2021)
62. A. Haleem, M. Javaid, R.P. Singh, R. Suman, S. Rab, Biosensors applications in medical field: a brief review. Sens. Int. **2**, 100100 (2021)

63. L.K. Povlich, J.C. Cho, M.K. Leach, J.M. Corey, J. Kim, D.C. Martin, Synthesis, copolymerization and peptide-modification of carboxylic acid-functionalized 3,4-ethylenedioxythiophene (EDOTacid) for neural electrode interfaces. Biochim. Biophys. Acta (BBA)-General Subj. **1830**(9), 4288–4293 (2013)
64. A. Markham, D. Faulds, Roxithromycin: an update of its antimicrobial activity, pharmacokinetic properties and therapeutic use. Drugs **48**, 297–326 (1994)
65. T. Nezakati, A. Seifalian, A. Tan, A.M. Seifalian, Conductive polymers: opportunities and challenges in biomedical applications. Chem. Rev. **118**(14), 6766–6843 (2018)
66. J.B. Kaushal, P. Raut, S. Kumar, Organic electronics in biosensing: a promising frontier for medical and environmental applications. Biosensors **13**(11), 976 (2023)
67. A.K. Poddar, S.S. Patel, H.D. Patel, Synthesis, characterization and applications of conductive polymers: a brief review. Polym. Adv. Technol. **32**(12), 4616–4641 (2021)
68. J.M. Marin-Beloqui, S. Gómez, H.I. Gonev, M. Comí, M. Al-Hashimi, T.M. Clarke, Truncated conjugation in fused heterocycle-based conducting polymers: when greater planarity does not enhance conjugation. Chem. Sci. **14**(4), 812–821 (2023)
69. S. Vallem, S. Venkateswarlu, S. Song, J. Kim, Y. Li, J. Bae, Flexible MXene-conjugated polymer nanoarchitectures: current developments and future frontiers in battery technology. Coord. Chem. Rev. **510**, 215778 (2024)
70. H. Kim, Y. Won, H.W. Song, Y. Kwon, M. Jun, J.H. Oh, Organic mixed ionic–electronic conductors for bioelectronic sensors: materials and operation mechanisms. Adv. Sci. **11**(27), 2306191 (2024)
71. M. Fuseini, M.M.Y. Zaghloul, M.F. Elkady, A.H. El-Shazly, Evaluation of synthesized polyaniline nanofibres as corrosion protection film coating on copper substrate by electrophoretic deposition. J. Mater. Sci. **57**(10), 6085–6101 (2022)
72. A.K. Nandi, A review on self-assembly driven optoelectronic properties of polythiophene-peptide and polythiophene-polymer conjugates. Langmuir **40**(18), 9385–9405 (2024)
73. H. Chen, J. Fraser Stoddart, From molecular to supramolecular electronics. Nat. Rev. Mater. **6**(9), 804–828 (2021)
74. P. Dubey, V. Shrivastav, T. Boruah, G. Zoppellaro, R. Zbořil, A. Bakandritsos, S. Sundriyal, Unveiling the potential of covalent organic frameworks for energy storage: developments, challenges, and future prospects. Adv. Energy Mater. **14**(24), 2400521 (2024)
75. S.B. Abel, E. Frontera, D. Acevedo, C.A. Barbero, Functionalization of conductive polymers through covalent postmodification. Polymers **15**(1), 205 (2022)
76. S. Dowlatshahi, M.J. Abdekhodaie, Electrochemical prostate-specific antigen biosensors based on electroconductive nanomaterials and polymers. Clin. Chim. Acta **516**, 111–135 (2021)
77. J. Kappen, M. Skorupa, K. Krukiewicz, Conducting polymers as versatile tools for the electrochemical detection of cancer biomarkers. Biosensors **13**(1), 31 (2023)
78. S. Sharma, A. Verma, S.M. Rangappa, S. Siengchin, S. Ogata, Recent progressive developments in conductive-fillers based polymer nanocomposites (CFPNC's) and conducting polymeric nanocomposites (CPNC's) for multifaceted sensing applications. J. Market. Res. **26**, 5921–5974 (2023)
79. X. Cao, Y. Han, Control over the aggregated structure of donor–acceptor conjugated polymer films for high-mobility organic field-effect transistors. Aggregate **5**(3), e501 (2024)
80. A. Marino, M. Battaglini, A. Desii, C. Lavarello, G. Genchi, A. Petretto, G. Ciofani, Liposomes loaded with polyphenol-rich grape pomace extracts protect from neurodegeneration in a rotenone-based in vitro model of Parkinson's disease. Biomater. Sci. **9**(24), 8171–8188 (2021)
81. X. Wang, Y. Tang, S. Cheng, Q. Gao, Y. Yuan, A. Li, S. Guan, PDMS-based conductive elastomeric composite with 3D reduced graphene oxide conductive network for flexible strain sensor. Compos. Part A: Appl. Sci. Manufact. **161**, 107113 (2022). https://doi.org/10.1016/j.compositesa.2022.107113
82. G. Nuroldayeva, M.P. Balanay, Flexing the spectrum: advancements and prospects of flexible electrochromic materials. Polymers **15**(13), 2924 (2023)

83. A. Mamadalimov, N. Khakimova, S.M. Norbekov, A.K. Yunusov, D. Rai, Study of electrophysical and optical properties of cotton fibers irradiated in γ-60Co radiation source. Fibers Polym. **25**(2), 535–541 (2024)
84. D.K. Pathak, H.C. Moon, Recent progress in electrochromic energy storage materials and devices: a minireview. Mater. Horiz. **9**(12), 2949–2975 (2022)
85. N. Singh, U. Riaz, Recent trends on synthetic approaches and application studies of conducting polymers and copolymers: a review. Polym. Bull. **79**(12), 10377–10408 (2022)
86. V. Harish, M. Ansari, D. Tewari, A.B. Yadav, N. Sharma, S. Bawarig, M.-L. García-Betancourt, A. Karatutlu, M. Bechelany, A. Barhoum, Cutting-edge advances in tailoring size, shape, and functionality of nanoparticles and nanostructures: a review. J. Taiwan Inst. Chem. Eng. **149**, 105010 (2023)
87. Y. Chen, W. Ren, T. Ma, N. Ren, S. Wang, X. Duan, Transformative removal of aqueous micropollutants into polymeric products by advanced oxidation processes. Environ. Sci. Technol. **58**(11), 4844–4851 (2024)
88. T.V. Nguyen, Q.V. Le, S. Peng, Z. Dai, S.H. Ahn, S.Y. Kim, Exploring conducting polymers as a promising alternative for electrochromic devices. Adv. Mater. Technol. **8**(18), 2300474 (2023)
89. Q. Liu, L. Wang, X. He, Toward practical solid-state polymer lithium batteries by in situ polymerization process: a review. Adv. Energy Mater. **13**(30), 2300798 (2023)
90. A. Buonerba, A. Grassi, Trends in sustainable synthesis of organics by gold nanoparticles embedded in polymer matrices. Catalysts **11**(6), 714 (2021)
91. K. Hema, A. Ravi, C. Raju, J.R. Pathan, R. Rai, K.M. Sureshan, Topochemical polymerizations for the solid-state synthesis of organic polymers. Chem. Soc. Rev. **50**(6), 4062–4099 (2021)
92. C.M.Q. Le, F. Morlet-Savary, A. Chemtob, Role of thiol oxidation by air in the mechanism of the self-initiated thermal thiol–ene polymerization. Polym. Chem. **12**(45), 6594–6605 (2021)
93. J. Park, A. Kim, B.-S. Kim, Anionic ring-opening polymerization of functional epoxide monomers in the solid state. Nat. Commun. **14**(1), 5855 (2023)
94. H. Yang, K. Shi, X. Dong, J. Zhang, Z. Du, P. Qiao, X. Zhang, J. Kong, S. Wang, R. Hu, Programmable compressing and decompressing for controllable polymerization: the case of methyl methacrylate. J. Phys. Chem. C **127**(34), 16929–16937 (2023)
95. P. Shokrollahi, Y. Omidi, L.X. Cubeddu, H. Omidian, Conductive polymers for cardiac tissue engineering and regeneration. J. Biomed. Mater. Res. B Appl. Biomater. **111**(11), 1979–1995 (2023)
96. Y. Gu, Y. Li, Q. Wu, Z. Wu, L. Sun, Y. Shang, Y. Zhuang, X. Fan, L. Yi, S. Wang, Chemical antifouling strategies in sensors for food analysis: a review. Compr. Rev. Food Sci. Food Saf. **22**(5), 4074–4106 (2023)
97. T. Dufour, From basics to frontiers: a comprehensive review of plasma-modified and plasma-synthesized polymer films. Polymers **15**(17), 3607 (2023)
98. P. Hariharan, S. Sundarrajan, G. Arthanareeswaran, S. Seshan, D.B. Das, A. Ismail, Advancements in modification of membrane materials over membrane separation for biomedical applications—review. Environ. Res. **204**, 112045 (2022)
99. M.G. Sumdani, M.R. Islam, A.N.A. Yahaya, S.I. Safie, Recent advancements in synthesis, properties, and applications of conductive polymers for electrochemical energy storage devices: a review. Polym. Eng. Sci. **62**(2), 269–303 (2022)
100. K. Lovato, P.S. Fier, K.M. Maloney, The application of modern reactions in large-scale synthesis. Nat. Rev. Chem. **5**(8), 546–563 (2021)
101. P.H. Reddy, J. Amalraj, S. Ranganatha, S.S. Patil, S. Chandrasekaran, A review on effect of conducting polymers on carbon-based electrode materials for electrochemical supercapacitors. Synth. Met. **298**, 117447 (2023)
102. R. Hsissou, R. Seghiri, Z. Benzekri, M. Hilali, M. Rafik, A. Elharfi, Polymer composite materials: a comprehensive review. Compos. Struct. **262**, 113640 (2021)
103. G. Heidari, M. Ghomi, Fabrication and morphology of composites based on conducting polymers, in *Electrically Conducting Polymers and Their Composites for Tissue Engineering* (ACS Publications, 2023), pp. 29–49

104. R.-Y. Li, X. Yang, J. Li, Y.-N. Shen, L.-X. Zhang, R. Lu, C. Wang, X.-H. Zheng, H.-S. Chen, T. Zhang, Review on polymer composites with high thermal conductivity and low dielectric properties for electronic packaging. Mater. Today Phys. **22**, 100594 (2022)
105. J.R. Sempionatto, J.A. Lasalde-Ramírez, K. Mahato, J. Wang, W. Gao, Wearable chemical sensors for biomarker discovery in the omics era. Nat. Rev. Chem. **6**(12), 899–915 (2022)
106. S.M. Naghib, Two-dimensional functionalised methacrylated graphene oxide nanosheets as simple and inexpensive electrodes for biosensing applications. Micro Nano Lett. **14**(4), 462–465 (2019). https://doi.org/10.1049/mnl.2018.5320
107. Z. Rahimzadeh, S.M. Naghib, E. Askari, F. Molaabasi, A. Sadr, Y. Zare, M. Afsharpad, K.Y. Rhee, A rapid nanobiosensing platform based on herceptin-conjugated graphene for ultrasensitive detection of circulating tumor cells in early breast cancer. Nanotechnol. Rev. **10**(1), 744–753 (2021)
108. M. Sadeghi, S. Kashanian, S.M. Naghib, E. Arkan, A high-performance electrochemical aptasensor based on graphene-decorated rhodium nanoparticles to detect HER2-ECD oncomarker in liquid biopsy. Sci. Rep. **12**(1), 3299 (2022). https://doi.org/10.1038/s41598-022-07230-3
109. S. Lanzalaco, B.G. Molina, Polymers and plastics modified electrodes for biosensors: a review. Molecules **25**(10), 2446 (2020)
110. M. Sadeghi, S. Kashanian, S.M. Naghib, F. Haghiralsadat, D. Tofighi, An efficient electrochemical biosensor based on pencil graphite electrode mediated by 2D functionalized graphene oxide to detect HER2 breast cancer biomarker. Int. J. Electrochem. Sci. **17**(4), 220459 (2022). https://doi.org/10.20964/2022.04.62
111. Y. Wang, A. Liu, Y. Han, T. Li, Sensors based on conductive polymers and their composites: a review. Polym. Int. **69**(1), 7–17 (2020)
112. S. Sharma, P. Sudhakara, A.A.B. Omran, J. Singh, R. Ilyas, Recent trends and developments in conducting polymer nanocomposites for multifunctional applications. Polymers **13**(17), 2898 (2021)
113. L. Guo, Z. Mu, B. Yan, J. Wang, J. Zhou, L. Bai, A novel electrochemical biosensor for sensitive detection of non-small cell lung cancer ctDNA using NG-PEI-COFTAPB-TFPB as sensing platform and Fe-MOF for signal enhancement. Sens. Actuators, B **350**, 130874 (2022). https://doi.org/10.1016/j.snb.2021.130874
114. S. Mukherjee, A. Mukherjee, Z. Bytesnikova, A.M. Ashrafi, L. Richtera, V. Adam, 2D graphene-based advanced nanoarchitectonics for electrochemical biosensors: applications in cancer biomarker detection. Biosens. Bioelectron. **250**, 116050 (2024). https://doi.org/10.1016/j.bios.2024.116050
115. K. Chen, H. Zhao, Z. Wang, M. Lan, Three-dimensional graphene-like homogeneous carbon architecture loaded with gold-platinum for the electrochemical detection of circulating tumor DNA. Mater. Today Chem. **24**, 100892 (2022)
116. G. Liu, X. Zhang, X. Chen, Y. He, L. Cheng, M. Huo, J. Yin, F. Hao, S. Chen, P. Wang, Additive manufacturing of structural materials. Mater. Sci. Eng. R. Rep. **145**, 100596 (2021)
117. S. Jose, A. Varghese, Design and structural characteristics of conducting polymer-metal organic framework composites for energy storage devices. Synth. Met. **297**, 117421 (2023)
118. S.W. Thomas, R.R. Khan, K. Puttananjegowda, W. Serrano-Garcia, Conductive polymers and metal oxide polymeric composites for nanostructures and nanodevices, in *Advances in Nanostructured Materials and Nanopatterning Technologies* (Elsevier, 2020), pp. 243–271
119. A. Shaver, K. Mallires, J. Harris, J. Kavner, B. Wang, R. Gottlieb, J. Lión-Villar, M.Á. Herranz, N. Martín, N. Arroyo-Currás, Survey of conductive polymers for the fabrication of conformation switching nucleic acid-based electrochemical biosensors. ACS Appl. Polym. Mater. **6**(1), 541–551 (2023)
120. H. Zhang, G. Rong, S. Bian, M. Sawan, Lab-on-chip microsystems for ex vivo network of neurons studies: a review. Front. Bioeng. Biotechnol. **10**, 841389 (2022)
121. P.K. Sharma, Kamini, A. Jain, V.K. Shukla, Innovative biomedical equipment for diagnosis of cancer, in *Targeted Cancer Therapy in Biomedical Engineering* (Springer, 2023), pp. 405–436

122. F. Karimi, H. Karimi-Maleh, J. Rouhi, N. Zare, C. Karaman, M. Baghayeri, L. Fu, S. Rostamnia, E.N. Dragoi, A. Ayati, Revolutionizing cancer monitoring with carbon-based electrochemical biosensors. Environ. Res. **239**, 117368 (2023)
123. V.K. Sarhadi, G. Armengol, Molecular biomarkers in cancer. Biomolecules **12**(8), 1021 (2022)
124. E. Demir, K.K. Kırboga, M. Işık, An overview of stability and lifetime of electrochemical biosensors, in *Novel Nanostructured Materials for Electrochemical Bio-sensing Applications* (2024), pp. 129–158
125. L. Vázquez-Iglesias, G.M.S. Casagrande, D. García-Lojo, L.F. Leal, T.A. Ngo, J. Pérez-Juste, R.M. Reis, K. Kant, I. Pastoriza-Santos, SERS sensing for cancer biomarker: approaches and directions. Bioactive Mater. **34**, 248–268 (2024)
126. Y. Xue, X. Feng, X. Fan, G. Zhu, J. McLaughlan, W. Zhang, X. Chen, Extracellular vesicles for the diagnosis of cancers. Small Struct. **3**(1), 2100096 (2022)
127. J.S. Stefano, L.R.G. e Silva, R.G. Rocha, L.C. Brazaca, E.M. Richter, R.A.A. Muñoz, B.C. Janegitz, New conductive filament ready-to-use for 3D-printing electrochemical (bio)sensors: towards the detection of SARS-CoV-2. Anal. Chim. Acta **1191**, 339372 (2022)
128. B.M. Kangarshahi, A. Beigi Javazm, S.M. Naghib, CRISPR-integrated metal-organic frameworks for biosensing applications: recent advances and future perspective. Sens. Bio-sens. Res. **47**, 100736 (2025). https://doi.org/10.1016/j.sbsr.2025.100736
129. B.M. Kangarshahi, S.M. Naghib, Nanogenosensors based on aptamers and peptides for bioelectrochemical cancer detection: an overview of recent advances in emerging materials and technologies. Discover Appl. Sci. **6**(2), 47 (2024). https://doi.org/10.1007/s42452-024-05681-z
130. B. Mikaeeli Kangarshahi, N.S. Morteza, N. and Rabiee, DNA/RNA-based electrochemical nanobiosensors for early detection of cancers. Crit. Rev. Clin. Lab. Sci. **61**(6), 473–495 (2024). https://doi.org/10.1080/10408363.2024.2321202
131. J. Luo, D. Liang, D. Zhao, M. Yang, Photoelectrochemical detection of circulating tumor cells based on aptamer conjugated Cu_2O as signal probe. Biosens. Bioelectron. **151**, 111976 (2020). https://doi.org/10.1016/j.bios.2019.111976
132. S.-T. Liu, X.-P. Liu, J.-S. Chen, C.-J. Mao, B.-K. Jin, Highly sensitive photoelectrochemical biosensor for microRNA159c detection based on a Ti_3C_2: CdS nanocomposite of breast cancer. Biosens. Bioelectron. **165**, 112416 (2020)
133. X. Zhang, X. Tan, P. Wang, J. Qin, Application of polypyrrole-based electrochemical biosensor for the early diagnosis of colorectal cancer. Nanomaterials **13**(4), 674 (2023)
134. S. Agrahari, R.K. Gautam, A.K. Singh, I. Tiwari, Nanoscale materials-based hybrid frameworks modified electrochemical biosensors for early cancer diagnostics: an overview of current trends and challenges. Microchem. J. **172**, 106980 (2022)
135. A. Khanmohammadi, A. Aghaie, E. Vahedi, A. Qazvini, M. Ghanei, A. Afkhami, A. Hajian, H. Bagheri, Electrochemical biosensors for the detection of lung cancer biomarkers: a review. Talanta **206**, 120251 (2020)
136. R. Chinnappan, T.A. Mir, S. Alsalameh, T. Makhzoum, A. Alzhrani, K. Alnajjar, S. Adeeb, N. Al Eman, Z. Ahmed, I. Shakir, Emerging biosensing methods to monitor lung cancer biomarkers in biological samples: a comprehensive review. Cancers **15**(13), 3414 (2023)
137. A. Singh, A. Sharma, A. Ahmed, A.K. Sundramoorthy, H. Furukawa, S. Arya, A. Khosla, Recent advances in electrochemical biosensors: applications, challenges, and future scope. Biosensors **11**(9), 336 (2021)
138. R. Salahandish, A. Ghaffarinejad, S.M. Naghib, K. Majidzadeh-A, A. Sanati-Nezhad, A novel graphene-grafted gold nanoparticles composite for highly sensitive electrochemical biosensing. IEEE Sens. J. **18**(6), 2513–2519 (2018)
139. R. Salahandish, A. Ghaffarinejad, S.M. Naghib, K. Majidzadeh-A, H. Zargartalebi, A. Sanati-Nezhad, Nano-biosensor for highly sensitive detection of HER2 positive breast cancer. Biosens. Bioelectron. **117**, 104–111 (2018). https://doi.org/10.1016/j.bios.2018.05.043
140. R. Salahandish, A. Ghaffarinejad, E. Omidinia, H. Zargartalebi, K. Majidzadeh-A, S.M. Naghib, A. Sanati-Nezhad, Label-free ultrasensitive detection of breast cancer miRNA-21 biomarker employing electrochemical nano-genosensor based on sandwiched AgNPs in PANI

and N-doped graphene. Biosens. Bioelectro. **120**, 129–136 (2018). https://doi.org/10.1016/j.bios.2018.08.025

141. A.-A. Zare, H. Naderi-Manesh, S.M. Naghib, M. Shamsipur, F. Molaabasi, Label-free electrochemical cancer cell detection leveraging hemoglobin-encapsulated silver nanoclusters and Cu-MOF nanohybrids on a graphene-assisted dual-modal probe. Sci. Rep. **13**(1), 21980 (2023). https://doi.org/10.1038/s41598-023-49418-1

Chapter 5
Nucleic Acid-Based Nanobiosensors for Geno-electrochemical Cancer Detection: A Summary of Current Developments in Cutting-Edge Materials and Technologies

Abstract Electrochemical biosensors based on nucleic acids, have shown significant promise for cancer detection due to their cheap cost, quick response, increased sensitivity and ease of use. This chapter will demonstrate the impact of current developments and uses of biosensors that rely on electrochemistry and nucleic acids for cancer identification. I present a few examples of electrochemical biosensors designed to detect cancer using nucleic acids, such as those that use DNA or RNA aptamers, carcinoembryonic antigens, microRNA-21 and prostate-specific antigens, and I compare them with antibody-based models. I go over the benefits and limitations of these biosensors as well as the difficulties they encounter including standardization, interference, stability and repeatability. I also outline some potential avenues for future research and development, including the creation of novel various RNA and DNA recognition elements, the investigation of novel transducer materials and configurations, the design of novel signal amplification techniques, the integration of electrochemical biosensors with portable instruments or microfluidic devices and the clinical evaluation of electrochemical biosensors using real samples from healthy donors or cancer patients. All things considered, I think that biosensors that use electrochemistry based on nucleic acids, provide a fortunate substitute for traditional cancer detection techniques and have significant promise for aiding in cancer early diagnosis and successful cancer therapy.

Keywords Nanomaterials · Genosensors · Peptides · Aptamers · Nucleic acids · Cancer detection · Electrochemical sensing · Biosensors · Biomarkers

5.1 Introduction

A major global issue that affects millions of people and contributes to substantial morbidity and death is cancer. According to WHO data, cancer remains among the primary causes of mortality, with an expected 10 million deaths in 2020. Many malignancies are still difficult to cure and treat because of their diversity, resistance, and spread, even with advancements in traditional treatments like surgery, chemotherapy,

S. M. Naghib, *Nano-Biosensors, Drug Delivery and Tissue Engineering*, Series in BioEngineering, https://doi.org/10.1007/978-981-95-1888-3_5

and radiation. Thus, developing new and efficient methods for cancer detection and treatment is essential [1–5]. Using aspects of molecular recognition that may bind to target molecules specifically linked to cancer cells, like biomarkers, antigens and receptors, is one potential cancer treatment strategy [6–8]. These components may be used for a number of tasks, including immune response modulation, anticancer drug delivery, cancer cell growth inhibition, and cancer cell detection [9–12]. Because of their great target accuracy, aptamers and antibodies have garnered a lot of interest among the many kinds of molecular recognition components [13–16]. The immune system proteins known as antibodies are able to identify and attach to a particular antigen. They have been extensively used, either alone or in conjunction with other modalities, as diagnostic and therapeutic techniques in the treatment of cancer [4, 17–20]. For example, we can use antibodies to deliver anticancer agents through antibody-drug conjugates or immunotoxins, identify cancer cells using flow cytometry or immunohistochemistry, and halt cancer cell growth. Oncoproteins or growth factors can be inhibited, and immune checkpoint activation or suppression can modify immune responses [2, 21, 22]. Antibodies do, however, have several drawbacks, including immunogenicity, complicated manufacture, high cost, toxicity, instability, and trouble entering tissues and tumors. Aptamers are molecules of RNA or DNA that possess the capacity to take on remarkable forms in three dimensions and firmly bind to a specific target molecule [1, 5, 23, 24]. They are made without the use of animals or cell cultures using the SELEX in vitro selection method. Contrary to antibodies, aptamers offer a number of benefits, including being tiny, stable, simple to change, low or non-immunogenic, cheap, and simple to manufacture. Aptamers are suitable for a number of uses in cancer detection and treatment because of their properties [25, 26].

A nanobiosensor is a nanoscaled analytical biodevice that detects the presence/concentration of biological or chemical substances by combining a biological component with a physicochemical detector [27–31]. A nanogenosensor is a specialized type of nanobiosensor designed to determine specific nucleic acid sequences such as DNA or RNA through the principle of molecular recognition [32–37]. Nucleic acid-based electrochemical biosensors are instruments that use the electrical characteristics of nucleic acids to identify certain biomolecules linked to illnesses like cancer. Nucleic acids like DNA and RNA, which may also serve as components of recognition that attach to target molecules with great specificity and affinity, are the building blocks of genetic information [38–42]. When various RNA and DNA attach to targets, biosensors that use electrochemistry may detect changes in current, voltage, or impedance. Examples of biosensors that use electrochemistry based on various RNA and DNA detection to identify cancer include RNA aptamers with graphene oxide to detect prostate-specific antigens, DNAzymes on gold nanoparticles to detect microRNA-21, and RNA aptamers with magnetic beads to detect carcinoembryonic antigen [43–47]. While these biosensors provide advantages like affordability, high sensitivity, fast reaction times, and ease of use, they also face challenges including standardization, interference, stability, and repeatability. Single strands of DNA or RNA known as various RNA and DNA aptamers have the ability to precisely bind to certain target molecules [36, 48–50]. Once found, an aptamer may be used as an identification element in cancer detection biosensors. When aptamers attach to their target molecule, they immobilize on a transducer surface and alter the transducer signal, which may be detected and measured. Aptamers offer a number of advantages over antibodies, including easier and quicker synthesis and modification, reduced immunogenicity and toxicity, improved stability and refoldability, and the ability to bind to sites that antibodies are unable to or have trouble recognizing. They are susceptible to nuclease degradation and clearance in vivo, have a lower binding affinity and specificity than antibodies, and have trouble passing through biological barriers and getting to intracellular targets, among other drawbacks [51, 52]. A schematic illustration of electrochemical nucleic acid-based genosensors based is shown in Fig. 5.1 [23].

Because of their great affinity and specificity for target molecules, aptamers hold great promise for cancer treatment. They may be employed in immunotherapy, cancer diagnostics, cancer treatment, and targeted drug delivery either by themselves or in combination with other molecules. Clinical studies have investigated AS1411's ability to cure a variety of malignancies by inducing apoptosis and inhibiting angiogenesis in cancer cells. Another such is the RNA aptamer NOX-A12, which binds to the chemokine CXCL12, which facilitates the spread and metastasis of cancer. NOX-A12 may strengthen the immune response against cancer cells by preventing CXCL12 from interacting with its receptor, CXCR4. Clinical trials of NOX-A12 are now being conducted to treat solid tumors and hematological malignancies.

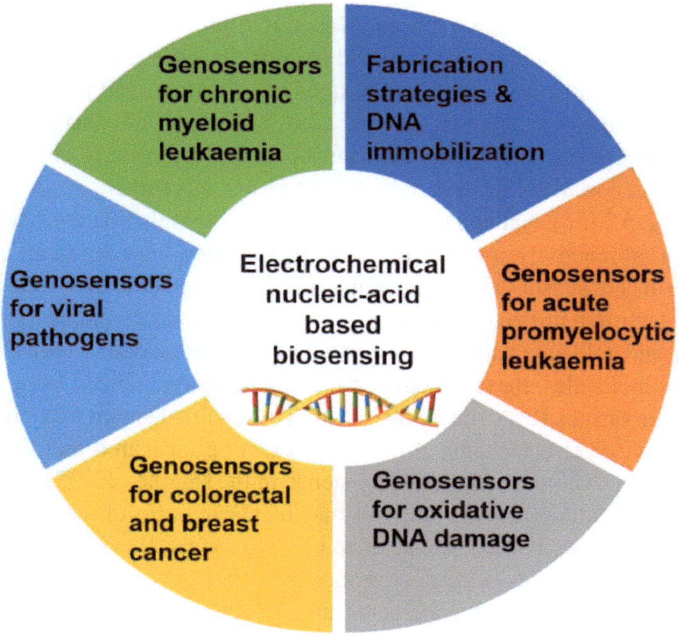

Fig. 5.1 Electrochemical nucleic acid-based genosensors: a promising approach for early cancer detection and biomolecular targeting [23]

Aptamer-drug conjugates (ApDCs), such as aptamer-chemical linker conjugates, aptamer-drug intercalation conjugates, and aptamer-nucleotide analog conjugates, are also used in cancer treatment. ApDCs may improve the pharmacokinetics and biodistribution of cancer medications, decrease their toxicity and adverse effects, and increase their specificity and effectiveness. For instance, AS1411-Dox is the result of conjugating AS1411 with the chemotherapy medication doxorubicin. In response to nucleolin binding, AS1411-Dox may release doxorubicin and specifically target cancer cells, increasing cytotoxicity and decreasing systemic toxicity [53, 54]. Antibodies and aptamers are molecular recognition components that have a high degree of precision in attaching to precise target molecules. Size, immunogenicity, stability, synthesis, binding affinity, and specificity are some of the variations that might impact their effectiveness and suitability for cancer treatment. Overall, there is a lot of promise for cancer diagnosis and treatment using electrochemical genosensors based on nucleic acids and aptamers, and more study in this area might result in new and better cancer patient diagnostic and treatment approaches. Immune system proteins known as antibodies are able to identify and attach to a particular antigen. They have been extensively used, either alone or in conjunction with other modalities, as both therapeutic and diagnostic techniques in the treatment of cancer. Antibodies do, however, have several drawbacks, including high cost, complicated manufacture, immunogenicity, toxicity, instability, and trouble entering tissues and tumors. Monostranded molecules of RNA or DNA known as aptamers possess the capacity to develop unique 3D structures that attach more specifically and with more affinity to a particular target molecule. They are created using a selection in vitro process called SELEX, which eliminates the need for cell cultures or animals. The advantages of aptamers over antibodies are many and include their low cost, ease of synthesis, little or no immunogenicity, stability, ease of modification, and tiny size. Because of these characteristics, aptamers may be used in a variety of cancer detection and treatment applications [55]. The nanomaterials often employed to develop bioelectrochemical genosensors to identify different tumor biomarkers are schematically shown in Fig. 5.2 [56]. Nanomaterials' unique qualities, including their huge surface area, enhanced reactivity, and tunable surface chemistry, have led to an increase in their usage in genosensor design in recent years. The image illustrates many nanomaterials that are often utilized to build genosensors, including graphene oxide, carbon nanotubes, and gold nanoparticles. To allow for the precise identification and binding of target biomolecules, these nanomaterials may be functionalized using aptamers, antibodies, or various RNA and DNA probes. The resultant signal may be converted into an optical or electrochemical output that can be measured and linked to the tumor biomarker's concentration and existence in the sample [23, 57].

This chapter will discuss the latest advancements and uses of biosensors that use electrochemistry based on nucleic acids to identify cancer. Electrochemical biosensors use the nucleic acids' electrical properties to detect certain proteins linked to illnesses like cancer. When nucleic acids attach to their targets, changes in current, voltage, or impedance may be detected using electrochemical biosensors. The advantages and disadvantages of nucleic acid-based electrochemical biosensors and those based on antibodies will be compared. Additionally, we provide several instances

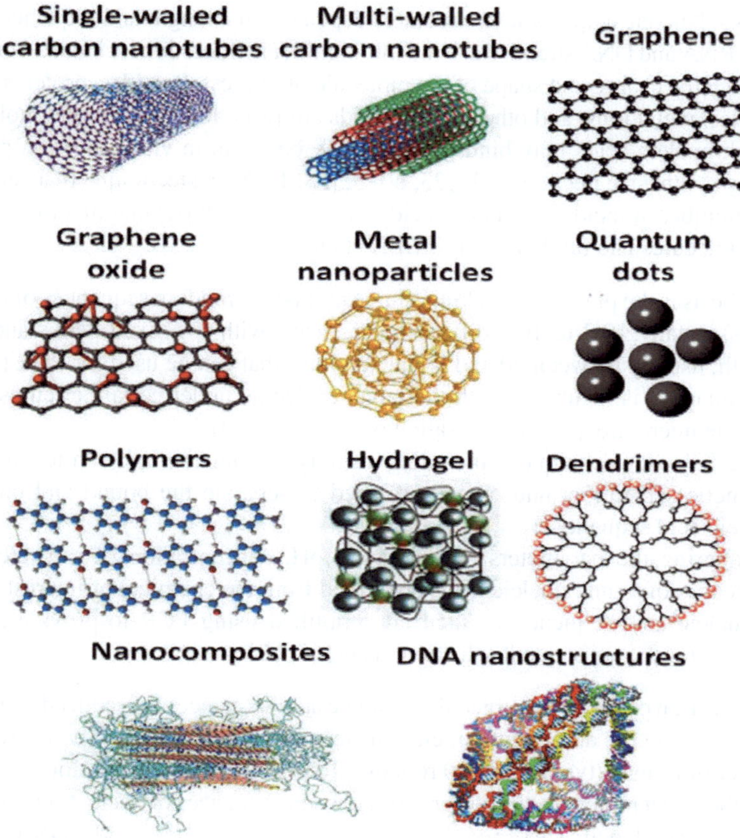

Fig. 5.2 An illustration of the nanoparticles often used in the development of bioelectrochemical genosensors for the detection of several tumor biomarkers [56]

of electrochemical reactions based on various RNA and DNA biosensors for cancer detection, including those that use DNA or RNA aptamers to detect carcinoembryonic antigen (CEA), miR-21 (microRNA-21), or prostate-specific antigen (PSA) [58].

5.2 Nucleic Acids as Components of Recognition

Nucleic acids may be used as components of recognizing that attach to molecules of interest and convey genetic information. DNA and RNA are the two categories into which nucleic acids are placed; the four nucleotides that make up DNA are ATCG. The four nucleotides are adenine (A), guanine (G), cytosine (C), and uracil (U), which combine to form RNA [59]. When C and G and T and A (or U) have complementary base couplings, RNA and DNA may form double- or single-stranded buildings. Nucleic acids, whose structures depend on the length and sequence of the various RNA and DNA strand, can bend into significant three-dimensional assemblies that match the charge and shape of the molecule of interest. Peptides, proteins, small molecules, metal ions, and other nucleic acids are just a few of the target molecules to which nucleic acids may bind, with SELEX being an in vitro selection process that may synthesize nucleic acids [23, 60–62]. SELEX is a technique that separates a huge number of incidental nucleic acid sequences from those that link to a specific target molecule. The fundamental SELEX stages are (Fig. 5.3):

- Synthesis is the process of making a huge number of random sequences of nucleic acids, usually 10^{15} to 10^{13} different sequences, with two fixed flanks and a set length, usually between 20 and 80 nucleotides, that can be used for PCR [63].
- The quantity is incubated with the target molecule under certain circumstances (e.g., temperature, pH, concentration of salt, etc.) [64].
- A variety of techniques, including affinity chromatography, filter binding, magnetic separation, and others, are used to separate the bound and unbound nucleic acid sequences.
- By altering the parameters (temperature, pH, concentration of salt, etc.), the sequences of bound nucleic acids are eluted from the chemical of interest.
- A nucleic acid sequence's eluted are amplified using PCR to provide a fresh amount for the subsequent selection cycle [65, 66].

A collection of the target molecule's nucleic acid sequences is produced by repeatedly performing the amplification, elution, separation, and selection processes over a number of rounds (typically 5–20 rounds). To determine which specific sequences bind to the target molecule, the last group of chosen sequences of nucleic acids may be sequenced and cloned. These sequences are known as aptamers, and a DNAzyme is another type of nucleic acid that can serve as a recognition element. A DNA sequence

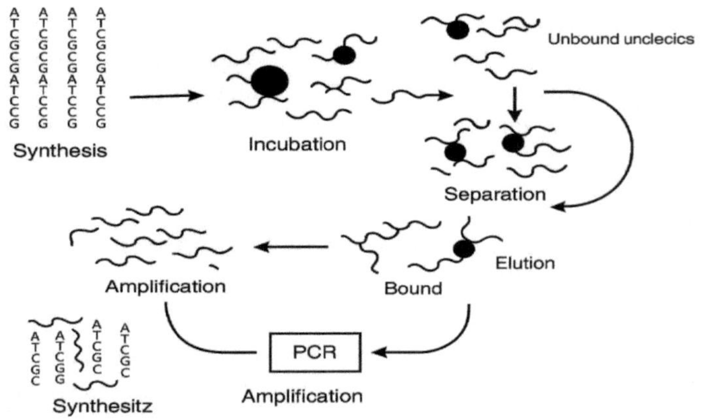

Fig. 5.3 SELEX process overview. Random nucleic acid sequences are incubated with a target molecule; bound sequences are separated, eluted, and amplified by PCR. Repeated cycles enrich high-affinity binders known as aptamers

known as a DNAzyme has the ability to catalyze a particular chemical process, such as the cutting or joining of another strand of DNA. One can construct DNAzymes using rational design or SELEX, and catalyze them when they bind to specific target molecules like nucleic acids or metal ions [67]. The genosensor recognizes certain DNA or RNA sequences linked to adenocarcinoma by using nucleic acids, which are molecules that contain genetic information. In order to identify and bind target DNA or RNA molecules, the genosensor fixes the nucleic acid probes on the surface of the electrode. The signal produced electrochemically by the following hybridization event may be measured and linked to the quantity and presence of adenocarcinoma in the sample. High sensitivity, specificity, and precision are only a few advantages of using nucleic acid-based genosensors [68].

5.3 Electrochemical Genosensing Based on Nucleic Acids

Electrochemical genosensors are instruments that use the electrical properties of nucleic acids to detect certain proteins linked to illnesses like cancer. The three primary parts of an electrochemical genosensor are a signal amplifier, a transducer, and a recognition section. The element that binds to the biomolecule of the detector with high affinity and specificity is known as the recognition moiety, where a nucleic acid or an antibody may act as the recognition element. The focus of this evaluation will be on genosensors that use electrochemistry and nucleic acids as components of recognition. The part that converts the connection event between the target and the recognition element into a quantifiable electrical signal is called a transducer [56, 69]. Cancer-associated biomarkers primarily include nucleic acids such as DNA and RNA sequences, proteins, exosomes and entire cells. Among these, circulating nucleic acids are particularly valuable because they can reveal the molecular characteristics of both primary tumor cells and micrometastatic cells. Consequently, they serve as highly effective biomarkers for the non-invasive screening of breast cancer through biological fluids. Figure 5.4 provides a schematic illustration of the electrochemical detection methods used to identify nucleic acid biomarkers for breast cancer detection from various body samples [56].

When connected to a voltmeter or power supply, an electrode or impedance sensor can function as a transducer, producing or measuring electrical current. When an alternating current is supplied to an electric circuit, a device known as an impedance sensor may detect the capacitance or resistance of the circuit. A signal amplifier is the component that enhances the electrical signal from the transducer to boost the sensor's accuracy and sensitivity. There are two possible methods for the signal amplifier: chemical and physical. Adding molecules or nanoparticles increases the current or impedance change brought on by the binding event in a chemical technique. Increasing the transducer's surface area or conductivity by structural or surface modification is known as a physical technique. Interactions between nucleic acids and their targets change in current, voltage, or impedance may be measured using electrochemical genosensors based on nucleic acids. These modifications may be

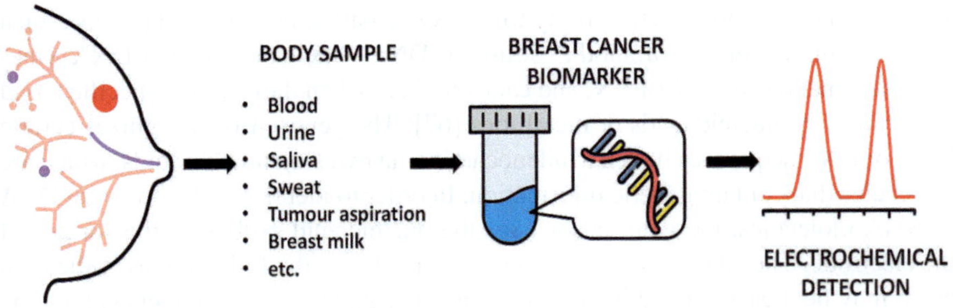

Fig. 5.4 Schematic representation of the electrochemical detection of nucleic acid breast cancer biomarkers that can be obtained via body samples [56]

brought about by a variety of processes, including displacement, catalysis, redox reactions, charge transfers, hybridization, conformational changes, and direct electron transfer [70]. Nucleic acid-based electrochemical genosensors provide many benefits, including affordability, high sensitivity, quick reaction times, and ease of use. But they also have to deal with issues like standardization, interference, repeatability, and stability. To increase their effectiveness and application, further study and development are thus required. A plan illustrating multiple methods for developing bioelectrochemical genosensors for the detection of miR-21, a microRNA biomarker having therapeutic importance in a variety of malignancies, is shown in Fig. 5.5 [71]. A variety of methods used for this purpose are shown in the figure, including electrochemiluminescence, surface plasmon resonance, and electrochemical impedance spectroscopy. These techniques make it possible to identify miR-21 in biological materials with high sensitivity and selectivity [72, 73].

5.3.1 Instances of Cancer Diagnosis Using Genosensors that Use Electrochemistry Based on Nucleic Acids

Recently, electroanalytical techniques have made significant strides in the assessment and identification of biomarkers for cancer using electrochemical DNA genosensors. These genosensors rely on the hybridization events that occur between the target nucleic acid sequence and a DNA or PNA capture probe, which result in a discernible shift in the electrochemical signal. These genosensors are a promising choice for early cancer detection because of their shown excellent sensitivity and specificity [74]. Specifically, a number of investigations have created nucleic acid-based electrochemical genosensors for the identification of breast biomarkers for cancer. Wang et al. identified miRNA-21 as a biomarker for bladder cancer and developed an electrochemical, label-free genosensor that uses DNAzyme and exhibits good selectivity against other miRNAs [75].

Two features related to an MOF (metal-organic framework) that enable simultaneous detection of multiple cancer markers. A simplified representation of the

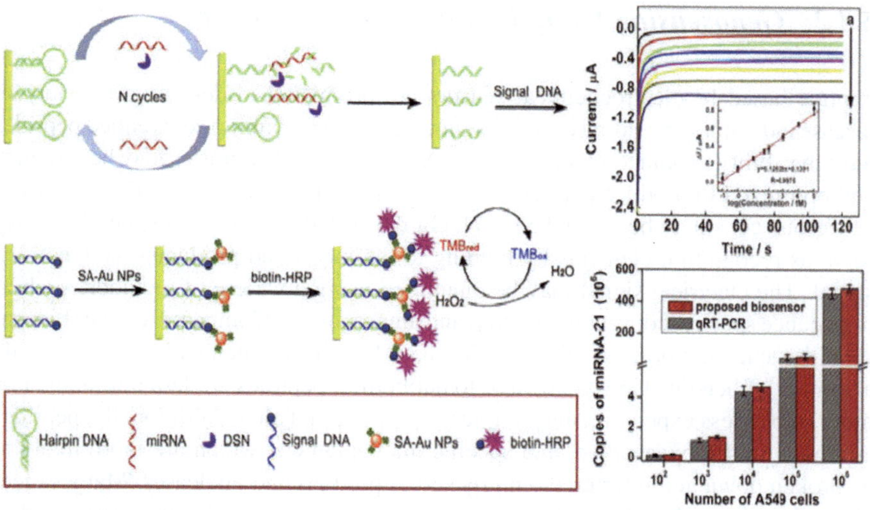

Fig. 5.5 Electrochemical genosensor strategies for miR-21 detection [71]

MOF's fabrication procedure, which includes adding nucleic acids to the framework to boost its affinity for target biomolecules, may be found in the first aspect (I). From complicated biological materials, the nucleic acid-biofunctionalized MOF may serve as a capturing mechanism to separate and enhance cancer indicators. The MOF-based method is a potential platform for cancer detection and monitoring since it has several advantages, including excellent sensitivity, selectivity, and specificity. The second aspect (II) of the figure is a MOF-assisted bioelectrochemical sensor, which allows the simultaneous identification of many cancer indicators in a single test. The MOF-based sensor amplifies the signal from electrochemistry produced by the collected indicators of cancer by using the special characteristics of the framework. By transducing and quantifying the resulting signal, one can quantitatively determine the quantity of each cancer indicator present in the specimen. Among the many advantages of the MOF-assisted bioelectrochemical sensor are its low detection limits, quick response, and high sensitivity [76].

The genosensor uses nucleic acid probes that are mounted on the electrode surface to detect certain DNA or RNA sequences that match miR-145. The electrochemical signal, which may be measured and linked to the quantity of miR-145 present in the sample, is changed by the hybridization between the target miR-145 and the fixed probes. High sensitivity, specificity, and selectivity are only a few advantages that make the bioelectrochemical genosensor a potential tool for cancer monitoring and detection [26].

5.3.2 Genosensing Using Peptides to Detect Cancer

Peptide-based bioelectrochemical sensors are promising in detecting a variety of disease indicators, including cancer, as short sequences of amino acids called peptides have the ability to bind to specific targets. Artificial peptide libraries may be screened and optimized to produce peptides, or natural peptides associated with enzymes or cancerous cells may be mimicked [77]. On the surface of electrodes, peptides may serve as recognition elements by binding to the target and producing an electrical signal. The electrical signal may be quantified using a variety of methods, such as impedance spectroscopy, voltammetry, and amperometry. Compared to other biosensors like aptamers or antibodies, peptide-based bioelectrochemical sensors provide a number of benefits [78]. Compared to antibodies, peptides are less immunogenic, more stable, less expensive, and simpler to produce and alter. Additionally, peptides are more versatile, sensitive, and specific than aptamers, which are more likely to be broken down and interfered with by serum proteins and nucleases. Many cancer biomarkers, including DNA mutations, proteins, hormones, enzymes, and metabolites, may be recognized by peptides. To improve the sensitivity, selectivity, and biocompatibility of the sensors, peptides may also be altered with other groupings that are functional, like polymers, redox mediators, or nanoparticles [79].

Bioelectrochemical cancer sensors based on peptide diagnosis include the following examples: a peptide-based sensor that imitates the VEGF binding site, a molecule that promotes tumor angiogenic processes. The sensor demonstrated high sensitivity and selectivity for VEGF in blood samples from individuals with breast cancer. A peptide-based sensor that binds to MMP-9 (matrix metalloproteinase-9), an enzyme that facilitates tumor invasion and metastasis by degrading the extracellular matrix [20]. The sensor accurately detected serum samples containing MMP-9 from patients with cancer of the colorectal region. EGFR, an overexpressed protein in several cancer types and a target for a number of anticancer medications, is recognized by a peptide-based sensor. The sensor was able to differentiate between EGFR in lung cancer cell line lysates that were wild-type and those that were mutant. Bioelectrochemical sensors based on peptides show promise in detecting a variety of illness indicators, including cancer. For the detection and tracking of cancer, they provide excellent sensitivity, selectivity, stability, and adaptability. Some of the constraints that still require attention are peptide sequence and structural optimization, enhancement of materials for electrodes and designs, mitigation of non-specific adsorption and interference, and clinical verification of the sensors. Peptide-based bioelectrochemical sensors are promising in detecting a variety of disease indicators, including cancer, because they are short chains of amino acids that can bind to specific targets, called peptides. Artificial peptide libraries may be screened and optimized to produce peptides, or natural interacting peptides with enzymes or cancerous cells can be mimicked. On the surface of electrodes, peptides may serve as recognition elements by binding to the target and producing an electrical signal. The electrical signal may be measured using a variety of methods, such as impedance spectroscopy, voltammetry, or amperometry [18, 80].

Compared to other biosensors like aptamers or antibodies, peptide-based bioelectrochemical sensors provide a number of benefits. Compared to antibodies, peptides are less immunogenic, more stable, less expensive, and simpler to produce and alter. Additionally, peptides are more versatile, sensitive, and specific than aptamers, which are more likely to be broken down and interfered with by serum proteins and nucleases. Many cancer biomarkers, including DNA mutations, proteins, hormones, enzymes, and metabolites, may be recognized by peptides. To improve the sensitivity, selectivity, and biocompatibility of the sensors, peptides may also be altered with other groupings that are functional, like polymers, redox mediators, or nanoparticles [16, 81].

Bioelectrochemical cancer sensors based on peptide diagnosis include the following examples: a peptide-based sensor that imitates the VEGF binding site, a molecule that promotes tumor angiogenesis. The sensor demonstrated high VEGF sensitivity and selectivity in blood samples from breast cancer cases. It is a peptide-based sensor that binds to MMP-9 (matrix metalloproteinase-9), an enzyme that facilitates tumor invasion and metastasis by degrading the extracellular matrix. The sensor accurately detected serum samples containing MMP-9 from patients with colorectal cancer. A peptide-based sensor detects EGFR, a protein overexpressed in many cancer types and a target for several anticancer medications [82, 83]. The sensor was able to differentiate between EGFR in lung cancer cell line lysates that were wild-type and those that were mutant. In addition to being helpful in cancer diagnosis, peptide-based bioelectrochemical sensors may also be used to track a patient's prognosis and response to therapy. For instance, variations in PSA levels during hormone treatment in individuals with prostate cancer were quantified using a peptide-based sensor that attaches itself to PSA, a prostate cancer biomarker [84]. The levels of CEA (carcinoembryonic antigen), an indicator of colorectal cancer, in blood samples previously obtained from individuals with colorectal cancer and after surgery were monitored using a different sensor that is according to a peptide that attaches itself to CEA. However, there are still some issues that need to be resolved before bioelectrochemical sensors based on peptides for the detection of cancer may be developed and used [17, 18]. Among these difficulties are: Peptide sequences and structures are optimized to provide minimal toxicity and immunogenicity for the host and high precision for the target. The development of electrode designs and materials to boost the sensors' stability, electrical conductivity, surface area, and biocompatibility. The decrease in adsorption and interference that are not specific caused by other chemicals or biomolecules seen in intricate biological specimens, including urine or serum [85]. Validation of sensors in medical environments using a large number of samples from different situations and sources suggests that peptide-based bioelectrochemical sensors have a potential future in cancer detection. More innovative and effective bioelectrochemical sensors grounded in peptides for different cancer biomarkers will be devised and manufactured with the advancements in electrochemical processes, bioinformatics tools, nanomaterials science, and modification approaches and peptide synthesis. For the first diagnostic, treatment monitoring, and prognosis assessment of cancer patients, these sensors will provide quick, precise, delicate, picky, transportable, and affordable instruments [86, 87].

(a) Cancer diagnosis via aptamer-based genomic sensing

Another cutting-edge technique for locating different disease indicators, including cancer, is aptamer-based bioelectrochemical sensors. Single-stranded RNA or DNA molecules called aptamers have the capacity to bind precisely to molecules of interest and fold into specific three-dimensional configurations, which SELEX may be used to select aptamers from vast combinatorial libraries. Compared to other detection systems such as peptides or antibodies, aptamers offer a number of advantages, including being more stable, reproducible, tunable, and compatible compared to antibodies. In addition, aptamers are simpler to regenerate, less immunogenic, and more thermally stable than peptides [18, 88].

Aptamers are a component of bioelectrochemical sensors that assist in identifying targets on the electrode surface; when they bind to a target, they transmit an electrical signal. The electrical signal may be measured using a variety of methods, including impedance spectroscopy, voltammetry, and amperometry. High sensitivity, selectivity, and speed may be attained using aptamer-based bioelectrochemical sensors for the detection of many indicators for cancer, encompassing proteins, nucleic acids, cells, or tissues. To improve the efficacy of the sensors, aptamers may also be used with other signal amplification techniques, like catalytic DNAzymes, enzymes, or nanomaterials [88].

Bioelectrochemical cancer sensors based on aptamer diagnoses include, for instance, an aptamer-based sensor that binds to VEGF, a molecule that promotes tumor angiogenesis. The sensor used a modified gold electrode with horseradish peroxidase (HRP) and gold nanoparticles as signal enhancers. With great sensitivity and selectivity, the sensor detected VEGF in blood samples from patients with breast cancer. An aptamer-based sensor that can identify the protein HER2, which is overexpressed in many cancer types and targets a number of anticancer medications [89]. The sensor used graphene oxide and hemin as signal amplifiers on a glassy carbon electrode. The sensor made a very accurate distinction between breast cancer cells that were HER2-positive and those that were HER2-negative. An aptamer-based sensor that attaches itself to the colorectal cancer biomarker mucin-1 glycoprotein (MUC1) [90]. Glucose oxidase (GOx) and gold nanorods were added to a gold electrode to act as signal amplifiers in the sensor. The sensor has great sensitivity and selectivity when measuring MUC1 in colorectal cancer patients' blood samples. Biochemical sensors based on aptamers can track a patient's prognosis and response to therapy in addition to helping with cancer diagnosis. For instance, PSA level variations during treatment in individuals with prostate cancer were quantified using an aptamer-based sensor that attaches to PSA, a prostate cancer biomarker [91]. An aptamer attaches itself to CEA, a different sensor that tracks the amounts of CEA (carcinoembryonic antigen), a colorectal cancer biomarker, in blood specimens taken from individuals with the disease both before and after surgery. Developing and using bioelectrochemical sensors based on aptamers for cancer detection requires overcoming some obstacles and restrictions. Among these difficulties are: aptamer sequences and architectures are optimized to ensure minimal toxicity and interference from other biomolecules or chemicals present in complicated biological specimens, as well as good precision for the target. The development of electrode designs and materials to boost the sensors' stability, electrical conductivity, surface area, and biocompatibility. The sensitivity and dynamic range of the sensors may be increased by using innovative signal amplification techniques without sacrificing simplicity or specificity [92].

Validation of sensors in medical settings using a large number of samples from different situations and sources suggests that aptamer-based bioelectrochemical sensors have a potential future in cancer detection. More innovative and effective aptamer-based bioelectrochemical sensors for a wide range of cancer biomarkers will be developed and produced thanks to advances in aptamer selection and modification methods, nanomaterials science, electrochemical methods, and bioinformatics tools. For the first diagnostic, treatment monitoring, and prognosis assessment of cancer patients, these sensors will provide quick, precise, delicate, picky, transportable, and affordable instruments [93].

5.3.3 Genosensing Using DNA to Detect Cancer

A new technique for identifying different illness indicators, including cancer, is DNA-based bioelectrochemical sensors. DNA, the molecule that contains the genetic information of living things, is made up of two entangled strands. In electrical and chemical biosensors, DNA may also serve as a recognition element, a signal converter, or a signal booster. By using its complementary base pairing characteristic, which enables it to hybridize with target nucleic acids, such DNA or RNA, with great specificity, DNA may be used as a recognition element. Utilizing its electrochemical characteristics, such as conductivity, charge transfer, or redox activity, which may produce an electrical signal upon contact with the target, DNA can also be employed as a signal transducer. By taking use of its enzymatic characteristics, such as polymerization, cleavage, or ligation, which may increase the electrical signal by creating additional DNA molecules, DNA can also be utilized as a signal amplifier [94].

When it comes to identifying cancer biomarkers like mutations, microRNAs, RNA expressions, or circulating tumor DNA, DNA-based bioelectrochemical sensors may achieve great sensitivity, selectivity, and speed. To improve their performance, DNA-based bioelectrochemical sensors may also be combined with other nanomaterials as graphene, quantum dots, or gold nanoparticles [95].

DNA-based bioelectrochemical sensors for cancer diagnosis include the following examples:

Changes in the KRAS gene, which are connected to colorectal cancer, are detected by a sensor that pairs DNA. The sensor made use of gold nanoparticles labeled with signal probes and a gold electrode covered with thiolated capture probes. With great accuracy and precision, the sensor detected KRAS alterations in clinical samples [96]. Telomerase activity, a marker of cancer cell proliferation, is detected via a sensor that employs DNA electricity. Glassy carbon served as the electrode for the sensor, and hairpin probes labeled with methylene blue and telomeric repeats tagged with ferrocene covered it. With great accuracy and precision, the sensor detected telomerase activity in extracts of cancer cells [97].

The effluence from tumor cells into the blood is detected using a sensor that employs DNA multiplication to detect ctDNA. The sensor used exonuclease III (Exo III), a multiplication enzyme, and a gold electrode covered with capture probes. With great accuracy and precision, the sensor detected ctDNA in plasma models from instances of lung cancer. DNA-bioelectrochemical sensors are useful not only for cancer detection but also for tracking patient prognosis and response to therapy [98]. For instance, EGFR gene alterations, which are linked to lung cancer and impact the responsiveness to tyrosine kinase inhibitors, are detected by a DNA hybridization-based sensor. The sensor made use of silver nanoparticles labeled with signal probes and a gold electrode covered with capture probes [99]. With great accuracy and precision, the sensor detected EGFR alterations in plasma samples from patients with lung cancer both before and after treatment. Changes in the p53 gene, which are associated with colorectal cancer and impact survival, are detected by a separate sensor that employs DNA multiplication [100]. The sensor used rolling circle multiplication (RCA) as a multiplication technique and a gold electrode covered with capture probes. With great accuracy and precision, the sensor detected p53 alterations in stool samples from individuals with colorectal cancer. The development and deployment of DNA-based biosensors that use chemistry and electrical for cancer detection, however, still face some challenges and restrictions [101]. Among these challenges are: in complex biological samples, DNA sequences and structures are fine-tuned to obtain high target accuracy and minimal interference and toxicity from other biomolecules or chemicals. Improving the electrode designs and materials to boost the sensors' stability, electrical conductivity, surface area, and biocompatibility [102]. The sensitivity and dynamic range of the sensors may be increased without sacrificing specificity and ease of use by developing new nanomaterials and multiplication techniques. Validation of sensors in medical settings using a large number of samples from different situations and sources suggests that DNA-based bioelectrochemical sensors have a promising future in cancer detection. More innovative and effective DNA-based bioelectrochemical sensors for different cancer indicators will be created and manufactured thanks to developments in DNA electrochemical methods, nanomaterials science, synthesis and modification procedures, and tools for bioinformatics. These sensors will provide quick, precise, delicate, picky, transportable, and affordable instruments for the first diagnostic, treatment monitoring, and prognosis assessment of cancer patients [103, 104].

5.3.4 RNA-Based Genomics for the Identification of Cancer

A novel method for detecting several illness signs, including cancer, is the utilization of electricity and chemistry in RNA-based biosensors. Genetic information is transferred from DNA to proteins via the single-stranded molecule known as RNA. Additionally, in bioelectrochemical sensors, RNA may function as a signal amplifier, transducer, or recognition element. By taking use of its complementary base pairing characteristic, which enables it to hybridize with target nucleic acids like DNA or RNA with great specificity, RNA may be utilized as a recognition element. By using its electrochemical characteristics, such as conductivity, charge transfer, or redox activity, which may produce an electrical signal upon contact with the target, RNA can also be used as a signal transducer. By taking use of its enzymatic characteristics, such as polymerization, cleavage, or ligation, which may increase the electrical signal by generating additional RNA molecules, RNA can also be utilized as a signal amplifier [105]. High sensitivity, selectivity, and speed are achievable using RNA-based bioelectrochemical sensors for the identification of various cancer biomarkers, including long non-coding RNAs, microRNAs, and RNA expressions. To improve their performance, RNA-based bioelectrochemical sensors may also be combined with other nanomaterials as graphene, quantum dots, or gold nanoparticles [106].

RNA-based bioelectrochemical sensors for cancer diagnosis include, for instance:

An RNA hybridization-based sensor detects the mRNA expression of VEGF-A, a gene that promotes tumor angiogenesis. The sensor made use of silver nanoparticles labeled with signal probes and a gold electrode covered with capture probes. With great accuracy and precision, the sensor detected VEGF-A expression in serum models from instances of lung cancer. An RNA electricity sensor detects miR-21, a protein that promotes tumor angiogenesis and aids in the development of cancer [107]. Glassy carbon served as the electrode for the sensor, and methylene blue-labeled capture probes and ferrocene-labeled hairpin probes covered it. With great accuracy and precision, the sensor detected miR-21 in serum samples from instances of breast cancer. RNA multiplication is used by a sensor to identify lengthy non-coding RNA. A big non-coding RNA called HOTAIR (HOX transcript antisense intergenic RNA) regulates gene expression and is connected to the spread of cancer [108]. The sensor used rolling circle multiplication (RCA) as a multiplication technique and a gold electrode covered with capture probes. With great accuracy and precision, the sensor detected HOTAIR in plasma samples from individuals with colorectal cancer. Electricity and chemistry-based RNA-based biosensors are helpful for cancer diagnosis, therapy response tracking, and patient prognosis [109]. For example, an RNA pairing sensor could find the mRNA expression of HER2, a gene that many types of cancer overexpress and that is the target of several anticancer drugs. The sensor made use of gold nanoparticles labeled with signal probes and a gold electrode covered with capture probes. With great accuracy and precision, the sensor detected HER2 expression in plasma samples from patients with breast cancer during their treatment [110]. mRNA expression of KRAS, a gene altered in many types of cancer and influencing the response to targeted treatment, is detected using a separate sensor that employs RNA multiplication. The sensor used an altered gold electrode with capture probes and the exponential amplification reaction (EXPAR) as an amplification technique. With outstanding sensitivity and selectivity, the sensor detected KRAS mutations in plasma samples taken from individuals who have colorectal cancer both before and after therapy [111].

The evolution and use of RNA-based cancer bioelectrochemical sensor diagnosis, however, still face some obstacles and restrictions. Among these difficulties are:

RNA sequences and structures are fine-tuned to ensure minimal toxicity and meddling from additional biomolecules or compounds found in intricate biological specimens, as well as high precision for the target. Improving the electrode designs and materials to boost the sensors' stability, electrical conductivity, surface area, and biocompatibility. Without sacrificing simplicity and specificity, the sensitivity and dynamic range of the sensors may be increased by using new nanomaterials and amplification techniques.

Validation of sensors in medical settings using a large number of samples from different situations and sources shows that RNA-based bioelectrochemical sensors hold enormous promise for cancer detection. Developments in RNA bioinformatics tools, electrochemical processes, nanomaterials science, and synthesis and modification methods will lead to the creation and manufacturing of more innovative and effective RNA-based bioelectrochemical sensors for various cancer biomarkers. For the first diagnostic, treatment monitoring, and prognosis assessment of cancer patients, these sensors will provide quick, precise, delicate, picky, transportable, and affordable instruments [112].

Several cancer cell microRNAs may be detected in a single experiment using peptide biosensors based on nucleic acids that combine chemical and electrical techniques, as shown in Fig. 5.6 [113]. The biosensors identify specific RNA sequences using probes for peptide nucleic acids on the electrode. By measuring the change in the electrical signal that results from the probes' interaction with the target microRNAs, one can determine the presence and quantity of microRNAs in the sample. In order to increase the detection's specificity and sensitivity, the biosensors additionally use catalytic multiplication of hairpin assemblies, a method that allows the signal generated by the corresponding event to be amplified. Catalytic hairpin assembly multiplication, a self-templated mechanism, produces several copies of the targeted RNA sequences, significantly boosting the electrical signal. Biosensors based on peptide nucleic acids, which employ both electrical and chemical techniques, offer advantages such as high sensitivity, specificity, and selectivity. Catalytic hairpin assemblies multiply, making them a promising tool for cancer monitoring and diagnostics [114]. Adenine-enriched unmethylated ss-DNA is more prevalent than guanine-enriched methylated ss-DNA when the two types of DNA are adhered to a solid-phase extraction (SPE) gold electrode. The test's electrical signal is generated by the (CV) probing of $[Ru(NH_3)_6]^{3+}$ complexes that are attached to DNA. The CV technique measures the current that results from applying a voltage to the DNA-bound $[Ru(NH_3)_6]^{3+}$ complexes. The relative quantities of guanine-enriched methylated ss-DNA and adenine-enriched unmethylated ss-DNA in the sample may be determined using the current that is acquired, which is correlated with the quantity of DNA on the electrode. A normal CV signal is shown in the figure's inset, where adenine-enriched unmethylated DNA has a greater CV charge than guanine-enriched methylated DNA. This demonstrates that the test may produce a detectable electrical signal and differentiate between the two types of DNA based on their attachment characteristics. There are several advantages to using an electrical test to measure DNA methylation specific to a gene [115]. The first-generation approach multiplies the DNA and two distinct DNA probes, each marked with digoxigenin (DIG) and horseradish peroxidase (HRP), respectively, using the polymerase chain reaction (PCR). An electrical signal and an anti-DIG antibody produced by HQ (hydroquinone) oxidation to BQ (benzoquinone) are then used to identify the resultant complex once the multiplied DNA and probes have been coupled [116]. LAMP (loop-mediated multiplication), which uses only one DNA probe and increases the DNA's number at a steady temperature, takes the role of PCR in the second-generation technique. After marking the probe with biotin, the amplified DNA and probe are captured by a magnetic bead coated with streptavidin. The signal from an electrical source produced via HQ's oxidation to BQ is then used to identify the collected complex. The comparison of the two approaches demonstrates that the second-generation approach has a number of benefits, such as a streamlined procedure, lower costs and time, and enhanced sensitivity and specificity [117]. This schematic illustrates a label-free electrochemical DNA genosensor for breast cancer gene detection. It involves functionalizing a gold electrode with cysteamine, glutaraldehyde, and Fc-PAMAM dendrimers (G1, G2, G3) to immobilize ssDNA probes. Hybridization with target DNA leads to measurable electrochemical signals for detection (Fig. 5.7) [118].

DNA molecules promote the oxidation or reduction of electroactive substances, which produces electrical current. Rolling circle multiplication (RCA), which creates lengthy single-stranded DNA products using a circular DNA template and a DNA polymerase, may be used to multiply the electrical signal. In order to electrically identify circulating tumor cells, the cells must be captured and isolated using a variety of methods, such as magnetic beads coated with antibodies against tumor cell surface

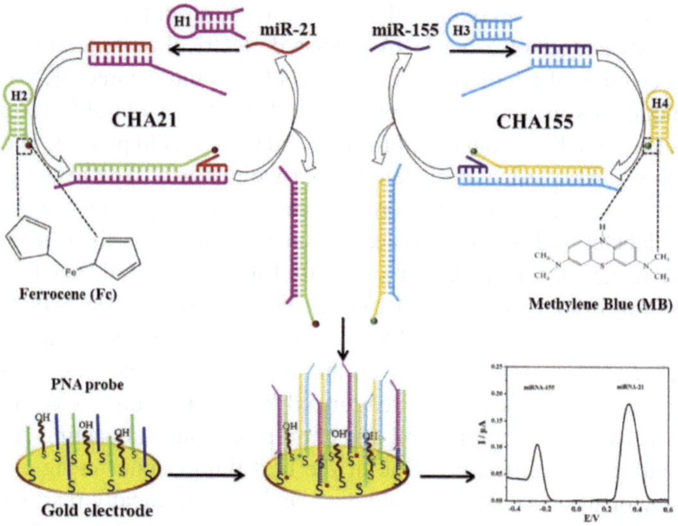

Fig. 5.6 Multiple-microRNA detection using a bioelectrochemical genosensor mediated by peptide nucleic acids [113]

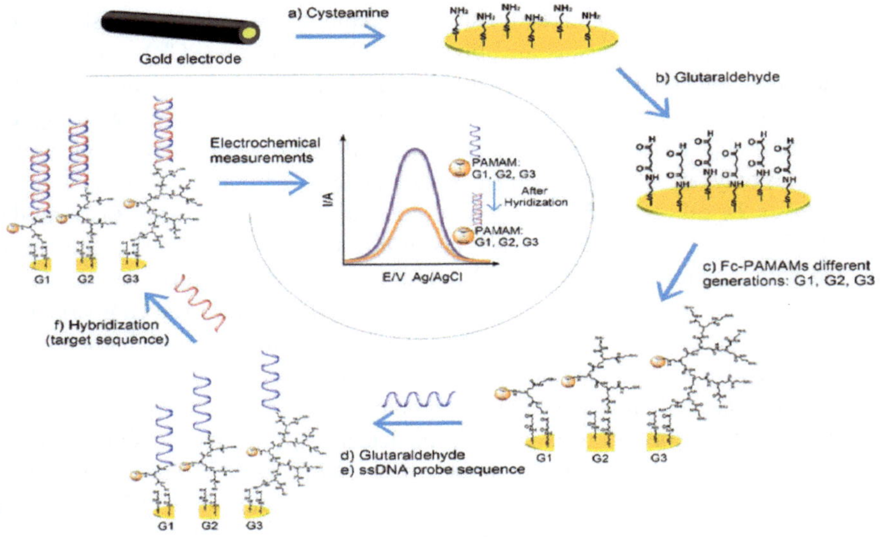

Fig. 5.7 Label-free detection of breast cancer gene markers using electrochemical DNA genosensors [118]

antigens or microfluidic devices. Following the lysis of the collected cells, the intracellular DNA is extracted and subjected to RCA multiplication in conjunction with an electroactive probe. An electrical signal generated by the electroactive probe's oxidation or reduction may be used to identify the lengthy single-stranded DNA products that are the consequence. High sensitivity, specificity, and selectivity are among the advantages of using DNA-mediated electrical current and RCA for the electrical detection of circulating tumor cells [119]. A unique microRNA identification technique uses an electrically reprogrammable network of gold-coated magnetic nanoparticles to enable signal multiplication and nucleic acid pairing. The technique depends on using a particular nucleic acid probe that is affixed to the gold-coated magnetic nanoparticles' surface in order to capture the target microRNA. A detection probe labeled redox-active molecule, such ferrocene, is then linked with the microRNA that was caught. CV may be used to identify and quantify the sandwich structure formed by the pairing event. With the help of an outside magnetic field, the electrically reprogrammable network of gold-coated magnetic nanoparticles can change how well they pair up and how many signals they send at once. The nanoparticle network may be rearranged by the magnetic field, changing the separation between the capture and detection probes and impacting signal multiplication and pairing efficiency. One of the best things about using a network of magnetic nanoparticles coated in gold that can be electronically rearranged to find microRNA is that it is very sensitive, very specific, and very selective. The technique is a powerful tool for illness monitoring and diagnosis as it may be adjusted to detect numerous microRNAs simultaneously [120]. An electrical genosensor that employs nuclease and quantum dot-DNA nanocomposites to detect a single gene alteration in breast cancer has been created using a revolutionary twofold signal multiplication technique. The genosensor aims to detect specific alterations in the BRCA1 gene, linked to an increased risk of ovarian and breast cancer. The genosensor may match the target DNA sequence in the sample using a specific DNA probe on the electrode's surface. To increase its sensitivity and accuracy, the genosensor uses two signal multiplication techniques. A nuclease enzyme is used in the initial signal multiplication approach to cut the DNA probe and release the target DNA sequence for further multiplication [121]. A quantum dot-DNA nanocomposite that serves as a second signal multiplier is then combined with the released DNA. When the nanocomposite binds to the target DNA sequence, it is intended to produce a powerful electrical impulse that is capable of being detected and connected to the quantity and intensity of the specimen alteration. High sensitivity, specificity, and selectivity are only a few advantages of the electrical genosensor that uses the double signal multiplication technique. The genosensor is a powerful tool for the early detection and diagnosis of breast and ovarian cancer since it can identify BRCA1 gene changes at low levels [122]. However, fouling the nonspecific adhesion of proteins and other biomolecules on the sensor surface can impede the identification process. A technique that combines DNA probes and antifouling peptides for the ultralow fouling electrical detection of cancer markers has been developed to get around this problem. The technique makes use of certain DNA probes that are affixed to the electrode surface and are able to identify and attach to the target cancer markers present in the sample. Antifouling peptides are added to the electrode surface to avoid fouling by preventing nonspecific biomolecules from adhering. The antifouling peptides use electrostatic repulsion and steric hindrance to reject proteins and other macromolecules. High sensitivity, specificity, and selectivity are among the advantages of electrically detecting cancer signs utilizing antifouling peptides and identifying DNA probes. Without being hampered by nonspecific biomolecules, the technique may identify cancer markers at low concentrations in complex biological fluids including blood and urine [123]. Through the identification of certain long non-coding RNAs (lncRNAs), a unique dual electrical genosensor has been created in order to identify prostate cancer early. A particular probe created to locate and bind to the target lncRNAs in the sample is used by the genosensor. After attaching itself to the electrode surface, the probe may couple with the target lncRNAs to produce a change in the signal from the electrical. This change can be quantified and linked to the concentration and presence of the lncRNAs in the sample. Among the many benefits of the dual electrical genosensor are its excellent sensitivity, selectivity, and specificity. Two distinct lncRNAs may be identified simultaneously thanks to the genosensor's integration of two electrical detection techniques. This characteristic improves the genosensor's precision and dependability for prostate cancer early detection. A non-invasive and economical technique for the early diagnosis of prostate cancer is the dual electrical genosensor, which detects lncRNAs. The genosensor may be used in clinical settings to test for and track instances of prostate cancer [124]. To find RNA markers in prostate cancer, a new electrical test based on loop-mediated isothermal multiplication (LAMP) has been created. A particular probe designed to identify and attach to the target RNA markers in the sample is used in the test. The electrical test may detect the huge number of DNA copies produced by the LAMP reaction, which multiplies the target RNA sequences. The test may detect many RNA markers at once and is both sensitive and specific. High sensitivity, quick identification, and simplicity of use are only a few benefits of the electrical LAMP-based test. The test is a powerful tool for early diagnosis and disease monitoring since it can identify RNA markers in prostate cancer at low levels. Real-time monitoring of the LAMP reaction is made possible by the electrical detection approach, which also yields quantitative findings that can be examined using a simple and inexpensive electrical instrument. An economical and non-invasive approach to early diagnosis and disease monitoring is provided by the electrical LAMP-based test for RNA markers in prostate cancer [125]. Optical genosensors, such as colorimetric and fluorescence sensors, have also been used to study the detection of cancer biomarkers. In the event that the target biomolecule attaches itself to the probe, these genosensors pick up changes in the optical signal. Fluorescent sensors are particularly appealing due to their ability to identify numerous targets for biomarkers using multiple fluorescent probes and their improved sensitivity of target identification through signal multiplication techniques [126]. The researchers examined new developments in optical and electrical biosensors for detecting various cancer markers. The significance of biosensors in early cancer detection and therapy is emphasized in the chapter. The authors list the benefits and drawbacks of every kind of biosensor and provide instances of how they are used in clinical settings. According to the study, the use of biosensors to identify cancer indicators shows promise as a means of early detection and therapy [127]. In their assessment of recent developments in electrical biosensors based on graphene for the detection of several biomolecules linked to illness. The researchers emphasized the benefits of these devices, such as their great selectivity and sensitivity as well as their capacity for use in clinical settings. But the authors also talk about the difficulties of using these biosensors, namely clinical translation and repeatability [128]. Li et al. combined a DNA probe with cancer-linked exosomes to create an electrical biosensor that can detect pancreatic cancer, demonstrating high specificity against healthy exosomes in people. Similarly, the researchers developed an electrical biosensor for lung cancer detection by combining circulating tumor DNA with a DNA probe. The biosensor demonstrated extraordinary specificity against healthy plasma DNA, highlighting the potential of graphene-based bioelectrical sensors for cancer marker detection [129]. There is potential for identifying cancer biomarkers thanks to recent developments in electrochemical biosensor technology. Graphene-based electrochemical biosensors, electrochemical DNA biosensors, and optical biosensors have demonstrated potential for early cancer detection and therapy. These biosensors are appealing for clinical usage because of their great sensitivity and specificity. Nevertheless, before being widely used in clinical settings, repeatability and clinical translation need to be addressed. The development of sensitive and targeted diagnostic instruments for cancer detection may result from further study and advancement in this field [130].

5.3.5 Genosensors Based on DNA Aptamers for PSA Identification

Genosensors based on DNA aptamers have shown promise in detecting PSA (prostate-specific antigen), a frequently-used indicator for prostate cancer detection. Blood levels of PSA, a glycoprotein secreted by the prostate gland, may reveal prostate cancer. However, the sensitivity and specificity of the PSA detection techniques used today, including ELISA, are limited. DNA aptamer-based genosensors, on the other hand, have shown excellent sensitivity and specificity, suggesting that they might be a way around these restrictions [131]. The invention of DNA aptamer-based biosensors for PSA and other cancer markers, such as CEA, HER2, and AFP, has also been documented in other circumstances. Numerous identifying principles, including colorimetric, electrical, and fluorescent, may be used to create the biosensors, which can then be tuned for sensitivity and specificity. The development of DNA aptamer-based biosensors is not without its difficulties, however, including the need to optimize biosensor performance and choose aptamers for the target biomolecule with high precision. However, DNA aptamers are appealing candidates for biosensor development due to their advantages, which include stability, high precision, and simplicity of synthesis. Therefore, a potential method for detecting cancer signs, such as PSA, is provided by DNA aptamer-based biosensors. They may be made to detect cancer markers using a variety of methods and have a number of benefits over conventional antibody-based biosensors. The creation of sensitive and targeted diagnostic instruments for early cancer detection and individualized therapy may result from the development of DNA aptamer-based biosensors for cancer diagnosis and treatment (Fig. 5.8) [132, 133].

5.3.6 Genosensors Based on RNA Aptamers for CEA Detection

A potential method for detecting cancer markers like carcinoembryonic antigen (CEA) is the use of RNA aptamer-based biosensors. CEA may be detected with great sensitivity and specificity thanks to RNA aptamers, which are tiny molecules of single-stranded RNA that attach to certain targets. RNA aptamer-based biosensors for CEA detection have been created in a number of studies; each has advantages and disadvantages [134]. One such biosensor is a label-free and user-friendly SPR biosensor for CEA detection that has been developed. Additional examples include the FP biosensor, the GO biosensor, the microcantilever biosensor, and the EIS biosensor. Though their detection techniques, detection limits, reagent needs, and manufacturing procedures may vary, these biosensors are likewise very sensitive and selective. The identification technique affects the sensitivity, specificity, and repeatability of these biosensors. Whereas FP, GO, and microcantilever biosensors are very sensitive and selective but could need extra reagents or intricate fabrication

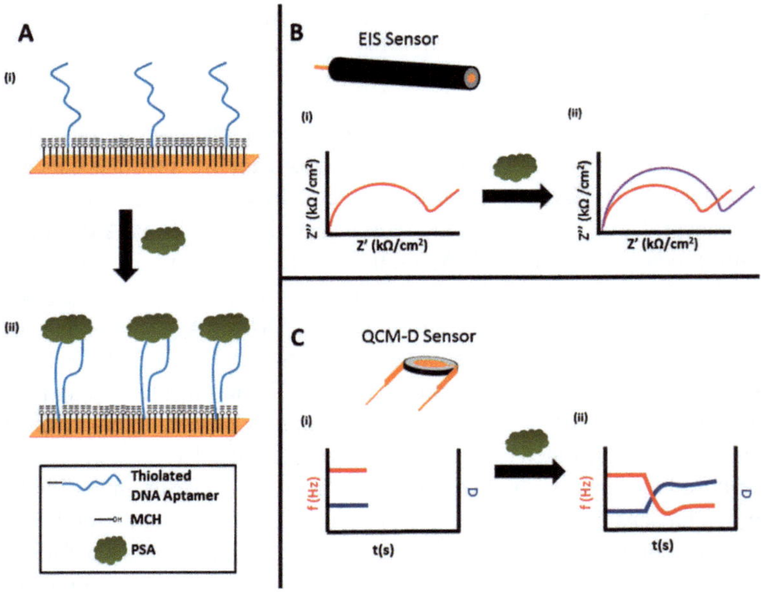

Fig. 5.8 **a** Biosensor detection principle with representation of the electrode gold surface; **b** EIS measurements showing a signal change in the Nyquist plot before (i) and after (ii) PSA binding; **c** QCM-D measurements showing frequency and dissipation responses before (i) and after (ii) PSA binding [133]

procedures, SPR and EIS biosensors are simple to use and have low detection limits [135].

5.4 Conclusion

The latest developments and uses of nucleic acid-based electrochemical genosensors for cancer detection have been covered in this chapter. Comparing the several research and showcasing this technology's potential for cancer detection were our goals. We have contrasted nucleic acid-based electrochemical genosensors and antibody-based ones, emphasizing the advantages and disadvantages of each. We have also shown instances of nucleic acid-based electrochemical genosensors for cancer detection, such as those that detect carcinoembryonic antigens, microRNA-21, or prostate-specific antigens using DNA or RNA aptamers. According to our study, electrochemical genosensors that rely on nucleic acids provide a viable substitute for conventional cancer detection techniques like qRT-PCR or ELISA. They are inexpensive, very sensitive, quick to react, and simple to use, among other benefits. They may also be able to identify different kinds of cancer biomarkers. However, issues with stability, repeatability, interference, and standardization plague genosensors that use electrochemistry based on various RNA and DNA. To maximize these genosensors' performance for clinical application, these issues need to be resolved. Notwithstanding these difficulties, electrochemistry based on various RNA and DNA have shown considerable promise as a cancer detection tool and may pave the way for the development of more precise, sensitive, and reasonably priced diagnostic systems. To increase their effectiveness and application, further study and development are thus required. We think there are a number of potential avenues and chances for further study and advancement in this field. One approach is to create new, more precise nucleic acid recognition elements for cancer biomarkers, including DNAzymes with increased catalytic activity or aptamers with altered bases or backbones. Investigating novel transducer designs and materials with enhanced electrical characteristics, like graphene, nanowires, carbon nanotubes, or nanocomposites, is an additional choice. Another option is to develop innovative methods for signal amplification, such as rolling circles, exonuclease-assisted amplification, or the use of nanoparticles with increased sensitivity and reduced background noise. An additional choice is to combine genosensors that use electrochemistry and portable tools or microfluidic instruments for tracking cancer biomarkers in vivo or point-of-care diagnostics. In order to confirm the accuracy and dependability of electrochemical genosensors, it is crucial to test them in medical environments using real samples taken from healthy donors or cancer patients. In conclusion, our study provides a comprehensive overview of the state-of-the-art and future prospects of nucleic acid-based electrical genosensors for cancer detection. We think this technology has a lot of promise to help with cancer therapy and early detection. To enhance their functionality, get over their drawbacks, and use them in therapeutic settings, further study and advancement are necessary. We hope that our study will lead to new discoveries in cancer diagnosis and treatment, as well as increased interest and creativity in this area.

References

1. M.A. Farzin, S.M. Naghib, Paper-based immunosensor integrated with bioinspired Cu-polydopamine nanozyme for voltammetric detection of CA-15-3 tumor marker. Sens. Actuators Rep. **8**, 100209 (2024). https://doi.org/10.1016/j.snr.2024.100209
2. M.A. Farzin, S.M. Naghib, Laccase-mimicking Cu@CDs-enabled bioinspired immunosensor using CuO nanoparticles for voltammetric detection of serum thyroxine level in patients with hyperthyroidism or hypothyroidism. Sens. Actuators Rep. **9**, 100291 (2025). https://doi.org/10.1016/j.snr.2025.100291
3. M.A. Farzin, S.M. Naghib, N. Rabiee, Bio-inspired and biomimetic composites based on biodegradable polymers for sensing applications with emphasis on early diagnosis of cancer. Chem. Eng. J. **493**, 152445 (2024). https://doi.org/10.1016/j.cej.2024.152445
4. M.A. Farzin, S.M. Naghib, N. Rabiee, Advancements in bio-inspired self-powered wireless sensors: materials, mechanisms, and biomedical applications. ACS Biomater. Sci. Eng. **10**(3), 1262–1301 (2024). https://doi.org/10.1021/acsbiomaterials.3c01633
5. M. Hosseine, A. Bakhshi, S.M. Naghib, N. Rabiee, Recent advancements in polyaniline-based biosensors for diagnosis of cancers: a comprehensive review. TrAC, Trends Anal. Chem. **181**, 118040 (2024). https://doi.org/10.1016/j.trac.2024.118040

6. A. Abi, Z. Mohammadpour, X. Zuo, A. Safavi, Nucleic acid-based electrochemical nanobiosensors. Biosens. Bioelectron. **102**, 479–489 (2018)
7. S.M. Naghib, B. Ahmadi, M. Mozafari, Stimuli-sensitive chitosan-based nanosystems-immobilized nucleic acids for gene therapy in breast cancer and hepatocellular carcinoma. Curr. Top. Med. Chem. **24**(17), 1464–1489 (2024)
8. O. Abazari, P. Dayati, M. Shahidi, J. ZavarReza, M. Rahmanian, S.M. Naghib, Exploring a desirable quadr-mRNAs panel for non-invasive and ultrasensitive bladder cancer diagnosis: in-silico and clinical studies. Curr. Med. Chem. (2024). https://doi.org/10.2174/0109298673302140240627053904
9. J. Zhao, K. Xia, P. He, G. Wei, X. Zhou, X. Zhang, Recent advances of nucleic acid-based cancer biomarkers and biosensors. Coord. Chem. Rev. **497**, 215456 (2023)
10. R. Salahandish, A. Ghaffarinejad, E. Omidinia, H. Zargartalebi, K. Majidzadeh-A, S.M. Naghib, A. Sanati-Nezhad, Label-free ultrasensitive detection of breast cancer miRNA-21 biomarker employing electrochemical nano-genosensor based on sandwiched AgNPs in PANI and N-doped graphene. Biosens. Bioelectron. **120**, 129–136 (2018). https://doi.org/10.1016/j.bios.2018.08.025
11. R. Salahandish, A. Ghaffarinejad, S.M. Naghib, K. Majidzadeh-A, H. Zargartalebi, A. Sanati-Nezhad, Nano-biosensor for highly sensitive detection of HER2 positive breast cancer. Biosens. Bioelectron. **117**, 104–111 (2018). https://doi.org/10.1016/j.bios.2018.05.043
12. Z. Mazidi, S. Javanmardi, S.M. Naghib, Z. Mohammadpour, Smart stimuli-responsive implantable drug delivery systems for programmed and on-demand cancer treatment: an overview on the emerging materials. Chem. Eng. J. **433**, 134569 (2022). https://doi.org/10.1016/j.cej.2022.134569
13. M. Sadeghi, S. Kashanian, S.M. Naghib, E. Arkan, A high-performance electrochemical aptasensor based on graphene-decorated rhodium nanoparticles to detect HER2-ECD oncomarker in liquid biopsy. Sci. Rep. **12**(1), 3299 (2022). https://doi.org/10.1038/s41598-022-07230-3
14. M. Sadeghi, S. Kashanian, S.M. Naghib, E. Askari, F. Haghiralsadat, D. Tofighi, A highly sensitive nanobiosensor based on aptamer-conjugated graphene-decorated rhodium nanoparticles for detection of HER2-positive circulating tumor cells. Nanotechnol. Rev. **11**(1), 793–810 (2022). https://doi.org/10.1515/ntrev-2022-0047
15. M. Sadeghi, S. Kashanian, S.M. Naghib, F. Haghiralsadat, D. Tofighi, An efficient electrochemical biosensor based on pencil graphite electrode mediated by 2D functionalized graphene oxide to detect HER2 breast cancer biomarker. Int. J. Electrochem. Sci. **17**(4), 220459 (2022). https://doi.org/10.20964/2022.04.62
16. N. Shahbazi, R. Zare-Dorabei, S.M. Naghib, Design of a ratiometric plasmonic biosensor for herceptin detection in HER2-positive breast cancer. ACS Biomater. Sci. Eng. **8**(2), 871–879 (2022). https://doi.org/10.1021/acsbiomaterials.1c01369
17. M. Sadeghi, S. Sadeghi, S.M. Naghib, H.R. Garshasbi, A comprehensive review on electrochemical nano biosensors for precise detection of blood-based oncomarkers in breast cancer. Biosensors **13**(4), 481 (2023)
18. B. Mikaeeli Kangarshahi, S.M. Naghib, N. Rabiee, DNA/RNA-based electrochemical nanobiosensors for early detection of cancers. Crit. Rev. Clin. Lab. Sci. **61**(6), 473–495 (2024). https://doi.org/10.1080/10408363.2024.2321202
19. M. Hosseine, S.M. Naghib, A. Khodadadi, Label-free electrochemical biosensor based on green-synthesized reduced graphene oxide/Fe_3O_4/nafion/polyaniline for ultrasensitive detection of SKBR3 cell line of HER2 breast cancer biomarker. Sci. Rep. **14**(1), 11928 (2024). https://doi.org/10.1038/s41598-024-62231-8
20. B.M. Kangarshahi, S.M. Naghib, Nanogenosensors based on aptamers and peptides for bioelectrochemical cancer detection: an overview of recent advances in emerging materials and technologies. Discover Appl. Sci. **6**(2), 47 (2024)
21. L. Sun, X. Wang, J. Saredy, Z. Yuan, X. Yang, H. Wang, Innate-adaptive immunity interplay and redox regulation in immune response. Redox Biol. **37**, 101759 (2020)

22. M. Hosseine, S.M. Naghib, N. Rabiee, Metal-organic frameworks: biosensor applications for diagnosis of cancers. Microchem. J. **208**, 112341 (2025). https://doi.org/10.1016/j.microc.2024.112341
23. B.M. Gunasekaran, S. Srinivasan, M. Ezhilan, N. Nesakumar, Nucleic acid-based electrochemical biosensors. Clin. Chim. Acta **559**, 119715 (2024). https://doi.org/10.1016/j.cca.2024.119715
24. B. Mikaeeli Kangarshahi, S. Sojdeh, H. Daneshgar, M. Bagherzadeh, S.M. Naghib, N. Rabiee, Electroconductive polymer-based biosensors for early cancer detection via liquid biopsy: advances, challenges, and future prospects. TrAC, Trends Anal. Chem. **183**, 118062 (2025). https://doi.org/10.1016/j.trac.2024.118062
25. A. Hashem, M.M. Hossain, A.R. Marlinda, M.A. Mamun, S. Sagadevan, Z. Shahnavaz, K. Simarani, M.R. Johan, Nucleic acid-based electrochemical biosensors for rapid clinical diagnosis: advances, challenges, and opportunities. Crit. Rev. Clin. Lab. Sci. **59**(3), 156–177 (2022)
26. R. D'Agata, M.C. Giuffrida, G. Spoto, Peptide nucleic acid-based biosensors for cancer diagnosis. Molecules **22**(11), 1951 (2017)
27. S. Morteza Naghib, S. Mahdi Katebi, S. Ghorbanzade, Introduction to electrochemical biosensors, in *Electrochemical Biosensors in Practice: Material and Methods* (Bentham Science Publishers, 2023)
28. S. Morteza Naghib, S. Mahdi Katebi, S. Ghorbanzade, Electrochemical biosensors design steps, in *Electrochemical Biosensors in Practice: Material and Methods* (Bentham Science Publishers, 2023)
29. S. Morteza Naghib, S. Mahdi Katebi, S. Ghorbanzade, Material and biomaterial for biosensing platform, in *Electrochemical Biosensors in Practice: Material and Methods* (Bentham Science Publishers, 2023)
30. S. Morteza Naghib, S. Mahdi Katebi, S. Ghorbanzade, Basic practical principles for studying electro-chemical biosensor, in *Electrochemical Biosensors in Practice: Material and Methods* (Bentham Science Publishers, 2023)
31. S. Morteza Naghib, S. Mahdi Katebi, S. Ghorbanzade, Biosensor application, in *Electrochemical Biosensors in Practice: Material and Methods* (Bentham Science Publishers, 2023)
32. N.A. Abtahi, S.M. Naghib, S.J. Ghalekohneh, Z. Mohammadpour, H. Nazari, S.M. Mosavi, S.M. Gheibihayat, F. Haghiralsadat, J.Z. Reza, B.Z. Doulabi, Multifunctional stimuli-responsive niosomal nanoparticles for co-delivery and co-administration of gene and bioactive compound: in vitro and in vivo studies. Chem. Eng. J. **429**, 132090 (2022). https://doi.org/10.1016/j.cej.2021.132090
33. N.A. Abtahi, S.M. Naghib, F. Haghiralsadat, M. Akbari Edgahi, Development of highly efficient niosomal systems for co-delivery of drugs and genes to treat breast cancer in vitro and in vivo. Cancer Nanotechnol. **13**(1), 28 (2022). https://doi.org/10.1186/s12645-022-00135-w
34. N.A. Abtahi, S.M. Naghib, F. Haghiralsadat, M. Akbari Edgahi, E. Askari, A comparative study on biopharmaceutical function of curcumin and miR-34a by multistimuli-responsive nanoniosome carrier: In-vitro and in-vivo. Front. Mol. Biosci. **9** (2022). https://doi.org/10.3389/fmolb.2022.1043277
35. N.A. Abtahi, S.M. Naghib, F. Haghiralsadat, J.Z. Reza, F. Hakimian, F. Yazdian, D. Tofighi, Smart stimuli-responsive biofunctionalized niosomal nanocarriers for programmed release of bioactive compounds into cancer cells in vitro and in vivo. Nanotechnol. Rev. **10**(1), 1895–1911 (2021)
36. N.A. Abtahi, S. Salehi, S.M. Naghib, F. Haghiralsadat, M.A. Edgahi, S. Ghorbanzadeh, W. Zhang, Multi-sensitive functionalized niosomal nanocarriers for controllable gene delivery in vitro and in vivo. Cancer Nanotechnol. **14**(1), 22 (2023). https://doi.org/10.1186/s12645-023-00175-w

37. S. Afereydoon, F. Haghiralsadat, N. Hamzian, A. Shams, M. Hemati, S.M. Naghib, M. Shabani, B. Zandieh-Doulabi, D. Tofighi, Multifunctional PEGylated niosomal nanoparticle-loaded herbal drugs as a novel nano-radiosensitizer and stimuli-sensitive nanocarrier for synergistic cancer therapy. Front. Bioeng. Biotechnol. **10**, 917368 (2022)
38. S.M. Naghib, S.M. Katebi, S. Ghorbanzade, *Electrochemical Biosensors in Practice: Materials and Methods* (Bentham Science Publishers, 2023)
39. F. Kazemi, S.M. Naghib, Z. Mohammadpour, Multifunctional micro-/nanoscaled structures based on polyaniline: an overview of modern emerging devices. Mater. Today Chem. **16**, 100249 (2020). https://doi.org/10.1016/j.mtchem.2020.100249
40. G.K. Kashani, S.M. Naghib, S. Soleymani, M.R. Mozafari, A review of DNA nanoparticles-encapsulated drug/gene/protein for advanced controlled drug release: current status and future perspective over emerging therapy approaches. Int. J. Biol. Macromol. **268**, 131694 (2024). https://doi.org/10.1016/j.ijbiomac.2024.131694
41. S.M. Naghib, B. Ahmadi, M.R. Mozafari, Smart physicochemical-triggered chitosan-based nanogels for siRNA delivery and gene therapy: a focus on emerging strategies and paradigms for cancer therapy. Curr. Med. Chem. (2024). https://doi.org/10.2174/0109298673286052240426044945
42. M.A. Khorasani, S.M. Naghib, A review of chitosan-based multifunctional nanocomposites for drug/gene/protein delivery and gene therapy in cancer treatments: promises, challenges and outlooks. Int. J. Biol. Macromol. **306**, 141394 (2025). https://doi.org/10.1016/j.ijbiomac.2025.141394
43. R. Pandey, Y. Lu, E.M. McConnell, E. Osman, A. Scott, J. Gu, T. Hoare, L. Soleymani, Y. Li, Electrochemical DNAzyme-based biosensors for disease diagnosis. Biosens. Bioelectron. **224**, 114983 (2023)
44. D.-B. Seo, T.N. Trung, D.-O. Kim, D.V. Duc, S. Hong, Y. Sohn, J.-R. Jeong, E.-T. Kim, Plasmonic Ag-decorated few-layer MoS_2 nanosheets vertically grown on graphene for efficient photoelectrochemical water splitting. Nano-Micro Lett. **12**, 1–14 (2020)
45. P. Vosoughi, S.M. Naghib, B.M. Kangarshahi, M.R. Mozafari, A review of RNA nanoparticles for drug/gene/protein delivery in advanced therapies: current state and future prospects. Int. J. Biol. Macromol. **295**, 139532 (2025). https://doi.org/10.1016/j.ijbiomac.2025.139532
46. P. Vosoughi, S.M. Naghib, M. Rahmanian, M.R. Mozafari, A comprehensive review of mRNA-based vaccines for COVID-19, a new era in pharmaceuticals: unspecified and unknown aspects, effects and challenges. Curr. Top. Med. Chem. (2025). https://doi.org/10.2174/0115680266325847241121034100
47. O. Abazari, P. Dayati, M. Shahidi, J. ZavarReza, M. Rahmanian, S.M. Naghib, Exploring a desirable quadr-mRNAs panel for non-invasive and ultrasensitive bladder cancer diagnosis: in-silico and clinical studies. Curr. Med. Chem. (2025)
48. H. Wang, J. Gao, C. Chen, W. Zhao, Z. Zhang, D. Li, Y. Chen, C. Wang, C. Zhu, X. Ke, PtNi-W/C with atomically dispersed tungsten sites toward boosted ORR in proton exchange membrane fuel cell devices. Nano-Micro Lett. **15**(1), 143 (2023)
49. S.M. Naghib, B. Ahmadi, B. Mikaeeli Kangarshahi, M.R. Mozafari, Chitosan-based smart stimuli-responsive nanoparticles for gene delivery and gene therapy: recent progresses on cancer therapy. Int. J. Biol. Macromol. **278**, 134542 (2024). https://doi.org/10.1016/j.ijbiomac.2024.134542
50. S.M. Naghib, B. Ahmadi, M.R. Mozafari, Stimuli-sensitive chitosan-based nanosystems-immobilized nucleic acids for gene therapy in breast cancer and hepatocellular carcinoma. Curr. Top. Med. Chem. **24**(17), 1464–1489 (2024). https://doi.org/10.2174/0115680266293173240506054439
51. Z. Wang, R. Yu, H. Zeng, X. Wang, S. Luo, W. Li, X. Luo, T. Yang, Nucleic acid-based ratiometric electrochemiluminescent, electrochemical and photoelectrochemical biosensors: a review. Microchim. Acta **186**, 1–19 (2019)
52. B. Sequeira-Antunes, H.A. Ferreira, Nucleic acid aptamer-based biosensors: a review. Biomedicines **11**(12), 3201 (2023)

53. J.-O. Jin, G. Kim, J. Hwang, K.H. Han, M. Kwak, P.C. Lee, Nucleic acid nanotechnology for cancer treatment. Biochim. Biophys. Acta (BBA) Rev. Cancer **1874**(1), 188377 (2020)
54. G. Sharma, M. Singh, K. Sharma, P. Sharma, S. Mehra, Role of nanotechnology in cancer treatment, in *Importance & Applications of Nanotechnology* (MedDocs Publishers, 2021)
55. A. Parihar, A. Singhal, N. Kumar, R. Khan, M.A. Khan, A.K. Srivastava, Next-generation intelligent MXene-based electrochemical aptasensors for point-of-care cancer diagnostics. Nano-Micro Lett. **14**(1), 100 (2022)
56. A.-M. Chiorcea-Paquim, Advances in electrochemical biosensor technologies for the detection of nucleic acid breast cancer biomarkers. Sensors **23**(8), 4128 (2023)
57. L. Tessaro, A. Aquino, P. de Almeida Rodrigues, N. Joshi, R.G. Ferrari, C.A. Conte-Junior, Nucleic acid-based nanobiosensor (NAB) used for Salmonella detection in foods: a systematic review. Nanomaterials **12**(5), 821 (2022)
58. L. Carvajal Barbosa, D. Insuasty Cepeda, A.F. León Torres, M.M. Arias Cortes, Z.J. Rivera Monroy, J.E. Garcia Castaneda, Nucleic acid-based biosensors: analytical devices for prevention, diagnosis and treatment of diseases. Vitae **28**(3) (2021)
59. T. Kawasaki, T. Kawai, S. Akira, Recognition of nucleic acids by pattern-recognition receptors and its relevance in autoimmunity. Immunol. Rev. **243**(1), 61–73 (2011)
60. C. Teller, I. Willner, Functional nucleic acid nanostructures and DNA machines. Curr. Opin. Biotechnol. **21**(4), 376–391 (2010)
61. A. Beigi, S.M. Naghib, A. Matini, M. Tajabadi, M.R. Mozafari, Lipid-based nanocarriers for targeted gene delivery in lung cancer therapy: exploring a novel therapeutic paradigm. Curr. Gene Ther. **25**(2), 92–112 (2025). https://doi.org/10.2174/0115665232292768240503050508
62. A. Shamaeizadeh, A. Beigi, S.M. Naghib, M. Tajabadi, M. Rahmanian, M.R. Mozafari, Smart nanobiomaterials for gene delivery in localized cancer therapy: an overview from emerging materials and devices to clinical applications. Curr. Cancer Drug Targets (2024). https://doi.org/10.2174/0115680096288917240404060506
63. S. Qian, D. Chang, S. He, Y. Li, Aptamers from random sequence space: accomplishments, gaps and future considerations. Anal. Chim. Acta **1196**, 339511 (2022)
64. Y. Du, S. Dong, Nucleic acid biosensors: recent advances and perspectives. Anal. Chem. **89**(1), 189–215 (2017)
65. M. Mascini, I. Palchetti, S. Tombelli, Nucleic acid and peptide aptamers: fundamentals and bioanalytical aspects. Angew. Chem. Int. Ed. **51**(6), 1316–1332 (2012)
66. B.B. Mendes, J. Conniot, A. Avital, D. Yao, X. Jiang, X. Zhou, N. Sharf-Pauker, Y. Xiao, O. Adir, H. Liang, Nanodelivery of nucleic acids. Nat. Rev. Methods Primers **2**(1), 24 (2022)
67. R. Micura, C. Höbartner, Fundamental studies of functional nucleic acids: aptamers, riboswitches, ribozymes and DNAzymes. Chem. Soc. Rev. **49**(20), 7331–7353 (2020)
68. M. Moccia, V. Caratelli, S. Cinti, B. Pede, C. Avitabile, M. Saviano, A.L. Imbriani, D. Moscone, F. Arduini, Based electrochemical peptide nucleic acid (PNA) biosensor for detection of miRNA-492: a pancreatic ductal adenocarcinoma biomarker. Biosens. Bioelectron. **165**, 112371 (2020)
69. E.E. Ferapontova, DNA electrochemistry and electrochemical sensors for nucleic acids. Annu. Rev. Anal. Chem. **11**(1), 197–218 (2018)
70. S. Campuzano, M. Pedrero, J.M. Pingarrón, Electrochemical nucleic acid-based biosensing of drugs of abuse and pharmaceuticals. Curr. Med. Chem. **25**(33), 4102–4118 (2018)
71. H. Zhang, M. Fan, J. Jiang, Q. Shen, C. Cai, J. Shen, Sensitive electrochemical biosensor for MicroRNAs based on duplex-specific nuclease-assisted target recycling followed with gold nanoparticles and enzymatic signal amplification. Anal. Chim. Acta **1064**, 33–39 (2019)
72. S.M. Mousavi, S.A. Hashemi, M.Y. Kalashgrani, V. Rahmanian, A. Gholami, W.-H. Chiang, C.W. Lai, Biomedical applications of an ultra-sensitive surface plasmon resonance biosensor based on smart MXene quantum dots (SMQDs). Biosensors **12**(9), 743 (2022)
73. B. Deiminiat, G.H. Rounaghi, A novel visible light photoelectrochemical aptasensor for determination of bisphenol A based on surface plasmon resonance of gold nanoparticles activated g-C3N4 nanosheets. J. Electroanal. Chem. **886**, 115122 (2021)

74. A. Sabahi, R. Salahandish, A. Ghaffarinejad, E. Omidinia, Electrochemical nano-genosensor for highly sensitive detection of miR-21 biomarker based on SWCNT-grafted dendritic Au nanostructure for early detection of prostate cancer. Talanta **209**, 120595 (2020)
75. W. Zhang, G. Xiao, J. Chen, L. Wang, Q. Hu, J. Wu, W. Zhang, M. Song, J. Qiao, C. Xu, Electrochemical biosensors for measurement of colorectal cancer biomarkers. Anal. Bioanal. Chem. **413**, 2407–2428 (2021)
76. J. Chang, X. Wang, J. Wang, H. Li, F. Li, Nucleic acid-functionalized metal-organic framework-based homogeneous electrochemical biosensor for simultaneous detection of multiple tumor biomarkers. Anal. Chem. **91**(5), 3604–3610 (2019). https://doi.org/10.1021/acs.analchem.8b05599
77. L. Yuan, L. Liu, Peptide-based electrochemical biosensing. Sens. Actuators, B Chem. **344**, 130232 (2021)
78. E.G. Ramirez-Valles, A. Rodríguez-Pulido, M. Barraza-Salas, I. Martínez-Velis, I. Meneses-Morales, V.M. Ayala-García, C.A. Alba-Fierro, A quest for new cancer diagnosis, prognosis and prediction biomarkers and their use in biosensors development. Technol. Cancer Res. Treat. **19**, 1533033820957033 (2020)
79. S. Sargazi, I. Fatima, M.H. Kiani, V. Mohammadzadeh, R. Arshad, M. Bilal, A. Rahdar, A.M. Díez-Pascual, R. Behzadmehr, Fluorescent-based nanosensors for selective detection of a wide range of biological macromolecules: a comprehensive review. Int. J. Biol. Macromol. **206**, 115–147 (2022)
80. J. Kappen, M. Skorupa, K. Krukiewicz, Conducting polymers as versatile tools for the electrochemical detection of cancer biomarkers. Biosensors **13**(1), 31 (2022)
81. T. Dahiya, S. Yadav, N. Yadav, A. Mann, M. Sharma, J. Rana, Monitoring of BNP cardiac biomarker with major emphasis on biosensing methods: a review. Sens. Int. **2**, 100103 (2021)
82. E. Askari, S.M. Naghib, A. Seyfoori, A. Maleki, M. Rahmanian, Ultrasonic-assisted synthesis and in vitro biological assessments of a novel herceptin-stabilized graphene using three dimensional cell spheroid. Ultrason. Sonochem. **58**, 104615 (2019)
83. W. Song, Y. Song, Q. Li, C. Fan, X. Lan, D. Jiang, Advances in aptamer-based nuclear imaging. Eur. J. Nucl. Med. Mol. Imaging **49**(8), 2544–2559 (2022)
84. S. Liu, Y. Xu, X. Jiang, H. Tan, B. Ying, Translation of aptamers toward clinical diagnosis and commercialization. Biosens. Bioelectron. **208**, 114168 (2022)
85. S. Aiman, Y. Alhamhoom, F. Ali, N. Rahman, L. Rastrelli, A. Khan, Q.u.A. Farooq, A. Ahmed, A. Khan, C. Li, Multi-epitope chimeric vaccine design against emerging Monkeypox virus via reverse vaccinology techniques-a bioinformatics and immunoinformatics approach. Front. Immunol. **13**, 985450 (2022)
86. S.-M. Yang, S. Lv, W. Zhang, Y. Cui, Microfluidic point-of-care (POC) devices in early diagnosis: a review of opportunities and challenges. Sensors **22**(4), 1620 (2022)
87. A. Haleem, M. Javaid, R.P. Singh, R. Suman, S. Rab, Biosensors applications in medical field: a brief review. Sens. Int. **2**, 100100 (2021)
88. B. Guan, X. Zhang, Aptamers as versatile ligands for biomedical and pharmaceutical applications. Int. J. Nanomed. 1059–1071 (2020)
89. N. Pandey, D. Biswas, N. Dutta, A. Hansda, G. Dutta, G. Mukherjee, Sensing soluble immune checkpoint molecules and disease-relevant cytokines in cancer: a novel paradigm in disease diagnosis and monitoring. Front. Sens. **3**, 789771 (2022)
90. P. Lakhera, V. Chaudhary, A. Jha, R. Singh, P. Kush, P. Kumar, Recent developments and fabrication of the different electrochemical biosensors based on modified screen printed and glassy carbon electrodes for the early diagnosis of diverse breast cancer biomarkers. Mater. Today Chem. **26**, 101129 (2022)
91. C. Dejous, U.M. Krishnan, Sensors for diagnosis of prostate cancer: looking beyond the prostate specific antigen. Biosens. Bioelectron. **173**, 112790 (2021)
92. M. Daniel, G. Mathew, M. Anpo, B. Neppolian, MOF based electrochemical sensors for the detection of physiologically relevant biomolecules: an overview. Coord. Chem. Rev. **468**, 214627 (2022)

93. U. Anand, A.K.S. Chandel, P. Oleksak, A. Mishra, O. Krejcar, I.H. Raval, A. Dey, K. Kuca, Recent advances in the potential applications of luminescence-based, SPR-based, and carbon-based biosensors. Appl. Microbiol. Biotechnol. **106**(8), 2827–2853 (2022)
94. A. Hashem, M.M. Hossain, M. Al Mamun, K. Simarani, M.R. Johan, Nanomaterials based electrochemical nucleic acid biosensors for environmental monitoring: a review. Appl. Surf. Sci. Adv. **4**, 100064 (2021)
95. N. Bidar, M. Amini, F. Oroojalian, B. Baradaran, S.S. Hosseini, M.-A. Shahbazi, M. Hashemzaei, A. Mokhtarzadeh, M.R. Hamblin, M. de la Guardia, Molecular beacon strategies for sensing purpose. TrAC, Trends Anal. Chem. **134**, 116143 (2021)
96. J. Das, S.O. Kelley, High-performance nucleic acid sensors for liquid biopsy applications. Angew. Chem. **132**(7), 2574–2584 (2020)
97. D.M. Kim, S.M. Yoo, DNA-modifying enzyme reaction-based biosensors for disease diagnostics: recent biotechnological advances and future perspectives. Crit. Rev. Biotechnol. **40**(6), 787–803 (2020)
98. H. Yang, J. Ren, M. Zhao, C. Chen, F. Wang, Z. Chen, Novel electrochemical immunosensor for O^6-methylguanine-DNA methyltransferase gene methylation based on graphene oxide-magnetic nanoparticles-β-cyclodextrin nanocomposite. Bioelectrochemistry **146**, 108111 (2022)
99. A. Atapour, H. Khajehzadeh, M. Shafie, M. Abbasi, S. Mosleh-Shirazi, S.R. Kasaee, A.M. Amani, Gold nanoparticle-based aptasensors: a promising perspective for early-stage detection of cancer biomarkers. Mater. Today Commun. **30**, 103181 (2022)
100. B. Zhang, K. Yao, E. Zhou, L. Zhang, C. Cheng, Chr20q amplification defines a distinct molecular subtype of microsatellite stable colorectal cancer. Can. Res. **81**(8), 1977–1987 (2021)
101. J. Gačanin, C.V. Synatschke, T. Weil, Biomedical applications of DNA-based hydrogels. Adv. Func. Mater. **30**(4), 1906253 (2020)
102. I.-H. Cho, D.H. Kim, S. Park, Electrochemical biosensors: perspective on functional nanomaterials for on-site analysis. Biomater. Res. **24**(1), 6 (2020)
103. Y. Hua, J. Ma, D. Li, R. Wang, DNA-based biosensors for the biochemical analysis: a review. Biosensors **12**(3), 183 (2022)
104. A.A. Saeed, J.L.A. Sánchez, C.K. O'Sullivan, M.N. Abbas, DNA biosensors based on gold nanoparticles-modified graphene oxide for the detection of breast cancer biomarkers for early diagnosis. Bioelectrochemistry **118**, 91–99 (2017)
105. Y. Lotan, J. de Jong, J.A. Proudfoot, S. Daneshmand, R.S. Svatek, V.M. Narayan, E. Gibb, E. Davicioni, S. Joshi, R. Li, A non-coding RNA based classifier for favorable outcomes in clinically organ confined bladder cancer. Am. Soc. Clin. Oncol. J. (2025)
106. J. Ouyang, X. Zhan, S. Guo, S. Cai, J. Lei, S. Zeng, L. Yu, Progress and trends on the analysis of nucleic acid and its modification. J. Pharm. Biomed. Anal. **191**, 113589 (2020)
107. P.B. Tiwade, V. Fung, R. VanKeulen-Miller, E.A. Narasipura, Y. Ma, O.S. Fenton, Non-viral RNA therapies for non-small cell lung cancer and their corresponding clinical trials. Mol. Pharm. (2025)
108. R. Prakash, K. Pande, A. Kannan, Exosomal non-coding RNAs: a future of therapeutics in regulating oxidative stress induced breast cancer, in *Handbook of Oxidative Stress in Cancer: Therapeutic Aspects* (Springer, 2021), pp. 1–34
109. V.K. Sarhadi, G. Armengol, Molecular biomarkers in cancer. Biomolecules **12**(8), 1021 (2022)
110. F. Tian, S. Zhang, C. Liu, Z. Han, Y. Liu, J. Deng, Y. Li, X. Wu, L. Cai, L. Qin, Protein analysis of extracellular vesicles to monitor and predict therapeutic response in metastatic breast cancer. Nat. Commun. **12**(1), 2536 (2021)
111. T.-M. Pan, P.-Y. Liao, High sensitivity and rapid detection of KRAS and BRAF gene mutations in colorectal cancer using $YbTi_xO_y$ electrolyte-insulator-semiconductor biosensors. Mater. Today Chem. **25**, 100979 (2022)
112. D.-G. Macovei, M.-B. Irimes, O. Hosu, C. Cristea, M. Tertis, Point-of-care electrochemical testing of biomarkers involved in inflammatory and inflammatory-associated medical conditions. Anal. Bioanal. Chem. **415**(6), 1033–1063 (2023)

113. P. Fu, S. Xing, M. Xu, Y. Zhao, C. Zhao, Peptide nucleic acid-based electrochemical biosensor for simultaneous detection of multiple microRNAs from cancer cells with catalytic hairpin assembly amplification. Sens. Actuators, B Chem. **305**, 127545 (2020). https://doi.org/10.1016/j.snb.2019.127545
114. R.R. Kumar, A. Kumar, C.-H. Chuang, M.O. Shaikh, Recent advances and emerging trends in cancer biomarker detection technologies. Ind. Eng. Chem. Res. **62**(14), 5691–5713 (2023)
115. D.K. Ban, Y. Liu, Z. Wang, S. Ramachandran, N. Sarkar, Z. Shi, W. Liu, A.G. Karkisaval, E. Martinez-Loran, F. Zhang, Direct DNA methylation profiling with an electric biosensor. ACS Nano **14**(6), 6743–6751 (2020)
116. M. Li, F. Yin, L. Song, X. Mao, F. Li, C. Fan, X. Zuo, Q. Xia, Nucleic acid tests for clinical translation. Chem. Rev. **121**(17), 10469–10558 (2021)
117. B.M. Felsheim, A. Fernandez-Martinez, C. Fan, A.D. Pfefferle, M.C. Hayward, K.A. Hoadley, N.U. Rashid, S.M. Tolaney, G. Somlo, L.A. Carey, Prognostic and molecular multi-platform analysis of CALGB 40603 (Alliance) and public triple-negative breast cancer datasets. NPJ Breast Cancer **11**(1), 24 (2025)
118. M. Senel, M. Dervisevic, F. Kokkokoğlu, Electrochemical DNA biosensors for label-free breast cancer gene marker detection. Anal. Bioanal. Chem. **411**, 2925–2935 (2019)
119. T.-T. Liang, X. Qin, Y. Xiang, Y. Tang, F. Yang, Advances in nucleic acids-scaffolded electrical sensing of extracellular vesicle biomarkers. TrAC, Trends Anal. Chem. **148**, 116532 (2022)
120. N.B. Aziz, R.G. Mahmudunnabi, M. Umer, S. Sharma, M.A. Rashid, Y. Alhamhoom, Y.-B. Shim, C. Salomon, M.J. Shiddiky, MicroRNAs in ovarian cancer and recent advances in the development of microRNA-based biosensors. Analyst **145**(6), 2038–2057 (2020)
121. K. Ondraskova, R. Sebuyoya, L. Moranova, J. Holcakova, P. Vonka, R. Hrstka, M. Bartosik, Electrochemical biosensors for analysis of DNA point mutations in cancer research. Anal. Bioanal. Chem. **415**(6), 1065–1085 (2023)
122. A. Mohammadpour-Haratbar, Y. Zare, K.Y. Rhee, Electrochemical biosensors based on polymer nanocomposites for detecting breast cancer: recent progress and future prospects. Adv. Coll. Interface. Sci. **309**, 102795 (2022)
123. D. Chakraborty, D. Ghosh, S. Kumar, D. Jenkins, N. Chandrasekaran, A. Mukherjee, Nano-diagnostics as an emerging platform for oral cancer detection: current and emerging trends. Wiley Interdiscip. Rev. Nanomed. Nanobiotechnol. **15**(1), e1830 (2023)
124. S. Prajapati, R. Malviya, P. Goyal, Treatment and detection of oral cancer using biosensors: advances and prospective, in *Soft Materials-Based Biosensing Medical Applications* (2025), pp. 449–466
125. Y. Shang, J. Sun, Y. Ye, J. Zhang, Y. Zhang, X. Sun, Loop-mediated isothermal amplification-based microfluidic chip for pathogen detection. Crit. Rev. Food Sci. Nutr. **60**(2), 201–224 (2020)
126. S. Kumar, R. Singh, Recent optical sensing technologies for the detection of various biomolecules. Opt. Laser Technol. **134**, 106620 (2021)
127. L.A. Stanciu, Q. Wei, A.K. Barui, N. Mohammad, Recent advances in aptamer-based biosensors for global health applications. Annu. Rev. Biomed. Eng. **23**(1), 433–459 (2021)
128. S. Akgönüllü, A. Denizli, Recent advances in optical biosensing approaches for biomarkers detection. Biosens. Bioelectron. X **12**, 100269 (2022)
129. S. Li, Q. Ma, Electrochemical nano-sensing interface for exosomes analysis and cancer diagnosis. Biosens. Bioelectron. **214**, 114554 (2022)
130. K.E. Keenan, J.G. Delfino, K.V. Jordanova, M.E. Poorman, P. Chirra, A.S. Chaudhari, B. Baessler, J. Winfield, S.E. Viswanath, N.M. deSouza, Challenges in ensuring the generalizability of image quantitation methods for MRI. Med. Physi. **49**(4), 2820–2835 (2022)
131. D. Wachholz Junior, P.B. Deroco, B.M. Hryniewicz, L.T. Kubota, Strategies for electrochemical point-of-care biosensors. Ann. Rev. Anal. Chem. **18** (2025)
132. S. Gupta, V. Mishra, A.A. Aljabali, A. Albutti, R. Kanday, M. El-Tanani, Y. Mishra, Breaking barriers in cancer diagnosis: unveiling the 4Ms of biosensors. RSC Adv. **15**(10), 8019–8052 (2025)

133. N. Formisano, P. Jolly, N. Bhalla, M. Cromhout, S.P. Flanagan, R. Fogel, J.L. Limson, P. Estrela, Optimisation of an electrochemical impedance spectroscopy aptasensor by exploiting quartz crystal microbalance with dissipation signals. Sens. Actuators, B Chem. **220**, 369–375 (2015)
134. S. Mukhtar, J.A. Bhat, M. Chesti, M. Bhat, S.H. Ul Baligah, T. Dolker, T. Mansoor, D. Sultan, S.A. Jan, Application of sensors and biosensors in soil contamination detection, in *Integrated Land and Water Resource Management for Sustainable Agriculture*, vol. 2 (Springer, 2025), pp. 167–184
135. A. Sarsenbayeva, S. Sadak, I. Kucuk, L. Kudreyeva, A.M. Bakytzhanovna, B. Uslu, Molybdenum-based electrochemical sensors for breast cancer biomarker detection: advances and challenges. Crit. Rev. Anal. Chem. 1–21 (2025)

Chapter 6
Stimuli-responsive Hydrogels for Tissue Engineering and Targeted Drug Administration via Multi-dimensional Printing: An Encouraging Strategy for Innovative Smart Materials and Technologies

Abstract One of the most groundbreaking developments in the biomedical industry's recent history, is the creation and spread of 3D printing. Nevertheless, there are several drawbacks in 3D printing technology, such as its inability to produce dynamic human tissues efficiently. Considered the fourth dimension at the time, the 4D printing technique has been seen as the next revolution in 3D printing since it involves the fourth-dimension parameter, solves this problem and allows for the creation of dynamic and living items. Consequently, the 3D additive manufacturing method is replaced with a 4D one, and stimuli-responsive materials are highly valued and recognized as one of the most crucial materials in this four-dimensional production process because of their characteristics. Investigating the developments in incorporating responsive material behavior into the 4D printing process, has been attempted in this chapter. Furthermore, current studies that investigate the relationships between various stimuli (categorized as chemical and biological signals) and associated stimulus-responsive compounds will be the primary emphasis of this study. Following a discussion of the process's potential capabilities, technical issues and opportunities for future research, the tremendous potential of the 4D printing of smart materials in medical applications particularly tissue engineering and localized drug delivery, will be covered.

Keywords 4D printing · Hydrogels · Tissue engineering · Targeted medication administration · Chemical and biological stimuli-responsive

6.1 Introduction

Hydrogels are 3D polymer networks capable of absorbing and retaining large amounts of water or biological fluids typically at least 10% by weight while maintaining their structural integrity [1–3]. Stimuli-responsive biological hydrogels are advanced smart biomaterials that dynamically respond to environmental cues, making them highly promising for tissue engineering, regenerative medicine, drug delivery and biomedical applications [4–9]. Although 3D printing was not formally

S. M. Naghib, *Nano-Biosensors, Drug Delivery and Tissue Engineering*, Series in BioEngineering, https://doi.org/10.1007/978-981-95-1888-3_6

introduced until the end of the twentieth century, the idea behind it was developed the year before. An estimated 0.2–0.6 trillion dollars are produced worldwide as a result of 3D printing, which is a substantial sum [10]. The main idea behind 3D printing technology is that intricate 3D structures with the required geometry are produced by precisely depositing the base material layer by layer using a nozzle [11–16]. The advent of additive printing techniques, such as 3D and 4D printing, has accelerated the growth and development of technologies, especially those in the biomedical sector [17–21]. The process's great capacity to turn preliminary plans and ideas into the original concept is one of the primary justifications for this approach. Biomedical devices may be prepared and manufactured using additive printing, which encompasses 3D and 4D printing, offering novel treatment options and solutions [3, 11, 16, 22–24]. Damaged tissues or organs may be repaired or replaced using biological materials and cell-based therapies, tissue engineering being the term used for this approach. The three fundamental building blocks of tissue engineering are a matrix (scaffold), cells (primary lineage or stem cells), and signal (mechanical, physical, electrical, and/or molecules: proteins, peptides, and cytokines) tissue. To accomplish the intended outcome in tissue engineering, these three bases must be taken into account [11, 16, 20, 22–24]. In addition to being crucial for cell proliferation, differentiation, and vaccination, scaffolding plays a significant role in tissue engineering [25]. The size and porosity of the generated scaffold have a direct impact on angiogenesis's oxygen supply and tissue development. Due to their inability to control the creation of pores and porosity, common biological scaffold manufacturing techniques such as the foam gas method, electrospinning, phase separation, solvent casting/particle washing, and dry ice emulsion cannot produce optimal 3D scaffolds. The 3D printing industry produces scaffolds with complicated geometries and excellent porosity, but it also slows or disturbs the delivery of nutrients to the cell and inhibits cell development within the scaffold due to its 3D structure or lack of linked pores. Because of this, 3D printing is regarded as a great way to create scaffolds in tissue engineering [17, 26]. Although complicated and precise biomedical devices may be produced with the use of 3D printing technology, the pieces produced by this technique are unable to mimic the tissue of living things, including people, since they lack dynamics [8]. This technique restricts the capacity to generate numerous biological goods associated to complicated bodily organs since human tissues continually interact with their surroundings, making static 3D printed objects unable to adapt to organs and tissues over time. For instance, mechanotransduction, a significant dynamic modification of the cellular environment that routinely and constantly results from the contractile force, takes place throughout the healing process of live tissue. It has an impact on the tissue, altering gene expression and cell behavior. All vital tissue activities are regulated by these biological dynamics [27–29]. The printed product's structure is thus perfect as it can react to biological dynamics, doesn't interfere with bodily functions, and adjusts to the specific biological organ. 4D printing is a crucial procedure that enhances the capacity to generate dynamic and reactive biological products in the human body, and it is one of the fundamental remedies for 3D printing's fundamental flaw, which is the result of static products against biological and physical reactions. In tissue regeneration, it shows promise [30]. The fourth dimension, time,

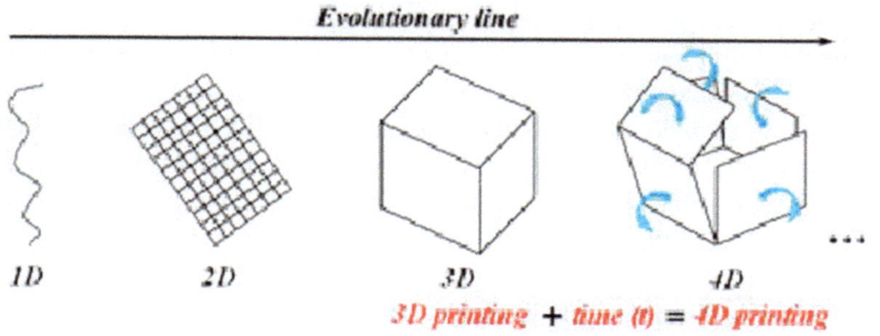

Fig. 6.1 Schematic illustration of 1D, 2D, 3D, and 4D concepts [31]

has been employed for manufacturing dynamics in this work in order to break free from the static state of 3D printed items and apply dynamic behavior in produced products, taking into account this change over time as another facet of production. It offers a tool for developing more intelligent and compatible devices, particularly for biomedical applications. This may be accomplished by the clever manipulation of certain materials' characteristics, which can be drastically changed or modified over time by physical, chemical, biological, or a variety of external stimuli [19–21]. For the 4D printing business to advance significantly and produce intelligent and optimum items, time-dependent change and transformation must be predictable (Fig. 6.1) [31]. A hydrogel is among the most crucial substances that might result in the creation of dynamic samples. One of the most crucial choices for creating or repairing tissue in tissue engineering is hydrogel. To put it simply, hydrogels are networks of three-dimensional polymers joined by cross-links. The "soft" aspect of hydrogels, which results from their hydrophilic and insoluble qualities, is its main attribute [32, 33]. By absorbing water, hydrogels achieve thermodynamic equilibrium and form a solid, hydrated substance with characteristics and functions similar to those of cells. Hydrogels may be employed in a variety of therapeutic applications due to their hydrophilic nature and structure, which provide advantageous biocompatibility qualities. Materials like agarose, chitosan, alginate, fibrin, poly (ethylene glycol) (PEG), poly (vinyl alcohol) (PVA), or poly (2-hydroxyethyl methacrylate) (PHEMA) are among the often-utilized hydrogels. When hydrogels transform into a material that reacts to stimuli, it is one of their most significant uses. The necessity for thorough data on the long-term behavior of stimulus-responsive materials in response to chemical and biological stimuli is covered in this study [34–42].

6.2 Hydrogels that Respond to Stimuli

The soft material class that includes hydrogels and gels has a high-water content in its structure. In other words, water makes up around 95% of the structure of hydrogels and gels [43]. Natural or artificial materials joined by connections including hydrogen bonds, receptor-receptor interactions, and metal ligands may serve as gelling agents in hydrogels. Important categories of natural polymers include polysaccharides, polypeptides, and nucleic acids [44]. Because of their high water content and unique bonding properties, hydrogels have a flexible structure [45]. These connections may be covalent or the result of host-guest interactions. A flexible three-dimensional network may sometimes be created using both kinds of bonds. Because hydrogels are soft, flexible, and have the ability to change their properties, they can be used in a wide range of pharmaceutical applications where they can easily swell and deform. This class of materials has garnered a lot of interest in novel medicinal applications because of its great potential [46]. Hydrogels that respond to specific environmental stimuli or a combination of many stimuli are known as stimulant-responsive hydrogels [3, 4, 6, 8, 9, 47, 48]. The hydrogel may undergo structural or mechanical changes as a consequence of the response [49]. Stimuli may be broadly divided into three categories: those pertaining to the body, those involving chemicals, and those involving living things (Fig. 6.2) [50]. Researchers have faced challenges in developing smart or intelligent hydrogels by modifying their physicochemical properties. These advanced hydrogels can respond to a range of stimuli, including physical factors like temperature, light, electromagnetic fields, pressure and ultrasound radiation, chemical triggers such as pH, glucose levels and ionic strength, and biological signals including enzymes and antigen–antibody interactions. Such materials are commonly referred to as stimuli-responsive hydrogels, as illustrated in Fig. 6.2. Sorting stimuli according to their origins inside or outside the body is another method. Special gels known as multi-stimulus responsive hydrogels have the ability to respond to several triggering stimuli [51–53]. Hydrogels that may respond to changes or modify their chemical makeup, for example, might detect odd signals in wounded tissues and initiate a natural process to carry out certain duties, such as drawing in cells [54, 55]. Modification and recombination of different gelling agents and cross-links within the hydrogel may result in a hydrogel's responsiveness to stimuli. This allows for the creation of varied features, such as controlled stiffness change or degradability [56]. The hydrogel may become stiffer or gel if the transverse connections between the gelling agents are increased; conversely, if the transverse connections are broken, the hydrogel may become less rigid or disintegrate. Certain cross-links of this problem, i.e., the appearance of certain changeable and reversible features, controlled hardness adjustment or sol-to-gel transition, or the facilitation and improvement of functional hydrogels for two hydrogels, are caused by the growth or reduction of these connections or several kinds of cross-links, such as permanent and stimuli-responsive cross-links; altering or breaking one of them causes stiffness to either increase or decrease [57, 58]. In general, the way several compounds interact with one another determines the response process. The hydrogels' structure and characteristics are impacted by this interaction, which may result in alterations such as size changes, shape changes, or structural changes in the water-based gel [3, 4, 6, 8, 9, 47, 48]. When creating and managing hydrogels that may react to changes, gel and its interactions with the liquid it is in are crucial. Three steps may be taken in the production of hydrogel. Primary bound water is first produced when the gel-forming agents absorb water molecules and become hydrated. The hydrophobic groups then make contact with the water molecules. A force known as the osmotic force is produced when the restricted water is placed in a certain location. A balanced system and the formation of the intended hydrogel are the end consequences of the crosslinks preventing the large amount of water that enters the network and attempts to dissolve the whole system in gelling agents. The force in the system pushes toward becoming less concentrated when gels that are joined together and dissolve readily in water are submerged in water. Nevertheless, the hydrogel maintains the proper volume and water content and may withstand dilution thanks to chemical or physical cross-linking. Its swelling ratio is determined by this. More connections between molecules in a hydrogel often translate into less swelling, greater strength, and fewer molecular gaps [59]. In other words, the characteristics of the gelling agents and the cross-links that form the sensitive hydrogel qualities determine the gelling-gelling and gelling-solvent interactions. SEM is a powerful tool for examining the appearance and surface characteristics of objects. It is also useful for determining the size and quantity of pores may monitor and determine the hydrogel's stability, stiffness, and thickness using a device called Rheometer [43, 60].

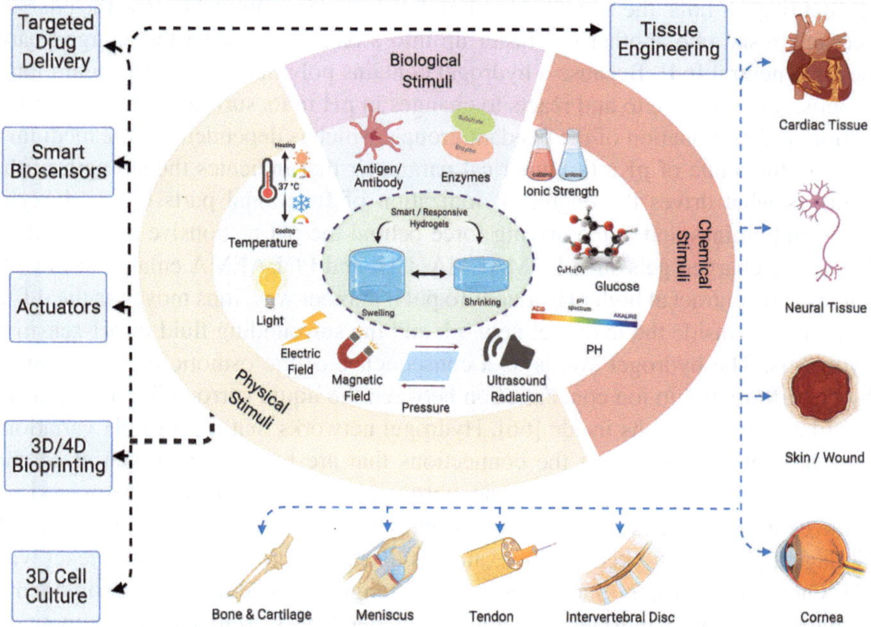

Fig. 6.2 Schematic illustration of different smart/stimuli-responsive hydrogels employed for different biomedical applications [50]

6.3 Hydrogels that Respond to Chemical Stimuli

Among the most prevalent features of many illnesses, such as cancer, heart disease, or even several infections in various bodily tissues, are changes in physiological conditions, such as variations in pH or ion concentrations [45, 61]. This feature may serve as a guide for the development of printed biological components and chemicals thanks to its demonstrated capabilities. It is useful in many medical applications, such as the creation of thermal systems for tissue growth and the targeted delivery of medications for the treatment of cancer [44, 62, 63].

6.3.1 Hydrogels that Respond to pH

Materials may weaken, expand, shrink, or disintegrate in response to pH variations. This occurs when the substance's bonds are disrupted by acid or when certain groups in the material acquire a positive charge. Materials that react to pH change form when the pH rises or falls: the long chains of molecules either coil up because the charged sections resist one another or cluster up into a ball shape when the charged parts become neutral [64]. Because a hydrogel contains polymer chains of pendant ionic groups, it is sensitive to and reacts to changes in pH in its surroundings. The protonation or deprotonation of the pendant groups, which is dependent on the medium's pH and the value of pKa (a numerical parameter that indicates the strength of the acid), is what drives the degree of ionization of functional parts like carboxylic acids and amines and is the driving force behind the pH-responsive property [65]. Positively charged gels such PDMAEMA, CH, and PDEAEMA enlarge at low pH values and contract at high pH values. To put it another way, ions move via the diffusion process inside the hydrogel network and the surrounding fluid in pH-sensitive hydrogels. The hydrogel swells as a consequence of the osmotic pressure caused by the differential in ion concentration between the liquid surrounding the squishy, gel-like network and its inside [66]. Hydrogel networks that react to pH variations may be created by varying the connections that are formed or broken at certain pH levels. Schiff base, acyl hydrazone, boronate-catechol, ionizable groups, and are some examples of these relationships. triplex DNA structures and i-motif [65, 67]. Collagen, gelatin, and keratin are examples of natural proteins that may react with pH. One crucial protein that supports and strengthens several bodily components, including our bones, muscles, and skin, is collagen [68]. Certain natural compounds, such as albumin, gelatin, CH, and Alg, may undergo size alterations in response to pH variations. Generally, polymers with unique chemical groups such as $-COOH$, $-OH$, $-NH_2$, $-N<$, $-CONH_2$, and $-SO_3H$ are used to create polymer platforms that react to pH variations [69]. Humans are naturally endowed with keratin, a protein that is abundant in hair and nails. Keratin is a protein that reacts to pH in contrast to collagen, keratin expands differentially in various pH conditions [70]. According to Ramos et al., keratin becomes smaller and less structured in acid. On the other hand, keratin expands significantly at pH values over 6, reaching its maximal expansion at pH values above 8 [71]. Research has used collagen and keratin's pH-responsiveness to safely release beneficial compounds like doxorubicin or antimicrobial zinc oxide nanosheets to treat cancer and prevent wound infections. Because natural proteins include certain components that aid in cell adhesion, they are generally believed to be chemicals that may enhance cell contact and development on materials [72]. Put more simply, Bear et al. discovered that at neutral pH, the degree of edema is minimal. A material is said to be "neutral" if its pH falls between 4 and 8. However, when things are subjected to slightly acidic or slightly alkaline environments, they swell the greatest. A reaction occurs when the pH falls between 2 and 4 or between 8 and 12. Swelling is reduced when the levels of acid or alkalinity are high. It states that pH 12 is very alkaline and pH 2 is extremely acidic [73]. Danan's idea states that the protein's surface becomes more positively or negatively charged when the pH is not at collagen's isoelectric point. More ions with opposing charges are created within the collagen gel as a result. Swelling results from the movement of osmotic forces, which are initiated by these ions with varying charges. The use of AM technology for protein printing has been the subject of several investigations. Though their special ability to react to pH has not yet been explored, the primary objective has been to develop printed structures that might enhance cell interactions [74]. For drug release applications, a broad range of pH-responsive synthetic polymers have been used in addition to natural polymers. Synthetic polymers including PAA, PMAA, PAAm, PDMAEMA, PDEAEMA, and their combinations are often utilized to make hydrogel systems that respond to pH variations. In this industry, these polymers are well-known and often used. Drug delivery systems with precise sizes and patterns that were impossible to produce with conventional manufacturing techniques were made possible by printing technology. To establish precise patterns for the release of medications, this precision is crucial. La Roche's approach included creating a unique gel from acrylic acid and turning it into tablets. In very acidic settings, these tablets may release medicine rapidly. Likewise, Okwuosa and associates tablets that are printed in a core-shell pattern, with the core and shell made of polyvinylpyrrolidone and methacrylic acid copolymer, respectively [75, 76]. These two trials' tablets were able to withstand stomach acid and release medication according to the degree of intestinal acidity. They might thus be used as a medicine delivery method via the digestive tract. To administer paracetamol, Goyans and his colleagues attempted to create pills using hypromellose acetate succinate. They discovered that only under certain acidic circumstances would the pills release the paracetamol gradually. To do this, they used a variety of enteric polymers. This greatly aids in the development of customized medications with varying dosages for the unique medical requirements of every patient [77]. pH response has proved useful in developing dressings with many uses, such as GelDerm, which is seen in Fig. 6.3, in addition to delivering medications. GelDerm is a unique kind of bandage that changes color in response to pH variations, indicating if you have a bacterial infection. Additionally, it has the ability to deliver medication directly at the incision to combat infection [72, 78]. A particular kind of cloth manufactured from a polymer known as alginate served as the study's material. There were unique color-changing beads in this cloth. A material known as mesoporous resin, which included a color that could react to pH variations, was used to fill these beads. Additionally, to make flow-controlling valves, employ synthetic polymers that react to pH changes. Nadgorny et al. discovered that a particular kind of gel composed of poly(2-vinyl pyridine) may alter in size in response to variations in the pH of its environment. The addition of acrylonitrile-butadiene-styrene gives this gel strength and flexibility. The flow of liquids through the gel may be regulated by this size change. Because natural pH-responsive polymers have worse mechanical qualities, their usage in practical settings has often been restricted. Because of their capacity to promote tissue development in the first healing phase after implantation, pH-sensitive hydrogels are often used in wound healing [24, 79–81]. In contrast to non-responsive and insensitive materials, scaffolds made of pH-responsive materials show swelling/expansion behavior in acidic environments, which has resulted in improved cell penetration and oxygen transfer. pH-sensitive hydrogels have also been investigated by researchers as potential materials for growth agents that release gradually. Certain chemicals known as GFs aid in regulating the growth, migration, multiplication, and gene usage of cells. The acidity of the surrounding environment may aid in the release of growth factors in a particular cell region, promoting tissue development. GFs are often added to hydrogels either during or after the hydrogel's creation. Covalent or non-covalent techniques, such as direct loading, enclosing them in a carrier, covalently joining them, or reversibly joining them, may be used to accomplish this. However, when GFs are delivered for tissue engineering purposes, the interactions that take place between them, their receptors, and other ECM components are crucial [82]. Developed incredible skin-inspired wearable technology that incorporates hydrogels, elastomers, electric circuits, and fluid channels. They followed a three-step process. To preserve their original structure, they first molded the hydrogels and elastomers before joining them. Then, by adding benzophenone, the researchers improved the procedure for combining hydrogels with hardened elastomers. In conclusion, tough hydrogels like PAAm-Alg, PAAm-hyaluronan, PAAm-CH, and PEGDA-Alg may be used to create surfaces that are durable and robust. The greatest features of elastomers and hydrogels were blended in these robust systems. Consequently, they may be used to a wide range of applications, including the prevention of dehydration, the development of flexible circuit boards using stretchy hydrogels, and the creation of flexible and interactive microfluidic devices may combine certain polysaccharides with heat-sensitive polymers to create hydrogels that react to two separate stimuli. When the body's temperature or pH fluctuates, this unique mixture of substances may also [43]. Researchers created hydrogels that can react to variations in temperature and pH by using a process known as graft polymerization in conjunction with chemical crosslinking. They accomplished this by affixing cellulose nanowhiskers to a substance known as AAm. The CNWs and AAm polymers that were made using NIPAAm were joined by MBAAm crosslinkers to produce the best hydrogel [83]. PH-sensitive hydrogels have significant promise for a variety of medicinal applications, including the transportation and blending of biomolecules. They were often developed as oral drug release devices that regulated the dosage of medication administered to the body. Additionally, they were used to administer insulin to particular body areas. Clinical delivery trucks are designed to provide medical equipment, supplies, and medications to patients or hospitals in need. Wu and his colleagues have recently developed unique gels that react rapidly to touch and may be injected into the body, the purpose of these gels is self-healing (Fig. 6.4) [84]. Certain specialists have created unique gels that react to variations in acidity. These gels may aid in wound healing and encourage the formation of new blood vessels in the skin since they have been improved with specific components derived from stem cells. A 3D printing technique was used to create these hydrogels. They functioned well as instruments with the ability to individually block capillaries. Future applications of specialized microrobots that can be supplied in 3D and 4D for lab investigations and organ-on-a-chip operations are made possible by the recent advancements in pH-sensitive hydrogels. The purpose of these microrobots is to precisely locate and address obstacles [85].

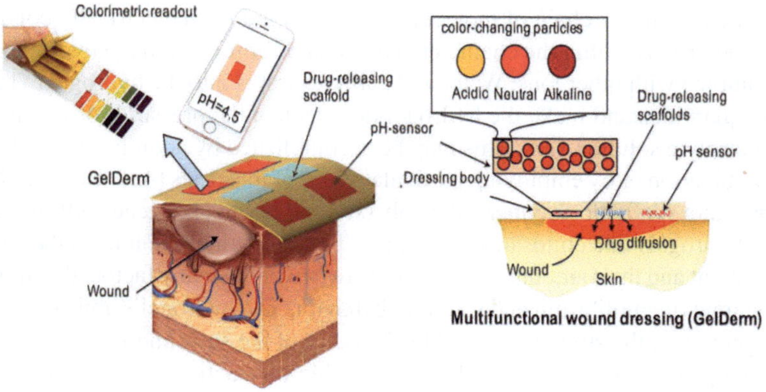

Fig. 6.3 Materials vulnerable to pH and 4D printing: this picture displays GelDerm, a unique bandage with many functions. In addition to its ability to inspect and arrange wounds, it has a marker that changes color according on the wound's acidity. Additionally, it releases medication to aid in the healing process [78]

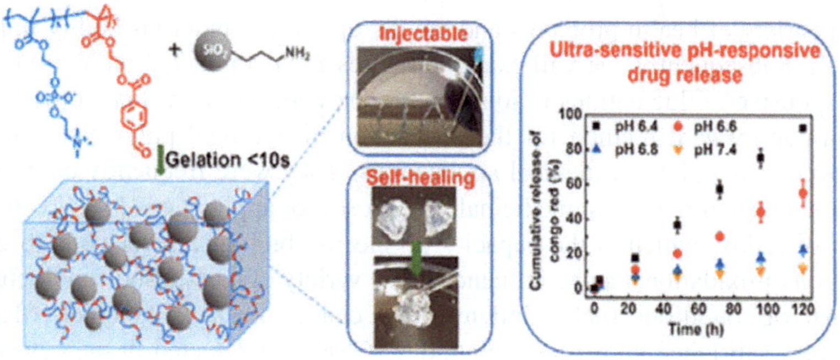

Fig. 6.4 Self-healing hydrogel that forms in under 10 s and enables pH-responsive drug release. Reprinted with permission from [84]. Copyright 2020 American Chemical Society

6.3.2 Hydrogels that Respond to Ionic Concentration and Redox

Ion-responsive hydrogels are a class of hydrogels that, depending on their kind, respond to certain ions to produce a particular behavior. Hydrogels' capacity to stretch and expand may be influenced by the kind of charged atoms and the quantity of tiny particles they contain. "Ionic quality" describes the quantity of chemicals that, when dissolved in water, have the ability to conduct electricity. Hydrogels may be created bigger by using tiny bits in a variety of ways, but the way they work relies on the substance used to form the hydrogel. First, ionic intelligence is often used in charged hydrogels to establish robust interactions. To join negatively charged polymers, employ unique ions with two or three positive charges. The ability of metal cations with two charges to bind together is a frequent example of alginate. The quantity of cations in the material determines how much swelling occurs in these hydrogels. When a lengthy chain of molecules is submerged in an electrically conductive liquid, the groups within the chain get charged and form ions. In the hydrogel, these ions combine with other ions. Water and particles travel into the hydrogel to balance out the quantities and make the hydrogel larger when the pressures of the hydrogel and electrolyte solution vary. This may be accomplished by adding certain elements to a combination or by employing particular types of materials [43]. These hydrogels may be either synthetic polymers like poly (styrene-divinyl benzene) sulfonic acid or natural hydrogels like chitosan and alginate. The strength of the ions in the swelling environment and their amount of ionization are two of the many factors that influence the dynamic of equilibrium and swelling behavior. The hydrogel's ability to swell is contingent upon the strength and acidity/basicity of the surrounding liquid [86, 87].

Developed unique hydrogels that are capable of reacting with reactive oxygen species (ROS). This serves to safeguard cells from the detrimental effects of ROS. ROS are chemical compounds that are unleashed during the energy production process of cells in the human body. Large quantities of them are present in the human organism. Furthermore, they are crucial for cell communication, despite the fact that they can cause oxidation, which can damage cell secretions, genetic material, and proteins. Consequently, the ongoing production of ROS is frequently associated with a variety of health problems and the disruption of numerous bodily functions [88, 89]. Subsequently, we will examine the research conducted on hydrogels that are responsive to fluctuations in ionic and redox concentration. Ionic cross-linking is a straightforward method for the production of hydrogel polymers, which are employed in a variety of medical applications. They were frequently employed in potential medical remedies, medicinal, and biosensor applications [48, 90, 91]. The power of redox, which is the capacity of specific biological processes to endure reduction or oxidation, can be influenced by a variety of factors and can fluctuate in certain toxic conditions, such as inflammation, cancer, or a lack of oxygen [61]. High levels of a substance known as glutathione (GSH) are detected in the cytosol and organelles, similar to pH gradients within cells, to determine redox potential. The high concentrations of glutathione in the cells are distinct from the adjacent area and are exclusively employed to activate medications within the cells [92]. The viscosity of redox-sensitive hydrogels decreases and they rapidly transform into a liquid solution when they are exposed to an oxidizing substance such as NaClO. Conversely, when they are exposed to reducing agents such as GSH, their viscosity increases and they rapidly form an emulsion. It appears that the same redox-responsive behavior occurred when it was exposed to an electrochemical stimulus. Incorporated redox-responsive structures into a heat-sensitive material to create redox-responsive hydrogels. The text pertains to a technique that is employed to impart unique properties to a polymer. The properties were added to the polymer using a thiol-yne reaction, which was produced using a distinct method known as ring-opening polymerization. The polymer's ability to adapt to changes in its environment is facilitated by the properties it has been endowed with. The plastic that has been produced may exhibit a reaction when it is exposed to a chemical known as H_2O_2. Redox-responsive hydrogels can also be produced by combining redox-responsive organometallic structures. Nakahata and his colleagues developed hydrogels that can adapt to fluctuations in oxidation and reduction through the utilization of specific polymers. They incorporated redox-responsive ferrocene (FE) and PAAc (PAAc/FE) as a unique form of polymer known as the guest polymer. This polymer interacts with another polymer, PAAc/β-CD, commonly referred to as the host polymer [93]. Additionally, the utilization of disulfide linkages to connect molecules can be a frequently employed approach to the development of hydrogels that are capable of adapting to environmental changes. Special polymers have been developed by Liang and Kiick to discharge drug-filled liposomes in regions with elevated levels of a chemical known as thiol in response to environmental changes [94]. This research demonstrated a straightforward approach to the development of improved drug carriers that degrade chemically in a unique manner. Researchers have observed that a substance known as ROS is frequently present in regions of tissue that are highly inflamed or deficient in oxygen, as well as in regions where infections are present. This has incited scientists to develop hydrogels that can respond to ROS, thereby facilitating the healing of lesions, inhibiting the proliferation of hazardous microorganisms, and reducing inflammation. Furthermore, researchers have created a form of gel known as ROS-responsive (PAAg-PGFe) nanocomposite hydrogel in order to enhance the efficacy of infection-fighting treatments. It is composed of hydrogels that utilize polyacrylic acid (PA). These hydrogels are coated with silver nanoparticles (AgNP) and contain iron (Fe^{2+}/Fe^{3+})-contained polyglutamic acid (PG) networks. These polymers have demonstrated substantial capacities to facilitate wound healing and combat infections [95]. In a separate investigation, researchers have developed hydrogels that are capable of responding to reactive oxygen species (ROS) and can be employed as a liquid wound dressing that is injected into the epidermis. In rodents with full-thickness lesions, these unique networks, which are composed of PEG, sodium Alg., and pectin, have been shown to effectively produce anti-inflammatory peptides, thereby facilitating the healing process. Synthetic polymers, natural peptides, and DNA are among the materials that are employed to produce responsive hydrogels. Additionally, these frameworks contain a variety of metal ions. Recently, a team developed a unique gel that can react to ions. They employed a substance known as polyacrylamide to enhance its strength, and they integrated sodium alginate to facilitate its reaction with ions and the retention of specific information. In this system, various metal ions were evaluated to determine whether they could alter the geometry of an object. After conducting tests on Fe^{3+}, Cu^{2+}, Ag^{+}, Ca^{2+}, Ni^{2+}, and Co^{2+}, it was determined that Fe^{3+} was the most effective in altering the morphology. The hydrogel was also discovered to be able to regain its shape at a faster rate when deposited in sodium citrate than in EDTA, sulfuric acid, or potassium carbonate. This investigation devised a method for fabricating a unique hydrogel that is capable of reacting to ions. It also provided a rule to enhance the hydrogel's ability to maintain its shape and regain its original shape [96]. Du and his colleagues developed a hydrogel system that has the ability to alter its configuration. The hydrogel has the ability to be contorted, formed into a tube, or shaped to resemble a plant. With the addition of Na^{+} or Ca^{2+} ions to the hydrogel, these geometries can be altered repeatedly. This gel-like system that responds to ions has significant potential for use in robotics and medicine. Metal ions are essential components of biological organisms. They can also be employed to modify the functionality of proteins, peptide derivatives, and nucleic acids [97]. In 2020, Chen and his team developed a novel gel that has the capacity to contain medication. They examined the impact of various metals on the morphology of the medicine. The detailed explanation of the peptides' assembly process is beneficial for the development of hydrogels for medical applications. Metal ions, such as potassium (K), mercury (Hg^{2+}), and silver (Ag^{+}), can react with nucleic acids, a type of biological material. These reactions can generate structures such as G-quadruplexes and duplexes, which are stabilized by metal ions [98]. Willner's team developed a collection of gel-like materials composed of DNA and stabilizing elements. A DNA hydrogel was generated that can be manipulated by ions to either become more solid or degrade. The Y-shape structure contained complementary pairings with an I-shape structure; however, there were discrepancies in the cytosine content. The Y-shape and I-shape DNA are connected to form a gel-like substance by hydrogen bonds and special ions (C–Ag^{+}–C) when Ag^{+} is present. The stabilized ions (C–Ag^{+}–C) were removed after the administration of cysteamine, resulting in the gel transitioning to a liquid state. Scientists incorporated man-made or natural materials (such as polyacrylamide, carboxymethylcellulose, and polyacrylic acid) into DNA hydrogels to provide structure and support, thereby making them more affordable. The permanent modification of DNA ligands on PNIPAM was facilitated by Ag^{+} ions [99]. After being treated with silver ions and cysteamine, the hydrogel networks can undergo a gel phase and subsequently decompose. The Ag^{+} stabilized hydrogel possessed a storage modulus (G') of 85 Pa, indicating that it is capable of storing a greater amount of energy. The loss modulus (G'') was 4 Pa, which denotes the quantity of energy that is lost. The hydrogel's strength decreased as cysteamine was employed for treatment. Similarly, a PNIPAM copolymer composed of G-quadruplexes was capable of separating and reassembling after being treated with kryptofix [2.2.2] (CP) and K^{+} (Fig. 6.5) [100].

Huang et al. investigated a hydrogel that reacts to changes in the ion concentration in its surroundings by changing its structure. A material known as hydroxyethyl acrylate is used to create the hydrogel. When making ink for printing, potassium

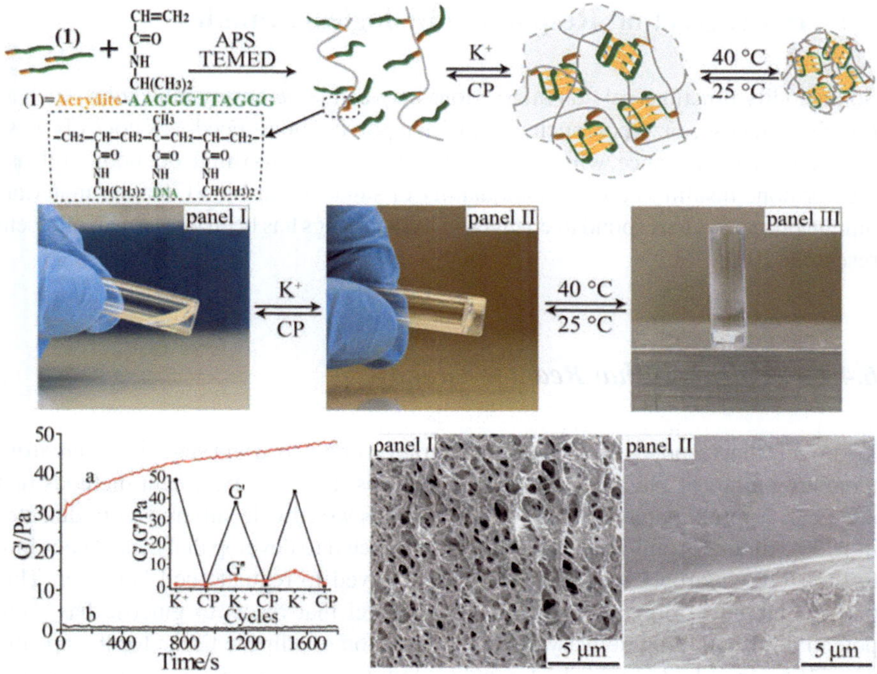

Fig. 6.5 Developing a hydrogel type that remains stable when exposed to potassium ions, and demonstrating how it is simple to build and degrade with the aid of potassium ions and kryptofix [2.2.2] (CP) [100]

3-sulfopropylmethacrylate is a component. It has the ability to alter form because it swells spontaneously in response to certain ion or salt concentrations in its environment [101]. Another example from the work of Kang and associates demonstrates that a certain kind of film composed of polystyrene and quaternate poly (2-vinyl pyridine) may be used as color sensors to track the skin's color. The researchers examined the expansion and contraction of the film. In this instance, the refractive index and the distance between domains may be successfully altered by ionic concentration [102]. The amount of electrolytes in our bodies may be a critical marker for a number of disorders, despite the fact that little is known about how materials respond to variations in ionic concentration. Lung and breast cancer, for instance, may result in hypercalcemia, or an excess of calcium in the blood. Furthermore, it was shown that patients with Alzheimer's disease had increased potassium (K^+) and sodium (Na^+) levels within certain brain areas [103].

6.4 Hydrogels that Respond to Biological Stimuli

Some bodily functions rely on interactions between large or tiny molecules, such as proteins, glucose, enzymes, nucleic acids, and polypeptides. Feedback controls these interactions. For instance, when our blood sugar levels are too high, our bodies release the hormone insulin to control the quantity of sugar in our blood. Giving materials unique properties to respond to changes in living beings has been the subject of recent research [104].

6.4.1 *Hydrogels that React to Glucose*

Hydrogels that respond to a certain concentration of glucose and sugar in the environment are known as glucose-responsive hydrogels. Diabetes treatment methods that are clear. Approximately 171 million individuals worldwide suffered from diabetes in 2000. About 366 million individuals are predicted to develop diabetes. According to the study, hydrogel scaffolds may be employed to regulate cell behavior. This particular kind of polymer-based smart hydrogel that reacts to glucose has great potential for both drug delivery and the production of adipose tissue for therapeutic uses [105]. In order to have a better understanding of this issue, it is vital to note that diabetes is a prevalent condition in society and that treatment is crucial. One chemical in our bodies that influences diabetes is glucose. Due to the rising prevalence of diabetes, researchers are working to develop materials that react to glucose and aid in the delivery and monitoring of insulin, a drug used to treat the disease [106]. In general, both type I and type II diabetes requires regular blood sugar checks. As a result, researchers are developing tools that can identify glucose and print them out for usage. Insulin injections are among the most straightforward methods of treating diabetes. However, the difficult and painful injection procedure, the need for regular testing, and the necessity for numerous injections have prompted researchers to look for more effective insulin delivery methods. A novel substance called hydrogels has the ability to hold insulin for an extended period of time. When our blood sugar levels are too high, these hydrogels can detect it and release insulin when it's time. We may no longer need to repeatedly administer insulin injections to ourselves, which is a very exciting development. These hydrogels might be used as a more practical therapy alternative instead [107]. Adams et al. used a method based on a substance's electrical conductivity to determine its concentration. A graphene and polylactic acid filament combination was used to print a glucose monitor. Glucose dehydrogenase and flavin adenine dinucleotide were used to produce the filament, discovered that the test is capable of measuring glucose levels between 0 and 400 mg/dL [108]. Blood sugar control may benefit from knowing this information. Similarly, as seen in the image. As seen in Fig. 6.6, platinum nanoparticles (PtNP) and glucose oxidase (GOx) were placed between two electrically conducting carbon-nanotube (CNT) electrodes to create a glucose sensor. An inkjet printer was then used to print a network of polyaniline nanowires on the region where CNT/GOx/Pt-NP was deposited. This was done in order to build a bridge between the CNT electrodes that can withstand electrical flow. The sensor receives a boost from GOx to produce hydrogen peroxide (H_2O_2) as a byproduct when glucose is present. To raise the acidity locally, PtNP will catalyze the H_2O_2. One substance that may be used to gauge how acidic a solution is polyaniline. It serves as a sensor to gauge blood glucose levels in order to manage diabetes. As low as two millimolar glucose concentrations are detectable by it. People with diabetes must inject themselves with insulin under the skin many times a day in addition to monitoring their blood sugar levels. Nevertheless, it is difficult to continuously monitor the patient and ensure that they adhere to all guidelines, which raises the risk of diabetic complications. As a result, drugs that can automatically react to variations in blood sugar levels by releasing insulin in a regulated manner are now required. Two strategies for blood sugar regulation are the main focus of the present insulin development process, which reacts to the body's glucose levels, insulin is released in two ways. In the first, insulin is released when swelling is brought on by an interaction between glucose and an enzyme known as glucose oxidase [109–111]. This occurs when glucose competes with glycopolymer-lectin complexes and causes the release. There haven't been any research on exploiting these processes to print a device that can react to glucose levels and release insulin, despite the fact that we already know how they function. Rather, Song and associates. Researchers have shown a novel insulin delivery method that uses beta cell release in a high-sugar environment. This is accomplished by printing a structure out of polylactic acid, a kind of plastic then cultivate stem cell clusters on this framework. This facilitates the response of the cell clusters to glucose. Researchers found that this cell-filled printed graft might increase the body's production of insulin in response to glucose. The usage of enzymes throughout the manufacturing process is the primary issue with developing materials that may react to glucose. Therefore, it is crucial to create enzymes under favorable circumstances to ensure their continued functionality. Furthermore, only substances that are practically safe for the human body to employ should be used to supply insulin and react to glucose [112].

For example, Lu used Ag-Lignin to establish a consistent environment. A persistent stickiness and gel formation resulted from the constant production of catechol groups and radicals in this environment [113]. Thordarson developed a unique kind of gel that, in the presence of a certain reductive material known as TCEP, may gradually decompose. In this system, a particular material and another substance that can respond to changes in its surroundings were combined to create a gel. The system changed from a liquid to a gel and back to a liquid at various initial TCEP concentrations. Because the system's disulfide crosslinks were deteriorating over time, this transformation occurred at various periods [114]. Two hydrogel systems were developed by Grinstaff that decompose as required. Thioester linkages are unique connections that hold these systems together. When cysteine methyl ester was present, a process known as thio-thioester exchange disrupted the crosslinks, causing the hydrogel to change form rapidly. Wounds might be healed with the aid of the two gel systems (Fig. 6.7a, b) [115, 116]. A unique network known as a constitutional dynamic network (CDN) was developed by Willner's group. The

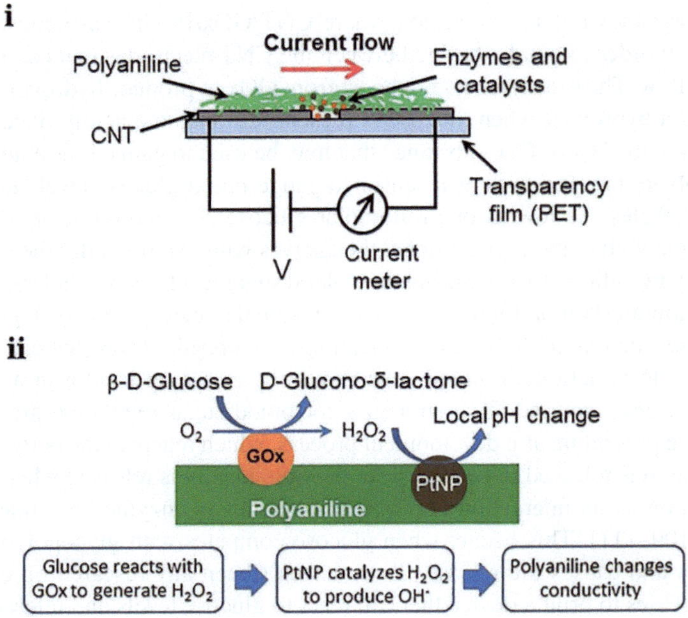

Fig. 6.6 Using 4D printing to create gadgets that react to the body's glucose levels. **i** Developing a glucose sensor that uses specific molecules to detect glucose. Measure the current flowing through the polyaniline nanowires using a current meter that is attached to a sensor. A unique procedure known as a catalytic reaction is used by the chemoreceptive sensor. When glucose is present, a chemical known as H_2O_2 is a byproduct that aids in the process. The H_2O_2 is then accelerated by a material known as PtNP, making the surrounding environment more acidic can measure the resistance of polyaniline nanowires to determine if a change has occurred [111]

network is divided into four sections: AA′, BB′, AB′, and BA′. Following E1 treatment, hydrogel X stabilized and underwent a process that hardened hydrogel Y. This occurred as a result of hydrogel Y having more tetradentate units AB′ and BA′ than before. Hydrogel X is transferred to hydrogel Z by the stabilized BB′ in the CDN hydrogel system after treatment of E2. The drop in AB′ and BA′ causes Hydrogel Z to become less rigid. This unique hydrogel structure is used for biocatalytic cascade operation, regulated drug release, and self-healing (Fig. 6.7c) [117]. Enzymatic reactions have been used by researchers to create intelligent hydrogels. Bio-chemically responsive gels that can identify biomolecules have shown to be very helpful in theranostics when compared to standard chemical gels. A unique kind of gel that reacts to proteins without the need of enzymes was recently developed by Hamachi's group. The protein signal in this system can be translated into mechanical hardness and enzyme activity. When the target protein is treated, enzyme-activity triggers (EATs) that prefer to bind to the protein are activated. Bovine carbonic anhydrase II (bCAII) is activated as a result. The gel transformed from a solid to a liquid and discharged its contents because enzymes have the ability to break the bonds between molecules. This gel shows great promise as an ingenious medication delivery method. It has

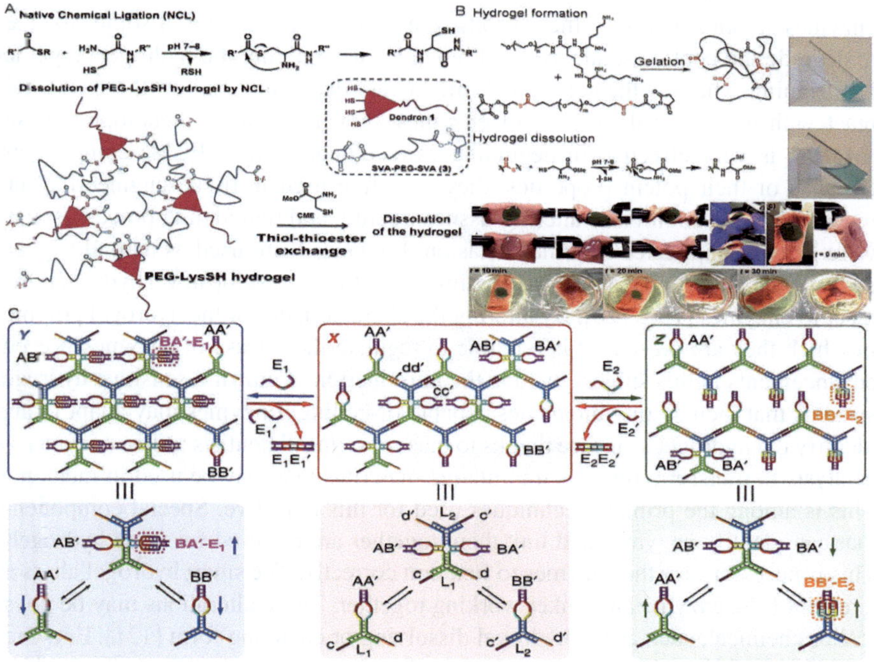

Fig. 6.7 **a** Used a unique chemical process to create a hydrogel created two distinct hydrogels, one pink and one green. When a particular solution was administered to human skin, they did not dissolve [115]. **b** This demonstrates how a gel is created and dissolved by a unique chemical process; images of the gel dissolving when desired are included [116]. **c** This figure pertains to a particular hydrogel. It describes how the hydrogel is composed of four distinct components that have undergone certain modifications. The text also includes a graphic or schematic of the hydrogel [117]

the ability to detect dangerous compounds in the body and deliver the medication precisely where it is required [118]. When a protein known as flap endonuclease 1 (FEN1) is present, a smart DNA nanogel system developed by Li and his colleague breaks down and releases chemicals. Many cancer cells have high amounts of FEN1 [119].

6.4.2 Hydrogels that React to Enzymes

Special molecules called enzymes regulate a wide range of vital bodily processes, including the synthesis of proteins, the adhesion of cells, and the contraction of muscles. Since the body already contains a wide variety of enzymes, enzyme-responsive materials have gained a lot of popularity. The biological environment of the body may naturally initiate enzyme stimulation, and the degree of stimulation is naturally modulated according to the state of the body. Hydrogels that respond to enzymes are known as enzyme-responsive hydrogels. The ability of some enzymes to degrade the peptide sequences and biomolecules holding the hydrogels together is intriguing. The way the gel expands might vary depending on how these enzymes react with the material. A class of enzymes known as matrix metalloproteinases (MMPs) is very effective in degrading extracellular matrix (ECM) components. Because of their potent properties, they are often used in tissue engineering and regeneration. In addition to altering tissue structure and remodeling processes, they were heavily implicated in tumor invasion. Enzymes were used as natural triggers and enzyme-catalyzed reactions as possible activators to develop novel hydrogels that react to enzymes [106, 120]. By altering the characteristics of the hydrogel polymer in which they are present, it is possible to regulate the release of enzymes. Recent advancements in this subject include the introduction of enzyme-sensitive hydrogel systems that include tiny molecules capable of gelling. Enzymes may adapt to and identify the hydrogel's surface thanks to these systems. Scientists utilize enzymes as catalysts to transform the gels into unique materials that may be used in medicine. This is among the primary techniques used for this objective. Special components that can identify enzymes and link them together are required for these hydrogels. This is necessary for the enzymes to function correctly. The smart hydrogel alters as a result of the enzyme and linker working together. These alterations may be physical or chemical, such as the hydrogel dissolving or changing form [121]. Enzymes have the ability to dissolve some hydrogels. There are two methods for making these unique hydrogels that respond to stimuli: The first technique involves simply joining the hydrogel polymer together. The second concept involves assembling a large number of tiny particles known as hydrogelators to form a structure. These particles join together to form nanofibers, which then come together to create the hydrogel. Hydrogel chemicals are typically made by connecting polymers together using strong bonds. They are put together using two main steps; the materials that can break down naturally are dissolved. Physical hydrogels are made by physically linking together biological components. The chemicals in the enzyme-responsive hydrogels that cause changes in their structure and composition are called reactive entities. These reactive entities are located either in the center or the side chain of the hydrogel. They form weak bonds that can easily break and cause the hydrogel to change. These changes happen because of various reactions that occur between the reactive entities and other substances in the hydrogel, but these reactions do not involve the sharing of electrons [122]. The molecules interact with each other by forming different types of chemical bonds such as electric charges, water repulsion, hydrogen connections, weak interactions, stacking of atoms, or a combination of these bonds (Fig. 6.8a) Cause changes in the outer layer of hydrogels, such as how they're built, how they stick together, and how they expand or shrink as illustrated in Fig. 6.8b [50].

Recently, Maki T. et al. were able to combine a chemical known as rhodamine 6G with urea to generate a unique gel. Enzymes may react with this gel. When the hydrogel is exposed to the β-galactosidase (β-Gal) enzyme, it may transform from a liquid to a gel [123]. In their research, Shen X. and associates found a way to combine chitin with β-(1 → 4) D-glucose units to form a hydrogel. When this

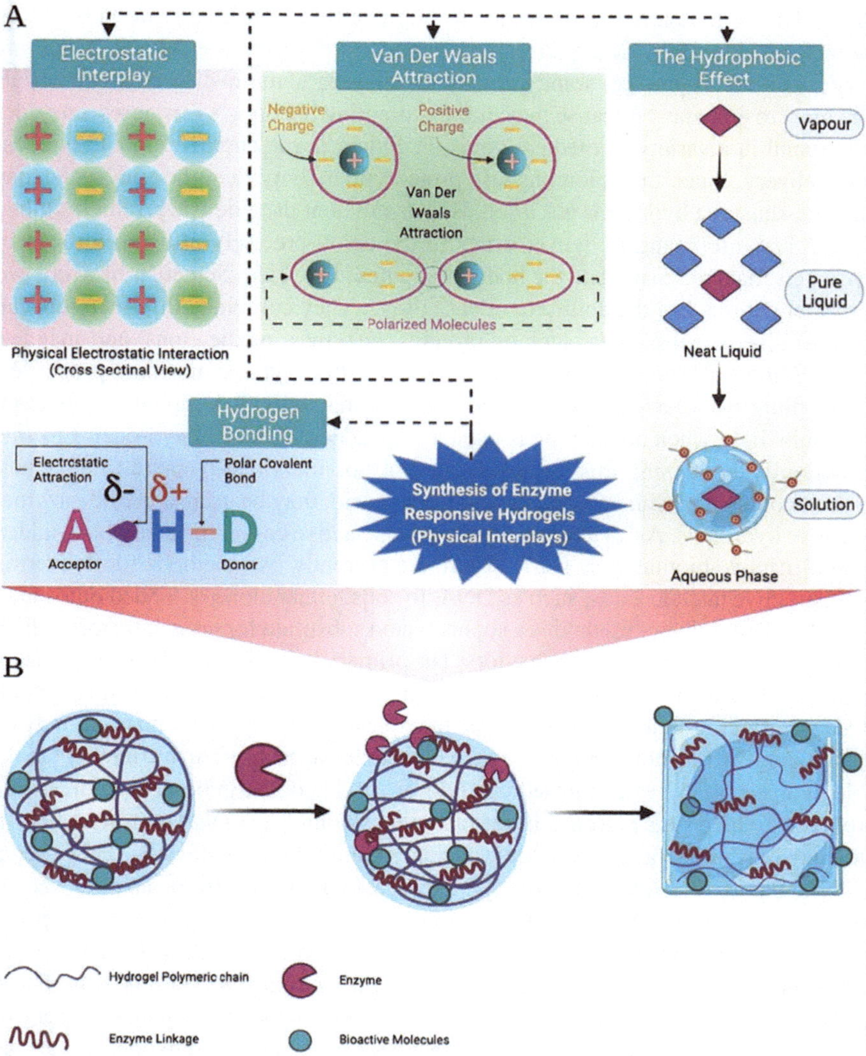

Fig. 6.8 The technique of making hydrogels that respond to enzymes is shown in this image. It also illustrates how these hydrogels behave in the presence of enzymes. **a** Researching how various materials combine to create unique gels that react with enzymes. **b** The way hydrogels react to enzymes and then release compounds that support biological activity [50]

unique gel detects a certain sugar, it may deliver medication. Drug delivery using this hydrogel is possible in response to 2-acetamido-2-deoxy-β-D-glucose [122]. Scientists have recently created unique gels that the body can decompose. Substances that aid in healing at the site of damage may be carried and released by these gels. Furthermore, compared to releasing the growth factors in the opposite order or without any control, it was likely to successfully demonstrate continuous release of VEGF and

PDGF-BB in this particular sequence to improve tissue healing in wounds caused by lack of blood flow. Many biological processes depend on enzymes, which are very sensitive. Therefore, some compounds may be combined to create materials that react to enzymes. Because they aid in promoting the body's responses, enzymes are crucial in a variety of medical fields, including tissue engineering and medication delivery. Since the colon has less proteolytic activity than the small intestine, enzyme-sensitive hydrogels are often used as efficient drug delivery vehicles there. Researchers are trying to deliver drugs to the colon precisely utilizing a range of hydrogels that are sensitive to the body's enzymes [124, 125]. Scientists have created a special gel that has the ability to release drugs. They combined different gels that react to diverse substances, such as proteins, enzymes, medications, and magnets [126]. Wang and Esker created tiny chitin films with enzyme-controlling properties. These films are referred to as enzyme-sensitive materials. Additionally, a material that emits light when an enzyme is present was added to the CH hydrogel. Enzyme detection is done using this hydrogel. Natural polymers that respond to enzymes, such as collagen, fibrin, gelatin, and hyaluronic acid, may be used to create enzyme-sensitive hydrogels. As an alternative, enzyme-sensitive components might be added to non-reactive biomaterials [127]. There are currently two methods for delivering bioinstructive nucleic acids, such as DNA, to cells: either alone or loaded onto small particles. This delivery method is a sophisticated substitute for growth factors, which are used to induce certain cell behaviors. The primary issue with this strategy is that in hydrogels based on HA or fibrin, DNA polyplexes have a tendency to group together. Because of this, efficiently loading gene therapies into the hydrogel matrices may be challenging. Strong nono-combined pDNA particles were inserted into readily breakable hydrogel matrices composed of fibrin or acrylated HA in order to address this problem. To help stem cells proliferate and mend wounds in diabetics, researchers in another study aimed to enhance the way genes communicate with cells. They used a material known as HA to make soft hydrogel scaffolds. They introduced microscopic particles containing DNA and added unique connections that degrade when certain enzymes are present. Researchers have recently shown that hydrogels that react to certain enzymes may be made using the gene-editing tools CRISPR and Cas. These hydrogels are more adaptable and practical for a range of uses as they may be made to carry living cells and biomolecules. Alp enzymes have been used to convert various materials, including gellan gum, chitosan, collagen-peG, and platelet-rich fibrin, into bone tissue [128]. Furthermore, the function of the enzyme is to convert the side chains of a polymer backbone into more active compounds. Parts of other neighboring polymer chains may then react with these active ingredients, joining them together. Tyrosine in gelatin, for example, is converted to o-quinone by the enzyme tyrosinase, which may then react with chitosan. According to some accounts, an enzyme called horseradish peroxidase may be used to turn a material known as hyaluronic acid-tyramine into a hydrogel. A chemical process brought on by this enzyme binds some of the substance's components together. Small peptides are often used as a binding material in hydrogels that may mimic the extracellular matrix, respond to enzymes generated by cells, and permit cell migration. Protein fragments may be broken down by enzymes known as matrix metalloproteinases (MMps) to form a gel that facilitates cell entry. During tissue change, certain enzymes known as MMp may also break down the small particles in the eCM. Therefore, the fact that hydrogels include components that MMPs may break down explains why they can be impacted by enzymes [129]. In another study, researchers created a gel with a particular kind of sugar and a unique protein that inhibits dangerous enzymes. The hydrogel's trapped MP3 was released as connections and then degraded by MMps, which are active enzymes. A smart vehicle that can carry growth factors to cells and improve their development has been the subject of much effort. GdnF, derived from glial cells, seems to have great potential for healing nerve damage. GdnF and other growth factors may now be delivered using nerve guidance conduits (nGCs). GdnF levels that are bound and released repeatedly inside peG microsphere scaffolds with heparin have been seen to exhibit almost straight lines. For certain particular applications, this discovery is significant. The GdnF compounds may have aided nerve recovery in a nerve guide channel since they remained on the hydrogel microspheres for more than a week. It was determined that these hydrogel scaffolds for nerve regeneration were safe for the body and did not result in any adverse effects, such as infection, tissue death, or an immune system reaction. Tumor invasion and metastasis in specific body districts have been linked to the production of specific proteins, such as lattice metalloproteinases (MMP). Therefore, MMP production acted as a site-specific natural trigger for the targeted release of anti-inflammatory drugs from hydrogel that was negatively charged. This trigger works by using the body's enzymes to break down the drug-filled hydrogel. This enables the hydrogel to release the medication where it is required [130]. Ceylan and his researchers, and Wang and his associates. Additionally, gelatin methacryloyl's capacity to degrade proteins was used to create small robots that can provide medication to certain bodily parts. (1) Magnets are used to operate these tiny robots. (2) As can be observed in the image, they are entirely eliminated by magnets after collagenase and MMP2 enzymes have been added, leaving no toxic materials behind. Their research has looked at frequent issues with additively manufactured micro-robots, such their inability to decompose naturally or their production of hazardous materials. There is a lot of interest in employing materials that can be broken down by enzymes for tissue regeneration in addition to using medications. These substances may be inserted into the body, and after serving their purpose, they will decompose and fall out on their own [131]. Researchers set out to do this created microscopic blood vessel-growing passageways on a biodegradable support framework. Enzymes ripped away the support structure that held the blood vessels together, allowing the cells inside of them to create new blood vessels [132]. Enzymes may result in a variety of additional responses than wear and tear. For example, the construction of buildings, their expansion or contraction, and the modifications to the way surfaces function. However, printed enzyme-responsive materials are now only useful because enzymes can break them down. The issue with employing materials that react to enzymes is that they may react with other enzymes in the body, making it hard to manage how they react. This particular issue is especially crucial when working with larger peptides or DNA sequences [133].

6.5 Conclusions, Limitations, and Future Outlook

6.5.1 Conclusion

Stimulus-responsive hydrogels produced by 4D printing have become a novel and exciting method for tissue engineering and medication delivery applications. These hydrogels' capacity to react to a range of stimuli, including chemical and biological signals, enables the development of dynamic structures that can change with their surroundings. They are thus perfect for creating implanted technology that can replicate the functions of real tissues and organs. More accuracy, speed, and customisation are just a few benefits of using 4D printing technology over conventional production techniques. This implies that hydrogels may be customized to meet the demands of individual patients, increasing the likelihood that medical operations will be successful and lowering the possibility of negative responses. Furthermore, 4D printing hydrogels has potential uses outside of biomedicine. Soft robotics, smart packaging, and other industries that need for flexible and responsive materials may also make use of these materials. All things considered, the advancement of 4D printing technology has created fascinating new opportunities for producing dynamic, useful materials with a variety of uses. The use of 4D printing hydrogels in the domains of medication delivery and tissue engineering is anticipated to grow in popularity as more research is done in this area, improving patient outcomes and quality of life.

6.5.2 Limitations

The restricted selection of materials available for use in hydrogel 4D printing is one of its drawbacks. The majority of recent studies have concentrated on the use of temperature-responsive synthetic hydrogels, such as poly(N-isopropylacrylamide) (PNIPAAm). The creation of novel materials that can react to different stimuli, such pH, light, or magnetic fields, is necessary even if these materials have shown encouraging outcomes. The low resolution of 4D printing methods is another drawback. Even while modern methods may produce intricate structures with great precision, they nevertheless have a lower resolution than more conventional manufacturing processes like lithography or injection molding. Future studies should concentrate on creating novel 4D printing methods that can produce prints with greater accuracy and resolution. The technology's scalability is another drawback of 4D printing hydrogels. The development of 4D printing processes has advanced significantly, yet producing large-scale buildings remains a difficulty. This is mostly because of the printing process's restrictions and the accessibility of appropriate materials. The development of scalable 4D printing methods and materials for the construction of large-scale structures should be the main goal of future research. Furthermore, 4D printing technology is now somewhat expensive. The expense of the tools, supplies, and labor needed for hydrogel 4D printing may prevent it from being widely used. However, it is anticipated that as the technology advances and is used more often, its prices will fall down, making it more affordable for researchers and medical professionals. Lastly, there is still much to learn about the stability and durability of 4D printed hydrogels over the long run. Despite the materials' encouraging responses to external stimuli, there are still concerns about their stability over time and in various environmental settings. In order to successfully translate these materials into therapeutic applications, future research should concentrate on examining their long-term stability and endurance. In conclusion, there are still a number of issues that need to be resolved even though 4D printing stimuli-responsive chemical and biological hydrogels has shown tremendous potential for tissue engineering and medication delivery applications. These include the scalability, affordability, and long-term stability and durability of printed hydrogels, as well as the restricted selection of materials and printing resolution. It will need more study and advancement in this area to overcome these constraints.

6.5.3 Future Prospects

Stimulus-responsive chemical and biological hydrogels printed using 4D printing have a bright future. Complex structures with exact forms and qualities may now be produced because to advancements in printing processes and novel materials. By making it possible to create specialized structures that react to certain stimuli, this technique has the potential to completely transform tissue engineering and medication delivery. Regenerative medicine is one area where 4D printing hydrogels is expected to find use in the future. It is feasible to encourage tissue regeneration and restore damaged tissues by developing scaffolds that can replicate the natural environment of cells. Additionally, medication delivery systems that target certain cells may be developed using this approach. The potential for 4D printing chemical and biological hydrogels that respond to stimuli is enticing. This technology has the potential to completely transform the fields of drug delivery and tissue engineering via the creation of novel materials, printing methods, and applications. The creation of organoids is one of the exciting potential uses for 4D printing hydrogels. Organoids are three-dimensional models that replicate the composition and capabilities of real organs. They may be used to personalized medicine, illness modeling, and medication screening. Complex organoid structures with particular forms and characteristics that may alter in response to outside stimuli can be produced using 4D printing of hydrogels. The discovery of novel medications and treatments for a variety of illnesses might be accelerated by this technology. The creation of smart implants is another possible use for hydrogels printed using 4D technology. In order to improve patient outcomes, smart implants are medical devices that can track and react to changes in the body. Hydrogels may be printed using 4D technology to make implants that can alter their form and characteristics in response to physiological changes like pH or temperature. Drug delivery, vital sign monitoring, and focused treatments for certain illnesses are all possible with these implants. Furthermore, artificial tissues and organs may be produced using 4D printing of hydrogels. It is possible to develop organs and tissues that can replace sick or damaged tissues by using the capacity to generate complex structures with certain forms and characteristics. Organ transplantation may no longer be necessary thanks to this technique, which also lowers the possibility of rejection. Lastly, hydrogels may be printed using 4D technology to provide customized implants and medical equipment. Customized implants and medical devices that match each patient's distinct anatomy and physiology may be made by using patient-specific data, such as genetic information and medical imaging. Both patient outcomes and healthcare expenses might be improved by this technology. To sum up, 4D printing of chemical and biological hydrogels that respond to stimuli is a quickly developing technology that has enormous promise for tissue engineering and medication administration. In the next years, biomedical research and healthcare are anticipated to be significantly impacted by the development of novel materials, printing methods, and applications. This technology has significant potential for the future and is anticipated to be crucial in the creation of novel cures and treatments for a variety of illnesses.

References

1. E. Behjat, M. Tajabadi, S.M. Naghib, H.R. Rezaie, Multifunctional stimuli-responsive self-healing polyvinyl alcohol/tannic acid/thyme-honey/aloe vera hydrogel as promising antibacterial drug delivery platforms: physicochemical and in vitro biological assays. Polymer **317**, 127897 (2025). https://doi.org/10.1016/j.polymer.2024.127897
2. R. Gholami, S.M. Naghib, Composites for BTE, in *Nanobioceramics for Bone Tissue Engineering and Regenerative Biomedicine: Tunable Biological Characteristics*, ed. by R. Gholami, S.M. Naghib (Springer Nature, Singapore, 2025), pp. 251–282. https://doi.org/10.1007/978-981-96-0041-0_9
3. S. Salehi, S.M. Naghib, H.R. Garshasbi, S. Ghorbanzadeh, W. Zhang, Smart stimuli-responsive injectable gels and hydrogels for drug delivery and tissue engineering applications: a review. Front. Bioeng. Biotechnol. **11**, 1104126 (2023)
4. N.A. Abtahi, S.M. Naghib, S.J. Ghalekohneh, Z. Mohammadpour, H. Nazari, S.M. Mosavi, S.M. Gheibihayat, F. Haghiralsadat, J.Z. Reza, B.Z. Doulabi, Multifunctional stimuli-responsive niosomal nanoparticles for co-delivery and co-administration of gene and bioactive compound: in vitro and in vivo studies. Chem. Eng. J. **429**, 132090 (2022)
5. N.A. Abtahi, S.M. Naghib, F. Haghiralsadat, M. Akbari Edgahi, E. Askari, A comparative study on biopharmaceutical function of curcumin and miR-34a by multistimuli-responsive nanoniosome carrier: In-vitro and in-vivo. Front. Mol. Biosci. **9** (2022). https://doi.org/10.3389/fmolb.2022.1043277
6. N.A. Abtahi, S.M. Naghib, F. Haghiralsadat, J.Z. Reza, F. Hakimian, F. Yazdian, D. Tofighi, Smart stimuli-responsive biofunctionalized niosomal nanocarriers for programmed release of bioactive compounds into cancer cells in vitro and in vivo. Nanotechnol. Rev. **10**(1), 1895–1911 (2021)
7. S. Afereydoon, F. Haghiralsadat, N. Hamzian, A. Shams, M. Hemati, S.M. Naghib, M. Shabani, B. Zandieh-Doulabi, D. Tofighi, Multifunctional PEGylated niosomal nanoparticle-loaded herbal drugs as a novel nano-radiosensitizer and stimuli-sensitive nanocarrier for synergistic cancer therapy. Front. Bioeng. Biotechnol. **10**, 917368 (2022)

8. H. Garshasbi, S. Salehi, S.M. Naghib, S. Ghorbanzadeh, W. Zhang, Stimuli-responsive injectable chitosan-based hydrogels for controlled drug delivery systems. Front. Bioeng. Biotechnol. **10**, 1126774 (2023). https://doi.org/10.3389/fbioe.2022.1126774
9. H.R. Garshasbi, S.M. Naghib, Smart stimuli-responsive alginate nanogels for drug delivery systems and cancer therapy: a review. Curr. Pharm. Des. **29**(44), 3546–3562 (2023). https://doi.org/10.2174/0113816128283806231211073031
10. A. Sadraei, S.M. Naghib, 4D printing of physical stimuli-responsive hydrogels for localized drug delivery and tissue engineering. Polym. Rev. **65**(1), 104–168 (2025)
11. T. Jafari, S.M. Naghib, M. Rahmanian, M.R. Mozafari, 3D Printing of self-healing materials for drug delivery applications: promises, advances and outlooks. Bioprinting **42**, e00353 (2024). https://doi.org/10.1016/j.bprint.2024.e00353
12. R. Salahandish, A. Ghaffarinejad, S.M. Naghib, A. Niyazi, K. Majidzadeh-A, M. Janmaleki, A. Sanati-Nezhad, Sandwich-structured nanoparticles-grafted functionalized graphene based 3D nanocomposites for high-performance biosensors to detect ascorbic acid biomolecule. Sci. Rep. **9**(1), 1226 (2019). https://doi.org/10.1038/s41598-018-37573-9
13. B. Mikaeeli Kangarshahi, S.M. Naghib, Multicomponent 3D-printed collagen-based scaffolds for cartilage regeneration: recent progress, developments, and emerging technologies. Curr. Org. Chem. **28**(8), 576–594 (2024). https://doi.org/10.2174/011385272829343724022 7065230
14. K. Mohammad-Jafari, S.M. Naghib, 3D Printing of microfluidic-assisted liposomes production for drug delivery and nanobiomedicine: a review. Curr. Med. Chem. (2024). https://doi.org/10.2174/0109298673285199231210170549
15. S.M. Naghib, M. Zarrineh, M.R. Mozafari, 3D Printing chitosan-based nanobiomaterials for biomedicine and drug delivery: recent advances on the promising bioactive agents and technologies. Curr. Org. Chem. **28**(7), 510–525 (2024). https://doi.org/10.2174/011385272 8298168240222114449
16. S. Soleymani, S.M. Naghib, 3D and 4D printing hydroxyapatite-based scaffolds for bone tissue engineering and regeneration. Heliyon **9**(9) (2023). https://doi.org/10.1016/j.heliyon.2023.e19363
17. M. Aghajani, H.R. Garshasbi, S.M. Naghib, M. Mozafari, 3D Printing of hydrogel polysaccharides for biomedical applications: a review. Biomedicines **13**(3), 731 (2025)
18. B. Mikaeeli Kangarshahi, N.S. Morteza, N. Rabiee, 3D Printing and computer-aided design techniques for drug delivery scaffolds in tissue engineering. Expert Opin. Drug Deliv. **21**(11), 1615–1636 (2024). https://doi.org/10.1080/17425247.2024.2409913
19. A. Sadraei, S.M. Naghib, 4D Printing of physical stimuli-responsive hydrogels for localized drug delivery and tissue engineering. Polym. Rev. **65**(1), 104–168 (2025). https://doi.org/10.1080/15583724.2024.2427184
20. A. Sadraei, N.S. Morteza, N. Rabiee, 4D Printing biological stimuli-responsive hydrogels for tissue engineering and localized drug delivery applications—Part 1. Expert Opin. Drug Deliv. **22**(4), 471–490 (2025). https://doi.org/10.1080/17425247.2025.2466772
21. A. Sadraei, N.S. Morteza, N. Rabiee, 4D Printing chemical stimuli-responsive hydrogels for tissue engineering and localized drug delivery applications—Part 2. Expert Opin. Drug Deliv. **22**(4), 491–510 (2025). https://doi.org/10.1080/17425247.2025.2466768
22. H.S. Ghazali, E. Askari, Z.S. Ghazali, S.M. Naghib, T. Braschler, Lithography-based 3D printed hydrogels: from bioresin designing to biomedical application. Colloid Interface Sci. Commun. **50**, 100667 (2022). https://doi.org/10.1016/j.colcom.2022.100667
23. H.S. Ghazali, E. Askari, A. Seyfoori, S.M. Naghib, A high-absorbance water-soluble photoinitiator nanoparticle for hydrogel 3D printing: synthesis, characterization and in vitro cytotoxicity study. Sci. Rep. **13**(1), 8577 (2023). https://doi.org/10.1038/s41598-023-35865-3
24. S.M. Naghib, S.N. Hosseini, A. Beigi, 3D 4D Printing of chitosan-based scaffolds for wound dressing applications. Carbohydr. Polym. Technol. Appl. **8**, 100594 (2024). https://doi.org/10.1016/j.carpta.2024.100594
25. P. Vosoughi, S.M. Naghib, M. Mozafari, Stimuli-responsive nano/biomaterials for smart drug delivery in cardiovascular diseases: promises, challenges and outlooks. Curr. Med. Chem. (2024)

26. H.R. Garshasbi, S. Soleymani, S.M. Naghib, M. Mozafari, Multi-stimuli-responsive hydrogels for therapeutic systems: an overview on emerging materials, devices, and drugs. Curr. Pharm. Des. **30**(26), 2027–2046 (2024)
27. A. Behera, A.K. Nayak, R.K. Mohapatra, A.A. Rabaan, *Smart Micro-and Nanomaterials for Drug Delivery* (CRC Press, 2024)
28. H.R. Garshasbi, S.M. Naghib, Injectable hydrogel for drug delivery, in *Biomaterial-Based Hydrogels: Therapeutics Carrier and Tissue Regeneration* (Springer, 2024), pp. 285–311
29. M. Pradhan, S. Swain, T.R. Rautray, T.Y. Kwon, Polymers for smart drug delivery, in *Smart Micro- and Nanomaterials for Drug Delivery* (CRC Press), pp. 47–73
30. Y. Zhu, 3D Printing biocompatible polymers for biomedical applications, Arizona State University (2024)
31. M. Quanjin, M. Rejab, M. Idris, N.M. Kumar, M. Abdullah, G.R. Reddy, Recent 3D and 4D intelligent printing technologies: a comparative review and future perspective. Procedia Comput. Sci. **167**, 1210–1219 (2020)
32. P. Kumar, P. Suryavanshi, S.K. Dwivedy, S. Banerjee, Stimuli-responsive materials for 4D printing: mechanical, manufacturing, and biomedical applications. J. Mol. Liq. 125553 (2024)
33. M.N. Sithole, H. Mndlovu, L.C. du Toit, Y.E. Choonara, Advances in stimuli-responsive hydrogels for tissue engineering and regenerative medicine applications: a review towards improving structural design for 3D printing. Curr. Pharm. Des. **29**(40), 3187–3205 (2023)
34. H. Yu, R. Gao, Y. Liu, L. Fu, J. Zhou, L. Li, Stimulus-responsive hydrogels as drug delivery systems for inflammation targeted therapy. Adv. Sci. **11**(1), 2306152 (2024)
35. Y.S. Lui, W.T. Sow, L.P. Tan, Y. Wu, Y. Lai, H. Li, 4D Printing and stimuli-responsive materials in biomedical aspects. Acta Biomater. **92**, 19–36 (2019)
36. S.N. Hosseini, S.M. Naghib, G.K. Kashani, M.R. Mozafari, Chitosan-grafted graphene materials for drug delivery in wound healing. Curr. Pharm. Des. **31**(9), 691–715 (2025). https://doi.org/10.2174/0113816128333493241014134711
37. K. Khodaverdi, S.M. Naghib, M.R. Mozafari, Chitosan/Bioglass nanocomposites for bone tissue engineering and regenerative medicine: an overview of promising biomaterials. Curr. Org. Chem. **28**(18), 1437–1451 (2024). https://doi.org/10.2174/01138527283147062405290525 35
38. K. Khodaverdi, S.M. Naghib, M.R. Mozafari, M. Rahmanian, Chitosan/Hydroxyapatite hydrogels for localized drug delivery and tissue engineering: a review, Carbohydr. Polym. Technol. Appl. **9**, 100640 (2025). https://doi.org/10.1016/j.carpta.2024.100640
39. M.A. Khorasani, S.M. Naghib, A review of chitosan-based multifunctional nanocomposites for drug/gene/protein delivery and gene therapy in cancer treatments: promises, challenges and outlooks. Int. J. Biol. Macromol. **306**, 141394 (2025). https://doi.org/10.1016/j.ijbiomac.2025.141394
40. S.M. Naghib, B. Ahmadi, B. Mikaeeli Kangarshahi, M.R. Mozafari, Chitosan-based smart stimuli-responsive nanoparticles for gene delivery and gene therapy: recent progresses on cancer therapy. Int. J. Biol. Macromol. **278**, 134542 (2024). https://doi.org/10.1016/j.ijbiomac.2024.134542
41. S.M. Naghib, B. Ahmadi, M.R. Mozafari, Smart physicochemical-triggered chitosan-based nanogels for siRNA delivery and gene therapy: a focus on emerging strategies and paradigms for cancer therapy. Curr. Med. Chem. (2024). https://doi.org/10.2174/0109298673286052240426044945
42. S.M. Naghib, B. Ahmadi, M.R. Mozafari, Stimuli-sensitive chitosan-based nanosystems-immobilized nucleic acids for gene therapy in breast cancer and hepatocellular carcinoma. Curr. Top. Med. Chem. **24**(17), 1464–1489 (2024). https://doi.org/10.2174/0115680266293173240506054439
43. P. Lavrador, M.R. Esteves, V.M. Gaspar, J.F. Mano, Stimuli-responsive nanocomposite hydrogels for biomedical applications. Adv. Func. Mater. **31**(8), 2005941 (2021)
44. M. Vázquez-González, I. Willner, Stimuli-responsive biomolecule-based hydrogels and their applications. Angew. Chem. Int. Ed. **59**(36), 15342–15377 (2020)

45. X. Lin, X. Wang, L. Zeng, Z.L. Wu, H. Guo, D. Hourdet, Stimuli-responsive toughening of hydrogels. Chem. Mater. **33**(19), 7633–7656 (2021)
46. Z. Deng, R. Yu, B. Guo, Stimuli-responsive conductive hydrogels: design, properties, and applications. Mater. Chem. Front. **5**(5), 2092–2123 (2021)
47. S. Gooneh-Farahani, M.R. Naimi-Jamal, S.M. Naghib, Stimuli-responsive graphene-incorporated multifunctional chitosan for drug delivery applications: a review. Expert Opin. Drug Deliv. **16**(1), 79–99 (2019). https://doi.org/10.1080/17425247.2019.1556257
48. Z. Mazidi, S. Javanmardi, S.M. Naghib, Z. Mohammadpour, Smart stimuli-responsive implantable drug delivery systems for programmed and on-demand cancer treatment: an overview on the emerging materials. Chem. Eng. J. **433**, 134569 (2022). https://doi.org/10.1016/j.cej.2022.134569
49. D. Yang, Recent advances in hydrogels 1987–1989 (2022)
50. H.M. El-Husseiny, E.A. Mady, L. Hamabe, A. Abugomaa, K. Shimada, T. Yoshida, T. Tanaka, A. Yokoi, M. Elbadawy, R. Tanaka, Smart/stimuli-responsive hydrogels: cutting-edge platforms for tissue engineering and other biomedical applications. Mater. Today Bio **13**, 100186 (2022)
51. Z. Li, Y. Zhou, T. Li, J. Zhang, H. Tian, Stimuli-responsive hydrogels: fabrication and biomedical applications. View **3**(2), 20200112 (2022)
52. W. Shi, J. Huang, R. Fang, M. Liu, Imparting functionality to the hydrogel by magnetic-field-induced nano-assembly and macro-response. ACS Appl. Mater. Interfaces **12**(5), 5177–5194 (2020)
53. Y. Zhu, J. Jia, G. Zhao, X. Huang, L. Wang, Y. Zhang, L. Zhang, N. Konduru, J. Xie, R. Yu, Multi-responsive nanofibers composite gel for local drug delivery to inhibit recurrence of glioma after operation. J. Nanobiotechnol. **19**(1), 198 (2021)
54. J. Tang, C. Yao, Z. Gu, S. Jung, D. Luo, D. Yang, Super-soft and super-elastic DNA robot with magnetically driven navigational locomotion for cell delivery in confined space. Angew. Chem. Int. Ed. **59**(6), 2490–2495 (2020)
55. Y. Zhang, L. Dong, L. Liu, Z. Wu, D. Pan, L. Liu, Recent advances of stimuli-responsive polysaccharide hydrogels in delivery systems: a review. J. Agric. Food Chem. **70**(21), 6300–6316 (2022)
56. Y. Jiang, Y. Wang, Q. Li, C. Yu, W. Chu, Natural polymer-based stimuli-responsive hydrogels. Curr. Med. Chem. **27**(16), 2631–2657 (2020)
57. Z. Zhu, D.W.H. Ng, H.S. Park, M.C. McAlpine, 3D-printed multifunctional materials enabled by artificial-intelligence-assisted fabrication technologies. Nat. Rev. Mater. **6**(1), 27–47 (2021)
58. C. Wang, A. Fischer, A. Ehrlich, Y. Nahmias, I. Willner, Biocatalytic reversible control of the stiffness of DNA-modified responsive hydrogels: applications in shape-memory, self-healing and autonomous controlled release of insulin. Chem. Sci. **11**(17), 4516–4524 (2020)
59. M.A. Haq, Y. Su, D. Wang, Mechanical properties of PNIPAM based hydrogels: a review. Mater. Sci. Eng. C **70**, 842–855 (2017)
60. K. Yue, G. Trujillo-de Santiago, M.M. Alvarez, A. Tamayol, N. Annabi, A. Khademhosseini, Synthesis, properties, and biomedical applications of gelatin methacryloyl (GelMA) hydrogels. Biomaterials **73**, 254–271 (2015)
61. I. Willner, Stimuli-controlled hydrogels and their applications. Acc. Chem. Res. **50**(4), 657–658 (2017)
62. M.C. Koetting, J.T. Peters, S.D. Steichen, N.A. Peppas, Stimulus-responsive hydrogels: theory, modern advances, and applications. Mater. Sci. Eng. R. Rep. **93**, 1–49 (2015)
63. D. Lei, Y. Yang, Z. Liu, S. Chen, B. Song, A. Shen, B. Yang, S. Li, Z. Yuan, Q. Qi, A general strategy of 3D printing thermosets for diverse applications. Mater. Horiz. **6**(2), 394–404 (2019)
64. F. Andrade, M.M. Roca-Melendres, E.F. Durán-Lara, D. Rafael, S. Schwartz Jr., Stimuli-responsive hydrogels for cancer treatment: the role of pH, light, ionic strength and magnetic field. Cancers **13**(5), 1164 (2021)
65. L. Dai, M. Ma, J. Xu, C. Si, X. Wang, Z. Liu, Y. Ni, All-lignin-based hydrogel with fast pH-stimuli responsiveness for mechanical switching and actuation. Chem. Mater. **32**(10), 4324–4330 (2020)

66. S. Chatterjee, P. Chi-leung HUI, Review of stimuli-responsive polymers in drug delivery and textile application. Molecules **24**(14), 2547 (2019)
67. Z. Han, P. Wang, G. Mao, T. Yin, D. Zhong, B. Yiming, X. Hu, Z. Jia, G. Nian, S. Qu, Dual pH-responsive hydrogel actuator for lipophilic drug delivery. ACS Appl. Mater. Interfaces **12**(10), 12010–12017 (2020)
68. T. Thambi, J.M. Jung, D.S. Lee, Recent strategies to develop pH-sensitive injectable hydrogels. Biomater. Sci. **11**(6), 1948–1961 (2023)
69. T. Mateti, K. Likhith, A. Laha, G. Thakur, A critical analysis of the recent developments in multi-stimuli responsive smart hydrogels for cancer treatment. Curr. Opin. Biomed. Eng. **25**, 100424 (2023)
70. D.J. Hulmes, Building collagen molecules, fibrils, and suprafibrillar structures. J. Struct. Biol. **137**(1–2), 2–10 (2002)
71. M.L.P. Ramos, J.A. González, L. Fabian, C.J. Pérez, M.E. Villanueva, G.J. Copello, Sustainable and smart keratin hydrogel with pH-sensitive swelling and enhanced mechanical properties. Mater. Sci. Eng. C **78**, 619–626 (2017)
72. W.T. Sow, Y.S. Lui, K.W. Ng, Electrospun human keratin matrices as templates for tissue regeneration. Nanomedicine **8**(4), 531–541 (2013)
73. R.S. Bear, The structure of collagen fibrils. Adv. Protein Chem. 69–160 (1952)
74. A.D. Nocera, R. Comín, N.A. Salvatierra, M.P. Cid, Development of 3D printed fibrillar collagen scaffold for tissue engineering. Biomed. Microdevice **20**, 1–13 (2018)
75. T.C. Okwuosa, B.C. Pereira, B. Arafat, M. Cieszynska, A. Isreb, M.A. Alhnan, Fabricating a shell-core delayed release tablet using dual FDM 3D printing for patient-centred therapy. Pharm. Res. **34**, 427–437 (2017)
76. J.R. Richter, R.C. de Guzman, M.E. Van Dyke, Mechanisms of hepatocyte attachment to keratin biomaterials. Biomaterials **32**(30), 7555–7561 (2011)
77. A. Goyanes, F. Fina, A. Martorana, D. Sedough, S. Gaisford, A.W. Basit, Development of modified release 3D printed tablets (printlets) with pharmaceutical excipients using additive manufacturing. Int. J. Pharm. **527**(1–2), 21–30 (2017)
78. B. Mirani, E. Pagan, B. Currie, M.A. Siddiqui, R. Hosseinzadeh, P. Mostafalu, Y.S. Zhang, A. Ghahary, M. Akbari, An advanced multifunctional hydrogel-based dressing for wound monitoring and drug delivery. Adv. Healthcare Mater. **6**(19), 1700718 (2017)
79. M. Nadgorny, Z. Xiao, C. Chen, L.A. Connal, Three-dimensional printing of pH-responsive and functional polymers on an affordable desktop printer. ACS Appl. Mater. Interfaces **8**(42), 28946–28954 (2016)
80. N. Dehghani, F. Haghiralsadat, F. Yazdian, F. Sadeghian-Nodoushan, N. Ghasemi, F. Mazaheri, M. Pourmadadi, S.M. Naghib, Chitosan/silk fibroin/nitrogen-doped carbon quantum dot/α-tricalcium phosphate nanocomposite electrospinned as a scaffold for wound healing application: In vitro and in vivo studies. Int. J. Biol. Macromol. **238**, 124078 (2023). https://doi.org/10.1016/j.ijbiomac.2023.124078
81. S.M. Naghib, S. Amiri, M.R. Mozafari, Stimuli-responsive chitosan-based nanocarriers for drug delivery in wound dressing applications: a review. Carbohydr. Polym. Technol. Appl. **7**, 100497 (2024). https://doi.org/10.1016/j.carpta.2024.100497
82. M.M. Martino, F. Tortelli, M. Mochizuki, S. Traub, D. Ben-David, G.A. Kuhn, R. Müller, E. Livne, S.A. Eming, J.A. Hubbell, Engineering the growth factor microenvironment with fibronectin domains to promote wound and bone tissue healing. Sci. Transl. Med. **3**(100), 100ra89–100ra89 (2011)
83. L. Li, J.M. Scheiger, P.A. Levkin, Design and applications of photoresponsive hydrogels. Adv. Mater. **31**(26), 1807333 (2019)
84. M. Wu, J. Chen, W. Huang, B. Yan, Q. Peng, J. Liu, L. Chen, H. Zeng, Injectable and self-healing nanocomposite hydrogels with ultrasensitive pH-responsiveness and tunable mechanical properties: implications for controlled drug delivery. Biomacromol **21**(6), 2409–2420 (2020)
85. K.Y. Lee, J.A. Rowley, P. Eiselt, E.M. Moy, K.H. Bouhadir, D.J. Mooney, Controlling mechanical and swelling properties of alginate hydrogels independently by cross-linker type and cross-linking density. Macromolecules **33**(11), 4291–4294 (2000)

86. M. Rudko, T. Urbaniak, W. Musiał, Recent developments in ion-sensitive systems for pharmaceutical applications. Polymers **13**(10), 1641 (2021)
87. M. Champeau, D.A. Heinze, T.N. Viana, E.R. de Souza, A.C. Chinellato, S. Titotto, 4D printing of hydrogels: a review. Adv. Func. Mater. **30**(31), 1910606 (2020)
88. Y.E. Kim, J. Kim, ROS-scavenging therapeutic hydrogels for modulation of the inflammatory response. ACS Appl. Mater. Interfaces **14**(20), 23002–23021 (2021)
89. Z. He, Q. Xu, B. Newland, R. Foley, I. Lara-Sáez, J.F. Curtin, W. Wang, Reactive oxygen species (ROS): utilizing injectable antioxidative hydrogels and ROS-producing therapies to manage the double-edged sword. J. Mater. Chem. B **9**(32), 6326–6346 (2021)
90. F. Kazemi, S.M. Naghib, Z. Mohammadpour, Multifunctional micro-/nanoscaled structures based on polyaniline: an overview of modern emerging devices. Mater. Today Chem. **16**, 100249 (2020). https://doi.org/10.1016/j.mtchem.2020.100249
91. S. Soleymani, S.M. Naghib, M.R. Mozafari, PH/Temperature/redox and light-responsive polymersome structure and application in cancer therapy: smart drug delivery and targeted drug release. Curr. Med. Chem. (2024). https://doi.org/10.2174/0109298673303351240507113603
92. J.S. Kahn, Y. Hu, I. Willner, Stimuli-responsive DNA-based hydrogels: from basic principles to applications. Acc. Chem. Res. **50**(4), 680–690 (2017)
93. M. Nakahata, Y. Takashima, H. Yamaguchi, A. Harada, Redox-responsive self-healing materials formed from host–guest polymers. Nat. Commun. **2**(1), 511 (2011)
94. Y. Liang, K.L. Kiick, Liposome-cross-linked hybrid hydrogels for glutathione-triggered delivery of multiple cargo molecules. Biomacromol **17**(2), 601–614 (2016)
95. J. Hu, Z. Liu, Q. Yu, T. Ma, Preparation of reactive oxygen species-responsive antibacterial hydrogels for efficient anti-infection therapy. Mater. Lett. **263**, 127254 (2020)
96. M. Wang, L. Luo, L. Fu, H. Yang, Ion responsiveness of polyacrylamide/sodium alginate (PAM/SA) shape memory hydrogel. Soft Mater. **17**(4), 418–426 (2019)
97. X. Du, H. Cui, Q. Zhao, J. Wang, H. Chen, Y. Wang, Inside-out 3D reversible ion-triggered shape-morphing hydrogels. Research (2019)
98. W. Fu, Z.F. Sabet, J. Liu, M. You, H. Zhou, Y. Wang, Y. Gao, J. Li, X. Ma, C. Chen, Metal ions modulation of the self-assembly of short peptide conjugated nonsteroidal anti-inflammatory drugs (NSAIDs). Nanoscale **12**(14), 7960–7968 (2020)
99. W. Guo, X.-J. Qi, R. Orbach, C.-H. Lu, L. Freage, I. Mironi-Harpaz, D. Seliktar, H.-H. Yang, I. Willner, Reversible Ag^+-crosslinked DNA hydrogels. Chem. Commun. **50**(31), 4065–4068 (2014)
100. C.-H. Lu, W. Guo, X.-J. Qi, A. Neubauer, Y. Paltiel, I. Willner, Hemin–G-quadruplex-crosslinked poly-N-isopropylacrylamide hydrogel: a catalytic matrix for the deposition of conductive polyaniline. Chem. Sci. **6**(11), 6659–6664 (2015)
101. A.E. Broadus, M. Mangin, K. Ikeda, K.L. Insogna, E.C. Weir, W.J. Burtis, A.F. Stewart, Humoral hypercalcemia of cancer. N. Engl. J. Med. **319**(9), 556–563 (1988)
102. Y. Kang, J.J. Walish, T. Gorishnyy, E.L. Thomas, Broad-wavelength-range chemically tunable block-copolymer photonic gels. Nat. Mater. **6**(12), 957–960 (2007)
103. K. Lee, A study on the preparation of stimuli-responsive hydrogels and their applications, École Supérieure de l'Université Nationale de Séoul (2020)
104. M. Kumi, O. Ejeromedoghene, W.D. Sudane, Z. Zhang, Unlocking the biological response of smart stimuli-responsive hydrogels and their application in biological systems. Eur. Polymer J. **209**, 112906 (2024)
105. M.M. Rahman, E.D. Giol, G. Cama, S. Van Vlierberghe, P. Dubruel, Stimuli-responsive hydrogels for tissue engineering (2016)
106. M. Baghali, H. Ziyadi, A. Di Martino, Engineering smart hydrogels for intelligent drug delivery. Polym. Bull. **82**(7), 2287–2328 (2025)
107. J. Kanapathy, Illness perception of type 2 diabetic patients in Malaysia, City University London (2015)
108. A. Adams, A. Malkoc, J.T. La Belle, The development of a glucose dehydrogenase 3D-printed glucose sensor: a proof-of-concept study. J. Diabetes Sci. Technol. **12**(1), 176–182 (2018)

109. L. Huang, R. Jiang, J. Wu, J. Song, H. Bai, B. Li, Q. Zhao, T. Xie, Ultrafast digital printing toward 4D shape changing materials. Adv. Mater. (Deerfield Beach, Florida) **29**(7) (2016)
110. K. Podual, F.J. Doyle III., N.A. Peppas, Glucose-sensitivity of glucose oxidase-containing cationic copolymer hydrogels having poly (ethylene glycol) grafts. J. Control. Release **67**(1), 9–17 (2000)
111. E. Song, T.H. da Costa, J.-W. Choi, A chemiresistive glucose sensor fabricated by inkjet printing. Microsyst. Technol. **23**, 3505–3511 (2017)
112. O. Abazari, P. Dayati, M. Shahidi, J. ZavarReza, M. Rahmanian, S.M. Naghib, Exploring a desirable quadr-mRNAs panel for non-invasive and ultrasensitive bladder cancer diagnosis: in-silico and clinical studies. Curr. Med. Chem. (2024)
113. D. Gan, W. Xing, L. Jiang, J. Fang, C. Zhao, F. Ren, L. Fang, K. Wang, X. Lu, Plant-inspired adhesive and tough hydrogel based on Ag-lignin nanoparticles-triggered dynamic redox catechol chemistry. Nat. Commun. **10**(1), 1487 (2019)
114. J.P. Wojciechowski, A.D. Martin, P. Thordarson, Kinetically controlled lifetimes in redox-responsive transient supramolecular hydrogels. J. Am. Chem. Soc. **140**(8), 2869–2874 (2018)
115. C. Ghobril, K. Charoen, E.K. Rodriguez, A. Nazarian, M.W. Grinstaff, A dendritic thioester hydrogel based on thiol-thioester exchange as a dissolvable sealant system for wound closure. Angew. Chem. (Int. Ed. Engl.) **52**(52), 14070 (2013)
116. M.D. Konieczynska, J.C. Villa-Camacho, C. Ghobril, M. Perez-Viloria, K.M. Tevis, W.A. Blessing, A. Nazarian, E.K. Rodriguez, M.W. Grinstaff, On-demand dissolution of a dendritic hydrogel-based dressing for second-degree burn wounds through thiol-thioester exchange reaction. Angew. Chem. Int. Ed. **55**(34), 9984–9987 (2016)
117. L. Yue, S. Wang, V. Wulf, I. Willner, Stiffness-switchable DNA-based constitutional dynamic network hydrogels for self-healing and matrix-guided controlled chemical processes. Nat. Commun. **10**(1), 4774 (2019)
118. H. Shigemitsu, R. Kubota, K. Nakamura, T. Matsuzaki, S. Minami, T. Aoyama, K. Urayama, I. Hamachi, Protein-responsive protein release of supramolecular/polymer hydrogel composite integrating enzyme activation systems. Nat. Commun. **11**(1), 3859 (2020)
119. H. Zhang, S. Ba, J.Y. Lee, J. Xie, T.-P. Loh, T. Li, Cancer biomarker-triggered disintegrable DNA nanogels for intelligent drug delivery. Nano Lett. **20**(11), 8399–8407 (2020)
120. P. Lavrador, V.M. Gaspar, J.F. Mano, Stimuli-responsive nanocarriers for delivery of bone therapeutics–barriers and progresses. J. Control. Release **273**, 51–67 (2018)
121. R. Chandrawati, Enzyme-responsive polymer hydrogels for therapeutic delivery. Exp. Biol. Med. **241**(9), 972–979 (2016)
122. X. Shen, J.L. Shamshina, P. Berton, G. Gurau, R.D. Rogers, Hydrogels based on cellulose and chitin: fabrication, properties, and applications. Green Chem. **18**(1), 53–75 (2016)
123. T. Maki, R. Yoshisaki, S. Akama, M. Yamanaka, Enzyme responsive properties of amphiphilic urea supramolecular hydrogels. Polym. J. **52**(8), 931–938 (2020)
124. L.J. Del Valle, A. Díaz, J. Puiggalí, Hydrogels for biomedical applications: cellulose, chitosan, and protein/peptide derivatives. Gels **3**(3), 27 (2017)
125. M. Najafi, H. Asadi, J. Van Den Dikkenberg, M.J. Van Steenbergen, M.H. Fens, W.E. Hennink, T. Vermonden, Conversion of an injectable MMP-degradable hydrogel into core-cross-linked micelles. Biomacromol **21**(5), 1739–1751 (2020)
126. A.A. Aimetti, A.J. Machen, K.S. Anseth, Poly(ethylene glycol) hydrogels formed by thiol-ene photopolymerization for enzyme-responsive protein delivery. Biomaterials **30**(30), 6048–6054 (2009)
127. C. Wang, A.R. Esker, Nanocrystalline chitin thin films. Carbohyd. Polym. **102**, 151–158 (2014)
128. M. Dash, S.K. Samal, T.E. Douglas, D. Schaubroeck, S.C. Leeuwenburgh, P. Van Der Voort, H.A. Declercq, P. Dubruel, Enzymatically biomineralized chitosan scaffolds for tissue-engineering applications. J. Tissue Eng. Regen. Med. **11**(5), 1500–1513 (2017)
129. C. Lin, Y. Hu, Z. Lin, L. Du, Y. Hu, L. Ouyang, X. Xie, P. Cheng, J. Liao, L. Lu, MMP-9 responsive hydrogel promotes diabetic wound healing by suppressing ferroptosis of endothelial cells. Bioactive Mater. **43**, 240–254 (2025)

130. S. Jia, H. Zhou, J. Chen, J. Lin, X. Zhu, J. Weng, W. Li, F. Yu, A comprehensive review of hydrogel strategies for repairing peripheral nerve injuries. Brain-X **3**(1), e70012 (2025)
131. X. Wang, X.H. Qin, C. Hu, A. Terzopoulou, X.Z. Chen, T.Y. Huang, K. Maniura-Weber, S. Pané, B.J. Nelson, 3D printed enzymatically biodegradable soft helical microswimmers. Adv. Func. Mater. **28**(45), 1804107 (2018)
132. K.H. Song, C.B. Highley, A. Rouff, J.A. Burdick, Complex 3D-printed microchannels within cell-degradable hydrogels. Adv. Func. Mater. **28**(31), 1801331 (2018)
133. I.S. Protsak, Y.M. Morozov, Fundamentals and advances in stimuli-responsive hydrogels and their applications: a review. Gels **11**(1), 30 (2025)

Chapter 7
Drug Delivery System Based on Nucleic Acid Nanoparticles: A Promising Approach for Establishing a Therapeutic Paradigm

Abstract The development of innovative drug delivery systems (DDSs) has advanced significantly in the area of nanomedicine during the last 10 years. Using DNA and RNA nanoparticles (NPs) as encapsulation vehicles for drug, proteins or genes, enabling controlled release of drugs is an efficient technique. The use of NPs of DNA and their possible uses in regulated drug release techniques are examined in this chapter. Scientists have developed NPs especially for drug delivery by taking advantage of the unique properties of DNA molecules, like their ability to self-assemble and compatibility with living things. When compared to traditional drug carriers, DNA NPs provide a number of advantages, including remarkable stability, easily customizable features, and scalable size and shape. By carefully planning their content and structure, researchers have been able to encapsulate many medicinal substances in a very effective manner. This development makes it possible to distribute medications precisely. There are significant benefits to adding medications, proteins or genes to DNA nanoparticles in relation to increasing therapeutic efficacy and lowering side effects. The encapsulated payloads are shielded by DNA NPs, which prolongs their time in the body and stops them from degrading. For sensitive biologics, including gene or protein treatments, which could be susceptible to enzymatic breakdown or rapid removal, the protective effect is particularly crucial. Additionally, DNA NPs' surface may be modified to enable targeted distribution to specific tissues or cells, increasing delivery accuracy. The ability of DNA NPs to control drug release kinetics is a major advantage. By adjusting the structure of DNA nanoparticles, researchers may control the release rate of the encapsulated payload, allowing for a controlled and extended drug delivery. For drugs with narrow therapeutic ranges or those that must be administered continuously to get the best possible therapeutic outcomes, this control is essential. Furthermore, DNA NPs may respond to environmental stimuli like changes in light, pH or temperature, which can start the payload's release at certain times or places. The accuracy of drug release control is improved by this function. DNA nanoparticles possess a wide array of possible uses in the controlled release of drugs and biomolecules. Drugs, peptides, proteins, DNA, RNA and aptamers are only a few of the medicinal molecules that the NPs may convey. They have the potential to be used therapeutically to address a range of illnesses, including illnesses that spread, cancer and genetic abnormalities.

S. M. Naghib, *Nano-Biosensors, Drug Delivery and Tissue Engineering*, Series in BioEngineering, https://doi.org/10.1007/978-981-95-1888-3_7

DNA and RNA nanoparticles may also be used for targeted drug delivery, overcoming biological obstacles and the limitations of traditional administration of drug techniques.

Keywords Encapsulation · DNA · Controlled drug release · Medication delivery · Gene delivery · Nanoparticles

7.1 Introduction

According to the FDA, a medicinal component that is active is a material that is formally identified by pharmacopoeias and is meant to be used in the diagnosis, treatment, mitigation, cure, and prevention of illnesses [1]. Dosage form administration methods differ according to the drug's chemical and physical properties, the desired site, and the length of therapy. Common dose forms include injections, tablets, ointments, syrups, pills, and capsules [2, 3]. Figure 7.1 summarizes various drug delivery routes, associated challenges, essential design and formulation principles, and regulatory aspects. Therefore, a table of several medication management routes is shown in Fig. 7.1 [4].

DNA nanotechnology has made considerable progress quickly in the previous several decades. Seeman was the first to propose using DNA as building blocks

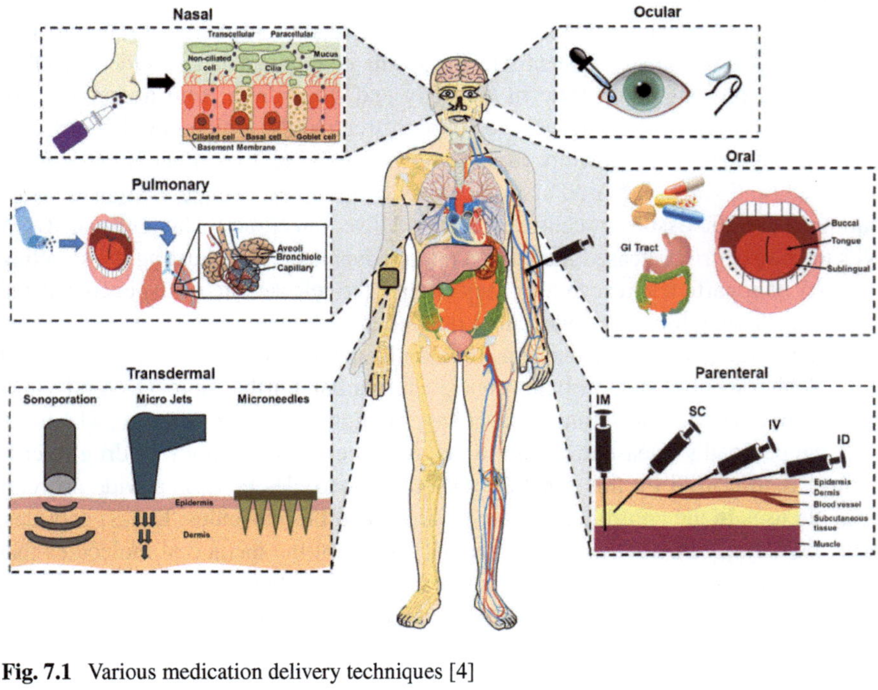

Fig. 7.1 Various medication delivery techniques [4]

by using the complementary base pairing concept. Origami assembly, tile-based bottom-up assembly, rolling circle replication, and composite material assembly are the four main techniques often used to produce DNA nanoscale structures [2, 5–8]. Numerous domains, including drug delivery [9–11], biomolecular analysis [12–15], biosensors [16–20], and molecular manipulation [21, 22], have benefited greatly from these strategies. Generally, DNA cannot readily enter the cell because of the electrostatic repulsion between the negative charges of DNA and the cell surface [3, 10, 23–25]. However, lysosomal avoidance is one of the many ways that DNA nanoscale structures may readily enter cells. Biological and chemical modifications of molecules, including proteins, aptamers, and tiny molecules, have also been used to enhance cell absorption and targeted distribution. Studies have shown that DNA NPs are more stable than DNA oligonucleotides in the physiological milieu of the nuclease [26, 27]. DNA materials are better than other materials for in-vivo investigations because of their inherent biocompatibility and degradability. Furthermore, DNA nanoscale structures facilitate the targeted delivery of drugs in response to specific stimuli [28–33]. The biomedical field has demonstrated numerous uses for nanoscale DNA structures. A brief summary of the use of living DNA nanoparticles in the field of biological uses was previously given by scientists [2, 34]. This research has conducted a thorough examination of medication administration by nanoscale structures of DNA and the associated response system. An overview of current developments in delivery strategies for responsive DNA nanostructures is given in this work. Among these systems are some that react to temperature, metal ions, Glutathione (GSH), ATP, light, pH, oligonucleotides, antigens, and enzymes [35, 36].

The processes that a medication goes through in the body are referred to as pharmacokinetics [2]. Pharmacokinetics is shown schematically in Fig. 7.2 [37].

The process by which medications are absorbed, distributed, metabolized, and excreted Within the body, together with their gradual entry and exit from the body is known as pharmacokinetics [38]. Pharmacokinetics' summary of medication metabolism and excretion is shown in Fig. 7.3 [39].

The capacity of DNA nanoparticles to shield the payload medication from immune system clearance and degradation is one of its main benefits. This is accomplished by the medication forming a protective shell that keeps the body from recognizing and getting rid of it. Furthermore, it is possible to engineer DNA nanoparticles to discharge the medication payload in a regulated way, enabling prolonged release and extended therapeutic benefits. Additionally, it is safe to employ DNA nanoparticles in medication applications for delivery since they are biocompatible and biodegradable. This is crucial for reducing the toxicity and any adverse effects of conventional DDSs. Furthermore, it is possible to design DNA nanoparticles to specifically target certain bodily tissues or cells, enabling accurate and effective medication delivery to the intended location of action [40, 41]. DNA nanoparticles provide a number of benefits over traditional DDSs in terms of pharmacokinetics. For instance, by encasing poorly soluble medications within the structure of the nanoparticle, they may increase their bioavailability. This may boost the drug's distribution and absorption throughout the body, improving therapeutic results. Furthermore, medications

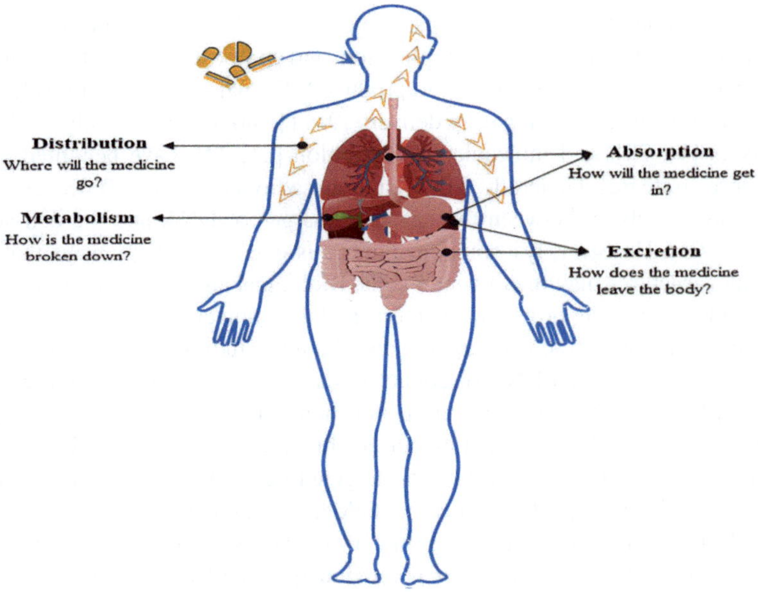

Fig. 7.2 A drug's pharmacokinetic stages include the following: 1. absorption, 2. dispersion, 3. metabolism, and 4. elimination [37]

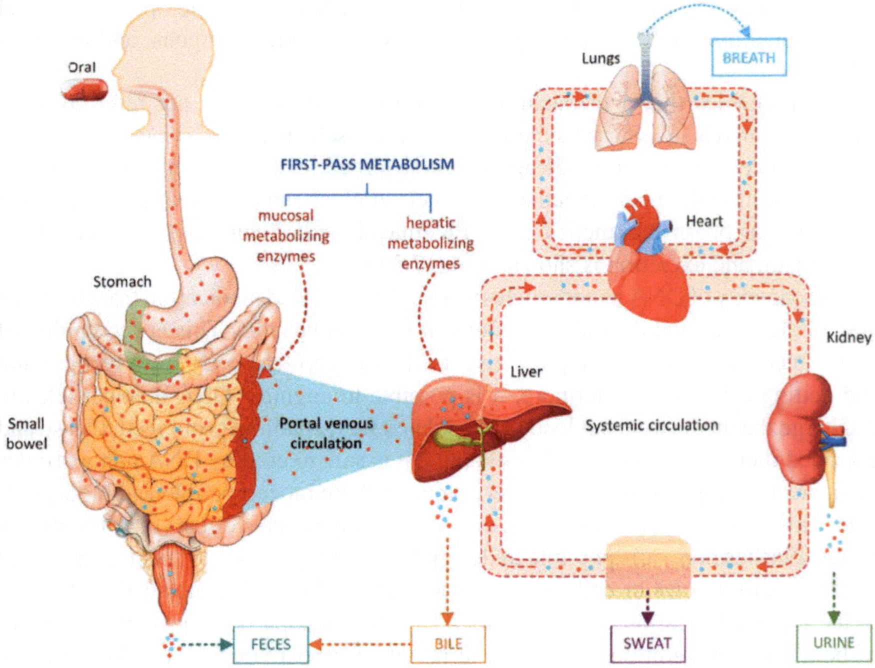

Fig. 7.3 Pharmacokinetics of drug schematic [39]

may be delivered to the central nervous system by using DNA nanoparticles that are engineered to go beyond biological barriers like the blood-brain barrier. This is especially crucial for treating neurological conditions, where conventional DDSs may not work because medications cannot cross the barrier between the brain and blood. Furthermore, scientists may design DNA thioparticles to respond to specific physiological stimuli, such as pH shifts or the presence of specific proteins. This lets us make smart DDSs that can release the medicine when the body goes through certain conditions, which improves the system's pharmacokinetics even more [42, 43]. The use of DNA nanoparticles to deliver nucleic acid (NA)-based therapies, including microRNA and small interfering RNA (siRNA), has gained popularity in recent years. These treatments have the ability to alter gene expression and cure a variety of illnesses, such as viral infections, cancer, and genetic problems. Because DNA nanoparticles can protect the fragile NA payload and make it easier for it to get into cells, they can be used to deliver NA-based therapies [43, 44]. Therefore, researchers have tried a variety of strategies to maintain the medication's blood level above the minimal effective concentration (MEC) but below the hazardous threshold. Although giving many dosages at consistent times can seem like a replacement for a single dosage, it might actually cause changes in the drug's blood levels. This often causes levels to rise beyond harmful levels and decrease below the necessary effectiveness. Patients who get many doses in a single day have less than ideal adherence. Another approach is giving a dose that is more than what is required, which may have unforeseen negative consequences. Therefore, to provide the intended therapeutic impact, controlled release systems (CRSs) of drug administration are required to keep the medication's circulation levels within the correct range for an extended length of time [45]. This DDS ensures that the tissues and circulatory system maintain a constant level of medication for an extended period. Drug levels in a standard delivery method vary below and above the MEC because successive dosages for injections or oral pills adhere to a normal pattern of bolus pharmacokinetics. A single dose of the controlled medication is released from a specific location in the controlled released organization, which, on the other hand, displays zero-order pharmacokinetics. Within the therapeutic range, the medication concentration is continuously maintained [46]. The process by which pharmaceuticals are released from the dosage form determines the classification of the controlled discharge system of drug release. These processes include chemically regulated, dissolution-controlled, diffusion-controlled, and NP-based systems [47].

7.2 Controlled Release Systems (CRS)

CRSs were created to increase medication efficacy, reduce adverse drug dose effects, and promote patient convenience and adherence. When controlled-release medicine is delivered, the dose stays at an effective level while allowing the drug's ingredients to be released in different ways depending on time or other variables (such as enzymes, pH, or temperature). Keeping physiologically active chemicals at appropriate levels within the therapeutic range helps minimize side effects and reduce medication toxicity. Because CRSs are specially intended to maintain the necessary medication plasma concentration for a long length of time, numerous doses are not required [48]. CRSs often use biodegradable synthetic polymers as delivery systems, and depending on the event scene and desired time-release profile, a variety of synthetic polymers with specific physicochemical properties have been widely used to develop CRSs [2]. Due to their benefits over biomaterials, particularly their mechanical characteristics that may be adjusted, polymers that biodegrade are often used in systems of supply to increase the efficacy of DDSs. Over the last 50 years, biodegradable synthetic polymers have emerged as the go-to option for sophisticated drug release systems [49]. This is because of their easily accessible and readily modifiable physicochemical characteristics. researchers developed polyester and its copolymer as biodegradable polymer-based artificial sutures [50]. Ever since, biodegradable polymers that are synthetic have been used mostly as biocompatible carriers that are simple to remove from the body when the drugs are released. Polycaprolactone and aliphatic polyesters are often used in biomedical applications because of their versatile degradability, broad availability, and biocompatibility. The substances fall within the category of biodegradable polymers. Due to their significant application in controlled-release delivery systems, various synthetic biodegradable polymers, including polyesters and several hydrogels, have been the focus of much study recently [51].

By changing the drug's pharmacokinetics, polymeric controlled-release solid dosage forms particularly stop medicines from being released right away. One has to be knowledgeable about the different processes of CRSs and their design considerations in order to choose the best delivery system for a given clinical scenario [25]. By developing a customized therapy that meets each patient's unique needs, the use of personalized medications in DDS has emerged as a way to increase treatment efficacy [28]. More specifically, an insulin dosage schedule that mimics the patient's normal daily cycle is thought to be advantageous as it enables innovative cancer treatment strategies. This emphasizes how important it is to design regulated release and delivery systems that can adjust the pace of medication discharge to different circumstances [52]. Diffusion, swelling, erosion, and degradation of polymers, osmosis, and release triggered by external stimuli are some examples of release processes. The fundamental methods for drug release systems with diffusion control are matrix and reservoir systems. Subcutaneous implants, ocular devices, oral pills, transdermal patches, vaginal implants, and drug-eluting stents are a few of the formulations (Fig. 7.4) [53, 54].

DNA shows a lot of promise as a practical choice for use as a building material in nanotechnology. The practical design of synthetic DNA receptor components is made possible by the exact Watson-Crick hydrogen bonding. The fundamental methods for drug release systems with diffusion control include matrix and reservoir systems, subcutaneous implants, ocular devices, oral pills, transdermal patches, vaginal implants, and drug-eluting stents, among others [2, 55]. The existence of a particular spot on the helix axis where branching may take place is a crucial need for using natural DNA as the basis for fascinating nanomaterials. Transient structures in DNA metabolism, such as repair and replication, produce DNA branch points. The

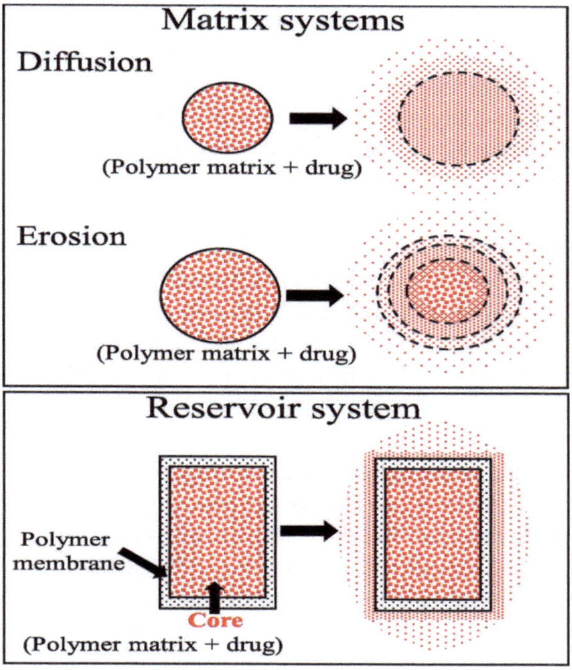

Fig. 7.4 Schematic representation of drug release mechanisms involved in different release systems [54]

Holliday framework, an intermediate DNA recombination process with four arms of branching, vividly illustrates them. DNA sequences can produce Holliday junctions in stable synthetic forms, which are branching molecules with varying numbers and more complex patterns [56, 57]. It also shows how a four-arm junction with sticky ends may be transformed into a 2D lattice. It is possible to create linked DNA structures, such as periodic and aperiodic arrays, by using branching intermediates [58]. In order to construct more complex structures, an alternative technique uses rigid tetrahedral organic vertices to which oligonucleotides are joined as connectors. As a result, different DNA structures may be created by joining different numbers of arms of oligonucleotides to the center organic tetrahedral linkers. Using DNA templates that self-assemble for the exact placement of some useful molecules is a useful way to make use of the extraordinary degree of architectural accuracy. This approach has been effectively used in several fields, encompassing enzyme networks, electronic molecular circuits, nanomechanical devices, and near-field optical devices [59]. Changes that can be switched between two distinct conformations make up the fundamental DNA nanoscale structures. Temperature variations, ionic conditions, and the involvement of a signaling molecule typically a DNA strand all contribute to the start of motion. To create even more complex structures, a different technique uses rigid organic tetrahedral vertices with connected oligonucleotides serving as connections. Therefore, joining a different number of oligonucleotide arms to the center of tetrahedral organic linkers can create different DNA structures [60]. Using DNA templates that self-assemble for the exact placement of some useful molecules is a useful way to make use of the extraordinary degree of architectural accuracy. Numerous applications, including near-field optical devices, enzyme networks, molecular electronic circuits, and nanomechanical devices, have effectively used this strategy. The fundamental active DNA nanoscale structures consist of switches that can transition between two distinct conformations. Temperature variations, ionic state changes, and interaction with a signaling chemical typically a DNA strand all cause motion [61].

Paul Rothmund was a pioneer in the field of DNA nanostructures when he developed a new method for creating unique DNA nanoscale structures by using many short "master" strands of DNA to control a long "scaffold" strand folded into a planar configuration of opposing helices. When compared to conventional assembly techniques, "scaffolded DNA origami" has several noteworthy advantages [62]. There are fewer mistakes made while folding a lengthy DNA strand, which leads to final structures with fewer flaws and more output. The procedure shortens the synthesis time, removes the need to purify the oligonucleotides, and lessens the dependence on stoichiometric ratios. Advanced processes can produce more complex structures with precise dimensions and perfect planning, which allows precise molecules to stick in specific places. Since the staples only need to bind to the scaffold and not to one another, the efficiency of origami scaffolded DNA is independent of the proportional amounts of the staples. Accurate staples are first attached, partly arranging the large structure so that the rest of the staples may be attached correctly. Strand invasion makes it possible to remove unwanted secondary structures and move erroneous or shortened staples, which leads to the right binding of staples. Since Rothemund's original study, numerous research groups have used scaffolded DNA origami to create a variety of structures and have demonstrated a number of notable applications. For the prior investigation, the circular, single-stranded (ss) M13mp18 genome was arranged into a framework made up of antiparallel helices. By using a particular pattern of recurring cross-overs, made possible using 200 staple strands to create helices, this arrangement was accomplished. Notably, the process of self-assembly produces very high yields of the desired structure. This procedure entails annealing the model for about one hour while it is surrounded by staple threads, which are usually one hundred times more numerous than are required. The assembly of DNA oligonucleotides into extended superlattices during the thermal annealing process typically takes 20 h. It is noteworthy, nevertheless, that the rate when the temperature drops from around room temperature at 95 °C is rather amazing. The entropic benefit of using one extended strand of scaffolding for folding is partly responsible for the remarkable performance of the DNA origami method [63]. Paul Rothemund was a trailblazer in the area of DNA nano-constructions. He came up with a new way to make unique DNA nanoscale structures by controlling a long "scaffold" strand folded into a flat shape of opposing helices using many short "staple" strands of DNA. When compared to conventional assembly techniques, "scaffolded DNA origami" has several noteworthy advantages [62]. There are fewer mistakes made while folding a lengthy DNA strand, which leads to final structures with fewer flaws and more output. The procedure shortens the synthesis time, removes the need to purify the oligonucleotides, and lessens the dependence on stoichiometric ratios. Advanced processes can produce more complex structures, resulting in nanoscale structures with precise dimensions and tunable programming, allowing precise molecules to adhere to specific locations. The efficiency of origami scaffolded DNA is independent of staple proportions because the staples only need to bind to the scaffold. Accurate staples are first attached, partly arranging the large structure so that the rest of the staples may be attached correctly. Strand invasion makes it possible to remove unwanted secondary structures and move erroneous or shortened staples, which leads to the right binding of staples. Because of Rothemund's original study, numerous research groups have used scaffolded DNA origami to create a variety of structures and shown a number of notable applications. For the prior investigation, the circular, single-stranded (ss) M13mp18 genome was arranged into a structure made up of antiparallel helices. By using a particular pattern of recurring cross-overs, made possible using 200 staple strands, this arrangement was accomplished. Notably, the process of self-assembly produces very substantial yields of the intended structure. This procedure involves surrounding the model with staple threads, typically one hundred times more than necessary, and annealing it for about an hour. The assembly of DNA oligonucleotides into extended superlattices during the thermal annealing process typically takes 20 h. It is noteworthy, nevertheless, that the rate when the temperature drops from around room temperature at 95 °C is rather amazing. The entropic benefit of using one extended strand of scaffolding for folding is partly responsible for the remarkable performance of the DNA origami method [64]. Building origami of 3D DNA on a lattice allows for a more thorough comprehension of force studies on single molecules and physical phenomena, such as plasmonics, protein interactions, and cascading reactions of enzymes. Furthermore, some investigations have shown that 3D origami of DNA may be used as nano-devices, nano-switches, and delivery systems by means of strand displacement. The effectiveness of lattice-engineered and scaffold-based 3D DNA origami has significantly increased with the advent of caDNAno, an easily navigable software application. This application makes it simple for non-specialists in a variety of professions to create and participate in 3D origami of DNA. The researchers used 3D DNA origami to simulate proteins with channels, which help move materials across cell membrane lipid layers, absorb loaded medication nanoparticles via cells, and administer them throughout living things. Top-down etching and lithography are commonly used to fabricate solid-state nanopores with the goal of improving their capacity by adding origami-like funnel-shaped structures to them, particularly inside silicon nitride filters. Scientists have found a new way to target and deliver the popular cancer drug carboplatin to cancerous cells in the uterus; they are doing this by using 3D DNA origami with a lattice structure [65]. Lattice engineering-based 3D DNA origami is more effective than non-targeted origami at specifically identifying and eliminating cells that overexpress certain receptors. According to the research, by using flexible DNA origami nanoscale structures, it will be easier to deliver mixtures of anticancer drugs to cancer cells that are resistant to them. In order to create 3D DNA origami, all of the strands must be combined, often in excess of two to ten times as many staple strands. After adjusting the concentrations of Mg^{2+} and Na^{+}, the blend is cooled progressively from high temperatures to promote renaturation. Nevertheless, several approaches rely on various denaturation procedures and use isothermal foldable ways [66]. Spin-filtration, gel electrophoresis, size-exclusion chromatography, and the use of ultracentrifugation at rate-zonal are used to remove the extra staples from the folded assemblies. The building's structural layout, future usage requirements, and size all influence the purifying method selection. AND and XOR are two examples of the many scaffold logic gates that may be built using tetrahedral DNA nanoscale structures. Various DNA tetrahedrons that can be programmed nanoscale structures have been developed, each having a distinct sequence that can be changed in response to certain protons, metal ions (Hg^{2+}), and small molecules (ATP). Logical calculations based on DNA nanostructures have the ability to detect illness biomarkers and regulate the release of tiny molecules inside living things. DNA-tetrahedron-linked hairpin probes may be used to develop a versatile optical sensing platform that enables a wide range of studies on endonucleases, small molecules, and miRNAs [58]. According to recent research, the same quantity of the drug doxorubicin (DOX) may be stored in all DNA origami nanoscale constructs. One DOX molecule may bind to each of the two to three DNA building units. Furthermore, the binding components are balanced in a matter of seconds; it is quicker than was first thought. This study investigated how drugs are broken down and released from complexes in the body during DNase I digestion, with the goal of understanding how drugs are released from complexes. Depending on the kind of DON, the quantity of released DOX molecules by the DONs might vary by up to 100 times. Furthermore, by adjusting DOX, magnesium, and pH levels, this study has shown how the light spectrum of DOX bunches together varies [67]. Irenicon is a sugar molecule that functions as an antibiotic and inhibits tumor growth by attaching an amino acid to a planar anthraquinone core. In DNA origami, irenicon, a semi-synthetic substitute for DOX, intercalates into the sequences of double-stranded (ds) DNA non-covalently, 5′-GC-3′ as well as 5′-CG-3′, to successfully cure breast cancer. With the exception of its stereochemistry, it is almost exactly the same as DOX (Fig. 7.5) [66].

To administer daunorubicin, an anthracycline, to human resistance of HL-60/ADR leukemia cells to numerous treatments, DNA origami rod at the nanoscale constructs that resemble Trojan "horses" was used. In the leukemia model, daunorubicin-loaded equine DNA nanoscale constructs have been shown to successfully combat medication resistance caused by multidrug resistance protein-1. By precisely intercalating into DNA origami, the antitumor drug aclarubicin is used to treat acute nonlymphocytic leukemia. When early chemotherapy is unsuccessful, idarubicin is used to treat advanced breast cancer [68]. For medication delivery, the origami double-stranded tetrahedron of folate-overhung DNA includes the anticancer drug mitoxantrone. The DNA origami mitoxantrone tetrahedra is around 25 nm in size and is decorated with extra folate molecules. Both in lab settings and in leukemia-affected animals, this nanostructure demonstrates the capacity to selectively attach to and breach the nuclear membrane of leukemic cells, trigger apoptosis, and improve the healing effectiveness against cells that are leukemic. One kind of flavonoid, quercetin, has anti-cancer, anti-inflammatory, immunomodulatory, antioxidant, and toxic actions, among other positive qualities that suggest its potential for medical use. Despite its many health benefits, quercetin has certain drawbacks, including restricted permeability, low bioavailability, poor solubility, and hydrophobicity. Incorporating

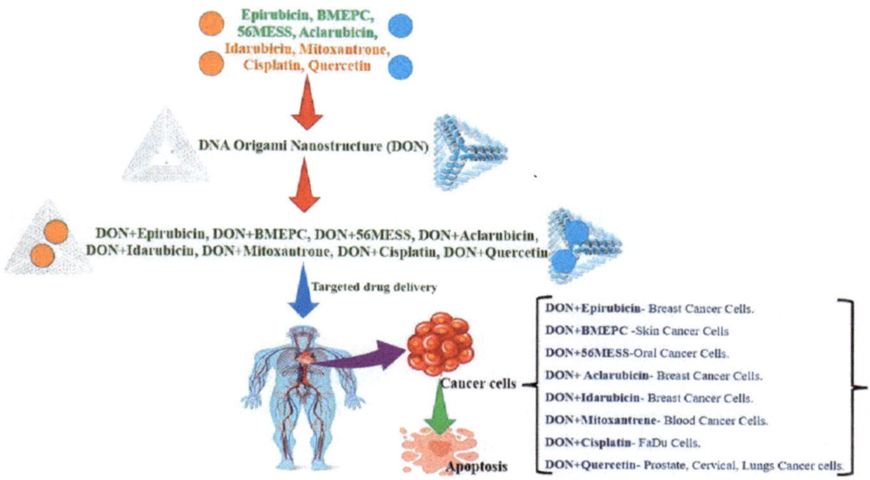

Fig. 7.5 An example showing how DNA-origami-NPs might be used to precisely deliver medications to tumor cells [66]

quercetin into DNA origami improves its absorption and solubility, which helps to get beyond some of the compound's drawbacks. Its water-repellent qualities, limited capacity for absorption by the body, poor mobility across membranes, and poor solubility in liquids are other drawbacks. By encasing luteolin in DNA origami, its bioavailability and solubility are improved, hence resolving some of the compound's drawbacks [69].

7.3 Drug-Encapsulated DNA Origami, Cellular Entry, and Cellular Targeting

DDSs must be delivered and loaded in a controlled manner, and they must contain sufficient amounts of medicine. The particular characteristics of the medicine, the kind of origami DNA used, and the need to preserve the drug's effectiveness after release all influence the choice of cargo integration technique. Thanks to continuous improvements in mass manufacturing and different techniques for maintaining DNA origami nanoscale structures in physiological fluids, the investigation of DNA origami nanoscale structures for bioapplications is now being justly conducted. Drug delivery is significantly hampered by the rigidity of DNA origami nanoscale structures due to their poor flexibility in the cellular environment, making DNA origami one of the most efficient methods for drug delivery [66, 70]. To enhance medication administration, delivery, and cell uptake, a variety of targeted ligands might be affixed to the origami structure of DNA. Aptamers, a class of NAs that can recognize membrane receptors, are readily included into triangle DNA origami. In HeLa cells, this may subsequently start aptamer-receptor-mediated endocytosis, which greatly increases the effectiveness of DOX delivery. Through careful manipulation of the density as well as the alignment of aptamers inside structures made using DNA origami, the effectiveness of cellular absorption may be adjusted. Proteins and small molecules have been used as targeted ligands in this application, where protein receptor-mediated endocytosis was used to attach transferrin, a protein that carries iron, to the outside of DNA origami structures. As a result, KB-3-1 cell lines showed a significant increase in internalization efficiency of up to 22-fold for DNA origami, which is used to treat prostate cancer with Z as a preclinical approach [71]. Ge et al. used DNA origami to insert 2-[3-(1,3-dicarboxy propyl)-ureido] acid pentanedioic, a ligand that targets tiny molecules. Because tailored ligands and antigens stick to tumor cells very well, those that show PSMA+ (the prostate-specific membrane antigen) worked best at ligand-modified DNA origami internalization. The interplay between ligand modifications and membrane receptors affects the specific efficiency of binding and the efficacy of targeting DNA origami structures [72]. Compared to studies carried out in lab settings, very few have been done on the dispersion of medicinally delivered origami DNA in living animals. However, many studies have been conducted on living DNA origami objects, with one of the first in vivo investigations on DOX distribution and DNA origami being conducted by Zhang and colleagues. Triangular DNA origami nanoscale constructs were used in this work to provide remarkable therapeutic effects in live creatures without endangering the system as a whole [63]. Additionally, by passively targeting tumors, these nanoscale devices demonstrated a significant buildup of DOX in malignant regions. Together with DNA intercalator medications, protein-based medications have been incorporated into nanoscale architectures of DNA for in vivo use testing. Researchers developed a self-governing DNA nanorobot to help transport a protein for medicinal purposes [73]. The robot's outer DNA aptamer forced the inner chamber to open and targeted the bound thrombin toward nucleolin, which caused necrosis, coagulation, and inhibition of tumor development in vivo. One important group of therapies that use DNA nanoscale structures to target cancer cells is called NA therapeutics. By combining DNA with antisense RNA nanoscale structures, self-assembly of NA is a promising way to make NA that is not affected by enzyme therapeutics. Rahman et al. reported that the first DNA origami structure with a rectangular and tubular shape was successfully made and then put into tumor cells for in vivo gene therapy. Short interfering RNA was used to effectively knock down BCL-2 mRNA. According to the results, up to 90% of BCL-2 was eradicated, and 70% of BCL-2 mRNA was successfully inhibited [74]. Pan and colleagues created rectangular DNA origami structures using antisense oligonucleotides to target the translation of glycoprotein P mRNA and two proteins associated with B-cell lymphoma, aiming to overcome the limitations of intercalator drugs. Together, these structures' healing effects in the HeLa/ADR and MCF-7/ADR cell lines were able to break down drug resistance [75]. Despite their study's limited coverage of RNA interference in vivo, Liu et al. created nanoscale constructs with consistent properties and carried out in vivo testing. DOX was used to treat drug-resistant MCF-7R cancers, together with a unique triangular nanostructure made using DNA origami as part of the research. DNA nanotechnology inhibited tumor development, showed no systemic toxicity, and may be used in the treatment of tumors with gene therapy [76]. In a follow-up study, the identical researchers used the same triangular DNA origami structure to deliver drugs called DOX and NA to combat tumor cells that are resistant to drugs. The NA medications selectively target the anti-apoptotic protein survivin and the P-glycoprotein by using two transcription templates for linear RNA. In order to effectively penetrate MCF-7R cells and limit tumor development in vivo, the nanocarriers used three complementary therapeutic approaches [77]. RNA transcription templates with short hairpins harboring the P-glycoprotein genes for multidrug resistance and endurance were encapsulated and distributed using the DNA origami nanoplatform. For the best tumor suppression in live beings, NA medications must often be used in combination with other therapeutic treatments [78]. To find out whether changes in internal structure affect the efficiency of DNA origami delivery systems, research was carried out to examine their penetration and absorption in cells and CSTMs, or cell spheroid tissue models. Despite having diverse inner designs, two buildings are built using similar molecular weights and geometrical features. One is small, while the other is built on lattice-based origami and has an open wireframe shape. In CSTMs, wireframe rods are more capable of penetrating deeper than closely spaced rods; moreover, wireframe rods loaded with DOX show elevated cytotoxicity in CSTMs. The two DNA origami design paradigms' variations in permeability, structural mechanics, local material density, and deformability pertaining to receptors in cells help to explain these results. Differences in how scavenger receptors interact are thought to be the main cause of the variations in penetration kinetics. In contrast to nanoscale polygonal formations of the same size and form, internalization of lattice-based structures is more likely to occur in a larger degree. Therefore, it is claimed that one of the most important factors in the use of origami DNA for drug delivery is the choice of structural design methods [79]. Origamis of DNA structure carrier types are often labeled using colors that glow to examine their cell transport capabilities. Because they can be seen under a microscope, infrared dyes are excellent molecular models for drug delivery systems that carry drugs into cells. Studies have shown that human tumor cells may be penetrated by Cy3 molecule-modified DNA nanoscale tubes. With the help of folate, tumor cells may internalize DNA nanotubes, which will cause folate receptors to develop. Ding and his colleagues have created a way to use luminescence to look at how DNA origami nanostructures are distributed and how long they last in living cells without the need for labeling. In a condensed DNA arrangement, a biscyanine dye based on carbazoles may bond to a duplex of DNA and create strong luminescence. Nevertheless, the intensity of their fluorescence decreased when the densely packed DNA structures were broken. Carbazole-based biscyanine was delivered to cells via DNA origami constructs. Surprisingly, during 60 h, this nanocarrier demonstrated exceptional resistance to cellular breakdown. This result provides strong support for the efficient management of cargo discharge [80].

7.4 Multipurpose Drug Carriers Based on DNA

When a drug carrier can perform many tasks, it is said to be multifunctional, because DNA origami is so versatile that it is useful for creating containers that can accommodate many drugs with different pharmacological effects. This makes it easier to provide medical treatment in a variety of methods, including gene therapy, immunotherapy, chemotherapy, photothermal therapy, photoacoustic diagnostics, and more. As a result, DNA origami influences the creation of flexible delivery methods. Successful attempts to use DNA origami to create adaptable drug delivery vehicles have recently been made [81]. One of the most important cancer management strategies is pharmacotherapy. However, as cancer patients get longer-term treatment, the problem of medication resistance has gradually increased in frequency. One of the main reasons why cancer therapy is unsuccessful is because resistance makes it more difficult for anti-cancer medications to kill cancer cells. Therefore, conquering cancer cells' resistance is a major problem in today's therapeutic practice. Because of their distinct photoelectric, chemical, and physical characteristics and their remarkable compatibility with biomaterials, gold nanoparticles (AuNPs) are an attractive option for modern cancer treatments. They are often used as DDSs to deliver medications to tumor cells and destroy cancer cells by photothermal effects [82]. Using DONs (DNA origami nanoscale structures) as gold nanorod transporters and medications is a unique way to treat tumors that are resistant to traditional treatments. This allows for accurate and focused distribution of the pharmaceuticals to the refractory tumor cells. Chemotherapy medications and targeted medications may work in concert to efficiently eliminate tumor cells that have become resistant to them [83]. Gu et al. targeted cancer cells deeply into tumor tissues by combining DOX, TDNs (tetrahedral DNA nanoscale structures), and gold nanoparticles (GNP) [84]. It was particularly intended to inflict varying degrees of tumor tissue damage, depending on pH variations; at first, the whole nanomaterial congregates and works together in tumor cells. The TDN connects DNA sequences to alter its shape from two helices to three helices at a pH of around 6.5 in the tumor microenvironment. Following its separation from GNP, it thoroughly penetrates the stroma of the tumor and within the cells. Photothermal GNP treatment is used in this procedure to successfully remove the tumor. In the end, TDN forms an i-motif to release DOX structures within the acidic lysosome (pH $= 5.0$), which causes the harmful effects of DOX to kill cancerous cells in the deepest areas (Fig. 7.6) [81, 85].

As a result, the DON-based drug delivery system can successfully produce the combined effects of physical, drug, gene, immunological, and other therapies that precisely target cancer cells, with drug-resistant cancer cells being more effectively eradicated by this method. As a result, using multifunctional drug carriers based on DNA origami presents a very promising and novel way to tackle the problem of treating tumors that are drug-resistant [81, 86–90].

Because of the Watson-Crick base pairing's very precise assembly ability, DNA may be programmed to create complex DNA machines. DNA origami, tile-based approaches, and DNA wireframe are the main methods used to assemble DNA

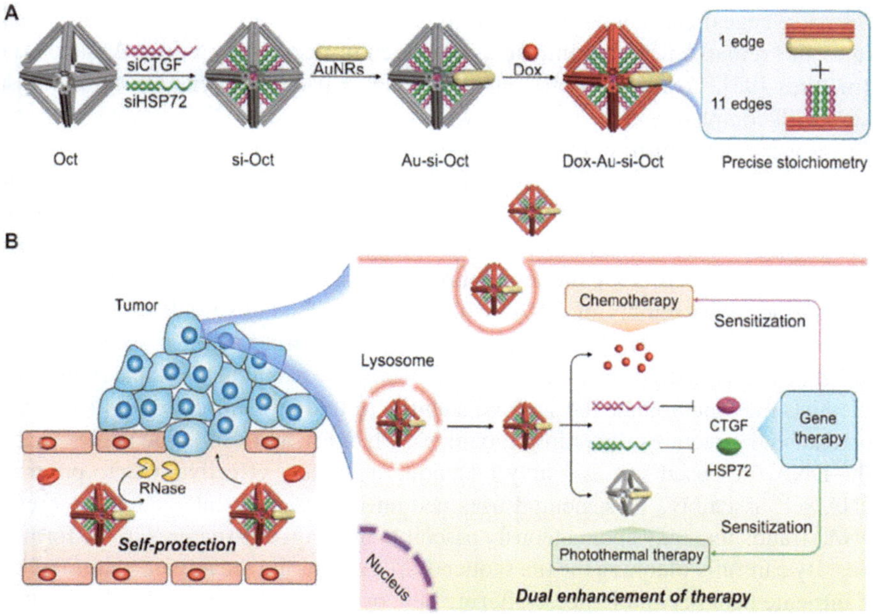

Fig. 7.6 DNA origami produces octahedral DNA nanoscale structures that are used in a variety of medication delivery techniques. **a** The octahedral DONs (diamond-shaped nanocrystals) were concurrently bonded to the small interfering RNA (siRNA) molecules (siCTGF and siHSP72), DOX, and AuNRs (gold nanorods). **b** When absorbed by cancer cells, the composite structure administers chemotherapy, gene therapy, and photothermal treatment while maintaining stability in the circulation [85]

nanoscale structures [91]. Four DNA strands come together to create the Holliday junction, which resembles the arm of a double helix. Brief S regions made up of five to ten nucleotides produce the projecting sections of NA strands, sometimes referred to as sticky parts. Consequently, the Holliday junction is used as the basic unit to generate further crossovers, resulting in the creation of complex and varied tiling patterns [92]. Rothemund's 2006 creation of DNA origami had a big influence on structural DNA nanotechnology. The "one-pot" method was used to fold long ssDNA with many short DNA strands (called staples) to create a variety of 2D structural forms, like triangles, triangles, stars, smiling faces, and more [62]. Complex 3D DNA nanoscale structures were created by further stacking numerous origamis. Other three-dimensional structures were produced with good yield and consistency [93]. Compared to 3D self-assembly using tiles, DNA origami displays an even more intricate and inflexible structure, mainly because of the large number of duplex intersection locations. In contrast to origami structures, which have a diameter of around 100 nm, tile-based structures are usually smaller, with a diameter in the tens of nanometers. Connecting a few short strands of DNA to make wireframe DNA nanoscale structures like icosahedrons, tetrahedrons, octahedrons, prismatic cages, and nanotubes is the third way to put DNA together. This illustrates the remarkable programmable self-assembly capability of DNA nanoscale structures [93]. DNA molecules have a special capacity to generate certain base pairs with complementary sequences, which is at the core of DNA nanotechnology. Researchers may construct DNA strands with certain sequences to produce building blocks that will self-assemble into predetermined forms and preferentially attach to one another. This process is similar to figuring out a jigsaw, with each DNA strand functioning as a component that can only be assembled with its corresponding piece [94]. The scaffolded DNA origami approach is among the many popular ways to instruct the self-assembly of DNA. With hundreds of small strands of DNA called staples to keep the scaffolding in place, a lengthy scaffold a single-stranded DNA molecule is folded into the required form. By carefully planning the staple sequences, scientists may manipulate the scaffold's ability to fold into intricate two- and three-dimensional structures, including tiny maps, triangles, and nanoscale smiley faces. The DNA tile-based self-assembly technique is another effective way to program DNA self-assembly. This method uses certain base pairing interactions to create DNA strands that may spontaneously assemble into bigger structures in the form of tiles. By carefully planning the tile sequences, scientists may manipulate the creation of intricate patterns and lattices, offering new opportunities for the development of materials and nanoscale devices with special qualities. Because DNA self-assembly is programmable, scientists may also add functional components to the nanostructures. In order to develop nanoscale devices for molecular sensing or targeted medication administration, for instance, researchers may bind certain compounds to the building components of DNA. For use in nanoelectronics and photonics, researchers may also develop tiny circuits and sensors by adding nanoparticles of metal or semiconductors to the structures [95]. Nanoscale DNA structures have the capacity to improve our basic knowledge of biological processes in addition to their potential uses in technology and medicine. Researchers may learn more about the rules guiding self-organization and molecular interactions in biological systems by simulating the DNA self-assembly processes seen in nature. To sum up, programmable self-assembling nanoscale structures of DNA provide a potent tool for building intricate nanoscale structures with specialized functions. The development of novel nanotechnologies with uses ranging from materials science to health is made possible by the capacity to precisely regulate the assembly of DNA molecules. In the near future, we may anticipate seeing much more complex and varied nanostructures that will transform a number of disciplines as scientists keep expanding the possibilities of DNA nanotechnology [96].

Many significant stimuli may cause nanocarriers to respond, including glutathione (GSH), light, PH, ATP, enzymes, oligonucleotides, metal ions, antigens, temperature, redox, toxins, and magnetic fields. Numerous studies have been carried out on responsive DNA nanoscale constructs with predictable and programmable architectures and functionalities. However, in the following areas, further research is needed. First, structural soundness. Stimuli-responsive DNA nanoscale constructs must maintain structural coherence until they arrive at their target place [97]. The difficulty with DNA nanostructures is that they are easily broken down by nucleases and removed from the bloodstream too soon. The construction of stiffer, more stable DNA molecules, structural encapsulation, chemical modification, and metal ion stabilization's use are some of the techniques that improve blood half-life. Second, low permeability; DNA struggles to get through biological barriers and escapes endosomes with less than ideal efficiency because of its size and significant negative electricity. As a result, even with great efforts, the DDS's capacity to enter tissues deeply is still severely limited [98]. Reducing the size by making the DNA origami structure smaller is one possible remedy. Incorporating biological ligands that are abundantly expressed into biofilms and adding materials or molecules that are positively charged may also aid in absorption and escape from endosomes [99]. Although a lot of research has been done on photoresponsive DNA nanoscale constructs, most of their light-responsive capabilities are restricted to the visible and ultraviolet spectrum. This makes it difficult to achieve the prerequisites for photodynamic treatment, which calls for deep invasion of tissue. Consequently, it is essential to integrate the infrared-responsive trigger into the nanostructure of DNA on a bigger scale. Furthermore, the previously noted problem is expected to be resolved by using the two-photon excitation technique. It is important to note that a lot of study has been done recently on the aberrant manifestation of tumor enzymes and antigens. However, there hasn't been much advancement in the creation of tumor indicators, especially for chemicals found within cells that react to vectors of DNA nanostructures. Also, drugs with DNA nanostructure carriers that respond to different environmental factors could be used in more complex settings, making it easier to give the right amount of medicine. The brain-blood barrier is still difficult for DNA nanoscale structures to cross at this time [100]. The development of environment-responsive DNA nanostructure DDSs that can operate efficiently in complicated systems should be the focus of future research. To improve the particularity, efficacy, and physiologic roles of medication administration, stimuli-responsive nanocarriers have been purposefully developed, taking into consideration a variety of pathological features in tumor microenvironments, intracellular spaces, and healthy tissues. In response to internal and external stimuli, nanocarriers may respond to magnetic fields, thermal temperature fluctuations, light, electric, and ultrasonic fields as examples of external stimuli; pH, hypoxia, redox potential, H2O2, ATP, and enzymes are examples of internal stimuli, which may occur inside cancer cells or in the tumor microenvironment [101]. Target cell delivery of functional pDNA seems to be less successful when the PEG layer is present. Yingyuad et al. looked into the use of pDNA-mediated enzyme activation treatment using a method called enzyme-assisted nanoparticle trigger ability. The researchers used peptidyl lipids, or PEG2000, with peptidoglyl components that react to the metalloproteinase-2 matrix or elastase present in tumor cells. Then, they created PEGylated pDNA nanoparticles that enzymes like elastase could use to break down the amide peptide bonds of elastin. Therapeutic NAs should be successfully delivered to tumor cells inside a living creature using this idea [102]. The system is made up of matrices of hydrogel that are charged with GOx and react to pH variations. There are two types of cooperative NAs that connect these matrices: pH-responsive "caged" i-motif duplexes and permanent duplex NAs. Acidification of the hydrogel is the consequence of gluconic acid being produced during the bioelectrocatalysis-induced oxidation of glucose. This acidity causes the i-motif's components to separate and the hydrogel's rigidity to diminish. The hydrogel-modified electrodes release insulin as a result of the matrices of hydrogel being filled with insulin. This release is regulated by the glucose levels and may be initiated by a possible alteration. Together with its ability to trigger the release of insulin in an "OFF or "ON" manner, the capacity of the pH-responsive hydrogel to adjust stiffness in response to variations in glucose concentration suggests that this system may serve for the exact and regulated release of insulin, both in terms of time and dose, as a self-regulating bioelectronic device [103]. Using metal-carboxylate bridges and NA duplexes to chemically link polymers together is a successful technique for producing hydrogels that respond to redox. The cooperative connecting of polymers via redox-active bridges between donors and acceptors and duplex NAs is the second technique for producing hydrogels that respond to redox. The development of polymer matrices supported by NA bridges that may go via trans/cis-photoisomerization and permanent crosslinking units like boronate ester/glucosamine is necessary for the introduction of photoresponsive hydrogels. As an alternative, this work presents photoresponsive hydrogels crosslinked with supramolecular complexes and NAs. Photoisomerizable donor-acceptor or host-guest complexes comprise these complexes. A hydrogel of DNA that displays a shift from gel to solid at 32 °C, which is near the average body temperature of 37 °C, was created by Guo et al. [104]. Drug delivery is significantly impacted by DNA hydrogels' temperature-dependent property; in particular, drugs may be added to certain hydrogels when they are solid. These hydrogels' duplex configuration changes become a gel and let go of the medication when they encounter a specific tissue spot whose temperature is lower (below 32 °C). By using the duplex of DNA breaking after being exposed to the appropriate temperature, which results when a medication is released in gel form, this allows pharmaceuticals to be localized to a particular spot [105]. Additionally, ATP-responsive DNA hydrogels were produced by the Qi group, and DNA strands were grafted onto polymer chains to create the hydrogels, which were then hybridized using aptamers for ATP to create a shell-core hydrogel. A nanogel with a comparatively high particle size was created when the researchers effectively added DOX, a chemotherapeutic drug, to the hydrogel. Both pH and ATP sensitivities were used in the Qi group's method. The medication was released when the hydrogel was compressed after being exposed to a tumor with a lower pH of 6.4. Furthermore, the drug's release was aided by the tumor locations with elevated ATP levels. An effective targeted DDS that was triggered by two different triggers was shown in this group's study [106]. By using magnetic nanoparticles with iron oxide as the center of a magnetic DNA nanogel, Yao et al. were able to accomplish a successful application. A chemotherapeutic drug called DOX was added to the nanogel matrix as part of the investigation. In order to guide the nanogel to the precise site of the tumor, an outside magnetic field was also used. The capacity of an electrically activated hydrogel of DNA to accurately find and deliver a medicine to a particular target region was successfully shown in this work. Further, it was shown that the nanogel was sensitive to pH and nuclease in addition to magnetic fields. This feature might lead to fine-grained control over the site and timing of medication release [107]. This work demonstrates that conjugates of DNA and magnetic nanoparticles may successfully stop the magnetic silica particles' pores from spreading when both DNA strands hybridize. Fluorescein, a model drug, was released from a pore that was created when the temperature was raised because the two strands of DNA gradually separated and opened as a result, and DNA linkage's reversibility leads to a release mechanism that is "on-off". Additionally, under the influence of an oscillating magnetic field, the system's magnetic component allows for the achievement of a temperature range of hyperthermia (42–47 °C). This feature makes it possible to trigger a medication delivery system from a distance. Furthermore, this adaptable device's ability to increase ambient temperature could significantly influence the development of advanced thermochemotherapy DDSs in the fight against cancer [108]. Along with the previously described molecule that responds to photon compounds, DNA hydrogels may also include photon-responsive DNA sequences, as a result of which the DNA hydrogels may exhibit photon-responsive properties. The development of the hydrogel memory system is advanced by combining photolithographic technology with photon-responsive DNA hydrogels to produce controlled form patterns in DNA hydrogels. A photolithographic method was created by Kasahara et al., to produce hydrogels made of DNA that were self-assemble in response to light. Y-shaped DNA nanostructures, or Y motifs, may be used to modify the shape of these hydrogels; two Y-shaped structures were created by the authors [109]. Based on the gating concept, Gu et al. reported a technique for selectively transporting macromolecular materials using DNA-gated hydrogel. Covalent connections are formed between DNA strands and the hydrogel in DNA-DNA interactions. The only cargo that can attach to these DNA strands that are gated reversibly and pass through obstacles is cargo labeled with complementary DNA sequences that correspond to the DNAs in the hydrogel. Two prerequisites must be satisfied for this technique to work: initially, the interactions between DNA must provide sufficient power of attraction to make it easier to put the cargo within the hydrogel. Additionally, the speed at which these interactions were place made it possible for the payload to travel and disperse throughout the hydrogel. In their further research, the authors looked into how cargo valency affects the selectivity of the delivery system and the strength of DNA's binding to it. In addition to being spread to additional scaffolds, this approach may be modified to provide field-mediated directional transport [110]. A hydrogel of DNA that can identify distinct miRNAs, was disclosed by Meng et al. AuNPs edited by DNA and their corresponding fluorescent sequences of DNA were used to accomplish this. By transfecting cells with certain miRNAs, the strand displacement reaction was started. This enhanced the intensity of fluorescence in imaging tasks and triggered the DNAzyme-assisted target recycling process. The results showed that this method can produce different levels of fluorescence by amplifying signals twice; this lets researchers figure out how many miRNAs are linked to tumors in different types of cells [111]. A DNA hydrogel that can detect AFB1 (aflatoxin B1) and deliver drugs using a digital balance was created by Tang et al. The polyethyleneimine and AFB1 aptamer were complementary ligands, and the hydrogel of DNA used linear grafted ssDNA with hyaluronic acid molecules as scaffolds. Recycling was made easier by

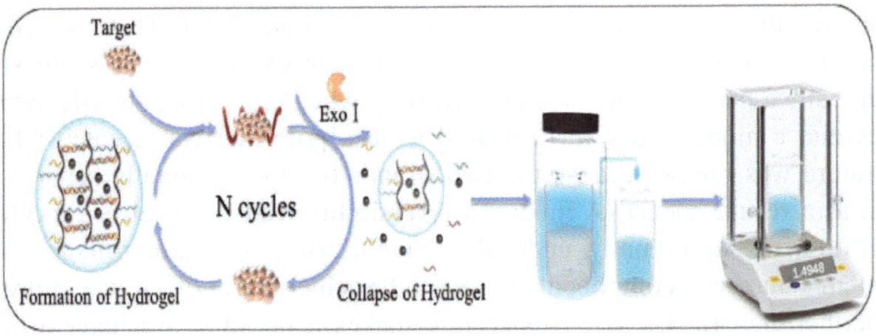

Fig. 7.7 A DNA hydrogel-based AFB1 detection system utilizes Exo I-assisted recycling and PtNPs for weight change measurement via a digital balance [112]

the introduction of Exonuclease I, which cleaved the AFB1 molecule and its aptamer; this allowed the released AFB1 to react using the hydrogel further. Additionally, by catalyzing hydrogen peroxide (H_2O_2), platinum nanoparticle-containing hydrogel (PtNPs) enabled the production of oxygen (O_2), which resulted in a decrease in the weight of the system, as determined by an electronic balance (Fig. 7.7) [112].

There are numerous possibilities for mixing NAs with nano/microcarriers to produce medication delivery systems that react to induce, including the capacity to modify the way NA structures are configured, techniques for attaching NAs to surfaces, and the possibility of using the data kept in NAs to generate complex 3D/2D structures. The evolution of stimuli-responsive nano/microcarriers with NA gates has advanced significantly. Nanoparticles of SiO_2, microcapsules of all-DNA, hydrogel microcapsules based on NA, and MOF nanoparticles functionalized with NA are a few examples of these carriers [113].

7.5 DNA-Based Drug Administration and Cancer Treatment

The use of DNA nanotechnology in biological domains, particularly in delivery systems and tumor treatment, has advanced significantly. Because of its exceptional structural adaptability, biocompatibility, low cytotoxicity, increased accessibility, and capacity to pass through membranes of cells without requiring transfection, DNA-based nanoscale structures are even more suitable materials than conventional drug nanocarriers such as gold and liposome nanoparticles. Multipurpose ligands may be integrated into structures by use of a variety of techniques, including conjugation, intercalation, encapsulation, and loading, depending on their intrinsic properties. As a result, several DNA NPs were produced and later used in a variety of biomedical fields, most notably medication delivery (Fig. 7.8). A thorough list and explanation of the several DNA NPs used in biomedical applications are given in the next section [97, 114].

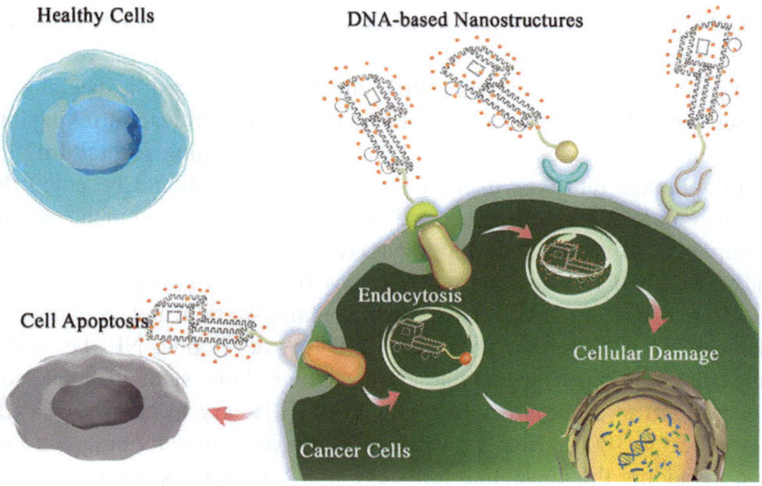

Fig. 7.8 Examining the use of DNA nanostructures in biology: current challenges and prospective developments [97]

A key strategy for treating diseases, targeted DDSs seek to maximize therapeutic effectiveness while reducing unwanted side effects. Nanoscale DNA constructs based on aptamers and aptamer-drug combinations are particularly noteworthy among these approaches because of the many previously highlighted benefits [115]. Because of their perfect biocompatibility, significant stability in biofluids, and a high systemic penetration rate, DNA nanoscale structures hold promise as efficient drug carriers for the treatment of a variety of disorders [2, 116].

Proof that the tetrahedron cage of DNA might effectively enter living cells in mammals and sustain its structural soundness for an extended period of time was shown by Turberfield and colleagues in 2011. Significant advancements have been made in the creation of aptamer-modified nanoDDSs. By reducing the likelihood of unwanted side effects, these systems seek to increase the efficacy of treatment [117]. This is seen in Fig. 7.9a, where Lin and his group used DNA aptamers (AS1411) and the counter-metabolite medication 5-fluorouracil [5-FU] to modify DNA tetrahedra in order to construct a nanomedicine. By looking at how breast cancer cells and healthy cells take in things differently, researchers found that this nanomedicine (AS1411-T-5-FU) meant to kill cancer cells worked better than free 5-FU at killing the cells it was meant to kill [118]. Additionally, using DNA nanoscale structures might significantly increase the amount of medication payload that can be carried. The desire to develop a DNA nanocentipede drove Wang and his associates. Two biotinylated hairpins are used in a process known as HCR to create the trunk that makes up the nanocentipede. Additionally, a streptavidin-biotin affinity interaction connects many legs, referred to as aptamers, to the trunk. Figure 7.9b provides an illustration of this. The proposed nanocentipede demonstrated enhanced cytotoxicity, selectivity, and binding affinity for cancer cells of interest due to the multivalent impact, while having no effect on negative control cells [119]. As shown in Fig. 7.9c, a new DNA nanostructure inspired by biotechnology known as DNA NF has also shown exceptional effectiveness of medication distribution. In the research, elongated DNA constructs created via rolling circular duplicates of a specially created model self-assembled to produce nanofibers (NFs). A significant drug payload and accurate cancer cell targeting may be achieved by integrating several drug loading sites and aptamers into NFs using a logical method. It should be noted that the tight arrangement of building pieces and liquid crystallization aid in the assembly of NF. Complex DNA sequence development is not required since this method does not depend on Watson-Crick base pairing. NFs have been shown to have a broad range of biological uses, like biosensing, targeted drug administration, and bioimaging [120]. The same team made further progress in 2019 by developing NFs-Fc-Sgc8 DNA nanostructure that is self-degradable and has controllable size; artificial sandwich bases were used to accomplish this (Fig. 7.9d). It is easy to modify NFs' dimensions (from 50 to 1000 nm) and give them the capacity to deteriorate via Fenton's reaction by adding the nucleotide ferrocene to DNA strands. In a xenograft tumor model, the modified NFs-Fc-Sgc8 showed remarkable tumor-targeting capacity and increased the cytotoxicity of DOX, highlighting its advantageous characteristics for drug administration [121]. Another benefit of using DNA nanoscale aptamer-based constructs is that they can be used to make tools that can be used for more than one purpose for targeted treatment. An adaptable and versatile DNA nanoassembly based on aptamers is an example of a sophisticated technology designed especially for targeted cancer treatment. Reduced drug resistance was the outcome of the DNA nano assembly based on aptamers focused on cytotoxic effects on cells of leukemia and suppression of P-gp expression (Fig. 7.9e). This was accomplished by mixing therapeutic antisense oligonucleotides with anticancer medications [122].

Pharmacological research and drug quality may benefit from the targeted ligand's conjugation with molecular medicines at a certain stoichiometric ratio. Aptamer-drug conjugates (ApDCs) are a novel kind of customized drug that has received a lot of interest lately. The drugs have cytotoxic effects for chemotherapy, and this arrangement includes aptamers used to identify their particular target molecules [123]. For example, Sullenger et al. linked aptamers E3 to powerful chemotherapy drugs, namely monomethyl auristatins F and E alone, to produce a modifiable cytotoxic ApDC [124]. Prostate cancer cells may be efficiently eradicated by internalizing these particular medications. ApPDCs (Aptamer-PolyproDrug Conjugates) with substantial loading of drugs were also created by Tan and associates. This was accomplished by mixing polyprodrugs with cell-targeting aptamers. Aptamer-PolyproDrug Conjugates have a number of advantages, including as strong cytotoxicity, a comparatively long half-life in the body, and remarkable nuclease resistance. Furthermore, combining two kinds of prodrugs via different pharmacological pathways may enable the exact combination of therapies to be achieved [125]. The same researchers previously showed how cb-apt, or circular bivalent aptamers, improved thermal stability, affinity for binding, and resistance to the breakdown of nucleases. Using the previously mentioned properties, they also added β-cyclodextrin to a cb-apt, which helps incorporate various therapeutic compounds, from tiny between molecules and macromolecules, using the chemistry between the host and the guest.

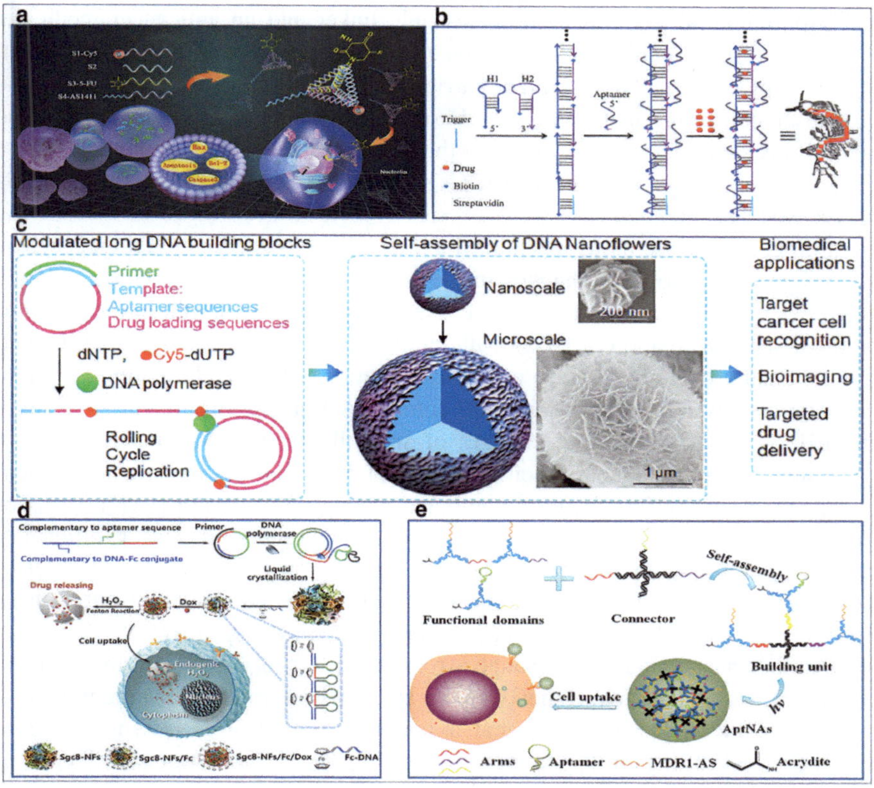

Fig. 7.9 The use of aptamer-integrated DNA nanoscale structures for targeted DDS is shown in this image. **a** In particular, it describes the use of a DNA tetrahedron modified by an aptamer and antimetabolite medication for the treatment of cancer. Reprinted with permission from [118]. Copyright 2019 American Chemical Society. **b** The DNA nano centipede is made up of many aptamer legs and a trunk put together using HCR. The multivalent cancer therapy is made possible by this design. Reprinted with permission from [119]. Copyright 2016 American Chemical Society. **c** Rolling circle replication produces aptamer-based DNA nanoflowers for targeted cancer therapy. Reprinted with permission from [120]. Copyright 2013 American Chemical Society. **d** The goal of this research is to create DNA nanoflowers that are self-degradable and scalable in size. These nanoflowers are intended for targeted cancer treatment and comprise synthetic sandwich bases. Reprinted with permission from [121]. Copyright 2019 American Chemical Society. **e** The goal of this work is to enhance a DNA nanoassembly that incorporates antisense oligonucleotides, aptamers, and anticancer medications. The objective is to specifically target leukemia cells, cause cell death, and suppress P-gp expression. Reprinted with permission from [122]. Copyright 2013 American Chemical Society

Furthermore, saporin, a cytotoxic agent, successfully inhibited the proliferation of tumor cells. The findings imply that the cb-apt-βCD could serve as a flexible framework for expanding and improving the capacities of aptamers [126]. The creation of a technique for automated modular synthesis is another important step forward in the area of ApDCs. This method focuses on creating a therapeutic module, namely a phosphoramidite that has a photocleavable linker and an anticancer medicinal component. This module's goal is to make solid-phase synthesis easier. As a result, the development of ApDCs would be carried out methodically, enabling the addition of many medications at specific sites. Furthermore, the produced ApDCs have the ability to release medications in a way that is regulated by light. According to the findings stated above and others, ApDCs seem to have potential as targeted DDSs for focused treatment [127].

In order to provide more specialized and effective treatments, intelligent DDSs are designed to recognize changes in the environment and then provide accurate quantities of medicine in reaction to certain environmental stimuli. In the realm of biomedicine, the creation of DDSs with intelligence more especially, those derived from the nanoscale architecture of DNA is now a fast-expanding subject. These methods usually take use of particular stimuli such as specific metal ions, light, and pH, as well as structural changes of specially made DNA nanotechnology, including the disintegration of their arrangement brought on by aptamer attachment to cancerous cells [128]. Using a combined strategy, Wang, along with his associate, created a nanotechnology known as a system of "sense-and-treat" to efficiently eradicate in-circulation tumor cells. The nanodevice was made out of a photosensitizer, a tetrahedron of DNA with DOX, and a hairpin switch aptamer. Both of these substances were used for phototherapy and chemotherapy, respectively. The nanodevices formed an aptamer hairpin shape to release them from their supporter when the target cells were identified. It then went on and used a combined treatment method to eradicate the cells [129]. Wang et al. assembled scaffolds of DNAzyme-substrate that underwent rolling circle polymerization to construct a non-toxic self-degrading nanosponge. Their efficient transport inside cancer cells was made possible by the encoded multivalent tandem aptamers. The encapsulated pH-sensitive ZnO nanoparticles may separate into Zn^{2+} ions within the cells. These ions then help DNAzymes break down DNA scaffolds, leading to intelligent delivery of anti-cancer drugs [130].

7.6 Clinical Studies of DNA Nanoparticles and DNA-Based Therapeutics in Drug Delivery

The use of structural nanotechnology of DNA in medicine and medication delivery has enormous promise. Numerous researches have proposed various structures and methods for loading, transporting, and releasing molecular drugs to treat cancer, including photodynamic therapy, immune system stimulation, and gene suppression. Static DNA structures are mostly used in treatment for anticancer medication delivery. In 2012, DOX was intercalated into DNA origami structures that were independently made by the Ding group and the Högberg group. Human breast cancer cells were exposed to and incorporated the drug-containing DNA constructs, demonstrating unmistakable apoptotic action. Depending on the dose, both groups demonstrated efficient administration of drugs and the efficacy of the anti-cancer treatment

[131]. Origami-based anticancer DDSs were investigated in small animals by Ding and associates (Fig. 7.10a). Quantum dots with fluorescent properties were used to decorate the origami structures, allowing researchers to study their dispersion over the body and rate of metabolism. Live organisms demonstrated the DOX's effectiveness, and no cytotoxic effects were detected [63]. DNA structures have been studied in cancer treatment and biomedical applications. The feasibility of using DNA to transfer immunostimulants into immune cells was examined by the Liedl group (Fig. 7.10b). The scientists created DNA nanotubes with CpG sequences and assessed how they affected spleen cells. The studies show that the delivery methods are not harmful and that their immune activity is as planned [132]. In a different study, Langer and associates used oligonucleotide nanoparticles of tetrahedral DNA (ONPs) as delivery vehicles for siRNA (Fig. 7.10c). They successfully demonstrated how some genes may be suppressed in tiny animal models [133]. One major benefit of dynamic DNA nanoscale structures over static delivery systems is their capacity to program the availability of the drug. A DNA origami box created by Gothelf and associates only opens or shuts its lid in reaction to certain outside stimuli. The contents of the box are visible when the lid is opened. Because dynamic structures may control the presence of medicinal molecules, they have significant advantages over static structures [134].

Innovative DDSs have shown great promise in the development of nanotechnology [136–144]. Using origami of DNA as a medium for drug delivery is an innovative approach. DNA origami is the process of using shorter staple strands of DNA to

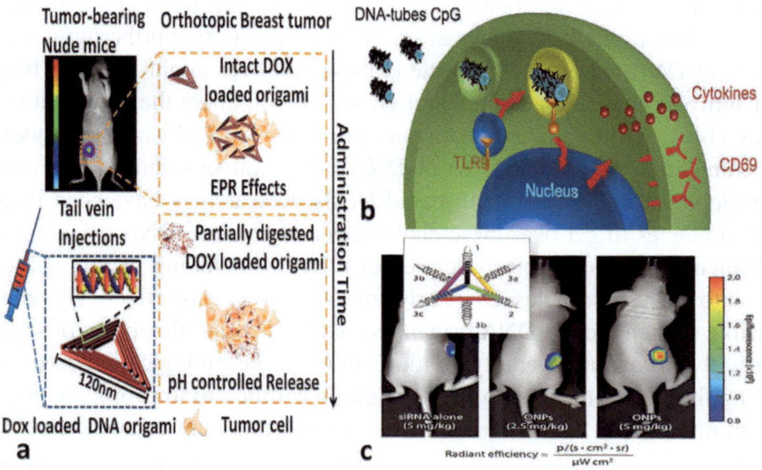

Fig. 7.10 Static DNA DDSs. **a** Triangular self-assembled DNA to cure cancer in vivo. Reprinted with permission from [63]. Copyright 2014 American Chemical Society. **b** Delivery of immunostimulants. Reprinted with permission from [135]. Copyright 2011 American Chemical Society. **c** Tetrapod DNA for in vivo delivery of siRNA. Toll-like receptor 9, or TLR9; DOX, doxorubicin; CpG, cytosine–phosphate–guanine; permeability, enhanced retention characteristic, and EPR impact; the acronyms ONP (oligonucleotide NP) and a cluster of differentiation 69, or CD69, are used [133]

carefully fold a long ssDNA molecule into a certain shape. This method has been used to create a number of nanostructures, such as nanocarriers and nanoparticles, for the delivery of pharmaceuticals. While DNA origami has great potential for accurate drug delivery, certain challenges must be addressed before its full potential can be realized. The intrinsic rigidity of DNA origami is a major challenge in its use for DDSs. The significant stiffness and lack of flexibility of DNA origami constructs may limit their ability to navigate the complex and dynamic living environment. These systems' rigidity may affect the diffusion and flow of DDSs based on DNA origami and hinder the efficient delivery of medicinal chemicals to their intended locations. Concerns about DNA origami's longevity and compatibility with living things may also arise from its rigidity. The therapeutic relevance of DNA origami constructs is limited due to their intrinsic stiffness, which may lead to increased immunogenicity and potential toxicity. Furthermore, DNA origami's rigidity may affect their ability to contain and release medicinal compounds in a controlled way, which is crucial for effective drug delivery [145]. The accumulation of DNA nanoparticles in the kidneys and liver is another barrier associated with utilizing DNA origami for the delivery of drugs. The reticuloendothelial system has the ability to remove DNA origami-based DDSs once they are introduced into the body, which causes the kidneys and liver to accumulate them. The build-up of these compounds may limit the movement of drug methods of distribution throughout the body and reduce their availability at the site of action, hence reducing their ability to effectively cure the ailment. Concerns about possible toxicity and long-term effects may arise from the accumulation of DNA nanoparticles in the kidneys and liver. Long-term accumulation of DNA origami in these organs may cause adverse reactions and damage to the kidneys and liver, placing patients at serious risk for injury. Furthermore, the accumulation of DNA nanoparticles in the kidneys and liver might limit the frequency and quantity of medication administration, which will affect the treatment's overall efficacy [146]. Addressing the challenges presented via DNA origami rigidity and the accumulation of nanoparticles of DNA in the kidneys and liver is crucial to the development of DNA origami-based DDSs. Researchers have been looking at several strategies to get over these challenges and enhance DNA origami's ability to deliver medications precisely and efficiently. By including pliable spacers and linkers into the nanoparticles, DNA origami's rigidity may be effectively reduced. It is possible to increase the DNA origami's adaptability and flexibility of structures, which will improve their ability to move through the biological environment, by including flexible elements into their design. In addition to lowering the possibility of an immunological reaction and toxicity issues, this may enhance the diffusion and flow of DDSs based on DNA origami. The development of systems that react to outside stimuli is an alternate method to overcome the rigidity of origami DNA. By encompassing stimuli-responsive elements in the DNA origami structures, like light-responsive triggers or pH-responsive switches, controlled and targeted drug release may be accomplished. This technique improves the efficacy of the DDSs, minimizes adverse effects, and makes it easier to accurately administer medication to the right place [147]. To deal with the accumulation of the liver and kidneys' DNA nanoparticles, researchers have been looking at several strategies to improve the biodistribution and systemic circulation of DDSs based on DNA origami. This entails using ligands that target to enhance the DDSs' intended buildup at the intended region and modifying the exterior properties of origami DNA to prevent their removal via the system of reticuloendothelium. Furthermore, the development of biocompatible coatings and modifications might help reduce the accumulation of DNA nanoparticles in the kidneys and liver, improving their long-term effects and safety profile. Enhancing DNA origami-based DDSs' interactions and effects on the body might boost their therapeutic efficacy and lessen any unfavorable side effects [148]. The limits of DNA nanoparticles in medication delivery have been the attention of nanotechnology scientists and researchers. The unique properties of DDSs based on DNA origami provide great promise for accurate and efficient medication administration. To fully use their potential, it is essential to resolve the inherent limitations. To get beyond these limitations, new approaches and multidisciplinary cooperation are needed to advance the area of drug delivery using DNA nanoparticles. Developing novel design strategies to increase the pliability and adaptability of DNA NPs is one field of research aimed at overcoming their limitations in drug delivery. By incorporating flexible spacers and linkers into DNA origami designs, researchers hope to decrease the rigidity of DNA nanoparticles and increase their mobility in complex biological environments [149]. This approach aims to improve the movement and dispersion of DDSs based on DNA NPs while simultaneously reducing any possible issues, given their rigidity, including toxicity and the ability to produce antibodies. Furthermore, research into the development of DNA nanoparticles that respond to stimuli is flourishing. By incorporating responsive-to-stimuli elements, like light-responsive triggers or pH-responsive switches, into DNA origami constructions, researchers hope to achieve precise and targeted drug release. This technique has great promise for improving the efficacy of DNA NP-based DDSs, decreasing adverse effects, and improving drug delivery precision. In addition to design strategies, research is focused on improving the exterior characteristics of NPs in DNA to prevent their removal via the system of reticuloendothelial. To increase their distribution over the body and reduce their accumulation in the kidneys and liver, DNA origami constructs must have their surfaces modified. Through the integration of targeting ligands and biocompatible coatings, researchers hope to enhance the exact accumulation of DNA NP-based DDSs at the intended area. This strategy seeks to reduce any adverse effects and accumulation in vital organs over time [150]. Additionally, a critical area of research focused on enhancing the security features of DDSs based on DNA nanoparticles is the development of biocompatible modifications and coatings. By enhancing the pharmacokinetics and pharmacodynamics of DNA origami-based DDSs, the investigators expect to increase their promise for therapy and decrease any possible side effects. Their clinical applicability will develop as a result of this; in order to overcome the limitations of medication delivery by DNA NP, it is also essential to look into other materials and hybrid nanostructures. In order to create hybrid nanostructures with enhanced properties and capabilities, researchers want to mix origami of DNA using additional nanomaterials, including lipids, inorganic nanoparticles, or polymers. This technique might expand the applications of DDSs based on DNA origami while addressing the limitations of nanoparticles of DNA, like their rigidity and diffusion in living things. It is essential to collaborate across disciplines and integrate advanced technologies, such as imaging with great resolution, computational modeling, and integrating high-volume screening, into the investigation to get over the restrictions of DNA NP-based drug delivery. By using these methods in conjunction with the information gained from other fields, scientists may get a deeper understanding of the complex interactions and effects of DDSs based on DNA NPs, which will eventually lead to the development of creative solutions to get beyond their limitations [151]. In the end, research into overcoming the limitations of medication delivery via DNA NP is an active and rapidly developing field with a tremendous deal of promise for expanding the field of nanomedicine. To improve the safety, versatility, and flexibility of DNA NP-based DDSs, scientists are employing innovative design strategies, surface alterations, systems that react to stimuli, and the creation of nanostructures that are hybrid. It is possible to overcome the limitations of medication delivery using DNA NPs and completely use their potential for accurate and efficient administration of drugs by carrying out more research and encouraging multidisciplinary cooperation. In conclusion, the inflexibility of the origami of DNA and the accumulation of nanoparticles of DNA in the kidneys and liver pose significant challenges for medication delivery. Ongoing research and advancements in technology, however, may provide answers to these issues. We can fully use the enormous potential of this state-of-the-art nanotechnology platform by recognizing and conquering the innate limitations of origami DNA and improving its properties for accurate and efficient medication delivery. DNA origami-based DDSs possess the capacity to greatly alter the area of nanomedicine and enhance the treatment outcomes for a range of illnesses by continuously developing and encouraging cooperation across several domains [152]. The stability of DNA NPs is one of the main issues they face; DNA NPs' circulation duration and effectiveness as medication carriers may be limited by their susceptibility to enzymatic breakdown by bloodstream nucleases. Researchers have looked at a number of approaches to deal with this problem, such as adding chemical moieties to the DNA backbone or applying shielding layers to increase the DNA nanoparticles' stability in biological settings; the possible immunogenicity of DNA NPs is another disadvantage. The body's immune system may get activated and experience negative reactions as a result of DNA molecules inciting immunological responses. This immunological reaction may restrict the therapeutic uses of DNA NPs and lessen their efficacy as medication carriers. Researchers are looking at approaches to create DNA NPs that reduce immunogenicity without sacrificing their ability to deliver drugs in order to address this problem [153]. Furthermore, DNA NPs' efficacy in drug delivery systems may also be influenced by their size and shape. Large DNA NPs may not be able to efficiently deliver therapeutic payloads to target areas due to inadequate tissue penetration and cellular absorption. Furthermore, DNA NPs' surface characteristics and form may impact how well they interact with cells and biological barriers, which in turn can affect how well they carry drugs overall. To enhance the dimensions, form, and surface properties using DNA nanoparticles for better delivery of drugs results, researchers are investigating various manufacturing processes and design approaches. The capacity to scale and repeatability of NP synthesis for DNA are major obstacles to their broad use in delivery of drug applications besides these difficulties. The translation of DNA NPs from bench to bedside is limited by the technical difficulty of producing them on a wide scale with uniform quality and characteristics. To guarantee DNA NPs' clinical viability and economic viability as delivery of drug platforms, it is essential to grow reliable and capable production techniques [154]. Furthermore, integrating DNA NP formulations into conventional drug delivery systems may be hampered by their intricacy and the need for specific knowledge in their design and characterization. To solve the many issues surrounding NPs of DNA and hasten their progress as efficient carriers of drugs, cooperation between interdisciplinary teams of chemists, biologists, engineers, and physicians is crucial. To sum up, DNA nanoparticles have a lot of potential for drug delivery applications, but they also possess a variety of problems that have to be fixed in order to improve their consistency, lower immunogenicity, maximize their size and shape, increase reproducibility and scalability, and simplify their description and formulation procedures. DNA NPs have the potential to transform medication delivery and open the door for the creation of cutting-edge treatment approaches for a range of illnesses if these obstacles are overcome by creative research and teamwork [155].

7.7 Conclusion

All things considered, DNA research nanoparticles containing medications, proteins, or genes for regulated medication release have become an exciting area of study with significant potential to change DDSs. The unique properties of DNA, together with developments in nanotechnology, have made it possible to develop very accurate and successful therapeutic treatments. To make NPs that can contain various medications, genes, and proteins, researchers have taken use of DNA molecules' innate capacity to self-assemble. At certain points inside the body, these NPs allow for the exact and regulated release of their contents. The ability of DNA NPs to overcome the limitations of conventional DDSs is a significant advantage. These NPs may be designed to increase the stability and prevent degradation of the payload, ensuring that the medicine, protein, or gene is unharmed until it arrives at its target. Furthermore, DNA-based carriers have the advantage of being able to decompose naturally and being compatible with living things, which lowers the possibility of adverse reactions and facilitates effective elimination from the body. The capacity of DNA NPs to precisely regulate the kinetics of medication release is a key benefit. Scientists have developed sophisticated methods to manage the rate of release of contained chemicals, enabling a controlled and extended medication dispensing. It is possible to achieve exact release patterns that correspond with the treatment requirements of different illnesses by customizing the NPs' features and design. Exact control minimizes the possible adverse effects associated with elevated medication concentrations while optimizing therapeutic effectiveness.

Additionally, DNA NPs are very versatile and may carry a wide range of therapeutic payloads. Therapeutic proteins, NAs, and small molecule medications are only a few of the pharmacological substances that DNA carriers may efficiently load. Because each patient's unique biological characteristics may be taken into account while developing treatment plans, the adaptability of this technology opens up new opportunities for personalized medicine. Combination treatment, in which several medications are delivered concurrently and potentially provide synergistic effects, is made possible by the ability to encapsulate many therapeutic chemicals inside a single DNA NP. DNA NPs' promise as a cutting-edge drug delivery technology is further enhanced by their precise delivery capabilities. It is possible to construct NPs to precisely recognize and adhere to target cells or tissues by altering them with specific ligands or aptamers. By using a targeted approach, the likelihood of unforeseen consequences is decreased and treatment efficacy is maximized, which benefits patients. Furthermore, DNA NPs may transcend biological boundaries, such as the blood-brain barrier, which makes it possible to deliver therapeutic agents to previously inaccessible areas. Even though the profession has seen significant progress, there are still challenges and opportunities for further improvement. To achieve optimal target specificity, release kinetics, and drug loading, DNA NPs' physicochemical characteristics like their surface charge, size, and stability must be optimized. Additionally, when manufacturing large-scale DNA NPs for therapeutic applications, the potential for growth and cost reduction must be taken into account. The science of nanotechnology has made tremendous strides in developing new medication delivery techniques during the last 10 years. As science has advanced, DNA nanoparticles have emerged as a novel means of encasing and delivering medicinal materials, including drugs, genes, and proteins, while carefully controlling their release. The use of nanoparticles of DNA to enclose medications, proteins, or genes for regulated drug release has enormous possibility to revolutionize patient care and medical treatment as researchers continue to explore the technology's possibilities. Let's now examine some key areas where we might expect significant advancements and exciting prospects. The ability of DNA nanoparticles to precisely target certain cells or tissues is one of its noteworthy features. Scientists may develop DNA nanoparticles with surface alterations that can recognize and interact with certain targets, such tumor cells or inflammatory regions, as our understanding of the molecular features of illnesses advances. Furthermore, by guaranteeing extended drug release, the DNA nanoparticles' controlled release properties provide a long-lasting and optimal therapeutic impact. Treatments are tailored to each patient's unique genetic profile in the rapidly developing area of personalized medicine. By containing gene therapies, DNA nanoparticles provide a unique opportunity to deliver customized treatments. Certain genetic defects associated to illnesses may be corrected or regulated by these medicines. It is conceivable that with further development, researchers may be able to produce DNA nanoparticles that include individuals' unique genetic information, enabling targeted and individualized treatment plans. Adopting this personalized medicine approach has the potential to revolutionize how we treat illness, leading to better outcomes for patients and a higher quality of life.

DNA nanoparticles' controlled release properties make it possible to precisely regulate and modify the speeds at which different drugs are made available, improving their communication and therapy outcomes. This method makes it easier to create innovative combination treatments that can more effectively treat complex illnesses. Certain illnesses need intrusive treatments, including surgery, which may be very difficult for sufferers. DNA nanoparticles may provide accurate and non-invasive delivery methods. Researchers are now looking at other delivery methods, including ingestion, inhalation, and skin absorption, where DNA nanoparticles may be designed to efficiently get past various biological barriers and arrive at targeted targets. Non-invasive delivery methods boost results and treatment compliance while also making patients more comfortable. Additionally, the use of DNA nanoparticles may improve the absorption and stability of therapeutic drugs, lowering the possibility of deterioration or loss of efficacy prior to reaching their intended sites. DNA nanoparticles may serve as very useful instruments in the domains of imaging and diagnostics in addition to being therapeutic agents. Scientists may develop adaptable systems that enable the ongoing monitoring of illness progression, therapy efficacy, and rates of medication release by incorporating imaging substances, like contrast agents or fluorescent dyes, within the particles of nanoparticles. Adding diagnostic and therapeutic features will expedite the development of platforms for theranostics, leading to more individualized treatment plans and improved patient care. There is much promise and excitement for the future of utilizing nanoparticles of DNA to enclose medications, proteins, or genes for regulated drug release. By leveraging the unique properties of DNA nanoparticles, clinicians and researchers may strive for more individualized and efficient therapeutic treatments. Improvements in combination medicines, non-invasive delivery methods, customized medicine strategies, improved targeting capabilities, and integrated imaging and diagnostics will all influence the direction of healthcare in the future. The potential to improve patient outcomes and revolutionize medical care becomes more promising as we develop nanotechnology and improve these strategies. The path towards a day when DNA nanoparticles will be used for regulated medication release excites us. We eagerly await the next discoveries and advancements in this area.

References

1. S. Adepu, S. Ramakrishna, Controlled drug delivery systems: current status and future directions. Molecules **26**(19), 5905 (2021)
2. G.K. Kashani, S.M. Naghib, S. Soleymani, M.R. Mozafari, A review of DNA nanoparticles-encapsulated drug/gene/protein for advanced controlled drug release: Current status and future perspective over emerging therapy approaches. Int. J. Biol. Macromol. **268**, 131694 (2024). https://doi.org/10.1016/j.ijbiomac.2024.131694
3. S.M. Naghib, B. Ahmadi, B. Mikaeeli Kangarshahi, M.R. Mozafari, Chitosan-based smart stimuli-responsive nanoparticles for gene delivery and gene therapy: recent progresses on cancer therapy. Int. J. Biol. Macromol. **278**, 134542 (2024). https://doi.org/10.1016/j.ijbiomac.2024.134542

4. D.S.W. Benoit, C.T. Overby, K.R. Sims Jr, M.A. Ackun-Farmmer, 2.5.12—Drug delivery systems, in *Biomaterials Science*, 4th edn., ed. by W.R. Wagner, S.E. Sakiyama-Elbert, G. Zhang, M.J. Yaszemski (Academic Press, 2020), pp. 1237–1266. https://doi.org/10.1016/B978-0-12-816137-1.00078-7
5. A. Beigi, S.M. Naghib, A. Matini, M. Tajabadi, M.R. Mozafari, Lipid-based nanocarriers for targeted gene delivery in lung cancer therapy: exploring a novel therapeutic paradigm. Curr. Gene Ther. **25**(2), 92–112 (2025). https://doi.org/10.2174/0115665232292768240503050508
6. B.M. Kangarshahi, A. Beigi Javazm, S.M. Naghib, CRISPR-integrated metal-organic frameworks for biosensing applications: recent advances and future perspective. Sens. Bio-sens. Res. **47**, 100736 (2025). https://doi.org/10.1016/j.sbsr.2025.100736
7. B.M. Kangarshahi, S.M. Naghib, Nanogenosensors based on aptamers and peptides for bioelectrochemical cancer detection: an overview of recent advances in emerging materials and technologies. Discover Appl. Sci. **6**(2), 47 (2024). https://doi.org/10.1007/s42452-024-05681-z
8. B. Mikaeeli Kangarshahi, N.S. Morteza, N. Rabiee, DNA/RNA-based electrochemical nanobiosensors for early detection of cancers. Crit. Rev. Clin. Lab. Sci. **61**(6), 473–495 (2024). https://doi.org/10.1080/10408363.2024.2321202
9. B. Mikaeeli Kangarshahi, S. Sojdeh, H. Daneshgar, M. Bagherzadeh, S.M. Naghib, N. Rabiee, Electroconductive polymer-based biosensors for early cancer detection via liquid biopsy: advances, challenges, and future prospects. TrAC Trends Anal. Chem. **183**, 118062 (2025). https://doi.org/10.1016/j.trac.2024.118062
10. S.M. Naghib, B. Ahmadi, M.R. Mozafari, Stimuli-sensitive chitosan-based nanosystems-immobilized nucleic acids for gene therapy in breast cancer and hepatocellular carcinoma. Curr. Top. Med. Chem. **24**(17), 1464–1489 (2024). https://doi.org/10.2174/0115680266293173240506054439
11. P. Vosoughi, S.M. Naghib, M. Rahmanian, M.R. Mozafari, A comprehensive review of mRNA-based vaccines for COVID-19, a new era in pharmaceuticals: unspecified and unknown aspects, effects and challenges. Curr. Top. Med. Chem. (2025). https://doi.org/10.2174/0115680266325847241121034100
12. M.A. Khorasani, S.M. Naghib, A review of chitosan-based multifunctional nanocomposites for drug/gene/protein delivery and gene therapy in cancer treatments: promises, challenges and outlooks. Int. J. Biol. Macromol. **306**, 141394 (2025). https://doi.org/10.1016/j.ijbiomac.2025.141394
13. S.M. Naghib, Two-dimensional functionalised methacrylated graphene oxide nanosheets as simple and inexpensive electrodes for biosensing applications. Micro Nano Lett. **14**(4), 462–465 (2019). https://doi.org/10.1049/mnl.2018.5320.
14. S.M. Naghib, M. Rabiee, E. Omidinia, P. Khoshkenar, Investigation of a biosensor based on phenylalanine dehydrogenase immobilized on a polymer-blend film for phenylketonuria diagnosis. Electroanalysis **24**(2), 407–417 (2012). https://doi.org/10.1002/elan.201100391
15. S.M. Naghib, M. Rabiee, E. Omidinia, P. Khoshkenar, D. Zeini, Biofunctionalization of dextran-based polymeric film surface through enzyme immobilization for phenylalanine determination. Int. J. Electrochem. Sci. **7**(1), 120–135 (2012). https://doi.org/10.1016/S1452-3981(23)13325-6
16. S. Morteza Naghib, S. Mahdi Katebi, S. Ghorbanzade, Introduction to electrochemical biosensors, in *Electrochemical Biosensors in Practice: Material and Methods* (Bentham Science Publishers, 2023)
17. S. Morteza Naghib, S. Mahdi Katebi, S. Ghorbanzade, Electrochemical biosensors design steps, in *Electrochemical Biosensors in Practice: Material and Methods* (Bentham Science Publishers, 2023)
18. S. Morteza Naghib, S. Mahdi Katebi, S. Ghorbanzade, Material and biomaterial for biosensing platform, in *Electrochemical Biosensors in Practice: Material and Methods* (Bentham Science Publishers, 2023)
19. S. Morteza Naghib, S. Mahdi Katebi, S. Ghorbanzade, Basic Practical principles for studying electro-chemical biosensor, in *Electrochemical Biosensors in Practice: Material and Methods* (Bentham Science Publishers, 2023)

20. S. Morteza Naghib, S. Mahdi Katebi, S. Ghorbanzade, Biosensor application, in *Electrochemical Biosensors in Practice: Material and Methods* (Bentham Science Publishers, 2023)
21. N. Dehghani, F. Haghiralsadat, F. Yazdian, F. Sadeghian-Nodoushan, N. Ghasemi, F. Mazaheri, M. Pourmadadi, S.M. Naghib, Chitosan/silk fibroin/nitrogen-doped carbon quantum dot/ α-tricalcium phosphate nanocomposite electrospinned as a scaffold for wound healing application: in vitro and in vivo studies. Int. J. Biol. Macromol. **238**, 124078 (2023). https://doi.org/10.1016/j.ijbiomac.2023.124078
22. H. Hasannezhad, A. Bakhshi, M.R. Mozafari, S.M. Naghib, A review of chitosan role in milk bioactive-based drug delivery, smart packaging and biosensors: recent advances and developments. Int. J. Biol. Macromol. **294**, 139248 (2025). https://doi.org/10.1016/j.ijbiomac.2024.139248
23. S.M. Naghib, M. Bakhshi, B. Ahmadi, A. Bakhshi, Gold nanoparticles based antibacterial and antiviral functional materials, in *Antibacterial and Antiviral Functional Materials*, vol. 2 (ACS Publications, 2024), pp. 221–256
24. S.M. Naghib, B. Ahmadi, M.R. Mozafari, Smart physicochemical-triggered chitosan-based nanogels for siRNA delivery and gene therapy: a focus on emerging strategies and paradigms for cancer therapy. Curr. Med. Chem. (2024). https://doi.org/10.2174/0109298673286052240426044945
25. S.M. Naghib, M. Zarrineh, M.R. Mozafari, 3D Printing chitosan-based nanobiomaterials for biomedicine and drug delivery: recent advances on the promising bioactive agents and technologies. Curr. Org. Chem. **28**(7), 510–525 (2024). https://doi.org/10.2174/0113852728298168240222114449
26. H. Zhang, G.S. Demirer, H. Zhang, T. Ye, N.S. Goh, A.J. Aditham, F.J. Cunningham, C. Fan, M.P. Landry, DNA nanostructures coordinate gene silencing in mature plants. Proc. Natl. Acad. Sci. **116**(15), 7543–7548 (2019)
27. P. Vosoughi, S.M. Naghib, M. Mozafari, Stimuli-responsive nano/biomaterials for smart drug delivery in cardiovascular diseases: promises, challenges and outlooks. Curr. Med. Chem. (2024)
28. S.M. Naghib, B. Ahmadi, Micro-and nanoscale oral drugs delivery platforms, in *Smart Micro- and Nanomaterials for Drug Delivery*, pp. 109–134
29. A. Shamaeizadeh, A. Beigi, S.M. Naghib, M. Tajabadi, M. Rahmanian, M.R. Mozafari, Smart nanobiomaterials for gene delivery in localized cancer therapy: an overview from emerging materials and devices to clinical applications. Curr. Cancer Drug Targets (2024). https://doi.org/10.2174/0115680096288917240404060506
30. P. Vosoughi, S.M. Naghib, B.M. Kangarshahi, M.R. Mozafari, A review of RNA nanoparticles for drug/gene/protein delivery in advanced therapies: current state and future prospects. Int. J. Biol. Macromol. **295**, 139532 (2025). https://doi.org/10.1016/j.ijbiomac.2025.139532
31. A.-A. Zare, H. Naderi-Manesh, S.M. Naghib, M. Shamsipur, F. Molaabasi, Label-free electrochemical cancer cell detection leveraging hemoglobin-encapsulated silver nanoclusters and Cu-MOF nanohybrids on a graphene-assisted dual-modal probe. Sci. Rep. **13**(1), 21980 (2023). https://doi.org/10.1038/s41598-023-49418-1
32. R. Salahandish, A. Ghaffarinejad, E. Omidinia, H. Zargartalebi, K. Majidzadeh-A, S.M. Naghib, A. Sanati-Nezhad, Label-free ultrasensitive detection of breast cancer miRNA-21 biomarker employing electrochemical nano-genosensor based on sandwiched AgNPs in PANI and N-doped graphene. Biosens. Bioelectron. **120**, 129–136 (2018). https://doi.org/10.1016/j.bios.2018.08.025
33. M. Yazdan, S.M. Naghib, M.R. Mozafari, Liposomal nano-based drug delivery systems for breast cancer therapy: recent advances and progresses. Anti-Cancer Agents Med. Chem. (Formerly Curr. Med. Chem. Anti-Cancer Agents) **24**(12), 896–915 (2024). https://doi.org/10.2174/0118715206293653240322041047
34. S.M. Naghib, S. Hoseinpour, S. Zarshad, *Localized Micro/Nanocarriers for Programmed and On-Demand Controlled Drug Release* (Bentham Science Publishers, 2022)

35. M. Yazdan, S.M. Naghib, Smart ultrasound-responsive polymers for drug delivery: an overview on advanced stimuli-sensitive materials and techniques. Curr. Drug Deliv. (2025)
36. S.N. Hosseini, S.M. Naghib, G.K. Kashani, M. Mozafari, Chitosan-grafted graphene materials for drug delivery in wound healing. Curr. Pharm. Des. (2024)
37. B. Chandrasekaran, S.N. Abed, O. Al-Attraqchi, K. Kuche, R.K. Tekade, *Computer-aided prediction of pharmacokinetic (ADMET) properties* (Elsevier, Dosage form Design Parameters, 2018), pp.731–755
38. S. Gooneh-Farahani, M.R. Naimi-Jamal, S.M. Naghib, Stimuli-responsive graphene-incorporated multifunctional chitosan for drug delivery applications: a review. Expert Opin. Drug Deliv. **16**(1), 79–99 (2019)
39. T.T.V. Tran, H. Tayara, K.T. Chong, Artificial intelligence in drug metabolism and excretion prediction: recent advances, challenges, and future perspectives. Pharmaceutics **15**(4), 1260 (2023)
40. B. Sun, X. Zhao, Y. Wu, P. Cao, F. Movahedi, J. Liu, J. Wang, Z.P. Xu, W. Gu, Mannose-functionalized biodegradable nanoparticles efficiently deliver DNA vaccine and promote anti-tumor immunity. ACS Appl. Mater. Interfaces **13**(12), 14015–14027 (2021)
41. W. Ho, M. Gao, F. Li, Z. Li, X.Q. Zhang, X. Xu, Next-generation vaccines: nanoparticle-mediated DNA and mRNA delivery. Adv. Healthcare Mater. **10**(8), 2001812 (2021)
42. P. Tiwari, K. Yadav, R.P. Shukla, S. Gautam, D. Marwaha, M. Sharma, P.R. Mishra, Surface modification strategies in translocating nano-vesicles across different barriers and the role of bio-vesicles in improving anticancer therapy. J. Control. Release **363**, 290–348 (2023)
43. D.V. Bhalani, B. Nutan, A. Kumar, A.K. Singh Chandel, Bioavailability enhancement techniques for poorly aqueous soluble drugs and therapeutics. Biomedicines **10**(9), 2055 (2022)
44. V.C. Vetter, E. Wagner, Targeting nucleic acid-based therapeutics to tumors: challenges and strategies for polyplexes. J. Control. Release **346**, 110–135 (2022). https://doi.org/10.1016/j.jconrel.2022.04.013
45. Y. Zhang, X. Liu, X. Li, Development and advantages of drug delivery systems, *Drug Delivery to Tumors: Recent Strategies and Techniques* (Springer, 2025), pp. 35–60
46. H.J. Vaughan, J.J. Green, S.Y. Tzeng, Cancer-targeting nanoparticles for combinatorial nucleic acid delivery. Adv. Mater. **32**(13), 1901081 (2020). https://doi.org/10.1002/adma.201901081
47. Q. Fan, Z. He, J. Xiong, J. Chao, Smart drug delivery systems based on DNA nanotechnology. ChemPlusChem **87**(3), e202100548 (2022)
48. S.M. Naghib, S. Amiri, M. Mozafari, Stimuli-responsive chitosan-based nanocarriers for drug delivery in wound dressing applications: a review. Carbohydr. Polym. Technol. Appl. 100497 (2024)
49. F. Kazemi, S.M. Naghib, *Smart controlled release of corrosion inhibitor from normal and stimuli-responsive micro/nanocarriers* (Elsevier, Corrosion Protection at the Nanoscale, 2020), pp.161–179
50. A.K. Grosskopf, A.A. Ginart, P. Spinosa, V. Shivva, Pharmacokinetics-based design of subcutaneous controlled release systems for biologics. CPT: Pharmacometrics Syst. Pharmacol. (2025)
51. X. Peng, F. Jiang, Z. Li, Enhancing compatibility of polylactic acid and lignin through acetylation for improving controlled release of pesticide. Pest Manag. Sci. (2025)
52. G. Sulli, M.T.Y. Lam, S. Panda, Interplay between circadian clock and cancer: new frontiers for cancer treatment. Trends Cancer **5**(8), 475–494 (2019)
53. Y.K. Sung, S.W. Kim, Recent advances in polymeric drug delivery systems. Biomater. Res. **24**(1), 12 (2020)
54. S. Borandeh, B. van Bochove, A. Teotia, J. Seppälä, Polymeric drug delivery systems by additive manufacturing. Adv. Drug Deliv. Rev. **173**, 349–373 (2021). https://doi.org/10.1016/j.addr.2021.03.022
55. B. Kou, Z. Wang, S. Mousavi, P. Wang, Y. Ke, Dynamic gold nanostructures based on DNA self assembly. Small **20**(23), 2308862 (2024)

56. J. Huang, W. Ma, H. Sun, H. Wang, X. He, H. Cheng, M. Huang, Y. Lei, K. Wang, Self-assembled DNA nanostructures-based nanocarriers enabled functional nucleic acids delivery. ACS Appl. Bio Mater. **3**(5), 2779–2795 (2020)
57. K.M. Abu-Salah, A.A. Ansari, S.A. Alrokayan, DNA-based applications in nanobiotechnology. Biomed. Res. Int. **2010**(1), 715295 (2010)
58. D.-X. Wang, J. Wang, Y.-X. Wang, Y.-C. Du, Y. Huang, A.-N. Tang, Y.-X. Cui, D.-M. Kong, DNA nanostructure-based nucleic acid probes: construction and biological applications. Chem. Sci. **12**(22), 7602–7622 (2021)
59. L.A. Lanier, H. Bermudez, DNA nanostructures: a shift from assembly to applications. Curr. Opin. Chem. Eng. **7**, 93–100 (2015)
60. Y. Ke, C. Castro, J.H. Choi, Structural DNA nanotechnology: artificial nanostructures for biomedical research. Annu. Rev. Biomed. Eng. **20**(1), 375–401 (2018)
61. R.K. Dubey, S. Shukla, K. Shah, H.K. Dewangan, A comprehensive review of self-assembly techniques used to fabricate as DNA origami, block copolymers, and colloidal nanostructures. Curr. Nanosci. **21**(3), 385–403 (2025)
62. P.W. Rothemund, Folding DNA to create nanoscale shapes and patterns. Nature **440**(7082), 297–302 (2006)
63. Q. Zhang, Q. Jiang, N. Li, L. Dai, Q. Liu, L. Song, J. Wang, Y. Li, J. Tian, B. Ding, Y. Du, DNA origami as an in vivo drug delivery vehicle for cancer therapy. ACS Nano **8**(7), 6633–6643 (2014). https://doi.org/10.1021/nn502058j
64. J. Nangreave, D. Han, Y. Liu, H. Yan, DNA origami: a history and current perspective. Curr. Opin. Chem. Biol. **14**(5), 608–615 (2010)
65. M.M.F.A. Baig, X.-H. Xia, The PA-receptor mediated internalization of carboplatin loaded poly-anionic DNA-nanowires for effective treatment of resistant hepatic-cancer HepG-2 cells. Appl. Nanosci. **10**, 1915–1926 (2020)
66. S. Ghosal, S. Bag, S. Bhowmik, Unravelling the drug encapsulation ability of functional DNA origami nanostructures: current understanding and future prospects on targeted drug delivery. Polymers **15**(8), 1850 (2023)
67. H. Ijäs, B. Shen, A. Heuer-Jungemann, A. Keller, M.A. Kostiainen, T. Liedl, J.A. Ihalainen, V. Linko, Unraveling the interaction between doxorubicin and DNA origami nanostructures for customizable chemotherapeutic drug release. Nucleic Acids Res. **49**(6), 3048–3062 (2021)
68. P.D. Halley, C.R. Lucas, E.M. McWilliams, M.J. Webber, R.A. Patton, C. Kural, D.M. Lucas, J.C. Byrd, C.E. Castro, Daunorubicin-loaded DNA origami nanostructures circumvent drug-resistance mechanisms in a leukemia model. Small **12**(3), 308–320 (2016)
69. Y.-Z. Bu, J.-R. Xu, Q. Luo, M. Chen, L.-M. Mu, W.-L. Lu, A precise nanostructure of folate-overhung mitoxantrone DNA tetrahedron for targeted capture leukemia. Nanomaterials **10**(5), 951 (2020)
70. S. Loescher, S. Groeer, A. Walther, 3D DNA origami nanoparticles: from basic design principles to emerging applications in soft matter and (bio-) nanosciences. Angew. Chem. Int. Ed. **57**(33), 10436–10448 (2018)
71. M. Cao, Y. Sun, M. Xiao, L. Li, X. Liu, H. Jin, H. Pei, Multivalent aptamer-modified DNA origami as drug delivery system for targeted cancer therapy. Chem. Res. Chin. Univ. **36**(2), 254–260 (2020)
72. Z. Ge, L. Guo, G. Wu, J. Li, Y. Sun, Y. Hou, J. Shi, S. Song, L. Wang, C. Fan, DNA origami-enabled engineering of ligand–drug conjugates for targeted drug delivery. Small **16**(16), 1904857 (2020)
73. S. Li, Q. Jiang, S. Liu, Y. Zhang, Y. Tian, C. Song, J. Wang, Y. Zou, G.J. Anderson, J.-Y. Han, A DNA nanorobot functions as a cancer therapeutic in response to a molecular trigger in vivo. Nat. Biotechnol. **36**(3), 258–264 (2018)
74. M.A. Rahman, P. Wang, Z. Zhao, D. Wang, S. Nannapaneni, C. Zhang, Z. Chen, C.C. Griffith, S.J. Hurwitz, Z.G. Chen, Systemic delivery of bc12-targeting siRNA by DNA nanoparticles suppresses cancer cell growth. Angew. Chem. Int. Ed. **56**(50), 16023–16027 (2017)

75. Q. Pan, C. Nie, Y. Hu, J. Yi, C. Liu, J. Zhang, M. He, M. He, T. Chen, X. Chu, Aptamer-functionalized DNA origami for targeted codelivery of antisense oligonucleotides and doxorubicin to enhance therapy in drug-resistant cancer cells. ACS Appl. Mater. Interfaces **12**(1), 400–409 (2019)
76. J. Liu, L. Song, S. Liu, Q. Jiang, Q. Liu, N. Li, Z.-G. Wang, B. Ding, A DNA-based nanocarrier for efficient gene delivery and combined cancer therapy. Nano Lett. **18**(6), 3328–3334 (2018)
77. Y. Zeng, R.L. Nixon, W. Liu, R. Wang, The applications of functionalized DNA nanostructures in bioimaging and cancer therapy. Biomaterials **268**, 120560 (2021)
78. X. Lu, J. Liu, X. Wu, B. Ding, Multifunctional DNA origami nanoplatforms for drug delivery. Chem. Asian J. **14**(13), 2193–2202 (2019)
79. Y. Wang, E. Benson, F. Fördős, M. Lolaico, I. Baars, T. Fang, A.I. Teixeira, B. Högberg, DNA origami penetration in cell spheroid tissue models is enhanced by wireframe design. Adv. Mater. **33**(29), 2008457 (2021)
80. J. Chao, H. Liu, S. Su, L. Wang, W. Huang, C. Fan, Structural DNA nanotechnology for intelligent drug delivery. Small **10**(22), 4626–4635 (2014)
81. Y. Zhang, X. Tian, Z. Wang, H. Wang, F. Liu, Q. Long, S. Jiang, Advanced applications of DNA nanostructures dominated by DNA origami in antitumor drug delivery. Front. Mol. Biosci. **10**, 1239952 (2023)
82. M. Alle, G. Sharma, S.-H. Lee, J.-C. Kim, Next-generation engineered nanogold for multimodal cancer therapy and imaging: a clinical perspectives. J. Nanobiotechnol. **20**(1), 222 (2022)
83. Z. Jin, K.T. Nguyen, G. Go, B. Kang, H.-K. Min, S.-J. Kim, Y. Kim, H. Li, C.-S. Kim, S. Lee, Multifunctional nanorobot system for active therapeutic delivery and synergistic chemo-photothermal therapy. Nano Lett. **19**(12), 8550–8564 (2019)
84. D. Gu, Y. Qiao, H. Fu, H. Zhao, X. Yue, S. Wang, Y. Yin, R. Xi, X. Fu, X. Zhao, Size-controllable DNA origami-stacked gold nanoparticles for deep tumor-penetrating therapy. ACS Appl. Mater. Interfaces **14**(33), 38048–38055 (2022)
85. T. Xu, S. Yu, Y. Sun, S. Wu, D. Gao, M. Wang, Z. Wang, Y. Tian, Q. Min, J.-J. Zhu, DNA Origami frameworks enabled self-protective siRNA delivery for dual enhancement of chemo-photothermal combination therapy. Small **17**(46), 2101780 (2021). https://doi.org/10.1002/smll.202101780
86. N.A. Abtahi, S.M. Naghib, S.J. Ghalekohneh, Z. Mohammadpour, H. Nazari, S.M. Mosavi, S.M. Gheibihayat, F. Haghiralsadat, J.Z. Reza, B.Z. Doulabi, Multifunctional stimuli-responsive niosomal nanoparticles for co-delivery and co-administration of gene and bioactive compound: in vitro and in vivo studies. Chem. Eng. J. **429**, 132090 (2022). https://doi.org/10.1016/j.cej.2021.132090
87. N.A. Abtahi, S.M. Naghib, F. Haghiralsadat, J.Z. Reza, F. Hakimian, F. Yazdian, D. Tofighi, Smart stimuli-responsive biofunctionalized niosomal nanocarriers for programmed release of bioactive compounds into cancer cells in vitro and in vivo. Nanotechnol. Rev. **10**(1), 1895–1911 (2021)
88. N.A. Abtahi, S. Salehi, S.M. Naghib, F. Haghiralsadat, M.A. Edgahi, S. Ghorbanzadeh, W. Zhang, Multi-sensitive functionalized niosomal nanocarriers for controllable gene delivery in vitro and in vivo. Cancer Nanotechnol. **14**(1), 22 (2023). https://doi.org/10.1186/s12645-023-00175-w
89. N.A. Abtahi, S.M. Naghib, F. Haghiralsadat, M. Akbari Edgahi, Development of highly efficient niosomal systems for co-delivery of drugs and genes to treat breast cancer in vitro and in vivo. Cancer Nanotechnol. **13**(1), 28 (2022). https://doi.org/10.1186/s12645-022-00135-w
90. N.A. Abtahi, S.M. Naghib, F. Haghiralsadat, M. Akbari Edgahi, E. Askari, A comparative study on biopharmaceutical function of curcumin and miR-34a by multistimuli-responsive nanoniosome carrier: in-vitro and in-vivo. Front. Mol. Biosci. **9** (2022). https://doi.org/10.3389/fmolb.2022.1043277
91. A. Bakhshi, S.M. Naghib, N. Rabiee, Antibacterial and antiviral nanofibrous membranes, in *Antibacterial and Antiviral Functional Materials*, vol. 2 (ACS Publications, 2024), pp. 47–88

92. T. Wang, Y. Liu, Q. Wu, B. Lou, Z. Liu, DNA nanostructures for stimuli-responsive drug delivery. Smart Mater. Med. **3**, 66–84 (2022)
93. W. Wang, S. Chen, B. An, K. Huang, T. Bai, M. Xu, G. Bellot, Y. Ke, Y. Xiang, B. Wei, Complex wireframe DNA nanostructures from simple building blocks. Nat. Commun. **10**(1), 1067 (2019)
94. H. Tanaka, T. Dotera, S.T. Hyde, Programmable self-assembly of nanoplates into bicontinuous nanostructures. ACS Nano **17**(16), 15371–15378 (2023)
95. J. Bernal-Chanchavac, M. Al-Amin, N. Stephanopoulos, Nanoscale structures and materials from the self-assembly of polypeptides and DNA. Curr. Top. Med. Chem. **22**(8), 699–712 (2022)
96. D. Bhatia, C. Wunder, L. Johannes, Self-assembled, programmable DNA nanodevices for biological and biomedical applications. ChemBioChem **22**(5), 763–778 (2021)
97. W. Ma, Y. Zhan, Y. Zhang, C. Mao, X. Xie, Y. Lin, The biological applications of DNA nanomaterials: current challenges and future directions. Signal Transduct. Target. Ther. **6**(1), 351 (2021)
98. L. Katrivas, N. Fardian-Melamed, D. Rotem, D. Porath, A. Kotlyar, Formation of novel octuplex DNA molecules from guanine quadruplexes. Adv. Mater. **33**(8), 2006932 (2021)
99. P.S. Kwon, S. Ren, S.-J. Kwon, M.E. Kizer, L. Kuo, M. Xie, D. Zhu, F. Zhou, F. Zhang, D. Kim, Designer DNA architecture offers precise and multivalent spatial pattern-recognition for viral sensing and inhibition. Nat. Chem. **12**(1), 26–35 (2020)
100. S. Shi, W. Fu, S. Lin, T. Tian, S. Li, X. Shao, Y. Zhang, T. Zhang, Z. Tang, Y. Zhou, Targeted and effective glioblastoma therapy via aptamer-modified tetrahedral framework nucleic acid-paclitaxel nanoconjugates that can pass the blood brain barrier. Nanomed. Nanotechnol. Biol. Med. **21**, 102061 (2019)
101. P. Mi, Stimuli-responsive nanocarriers for drug delivery, tumor imaging, therapy and theranostics. Theranostics **10**(10), 4557 (2020)
102. P. Yingyuad, M. Mével, C. Prata, S. Furegati, C. Kontogiorgis, M. Thanou, A.D. Miller, Enzyme-triggered PEGylated pDNA-nanoparticles for controlled release of pDNA in tumors. Bioconjug. Chem. **24**(3), 343–362 (2013)
103. M. Fadeev, G. Davidson-Rozenfeld, Z. Li, I. Willner, Stimuli-responsive DNA-based hydrogels on surfaces for switchable bioelectrocatalysis and controlled release of loads. ACS Appl. Mater. Interfaces **15**(30), 37011–37025 (2023)
104. C. Zhang, Y. Zhou, H. Han, H. Zheng, W. Xu, Z. Wang, Dopamine-triggered hydrogels with high transparency, self-adhesion, and thermoresponse as skinlike sensors. ACS Nano **15**(1), 1785–1794 (2021)
105. J.S. Kahn, Y. Hu, I. Willner, Stimuli-responsive DNA-based hydrogels: from basic principles to applications. Acc. Chem. Res. **50**(4), 680–690 (2017)
106. S.R. Shin, B. Migliori, B. Miccoli, Y.C. Li, P. Mostafalu, J. Seo, S. Mandla, A. Enrico, S. Antona, R. Sabarish, Electrically driven microengineered bioinspired soft robots. Adv. Mater. **30**(10), 1704189 (2018)
107. X. Zhao, H. Wu, B. Guo, R. Dong, Y. Qiu, P.X. Ma, Antibacterial anti-oxidant electroactive injectable hydrogel as self-healing wound dressing with hemostasis and adhesiveness for cutaneous wound healing. Biomaterials **122**, 34–47 (2017)
108. E. Ruiz-Hernandez, A. Baeza, M. Vallet-Regí, Smart drug delivery through DNA/magnetic nanoparticle gates. ACS Nano **5**(2), 1259–1266 (2011)
109. Y. Kasahara, Y. Sato, M.K. Masukawa, Y. Okuda, M. Takinoue, Photolithographic shape control of DNA hydrogels by photo-activated self-assembly of DNA nanostructures. APL Bioeng. **4**(1) (2020)
110. Z. Wang, R. Chen, S. Yang, S. Li, Z. Gao, Design and application of stimuli-responsive DNA hydrogels: a review. Mater. Today Bio **16**, 100430 (2022)
111. X. Meng, K. Zhang, W. Dai, Y. Cao, F. Yang, H. Dong, X. Zhang, Multiplex microRNA imaging in living cells using DNA-capped-Au assembled hydrogels. Chem. Sci. **9**(37), 7419–7425 (2018)

112. L. Tang, Y. Huang, C. Lin, B. Qiu, L. Guo, F. Luo, Z. Lin, Highly sensitive and selective aflatoxin B1 biosensor based on exonuclease I-catalyzed target recycling amplification and targeted response aptamer-crosslinked hydrogel using electronic balances as a readout. Talanta **214**, 120862 (2020)
113. Z. Li, C. Wang, J. Li, J. Zhang, C. Fan, I. Willner, H. Tian, Functional DNA structures and their biomedical applications. CCS Chem. **2**(5), 707–728 (2020)
114. K. Xia, H. Kong, Y. Cui, N. Ren, Q. Li, J. Ma, R. Cui, Y. Zhang, J. Shi, Q. Li, Systematic study in mammalian cells showing no adverse response to tetrahedral DNA nanostructure. ACS Appl. Mater. Interfaces **10**(18), 15442–15448 (2018)
115. C. Cerchia, A. Lavecchia, Small molecule drugs and targeted therapy for melanoma: current strategies and future directions. Curr. Med. Chem. **24**(21), 2312–2344 (2017)
116. A.H. Okholm, J. Kjems, The utility of DNA nanostructures for drug delivery in vivo. Expert Opin. Drug Deliv. **14**(2), 137–139 (2017)
117. A.S. Walsh, H. Yin, C.M. Erben, M.J. Wood, A.J. Turberfield, DNA cage delivery to mammalian cells. ACS Nano **5**(7), 5427–5432 (2011)
118. Y. Zhan, W. Ma, Y. Zhang, C. Mao, X. Shao, X. Xie, F. Wang, X. Liu, Q. Li, Y. Lin, DNA-based nanomedicine with targeting and enhancement of therapeutic efficacy of breast cancer cells. ACS Appl. Mater. Interfaces **11**(17), 15354–15365 (2019)
119. W. Li, X. Yang, L. He, K. Wang, Q. Wang, J. Huang, J. Liu, B. Wu, C. Xu, Self-assembled DNA nanocentipede as multivalent drug carrier for targeted delivery. ACS Appl. Mater. Interfaces **8**(39), 25733–25740 (2016)
120. G. Zhu, R. Hu, Z. Zhao, Z. Chen, X. Zhang, W. Tan, Noncanonical self-assembly of multifunctional DNA nanoflowers for biomedical applications. J. Am. Chem. Soc. **135**(44), 16438–16445 (2013)
121. L. Zhang, R. Abdullah, X. Hu, H. Bai, H. Fan, L. He, H. Liang, J. Zou, Y. Liu, Y. Sun, Engineering of bioinspired, size-controllable, self-degradable cancer-targeting DNA nanoflowers via the incorporation of an artificial sandwich base. J. Am. Chem. Soc. **141**(10), 4282–4290 (2019)
122. C. Wu, D. Han, T. Chen, L. Peng, G. Zhu, M. You, L. Qiu, K. Sefah, X. Zhang, W. Tan, Building a multifunctional aptamer-based DNA nanoassembly for targeted cancer therapy. J. Am. Chem. Soc. **135**(49), 18644–18650 (2013)
123. G. Zhu, G. Niu, X. Chen, Aptamer–drug conjugates. Bioconjug. Chem. **26**(11), 2186–2197 (2015)
124. B. Powell Gray, L. Kelly, D.P. Ahrens, A.P. Barry, C. Kratschmer, M. Levy, B.A. Sullenger, Tunable cytotoxic aptamer–drug conjugates for the treatment of prostate cancer. Proc. Natl. Acad. Sci. **115**(18), 4761–4766 (2018)
125. Z. Deng, Q. Yang, Y. Peng, J. He, S. Xu, D. Wang, T. Peng, R. Wang, X.-Q. Wang, W. Tan, Polymeric engineering of aptamer–drug conjugates for targeted cancer therapy. Bioconjug. Chem. **31**(1), 37–42 (2019)
126. Y. Jiang, X. Pan, J. Chang, W. Niu, W. Hou, H. Kuai, Z. Zhao, J. Liu, M. Wang, W. Tan, Supramolecularly engineered circular bivalent aptamer for enhanced functional protein delivery. J. Am. Chem. Soc. **140**(22), 6780–6784 (2018). https://doi.org/10.1021/jacs.8b03442
127. R. Wang, G. Zhu, L. Mei, Y. Xie, H. Ma, M. Ye, F.-L. Qing, W. Tan, Automated modular synthesis of aptamer-drug conjugates for targeted drug delivery. J. Am. Chem. Soc. **136**(7), 2731–2734 (2014). https://doi.org/10.1021/ja4117395
128. P. Sanchez-Moreno, J.L. Ortega-Vinuesa, J.M. Peula-Garcia, J.A. Marchal, H. Boulaiz, Smart drug-delivery systems for cancer nanotherapy. Curr. Drug Targets **19**(4), 339–359 (2018). https://doi.org/10.2174/1389450117666160527142544
129. N. Chen, S. Qin, X. Yang, Q. Wang, J. Huang, K. Wang, Sense-and-treat DNA nanodevice for synergetic destruction of circulating tumor cells. ACS Appl. Mater. Interfaces **8**(40), 26552–26558 (2016). https://doi.org/10.1021/acsami.6b08695
130. J. Wang, H. Wang, H. Wang, S. He, R. Li, Z. Deng, X. Liu, F. Wang, Nonviolent self-catabolic DNAzyme nanosponges for smart anticancer drug delivery. ACS Nano **13**(5), 5852–5863 (2019). https://doi.org/10.1021/acsnano.9b01589

131. Y.-X. Zhao, A. Shaw, X. Zeng, E. Benson, A.M. Nystrom, B. Hogberg, DNA origami delivery system for cancer therapy with tunable release properties. ACS Nano **6**(10), 8684–8691 (2012)
132. V.J. Schüller, S. Heidegger, N. Sandholzer, P.C. Nickels, N.A. Suhartha, S. Endres, C. Bourquin, T. Liedl, Cellular immunostimulation by CpG-sequence-coated DNA origami structures. ACS Nano **5**(12), 9696–9702 (2011). https://doi.org/10.1021/nn203161y
133. H. Lee, A.K. Lytton-Jean, Y. Chen, K.T. Love, A.I. Park, E.D. Karagiannis, A. Sehgal, W. Querbes, C.S. Zurenko, M. Jayaraman, Molecularly self-assembled nucleic acid nanoparticles for targeted in vivo siRNA delivery. Nat. Nanotechnol. **7**(6), 389–393 (2012)
134. E.S. Andersen, M. Dong, M.M. Nielsen, K. Jahn, R. Subramani, W. Mamdouh, M.M. Golas, B. Sander, H. Stark, C.L.P. Oliveira, J.S. Pedersen, V. Birkedal, F. Besenbacher, K.V. Gothelf, J. Kjems, Self-assembly of a nanoscale DNA box with a controllable lid. Nature **459**(7243), 73–76 (2009). https://doi.org/10.1038/nature07971
135. V.J. Schuller, S. Heidegger, N. Sandholzer, P.C. Nickels, N.A. Suhartha, S. Endres, C. Bourquin, T. Liedl, Cellular immunostimulation by CpG-sequence-coated DNA origami structures. ACS Nano **5**(12), 9696–9702 (2011). https://doi.org/10.1021/nn203161y
136. S. Afereydoon, F. Haghiralsadat, N. Hamzian, A. Shams, M. Hemati, S.M. Naghib, M. Shabani, B. Zandieh-Doulabi, D. Tofighi, Multifunctional PEGylated niosomal nanoparticle-loaded herbal drugs as a novel nano-radiosensitizer and stimuli-sensitive nanocarrier for synergistic cancer therapy. Front. Bioeng. Biotechnol. **10**, 917368 (2022)
137. Z. Gharehdaghi, S.M. Naghib, R. Rahimi, A. Bakhshi, A. Kefayat, A. Shamaeizadeh, F. Molaabasi, Highly improved pH-responsive anticancer drug delivery and T2-weighted MRI imaging by magnetic MOF CuBTC-based nano/microcomposite. Front. Mol. Biosci. **10**, 1071376 (2023)
138. Z. Gharehdaghi, R. Rahimi, S.M. Naghib, F. Molaabasi, Cu (II)-porphyrin metal–organic framework/graphene oxide: synthesis, characterization, and application as a pH-responsive drug carrier for breast cancer treatment. J. Biol. Inorg. Chem. **26**, 689–704 (2021)
139. Z. Gharehdaghi, R. Rahimi, S.M. Naghib, F. Molaabasi, Fabrication and application of copper metal–organic frameworks as nanocarriers for pH-responsive anticancer drug delivery. J. Iran. Chem. Soc. **19**(7), 2727–2737 (2022)
140. T. Jafari, S.M. Naghib, M. Rahmanian, M.R. Mozafari, 3D printing of self-healing materials for drug delivery applications: promises, advances and outlooks. Bioprinting **42**, e00353 (2024). https://doi.org/10.1016/j.bprint.2024.e00353
141. S. Kabiri, R. Rahimi, M.R. Mozafari, S.M. Naghib, Porphyrin-based metal–organic frameworks in drug delivery for cancer therapy: promises, advances and prospects. Discover Appl. Sci. **7**(4), 313 (2025)
142. Z. Mazidi, S. Javanmardi, S.M. Naghib, Z. Mohammadpour, Smart stimuli-responsive implantable drug delivery systems for programmed and on-demand cancer treatment: an overview on the emerging materials. Chem. Eng. J. **433**, 134569 (2022). https://doi.org/10.1016/j.cej.2022.134569
143. S.M. Naghib, K. Mohammad-Jafari, Microfluidics-mediated liposomal nanoparticles for cancer therapy: recent developments on advanced devices and technologies. Curr. Top. Med. Chem. **24**(14), 1185–1211 (2024). https://doi.org/10.2174/0115680266286460240220073334
144. S. Soleymani, S.M. Naghib, M.R. Mozafari, PH/Temperature/redox and light-responsive polymersome structure and application in cancer therapy: smart drug delivery and targeted drug release. Curr. Med. Chem. (2024). https://doi.org/10.2174/0109298673303351240507113603
145. W. Jorieke, B. Maartje, DNA origami nanostructures for controlled therapeutic drug delivery (2021)
146. J.F. Berengut, C.K. Wong, J.C. Berengut, J.P.K. Doye, T.E. Ouldridge, L.K. Lee, Self-limiting polymerization of DNA origami subunits with strain accumulation. ACS Nano **14**(12), 17428–17441 (2020). https://doi.org/10.1021/acsnano.0c07696
147. A. Jabbari, E. Sameiyan, E. Yaghoobi, M. Ramezani, M. Alibolandi, K. Abnous, S.M. Taghdisi, Aptamer-based targeted delivery systems for cancer treatment using DNA origami and DNA nanostructures. Int. J. Pharm. **646**, 123448 (2023). https://doi.org/10.1016/j.ijpharm.2023.123448

148. Q. Jiang, Y. Shang, Y. Xie, B. Ding, DNA origami: from molecular folding art to drug delivery technology. Adv. Mater. **36**(22), 2301035 (2024). https://doi.org/10.1002/adma.202301035
149. R. Li, A.S. Madhvacharyula, Y. Du, H.K. Adepu, J.H. Choi, Mechanics of dynamic and deformable DNA nanostructures. Chem. Sci. **14**(30), 8018–8046 (2023). https://doi.org/10.1039/D3SC01793A
150. M.J. Mitchell, M.M. Billingsley, R.M. Haley, M.E. Wechsler, N.A. Peppas, R. Langer, Engineering precision nanoparticles for drug delivery. Nat. Rev. Drug Discovery **20**(2), 101–124 (2021). https://doi.org/10.1038/s41573-020-0090-8
151. S.H. Kiaie, H. Salehi-Shadkami, M.J. Sanaei, M. Azizi, M. Shokrollahi Barough, M.S. Nasr, M. Sheibani, Nano-immunotherapy: overcoming delivery challenge of immune checkpoint therapy. J. Nanobiotechnol. **21**(1), 339 (2023)
152. J. Yang, C. Jia, J. Yang, Designing nanoparticle-based drug delivery systems for precision medicine. Int. J. Med. Sci. **18**(13), 2943–2949 (2021). https://doi.org/10.7150/ijms.60874
153. N. Maghsoudnia, R.B. Eftekhari, A.N. Sohi, A. Zamzami, F.A. Dorkoosh, Application of nano-based systems for drug delivery and targeting: a review. J. Nanopart. Res. **22**(8), 245 (2020). https://doi.org/10.1007/s11051-020-04959-8
154. Y. Dang, J. Guan, Nanoparticle-based drug delivery systems for cancer therapy. Smart Mater. Med. **1**, 10–19 (2020). https://doi.org/10.1016/j.smaim.2020.04.001
155. G.-M. Han, B. Liu, D.-M. Kong, L.-N. Zhu, DNA as highly biocompatible carriers for drug delivery. Mater. Chem. Front. **7**(24), 6345–6365 (2023). https://doi.org/10.1039/D3QM00395G

Chapter 8
Three-Dimensional Printing Methods for Designing Nanostructured Scaffolds in Tissue Engineering and Localized Drug Delivery

Abstract The use of computer-aided design (CAD) and additive manufacturing (AM) in tissue engineering, particularly for cartilage and bone regeneration, is examined in this chapter. Developing scaffolds that resemble the natural extracellular matrix is the difficult part of tissue engineering. Recent developments in additive manufacturing, such as 3D printing, allow scaffolds to be precisely customized with control over attributes like shape and bioactivity. Design utilizing patient-specific data is made easier by CAD tools. The benefits of materials like hydrogels and polymers for tissue engineering, are highlighted. The use of bioactive compounds such as growth factors in scaffolds to promote tissue regeneration, is covered in the chapter. To improve tissue growth, several cell sources including stem cells, are being investigated for incorporation into the scaffold matrix. The use of AM and CAD in cartilage and bone tissue engineering to treat issues including osteochondral abnormalities and osteoarthritis, are highlighted.

Keywords Cartilage regeneration · Bone regeneration · 3D printing · Tissue engineering · Regenerative medicine · Natural extracellular matrix (ECM) · Additive manufacturing

8.1 Introduction

Loss of cartilage brought on by bone diseases such as osteoarthritis (OA) is a serious health problem. The most prevalent joint condition, OA causes pain and suffering for those who have it and may be quite expensive for healthcare systems. One common characteristic of OA seems to be the gradual deterioration of the articular cartilage [1]. The ankle, hip, and knee are common weight-bearing joints where symptoms of articular cartilage damage manifest. These ailments often prevent the individual from doing sports. Five to ten percent of knees undergoing arthroscopy had full-thickness lesions [2, 3]. Around 10% of hips treated for femoroacetabular impingement (FAI) have full-thickness acetabular lesions, which are linked to cartilage damage in the hip and dysplasia. The difficulties in delivering treatments to the cartilage highlight

S. M. Naghib, *Nano-Biosensors, Drug Delivery and Tissue Engineering*, Series in BioEngineering, https://doi.org/10.1007/978-981-95-1888-3_8

the need for efficient drug-delivery devices in the treatment of cartilage degeneration [4, 5]. The administration of drugs in tissues is significantly hampered by the extracellular matrix's thick network and the absence of vascularization [6–8]. Present intra-articular drug delivery methods may not allow the medications to adequately penetrate tissue to reach cell and matrix targets at the quantities necessary to elicit the desired biological response [9, 10]. Therefore, finding effective treatment approaches for osteoarthritis, cartilage and bone loss is crucial [1, 11, 12]. A unique connective tissue called cartilage provides the joints with load-bearing capabilities and smooth articulation. However, since cartilage is avascular and low cellularity, its ability to self-repair is restricted [11]. Osteoarthritis is a persistent and crippling condition that affects millions of people worldwide and is caused by cartilage degradation or degeneration. Therefore, there is an urgent need to develop effective methods for regenerating cartilage tissue. Tissue engineering creates functional constructions for tissue regeneration by combining biomaterials, cells, and bioactive substances. Biomaterials are essential among these elements because they serve as scaffolding that facilitate cell adhesion, proliferation, delivery, and differentiation [13, 14]. 3D biomaterials may provide chemical and physical signals to alter tissue development and cell behavior. By layering materials, 3D printing is a developing technique that can create scaffolds with intricate and accurate architecture [2, 5, 15–17]. The form, size, mechanical, and porosity characteristics of the scaffolds may be altered via 3D printing to meet the unique needs of various tissues [15, 18, 19]. Moreover, bioactive compounds may be added to the scaffolds by 3D printing to improve their biological properties. A biomimetic scaffold that mimics the zonal design of natural cartilage which is made up of discrete layers with different pore sizes, mechanical characteristics, and drug-loading capacities was developed in order to achieve this goal [3, 16, 20]. The scaffold is made of gelatin methacrylate (GelMA) and poly(ε-caprolactone) (PCL), which were selected for their biocompatibility and biodegradability. Important for tissue formation and cell function, the scaffold may also improve cytoskeletal structure, adhesion, and spreading. The scaffold was shown in in vivo experiments to stimulate cartilage-like tissue formation with improved histological and biochemical properties (Fig. 8.1) [21, 22].

These improvements include increased expression of glycosaminoglycan (GAG), decreased production of fibrous tissue, and increased adhesion to the surrounding host tissue. Additionally, it clarified the fundamental processes of scaffold-mediated chondrogenesis by an examination of hMSC morphology, cytoskeleton, adhesion, and signaling pathways [23]. Based from the results, the scaffold might improve the chondrogenic capacity of hMSCs by modifying the interactions between cells and the microenvironment. The scaffold may specifically stimulate the GR pathway and FLNA-mediated signaling, which are implicated in ECM formation and chondrogenesis [24]. It includes information on the structure of articular cartilage, such as the shape of the chondrocytes and the orientation of the collagen fibers, in addition to a histological image that displays the distribution of chondrocytes. A schematic of the chondrogenesis process, which begins with MSC proliferation and concludes with endochondral ossification, is also shown in the picture, along with an emphasis on crucial ECM proteins. It also explains the ideal properties of bioinks for 3D

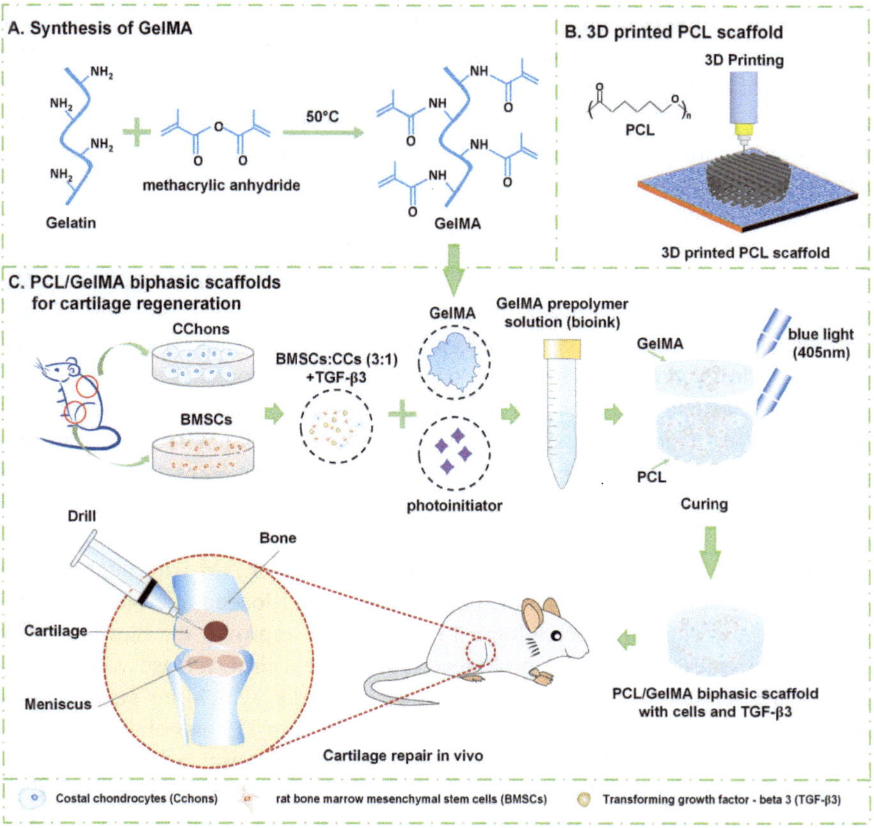

Fig. 8.1 Schematic illustration of GelMA synthesis, preparation of PCL/GelMA biphasic scaffolds and in vivo cartilage regeneration [22]

printing and in vivo cartilage tissue regeneration. The chapter concludes by discussing 3D bioprinted biomimetic scaffold examples and their potential to improve tissue regeneration in chondral defects [25].

One critical study presents a novel method for 3D printing biomimetic scaffolds that have the special capacity to deliver two drugs at once to promote cartilage regeneration. This scaffold architecture increases the chondrogenic capability of human mesenchymal stem cells (hMSCs) by actively modifying cell-microenvironment interactions. Additionally, it helps the regenerated tissue blend in seamlessly with the surrounding host tissue. The use of this scaffold has the potential to solve important health issues such osteoarthritis (OA) and cartilage abnormalities, which have major social and economic ramifications. Future directions for development include improving the scaffold's structure and drug release kinetics, carrying out thorough evaluations of its long-term safety and effectiveness, testing it in bigger animal models, and ultimately putting it through clinical trials [20, 26, 27]. This chapter's novelty in concentrating on the future of Additive Manufacturing (AM) in tissue engineering (TE) is defined by the creative fusion of several state-of-the-art technologies and a deeper comprehension of tissue biology, which together allow for the production of highly biomimetic and functional tissue constructs that were previously unattainable. The building of complex tissue microenvironments, which goes beyond conventional scaffold manufacturing to replicate the complicated signaling pathways that direct tissue growth, is a crucial component of this innovation. Researchers may better replicate the natural surroundings of tissues by manipulating the temporal and spatial presentation of biomolecules inside scaffolds. An important improvement over traditional methods is the ability to recreate the anisotropic phenotypes of original tissues, such the meniscus, by precisely placing growth factors inside a scaffold. The use of stimuli-responsive materials in "smart" scaffolds which may release therapeutic drugs in reaction to certain triggers, such ultrasound is another groundbreaking advancement. Bone grafts, which release bone morphogenetic proteins (BMP) only when required to stimulate localized bone development, are an example of this strategy's unparalleled control over drug delivery, enabling tailored therapy and reducing systemic adverse effects [28, 29]. Furthermore, new opportunities for creating complex microstructures and microparticles with highly controlled characteristics have been made possible by the combination of AM with microfluidics and soft lithography. By enabling the embedding of drug-loaded microparticles into 3D-printed scaffolds, this technological combination improves the biomimicry of engineered tissues by enabling prolonged and targeted drug release with various profiles. The creation of functioning vascular networks, which are necessary for tissue survival and functionality, is a recurring problem in TE, especially for bigger tissue constructions [30]. Future research must address this vascularization obstacle, and two such approaches include incorporating pro-angiogenic factors into scaffolds to promote blood vessel development and integrating bioprinting with endothelial cells. Driven by a stronger knowledge of tissue biology and biophysics, these improvements are really innovative in their synergistic integration, which has the potential to transform the field of AM in TE. It is anticipated that this technological convergence will result in the creation of more complex, useful, and clinically applicable tissue engineering solutions, opening the door for further advancements in regenerative medicine [31]. Researchers are investigating novel approaches to develop biomimetic constructions that have significant promise for tissue repair and regenerative medicine by imitating natural signals from native tissues. In order to reproduce the anisotropic phenotypes of different tissues, such as the meniscus in the knee, where the careful placement of growth factors may assist simulate its complicated shape, a crucial component of this study is the exact positioning and distribution of biomolecules inside scaffolds. For instance, a pioneering study successfully replicated the tissue's structural complexity and provided a platform for targeted regeneration strategies by using multi-tool biofabrication to create a meniscus scaffold with zone-specific properties by incorporating growth factor-loaded microparticles into distinct regions. In order to replicate the shift from the superficial to the deep zones in normal articular cartilage, a gradient-structured bioprinted scaffold with variable PCL fiber spacing was employed in another noteworthy attempt to promote anisotropic cartilage regeneration. The formation of cartilage tissue with spatially specified features was made possible by this design in conjunction with growth factor-loaded microcarriers, indicating the possibility of customized regeneration techniques. A significant breakthrough in scaffold manufacturing, the ability to precisely regulate biomolecule distribution both geographically and temporally opens up new possibilities for producing scaffolds that closely resemble natural tissue microenvironments. In addition to increasing the structural integrity of engineered tissues, this complexity has great potential for individualized regenerative treatments that are catered to the requirements of each patient [32]. With potential uses in organ regeneration, disease modeling, and customized therapy, the implications for tissue engineering and regenerative medicine are significant as this field of study develops. This multidisciplinary approach is essential to forming the future of healthcare and providing hope to patients who need advanced therapeutic solutions tailored to their specific biological contexts, as researchers are poised to revolutionize tissue engineering and lay the foundation for transformative clinical applications by perfecting the precise delivery of biomolecules [33].

Treatments and Drug Delivery on-Demand: It is possible to create "smart" scaffolds that may release therapeutic drugs in response to certain triggers by including stimuli-responsive materials [34–40]. Changes in temperature, pH, light exposure, ultrasound, magnetic fields, and even the presence of certain enzymes are examples of these stimuli. Scientists created hydrogel biomaterials that melt when exposed to near-infrared light-induced localized heating. The regulated release of encapsulated medicinal cargo was made possible by this on-demand liquefaction [41–45]. Iron oxide nanoparticles have been embedded in a biocompatible matrix to create ferrogels, which are magnetically sensitive scaffolds. These nanoparticles move when an external magnetic field is applied, which deforms the scaffold and speeds up the release of medications that are enclosed. Through the regulated release of growth factors such as TGF-β1 and PDGF, this approach has shown potential in encouraging tissue regeneration. A major advancement in TE is represented by this degree of control over drug release, which may lead to focused treatments with fewer adverse effects [46].

Combining AM with Soft Lithography and Microfluidics: Combining AM with soft lithography and microfluidics enables the production of intricate microparticles and microstructures with precisely regulated characteristics [47].

Microfluidic Emulsions: Using microfluidic devices, it is possible to create multicored microparticles that each contain a distinct drug or cell [48–50]. Engineering scaffolds that can transport numerous therapeutic agents with different release profiles is made possible by this capacity, which presents fascinating opportunities [51].

Lithographically Defined Microcontainers: Lithography-based methods, such those that use patterned molds or UV-photopolymerization, provide exact control over the creation of microcontainers that hold drugs and cells. To ensure stability and maximize loading efficiency, these microcontainers may be loaded with numerous bioactive compounds in powder form or by supercritical CO_2 processing. Researchers may get previously unheard-of control over the temporal and geographical release of medicinal medicines by integrating these carefully designed microparticles and microcontainers into 3D-printed scaffolds. The creation of highly biomimetic scaffolds that can replicate the intricate signaling contexts present in real tissues is made possible by this combination of technologies [52, 53].

Vascularization Challenge: Obtaining sufficient vascularization in TE is still very difficult, especially for bigger tissue structures. Without a working circulatory system, designed constructions' cells and tissues won't be able to get the oxygen and nutrients they need, which will eventually cause them to die. Although this difficulty is not specifically addressed as a uniqueness of future AM approaches, sources stress how important it is to overcome it. Future studies will probably concentrate on combining endothelial cells, which are the building blocks of blood vessels, with methods like bioprinting. In order to promote the creation of blood vessels from the surrounding host tissue, other strategies can include adding pro-angiogenic substances to scaffolds. Long-term survival and bodily integration of engineered tissues will depend critically on the effective creation of functioning circulatory networks inside those tissues [54]. Through the harmonic integration of cells, biomaterials, and bioactive agents, tissue engineering develops as a dynamic interdisciplinary field with the goal of creating completely functioning tissues or organs. Three broad approaches scaffold-based, scaffold-free, and cell-free approaches have emerged in the context of cartilage regeneration. Cells are strategically seeded into a well planned three-dimensional (3D) porous architecture using scaffold-based techniques 3D. This design serves as an essential framework, providing biochemical signals, mechanical reinforcement, and spatial guidance that coordinate critical cellular functions, such as extracellular matrix deposition, adhesion, proliferation, and differentiation. This scaffold's composition includes a range of natural, synthetic, and hybrid materials that exhibit unique properties such as mechanical robustness, porosity, biodegradability, and biocompatibility [55]. The next step up is to functionalize these scaffolds with medicines, growth factors, or genetic material to provide cells and scaffolds a higher biological stimulation. Significant examples of scaffold-based methods include matrix-assisted chondrocyte implantation (MACI), autologous chondrocyte implantation (ACI), and mesenchymal stem cell (MSC)-based treatments. Scaffold-free methods do not use any external scaffolds; instead, they depend on cells self-assembling into three-dimensional structures. Immunogenicity, infection, inflammation, and poor integration with the host tissue are some of the drawbacks of scaffold-based approaches that scaffold-free approaches can get over [56]. In addition to creating cartilage-like tissues with a high cell density and uniform distribution, scaffold-free techniques can maintain the natural connections between cells and between cells and the matrix. Spheroid, pellet, and cell sheet engineering are scaffold-free methods. By directing the migration and differentiation of endogenous cells toward the site of damage, cell-free techniques make use of biomaterials and biological factors either alone or in combination. Through avoiding some issues inherent in cell-based approaches, such as cell acquisition, growth, and viability as well as phenotypic stability, cell-free approaches provide a solution. The financial, logistical, and regulatory challenges related to cartilage tissue creation may also be lessened by these methods. Additionally, developing new methods that combine many techniques into complementary multimodal treatments is essential to pushing the boundaries of cartilage tissue engineering results [57]. Because they provide the critical physical and chemical environment required for cellular development, differentiation, and matrix production, biomaterials play a key role in cartilage tissue engineering. These biomaterials may be classified as either natural or synthetic, and each has unique advantages and disadvantages. Hyaluronic acid, collagen, alginate, and chitosan are examples of natural biomaterials that come from biological sources and have higher levels of biocompatibility, biodegradability, and a similarity to the extracellular matrix of natural cartilage. These materials do have some drawbacks, however, such as batch-to-batch variability, immunogenicity, low mechanical strength, and processing difficulties. Conversely, well crafted synthetic biomaterials like poly(lactic-co-glycolic acid), polyethylene glycol, and polyurethane provide tunable properties including stiffness, porosity, degradation kinetics, and molecular weight. Nevertheless, there are drawbacks to this type of materials, including poor bioactivity, poor cell adherence, and possible toxicity [58, 59]. Significant progress has recently been made in creating new biomaterials designed especially for cartilage tissue engineering. Figure 8.2 depicts two main topics: 3D bioprinting technology in cartilage tissue engineering, and articular cartilage and chondrocytes on the joint surface. This includes the distribution and shape of chondrocytes in articular cartilage as well as the production of cartilage tissue using 3D bioprinting technology [60].

8.2 Different Materials for Cartilage Scaffold 3D Printing

Biomimetic Materials: These work on many dimensions, including the molecular, cellular, and tissue levels, and are cleverly made to mimic the complex structure and functions of genuine cartilage tissue. Biomimetic materials have the revolutionary potential to improve the biological performance of manufactured cartilage by precisely mimicking these aspects. Their capacity to provide specific signals for essential cellular functions, including as attachment, proliferation, differentiation, and matrix deposition, is what makes this augmentation possible. Consider the creative development of biomimetic hydrogels. This entails introducing different components into the hydrogel network, such as bioactive chemicals (like growth factors, peptides, or genes), nanofibers (like collagen or silk), or microspheres (like calcium phosphate or hyaluronic acid). The artificial tissue environment may become even more favorable for successful cartilage regeneration with the use of such biomimetic constructions [25, 61].

Novel Materials: These are very sensitive to external stimuli, such as changes in temperature, pH, light, electric fields, or magnetic fields. These innovative materials have the unique ability to dynamically adjust their properties in response to changes

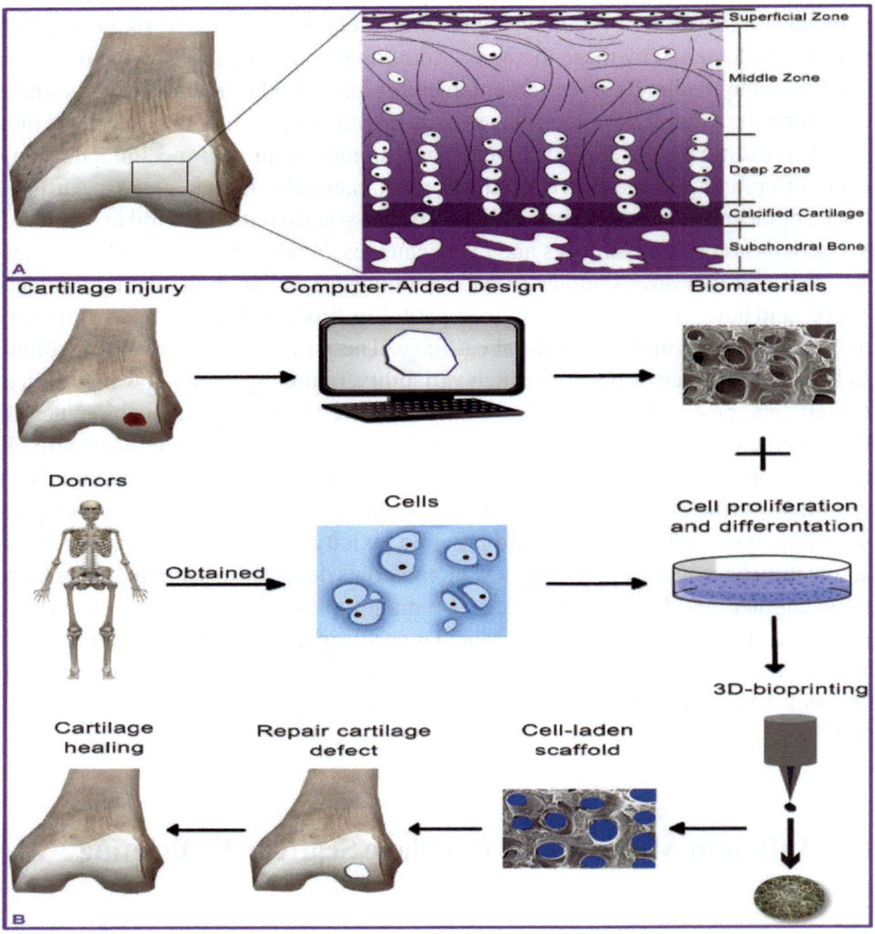

Fig. 8.2 **a** A diagram showing the composition of articular cartilage and the location of chondrocytes on the joint surface. **b** A schematic illustration of the use of 3D bioprinting technology in cartilage tissue engineering can be seen. Reproduced with permission from Ref. [60]. Copyright 2021 Frontiers in Bioengineering and Biotechnology

in the physiological environment or outside stimuli. As an example, think of the field of intelligent hydrogels, which may reversibly expand or contract in direct reaction to temperature or pH changes [62]. These cutting-edge materials have the capacity to direct the controlled release of bioactive substances, trigger mechanical reactions in cellular structures, and even operate as watchful monitors, continuously assessing the condition of the manufactured cartilage [63]. Figure 8.3 shows a complex scene of 3D-printed scaffolds that are artistically adapted to satisfy the various needs of various tissues and organs within the broad framework of regenerative medicine. These scaffolds, which are made of biocompatible polymers, are smooth combinations of materials, with each component working together to promote regenerative results.

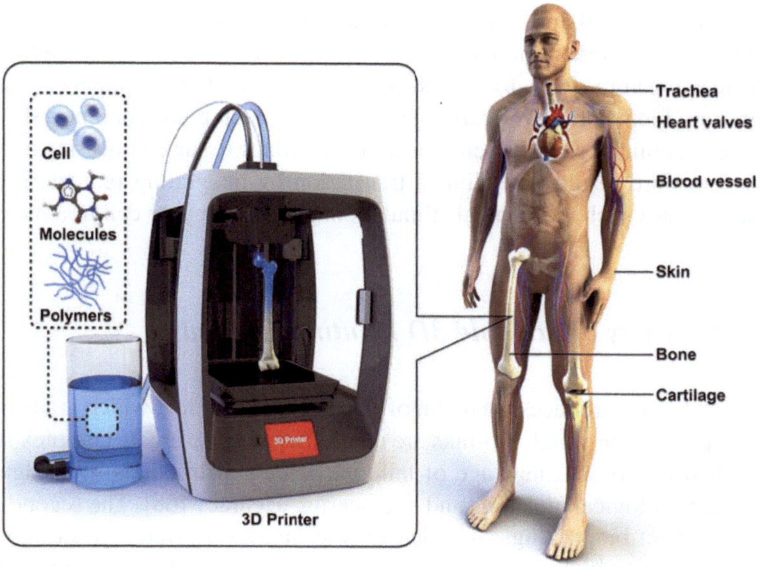

Fig. 8.3 Various scaffolds for tissues and organs have been 3D printed in the regenerative medicine field. 3D scaffolds are typically made of biocompatible polymers. Cells and bioactive molecules are often incorporated with the scaffolds to enhance bioactive properties. Reproduced with permission from Ref. [61]. Copyright 2020 Frontiers in Bioengineering and Biotechnology

These structures are far from static, however; their orchestration goes beyond their structural soundness. Rather, it develops as a symphony of intent, where the deliberate interaction of cells and bioactive compounds culminates, advancing these scaffolds onto a higher plane of enhanced bioactivity and therapeutic possibilities. The essence of these dynamic, regenerating scaffolds is vividly captured in this picture, which sheds light on this multifaceted story [61, 64].

8.3 3D Printing of Cartilage Regeneration Scaffolds

One flexible technique for creating scaffolds for cartilage tissue engineering is 3D printing. It provides careful control over shape, porosity, and mechanical properties via the creation of complex and customized designs. Interestingly, 3D printing goes beyond its potential by combining various materials, cells, and bioactive components into cohesive scaffolds that are positioned in space [65]. The complexities of creating, manufacturing, and evaluating 3D-printed scaffolds for cartilage regeneration will be covered in detail in this section. It will also analyze the current issues and provide a path for future developments. 3D With great accuracy and resolution, this manufacturing process may produce intricate 3D structures. With spatial control, 3D printing may also combine different cell types or biomaterials into a single build.

The combination of 3D printing and other techniques, such as bioprinting or electrospinning, may work in harmony to create hybrid structures with sophisticated and advanced features [66]. To address the difficulties and requirements of cartilage regeneration, biomaterials for cartilage tissue engineering are continuously developing and becoming better. It is envisaged that cartilage tissue engineering may soon achieve improved results and clinical translation by integrating biomaterials with other approaches such mechanical stimulation, gene therapy, or cell treatment [67].

8.3.1 A Variety of Scaffold 3D Printing Methods

The first category introduces the realm of extrusion bioprinting, which is a technique similar to painting in which bio-inks are carefully placed to create complex shapes. This method ushers in a new age of tailored tissue constructions by bridging the gap between biological materials and engineering accuracy [68]. The second aspect introduces inkjet bioprinting, an endeavor that mimics a printer's accuracy while paving the way for biological innovation. Bio-ink droplets form a symphony that orchestrates the creation of complex biological designs as they dance onto surfaces. The third group focused on laser-assisted bioprinting, which uses laser pulses to transfer tiny droplets of bio-ink onto target surfaces. The subtle skill that underlies tissue engineering is shown by this methodical approach [69]. Last but not least, it revealed stereolithography bioprinting, in which objects are given life layer by layer by photopolymerization. Bio-inks are carefully shaped into complex patterns by light, creating a canvas that combines the poetry of biology with the accuracy of technology [70]. These four bioprinting pillars are summed up in Fig. 8.4, which provides a thorough overview of the techniques that expand tissue engineering's potential. Illuminating these categories served as a compass for comprehending the many facets that support the development and expansion of this cutting-edge field [71, 72]. Inkjet bioprinting provides a significant advantage in the precise deposition of biological inks onto designated locations following a predetermined pattern. This technique enables the fabrication of multifunctional bionic structures while minimizing disruption to existing constructs. By delivering cells and biomaterials from small droplets with high spatial accuracy, inkjet bioprinting allows for intricate patterning and multilayered tissue construction, supporting complex and organized biological architectures for applications in regenerative medicine and tissue engineering (Fig. 8.4).

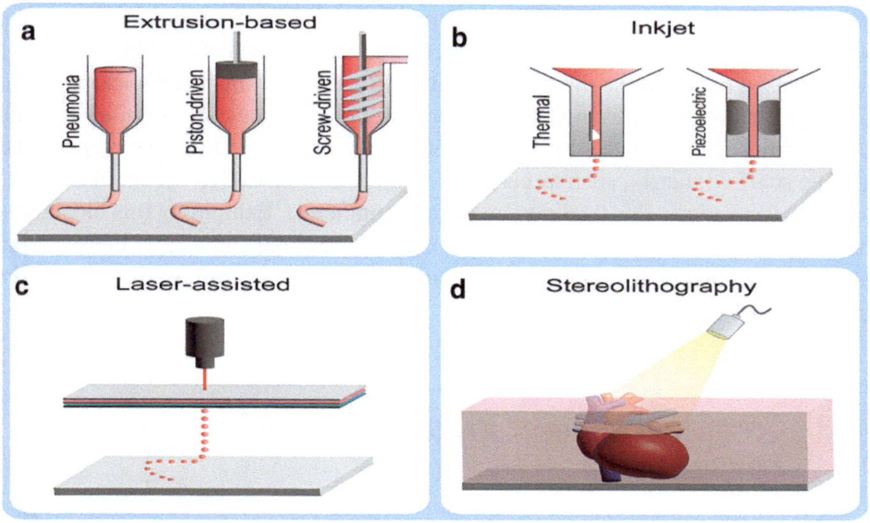

Fig. 8.4 Representative 3D bioprinting techniques. **a** Extrusion-based bioprinting; **b** inkjet bioprinting; **c** laser-assisted bioprinting; **d** stereolithography bioprinting [72]

8.3.2 Enhancing the Design Paradigm: Cartilage Regeneration Made Possible by 3D Printed Scaffolds

To guarantee thorough gap covering, 3D Precision in the scaffold's size and shape formulation is crucial, ensuring that it blends in perfectly with the defect site's contours. In addition, the scaffold's surfaces and edges should have a smooth profile to avoid causing any irritation or inflammation in the surrounding tissue. To accurately measure these characteristics, sophisticated imaging techniques such as magnetic resonance imaging (MRI) and computed tomography (CT) are used. By carefully creating a 3D model of the affected region, this data may be easily converted into a digital format that can be divided into printable layers with the aid of state-of-the-art 3D printing technologies [73]. Transference, cellular penetration, nutrition distribution, waste outflow, and mechanical properties are all orchestrated by the complex dance between a scaffold's porosity and pore size. A scaffold that allows for sufficient cell penetration and vascularization while maintaining the necessary mechanical strength and stability is the result of fine-tuning this interaction. By carefully adjusting printing parameters such as printing velocity, layer thickness, nozzle diameter, and inter-filament or droplet spacing, the equilibrium is achieved by carefully calibrating the parameters controlling porosity and pore size [74].

Mechanical Integrity and Adaptation: Maintaining the toughness of joint-related physiological stresses requires striking a balance between the scaffold's mechanical characteristics and the natural characteristics of cartilage tissue. The scaffold's mechanical profile is determined by a complicated interaction between porosity, pore diameters, microstructure, and material composition [75]. In order to quantify this profile, well-established methods such as compression, tension, bending, and torsion tests are used to determine the scaffold's ability to replicate the mechanical response of the original tissue. The bioactivity of the scaffold depends on its ability to stimulate or improve tissue and cellular responses. By including bioactive components into the scaffold matrix, such as ceramics, bioglass, or bioactive agents like growth hormones and genes, this liveliness may be strengthened. Examining this bioactivity involves using in vitro or in vivo tests to see if the scaffold can trigger these crucial biological behaviors by exploring cellular adhesion, proliferation, differentiation, matrix formation, and tissue fusion [76].

8.3.3 Creation of Cartilage Regeneration Scaffolds Using 3D Printing

The process of creating 3D-printed scaffolds for cartilage regeneration requires careful consideration of both printing methods and material selection. These elements, which are essential to the scaffold's construction, may be divided into two main groups: synthetic and natural. Natural materials having biological origins boast remarkable biocompatibility, biodegradability, and an innate affinity for the extracellular matrix of natural cartilage. Despite these advantages, natural materials have drawbacks, including difficult processability, poor mechanical robustness, possible immunogenicity, and batch-to-batch variations. Conversely, artificial materials, which are carefully created by chemical synthesis, exhibit modifiable characteristics such as molecular weight, rate of disintegration, porosity, and stiffness. However, there are other drawbacks to their story, most notably a lack of bioactivity, poor cell adherence, and possible toxicity [77].

Stem Cells: Potential cell sources for cartilage repair include mesenchymal stem cells, induced pluripotent stem cells, and embryonic stem cells. For these stem cells to develop into the diverse tissue types they will ultimately become, certain growth factors must be present in the right concentration range (Fig. 8.5) [78, 79].

Growth Factors: Growth factors are necessary for cartilage repair. They can improve cartilage regeneration by changing the local microenvironment. Members of the fibroblast growth factor family, insulin-like growth factor-I, platelet-derived growth factor, and transforming growth factor-β superfamily have all been investigated as possible adjuncts in the treatment of chondral injuries and early arthritis.

Microenvironment: It is important to understand the concentrations of morphogen gradients and signaling molecules in the cellular microenvironment during the early stages of embryonic development since these molecules are essential inducers of tissue growth [80].

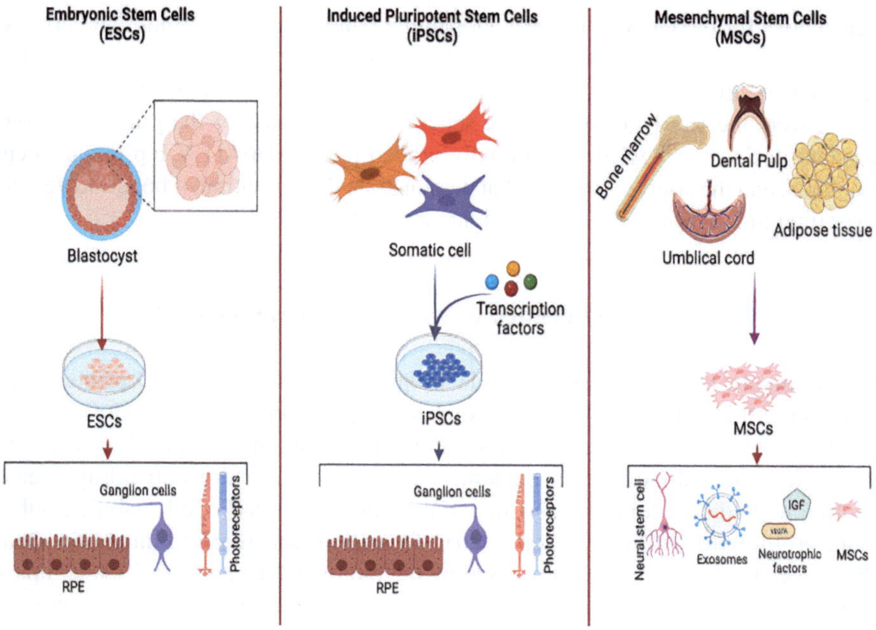

Fig. 8.5 Stem cell sources and differentiation pathways relevant to cartilage repair, including embryonic, induced pluripotent, and mesenchymal stem cells [79]

Chemokines: They also play a role in cartilage repair. For instance, the MCP-1–CCR2 axis is essential for the onset and progression of cartilage deterioration.

SOX9 Regulation: SOX9 regulation is necessary for the chondrocyte phenotype. Maintaining cell viability and extracellular matrix (ECM) homeostasis is also necessary for large-scale chondrocyte proliferation. These components work together to promote cartilage repair [81].

The development of these regenerative scaffolds is guided by the printing technique selection in addition to the material selection. These methods include paradigms that are based on extrusion, inkjet, laser, and light. Extrusion-based techniques use a nozzle to precisely extrude continuous filament or material droplets onto consecutive substrate layers. With inkjet-based techniques, tiny material droplets are ejected onto the substrate layer by layer using a printhead. Laser-based techniques use laser beams to sequentially melt or sinter powder bed layers. The foundation of light-based techniques is the careful layer-by-layer curing of liquid resin using a light source. The creation of customized cartilage scaffolds requires a careful tango between the benefits and drawbacks of each technique in terms of resolution, velocity, accuracy, and compatibility with various materials [82]. Figure 8.6 illustrates many facets of bone tissue engineering, including the creation of dual-modular scaffolds with functionalization to provide growth factors (GF), two-ink 3D printing of vascularized CaP scaffolds, and GelMA-based constructs. This includes gene expression data, microscope photos, micro-CT analysis, and descriptions of the manufacturing schematics for the GelMA/HA/HA-Dopamine bone implant scaffolds. The process of 3D printing vascularized CaP scaffolds, including the formation of endothelial monolayers, the elimination of sacrificial ink, and ink extrusion, is also covered. It also covers the LPFB technique, the schematic illustration of the gradient porous implant design strategy for functionally graded porous Ti6Al4 V hip replacement implants, and the confocal images that demonstrate the formation of bone and vessels over time in functionalized dual-modular scaffolds [83].

8.3.3.1 Evaluation of 3D-Printed Cartilage Regeneration Scaffolds

Examining the physical, chemical, and biological characteristics of 3D-printed scaffolds for cartilage regeneration is a complex process. Form, porosity, pore size, and mechanical characteristics are all included in an evaluation of physical features. As we get into the realm of chemical characteristics, the focus shifts to the drug release profile, degradation kinetics, and material composition [84]. Cell viability, adhesion, proliferation, differentiation, matrix formation, and scaffold integration into the tissue environment are all traversed by the biological assessment. This complex assessment is handled using a variety of characterisation techniques, which may be roughly divided into destructive and non-destructive methods. Spectroscopic methods and imaging parameters are included in non-destructive approaches. On the other hand, the destructive suite includes techniques that need the scaffold to be modified, such as molecular tests, immunohistochemistry, biochemical studies, and histological evaluations [85]. These two paths provide a comprehensive grasp of the scaffold's complexities, supporting its ability to function as an effective cartilage regenerating tool. Figure 8.7 highlights the colorful tapestry of 3D-printed scaffolds that have given regenerative medicine new life in the complex field of tissue engineering. Every picture in this figure is a doorway into a unique story that captures the development, inventiveness, and promise of these scaffolds. These images span the spectrum from the macrocosm to the microcosm: from the close-up view of a poly(l-lactide-co-ε-caprolactone) scaffold for adipose tissue engineering via an optical microscope to the enormous size of a pig condyle scaffold seen in an STL picture. A frontal view of a PCL scaffold that has been 3D printed recalls the technology's early days. In addition, a SEM picture provides a finely detailed view of a scaffold the size of a mouse that is intended for femoral mid-diaphysis regeneration. The lens's fine pore diameters create a microcosmic symphony as it zooms deeper, immersing it in the micro-porosity of a calcium phosphate-collagen composite scaffold [86]. Last but not least, a 3D printed scaffold design based on PU/HA hints to a possible route to in situ spontaneous chondrogenesis. Reproduced with permission from the contributing research, these striking images are a monument to the creativity and artistry shedding light on the way to improved regeneration results [61].

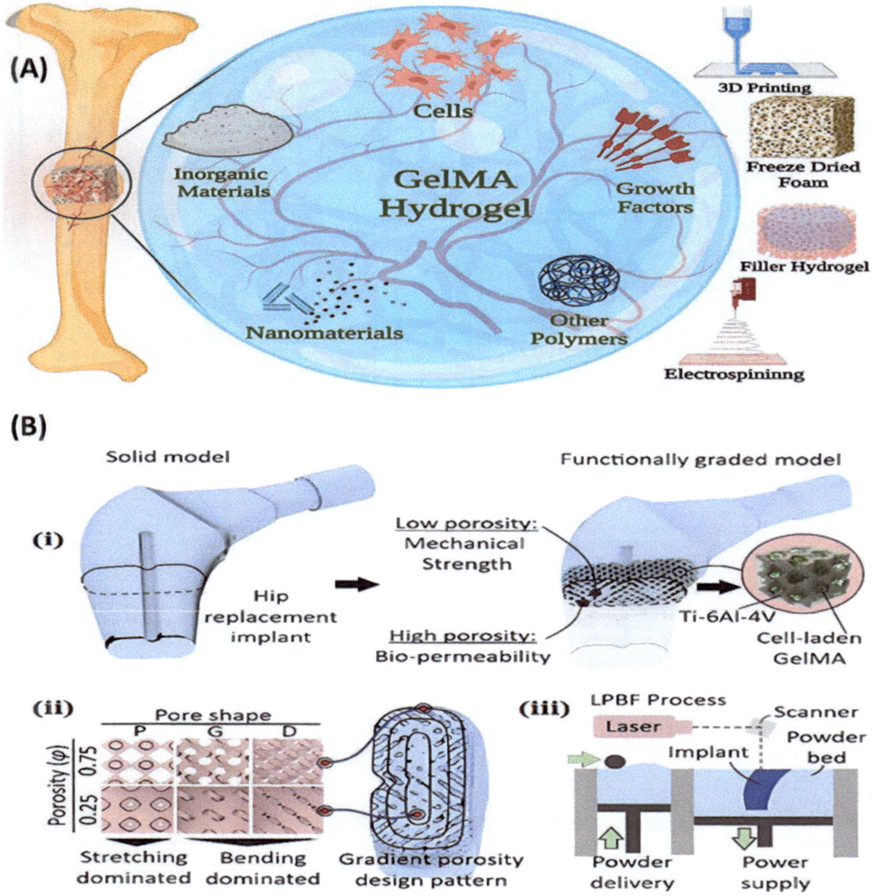

Fig. 8.6 Schematic illustrations and strategies in bone tissue engineering; **a** GelMA hydrogel system incorporating inorganic materials, growth factors, nanomaterials, and other polymers to support cell growth and regeneration, using advanced fabrication techniques such as 3D printing, freeze-dried foam, filler hydrogel, and electrospinning. **b** Functionally graded porous hip replacement implant model: (i) transformation of solid model to functionally graded porous implant, balancing mechanical strength (low porosity) and bio-permeability (high porosity) using Ti-6Al-4V and cell-laden GelMA. (ii) Influence of pore shape and porosity on mechanical properties, transitioning from stretching-dominated to bending-dominated behavior, and gradient porosity design strategy. (iii) Laser powder bed fusion (LPBF) process schematic, illustrating the gradient porosity design and the 3D printing approach for optimized mechanical and biological integration. Reprinted with permission from [83]. Copyright 2023 American Chemical Society

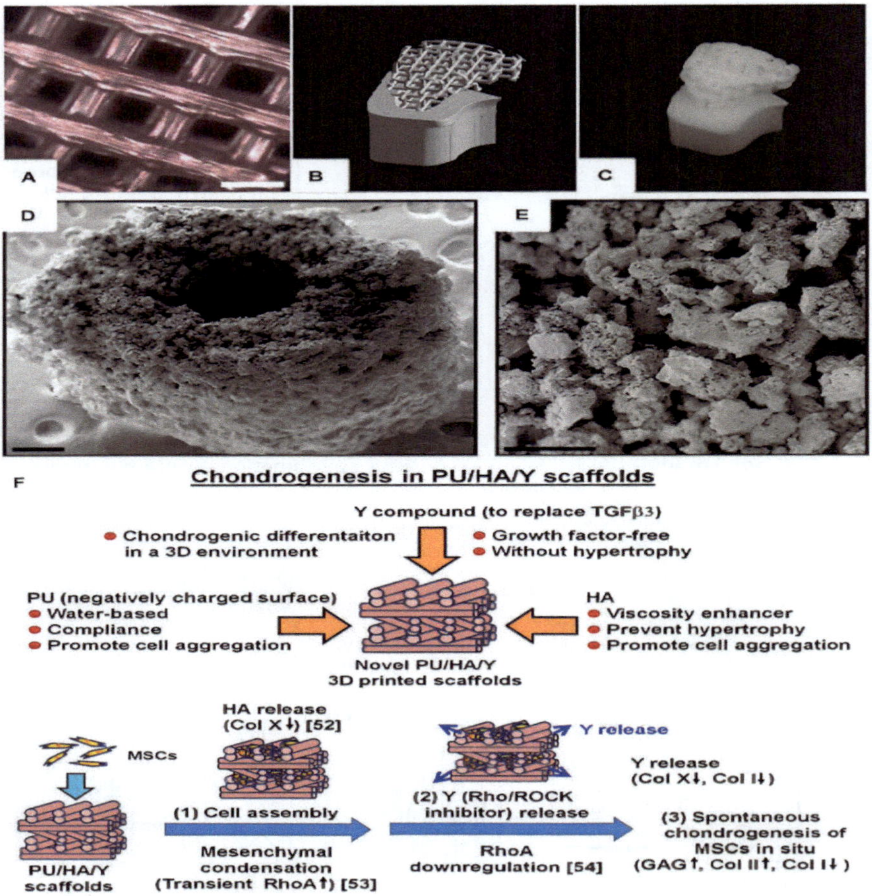

Fig. 8.7 **a** Visual exploration of a 3D printed scaffold for adipose tissue engineering made of poly(l-lactide-co-ε-caprolactone), as seen using an optical microscope (thanks to YJ, scale bar = 2 mm). **b** The complex artistry of 3D printing is shown by the world of porcine condyle scaffold as seen in an STL picture. **c** Honor the 3D-printed PCL scaffold's founding origins by seeing it from the front. **d** A murine-sized scaffold that has been painstakingly constructed for femoral mid-diaphysis regeneration can be seen in great detail using a scanning electron microscope (SEM) (scale bar = 250 mm). **e** The calcium phosphate-collagen composite scaffold's remarkable micro-porosity, with pore diameters ranging from 20 to 50 mm (scale bar = 100 mm), is revealed when we explore the microcosm. **f** The 3D printed PU/HA-based scaffold design brings an artistic endeavor to life and may open the door to spontaneous in situ chondrogenesis. Reproduced with permission from Ref. [61]. Copyright 2020 Frontiers in Bioengineering and Biotechnology

8.4 Cartilage Regeneration Using 3D-Printed Scaffolds: Challenges and Possible Future Directions

The diameters of nozzles, printheads, laser beams, and light sources, as well as the viscosity and rheology of materials, limit the resolution and accuracy of 3D printing. By improving these characteristics, it will be possible to create scaffolds with complex structures and features that closely mimic the structural subtleties and functions of natural cartilage tissue [15, 87]. In order to meet the needs of both printing and biological compatibility, the field of materials and bio-inks utilized in 3D printing requires a dual synergy. With increased biocompatibility, biodegradability, bioactivity, and mechanical robustness, the development of innovative materials and bio-inks fosters a jump in scaffold performance and functionality [88]. Multifunctional, hierarchically constructed platforms are the result of the skillful fusion of various materials, cells, and bioactive components inside 3D printed scaffolds. The environment created by this orchestration supports cartilage regeneration by finely influencing cellular dynamics. Printheads, a variety of printing modalities, or combining 3D printing with complementing processes like electrospinning or bioprinting may all help achieve this goal. The thorough evaluation of 3D printed scaffolds' long-term safety and efficacy in living creatures serves as the yardstick for determining their clinical viability. In addition to resolving ethical, legal, and financial issues, this crucial stage requires the incorporation of relevant animal models, standardized procedures, and reliable result metrics [77]. An Instigator of a Novel Era in Cartilage Restoration: There is no denying 3D printing's superiority as a scaffold manufacturing method in cartilage tissue engineering. With careful control over shape, porosity, and mechanical properties, this technology is capable of creating complex and customized structures. Beyond this, the real wonder is its ability to symphonize spatial distribution by integrating various materials, cells, and bioactive substances in a seamless manner. However, obstacles and constraints still exist, requiring a combination of 3D printing with complementary approaches, such as mechanical stimulation or gene and cell therapy. Cartilage tissue engineering hopes to accomplish both improved results and a quick transition to clinical reality via this symphony of innovation [89].

8.5 Cartilage Regeneration Using Bioactive Molecules: Mechanisms, Delivery Methods, and Uses

The biological functions of cells and tissues, including migration, differentiation, proliferation, and matrix production, may be influenced by bioactive compounds. Because they can promote the chondrogenic potential of different cell sources, improve the integration and repair of cartilage defects, and stop or reverse the degradation of cartilage tissue, bioactive chemicals may be employed for cartilage regeneration [90]. There are two types of bioactive molecule pathways for cartilage regeneration: direct and indirect. Through direct methods, some genes or proteins that control the chondrogenic phenotype and cell function are activated or inhibited by binding bioactive substances to certain receptors or signaling pathways on the cell membrane, in the cytoplasm, or in the nucleus. By altering the cartilage tissue's extracellular matrix or milieu, indirect methods may impact the tissue's mechanical characteristics, inflammatory state, nutrition transport, and cell-matrix interactions [91]. A symphony of cellular functions, from growth and survival to differentiation and migration, is orchestrated by pivotal proteins in the biological landscape. These influences can interact with certain cell surface receptors, which are often characterized as serine/threonine kinase or tyrosine receptors. Intricate intracellular signaling cascades that gracefully alter gene expression and protein synthesis are triggered by this union. Some growth factors operate as trailblazers in the context of cartilage regeneration, starting the process of chondrogenesis and the subsequent revitalization of cartilage tissue. Bone morphogenetic protein (BMP), fibroblast growth factor (FGF), insulin-like growth factor (IGF), platelet-derived growth factor (PDGF), transforming growth factor beta (TGF-β), and vascular endothelial growth factor (VEGF) are some of these trailblazers. Underpinning cartilage regeneration, these growth factors function as conductors, coordinating a symphony of cellular reactions [92]. Organic substances known as small molecules have the ability to alter the activity of signaling molecules, transcription factors, or enzymes involved in chondrogenesis and cartilage homeostasis. Targets in the cytoplasm or nucleus may interact with small compounds that can pass across the cell membrane. Vitamin C, rapamycin (RAPA), forskolin (FSK), kartogenin (KGN), and 5-azacytidine (5-AZA) are a few small molecules that have been shown to promote cartilage regeneration and chondrogenesis [93]. DNA segments called genes encode RNA or protein molecules that control various biological processes. Viral or non-viral vectors may be used to transfer genes into cells, which can lead to the overexpression or knockdown of certain genes involved in cartilage repair and chondrogenesis. Several genes, including Sox9, Runx2, Aggrecan, Collagen II, TGF-β, BMP-2, BMP-7, IGF-1, and VEGF, enhance chondrogenesis and cartilage regeneration [94].

Here are a few instances of bioactive compounds that work indirectly:

By reducing inflammation and joint pain, intervention's mainstays pave the way for alleviation. Inflammation plays a key role in the development of osteoarthritis (OA) and cartilage degeneration by triggering the production of pro-inflammatory cytokines, including matrix metalloproteinases (MMPs) and interleukin-1 beta (IL-1β) and tumor necrosis factor-alpha (TNF-α), which aid in the breakdown of cartilage matrix and prevent chondrocyte anabolism. Certain substances are important because they are effective in inhibiting the production or effects of certain inflammatory mediators, which promotes cartilage health and function. Non-steroidal anti-inflammatory medications (NSAIDs), corticosteroids, hyaluronic acid (HA), curcumin, resveratrol, and green tea polyphenols are among the notables that stand out among these substances. In essence, these substances provide a calculated defense against the damaging consequences of inflammation, working to preserve cartilage tissue health. Chondroprotective substances promote cartilage healing and shield it from harm. By altering the extracellular matrix of cartilage, chondroprotective drugs may improve its mechanical characteristics and resistance to deterioration. Additionally, chondrocytes or stem cells might be encouraged to synthesize or retain cartilage matrix components by chondroprotective agents. Glucosamine, chondroitin sulfate, collagen hydrolysate, HA, pentosan polysulfate (PPS), diacerein, avocado/soybean unsaponifiable (ASU), and strontium ranelate are a few chondroprotective compounds that have been employed for cartilage regeneration [95, 96].

8.6 Fabrication Process

The precise orchestration of scaffold construction, in which layers of biocompatible materials effortlessly weave together, opens the prologue. Similar to an artist's canvas, this scaffold creates the framework on which the next artwork will be built. As the story progresses, attention shifts to the scaffold's assimilation of cells that give it life. Cells are arranged deliberately, each playing a crucial role in the harmonic symphony of tissue regeneration, much like notes on a musical score. The story then switches to the infusion of medications and bioactive compounds that are carefully woven into the scaffold's fabric. These chemicals provide the scaffold the ability to coordinate biological reactions and healing processes, just as catalysts do. The story smoothly shifts to the incorporation of biomolecules when the scaffold is loaded with bioactive compounds. The scaffold gains the capacity to interact and direct cellular behavior via the interlacing of these molecular protagonists, which range from growth factors to signaling molecules. Scaffolds, cells, medications, and biomolecules all converge in a dynamic dance of interaction during the crescendo of creation. Each element plays a distinct part in the symphony, which comes together to form an ensemble that coordinates tissue regeneration. This display of scaffold-cell, drug, and biomolecule complexes comes together to form a cohesive whole that is more important than the sum of its parts. Under the direction of bioactive signals, this scaffold directs the complex composition of cellular connections, much as a conductor leads an orchestra Fig. 8.8 [61, 71].

8.7 Methods for Delivering Bioactive Compounds to Promote Cartilage Regeneration

Enhancing the bioactive molecules' localization, retention, and activity at the target region is the goal of the delivery methods for cartilage regeneration. Passive and active delivery systems are the two categories into which the delivery systems fall.

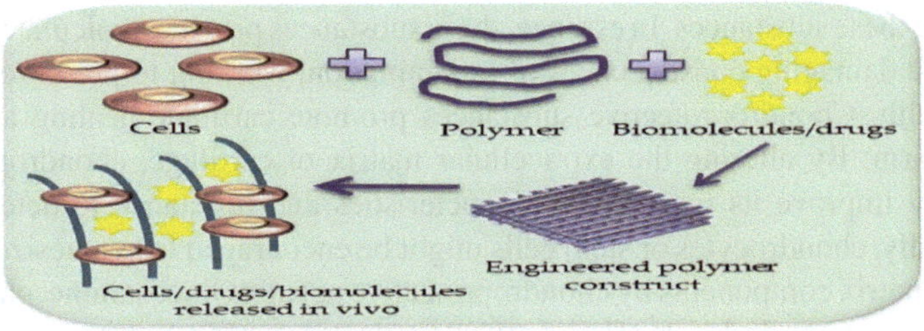

Fig. 8.8 Is a powerful illustration that reveals the fascinating process of scaffold-cell/drug/biomolecule assembly by removing the layers. This visual story invites us to see the complex dance of creation that takes place at the nexus of engineering, biology, and invention, going beyond simple illustration. This illustration perfectly depicts the process of creation, in which scaffold, cells, medications, and biomolecules come together in a complex dance to compose the narrative of regeneration. It challenges us to go beyond the picture and explore the intricate web of mechanisms that underpin tissue engineering and regenerative medicine [71]

The diffusion or convection of the bioactive chemicals from the delivery vehicle to the cartilage tissue is the basis of passive delivery methods. On the other hand, active delivery systems react to internal or external stimuli by releasing or activating the bioactive chemicals. The simplest and most popular way to distribute bioactive molecules for cartilage regeneration is by injections. Injections are the direct administration of liquid or gel bioactive substances into the cartilage defect or joint cavity. Rapid clearance, poor retention, systemic adverse effects, and the need for repeated injections are some of the disadvantages of injections, despite their ability to deliver large amounts of bioactive molecules rapidly. Scaffolds are complex 3D porous structures that play a crucial role in providing three essential roles: mechanical support, spatial orientation, and biochemical cues for a variety of cellular processes, including attachment, proliferation, differentiation, and matrix assembly. In addition to their architectural function, scaffolds have the ability to operate as bioactive agent reservoirs, facilitating their progressive release into the cartilage void. These scaffolds may be made from a variety of materials, including synthetic and natural ones, or even a tasteful blend of the two. The scaffold's potential is shaped in this area by characteristics including mechanical durability, porosity, biodegradability, and biocompatibility. Additionally, by adding growth factors, drugs, or genes, the scaffolds may be made to represent functionality, enhancing the coordination of biological processes in both the cells and the scaffold [97, 98].

8.7.1 Nanoparticles and Microparticles

Bioactive compounds may be encapsulated and shielded against deterioration or deactivation by micro- and nanoparticles. Additionally, the solubility, stability, and bioavailability of bioactive compounds may be improved by micro- and nanoparticles. For cartilage regeneration, micro- and nanoparticles may be added to scaffolds or injected into the joint cavity. Natural or artificial materials, including metals, polymers, lipids, ceramics, and bioglass, may be used to create micro- and nanoparticles of various sizes, shapes, and surface characteristics [99, 100].

8.7.2 Novel Materials

Temperature, light, pH, electric fields, and magnetic fields are examples of internal or external stimuli that may affect intelligent materials [101–103]. Novel materials have the ability to dynamically modify their characteristics in response to external signals or changes in the physiological environment [104–107]. The efficiency and specificity of bioactive compounds may be improved by using novel materials to regulate their release or activation in a spatiotemporal way. Thermosensitive, pH-sensitive, light-sensitive, electrically-sensitive, and magnetism-sensitive materials are a few types of novel materials [108].

8.7.3 Biomimetic Substances

At several levels, including the molecular, cellular, and tissue levels, biomimetic materials are made to resemble the composition and functionality of actual cartilage tissue. By offering particular signals for cell attachment, proliferation, differentiation, and matrix deposition, biomimetic materials might improve the biological performance of the synthetic cartilage. Collagen-, hyaluronic acid-, chitosan-, and silk-based materials are a few types of biomimetic materials [109].

8.7.4 Orthobiology

Orthobiologics are biological agents that have the ability to regulate musculoskeletal tissue regeneration and repair. Cells, platelet-rich plasma (PRP), and extracellular vesicles (EVs) are examples of orthobiologics. Cells may produce bioactive compounds in situ, acting as "factories" for the delivery of drugs. PRP is a concentrated platelet solution that aids in cartilage healing by incorporating a variety of cytokines and growth factors. Proteins, lipids, and nucleic acids make up membrane-bound vesicles called EVs, which mediate intercellular communication and alter cellular processes [110].

8.8 Bioactive Molecule Applications for Cartilage Regeneration

There are two types of bioactive molecule applications for tissue regeneration include restorative and regenerative applications [111–113]. Through the reduction of inflammation, discomfort, and matrix deterioration, restorative treatments seek to maintain or repair the structure and function of injured or degenerated tissues [12, 114]. By promoting cell proliferation, differentiation, migration, and matrix formation, regenerative applications seek to produce new cartilage tissue. By introducing chondroprotective or anti-inflammatory substances into the joint cavity, intra-articular injections help to relieve the symptoms of osteoarthritis. Corticosteroids, HA, glucosamine, chondroitin sulfate, curcumin, resveratrol, green tea polyphenols, PRP, and EVs are a few types of intra-articular injections. Although intra-articular injections might reduce pain and inflammation momentarily, there are dangers associated with them, including the possibility of infection, hemorrhage, or cartilage injury. Topical treatments use creams, gels, patches, or sprays to apply chondroprotective or anti-inflammatory substances to the joint surface. Although topical treatments have minimal penetration and bioavailability, they may prevent certain systemic adverse effects of oral or injectable medications. NSAIDs, capsaicin, menthol, camphor, and lidocaine are a few examples of topical treatments [115]. Scaffold-based methods use 3D porous constructs that provide mechanical support and biochemical signals to transfer cells and bioactive substances to the cartilage defect. Scaffold-based methods may improve the repair and integration of the defect with the host tissue by fostering an environment that is conducive to cell proliferation and differentiation. Autologous chondrocyte implantation (ACI), matrix-assisted chondrocyte implantation (MACI), and mesenchymal stem cell (MSC)-based treatments are a few instances of scaffold-based approaches [90, 116]. Figure 8.9 offers a thorough explanation of the complex electrospinning process via a well crafted schematic. Polymers are delicately woven into beautifully spun textiles using this method. Explore the many levels of this technology and discover the underlying mechanics and relationships that lead to its amazing results as you go on a visual voyage [71].

8.8.1 Techniques Without Scaffolding

Scaffold-free methods use self-assembled 3D structures without the need for external scaffolds to transfer cells and bioactive compounds to the cartilage lesion. Some of the drawbacks of scaffold-based approaches, including immunogenicity, infection, inflammation, and poor integration, may be addressed by scaffold-free approaches. Additionally, scaffold-free methods may create cartilage-like tissues with a high cell density and uniform distribution while maintaining the natural cell-cell and cell-matrix connections. Cell sheet engineering, spheroid culture, and pellet culture are a few instances of scaffold-free methods [117].

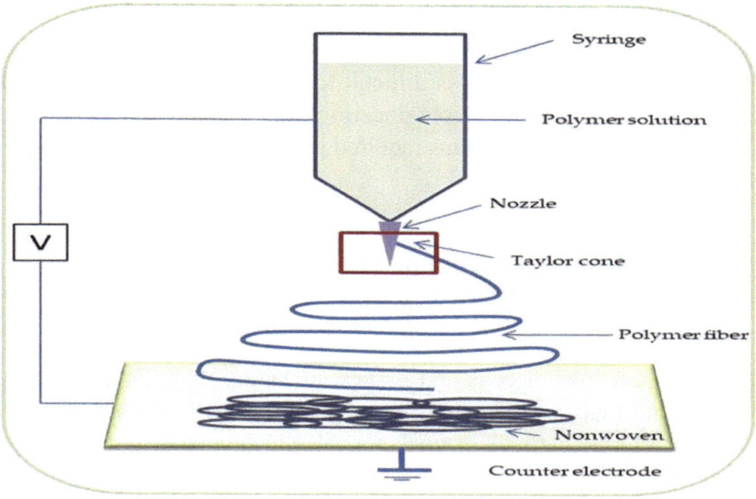

Fig. 8.9 A meticulously designed schematic, offers a comprehensive understanding of the electrospinning method, a complex procedure that transforms polymers into finely spun fibers [71]

8.8.2 Methods Without Cells

Using biomaterials or biological factors that cause endogenous cells to migrate and differentiate into the cartilage defect site, cell-free approaches are employed to deliver bioactive chemicals either alone or in combination to the defect site. Certain issues with cell-based approaches, such cell collection, growth, delivery, viability, and phenotypic stability, may be circumvented by cell-free approaches. The expense, complexity, and regulatory barriers associated with cartilage tissue creation may also be decreased by using cell-free methods. Microfracture, the osteochondral autograft transfer system (OATS), and platelet-rich plasma (PRP) injection are a few instances of cell-free methods [118].

8.9 Innovative Methods for Improving Mesenchymal Stem Cells' Chondrogenic Differentiation in Cartilage Tissue Engineering

When mesenchymal stem cells (MSCs) undergo chondrogenic differentiation, they develop into chondrocytes, which are the main cell type that make up cartilage tissue. For cartilage tissue engineering, which attempts to produce functional cartilage-like tissues for repairing or rebuilding damaged or deteriorated cartilage, chondrogenic differentiation of MSCs is crucial. However, since MSCs often lose their multipotency and experience hypertrophy or dedifferentiation in vitro or in vivo, chondrogenic differentiation of MSCs is difficult to attain and sustain. To enhance the results and uses of cartilage tissue engineering, new methods for promoting MSC chondrogenic differentiation are thus required [119].

8.9.1 Newly Discovered Methods to Improve MSC Chondrogenic Differentiation

The biological functions of cells and tissues, including migration, differentiation, proliferation, and matrix production, may be influenced by bioactive substances [6–10, 86]. By offering particular signals for cell attachment, proliferation, differentiation, and matrix deposition, bioactive substances may promote the chondrogenic differentiation of MSCs. Growth factors (like TGF-β, BMP, FGF, IGF, PDGF, and VEGF), small molecules (like KGN, RAPA, FSK, 5-AZA, and vitamin C), genes (like Sox9, Runx2, Aggrecan, Collagen II, TGF-β, BMP-2, BMP-7, IGF-1, and VEGF), anti-inflammatory agents (like NSAIDs, corticosteroids, HA, curcumin, resveratrol, and green tea polyphenols), and chondroprotective agents (like glucosamine, chondroitin sulfate, collagen hydrolysate, HA, PPS, diacerein, ASU, and strontium ranelate) are a few examples of bioactive factors that have been demonstrated to support chondrogenesis and cartilage regeneration. For MSC chondrogenic development, bioactive factor optimization include choosing the best kind, mix, dose, timing, and delivery method of bioactive factors [120]. In vitro or in vivo, cell behavior and function are influenced by the physical and chemical environment. Culture conditions that provide appropriate parameters for cell development and differentiation may improve MSC chondrogenic differentiation. Osmolarity (high or low), temperature (high or low), mechanical stimulation (static or dynamic), electrical stimulation (direct or alternating current), magnetic stimulation (static or dynamic field), light stimulation (visible or infrared light), oxygen tension (hypoxia or normoxia), pH (acidic or alkaline), and bioreactor systems (spinner flask, rotary wall vessel, perfusion chamber, or hollow fiber bioreactor) are a few examples of culture conditions that have been demonstrated to affect chondrogenesis and cartilage regeneration. The best level, duration, frequency, and order of culture factors for MSC chondrogenic development are all adjusted as part of the optimization of culture conditions. Choosing the best cell source for cartilage tissue engineering is known as cell type selection optimization. The quality and chondrogenic differentiation capability of MSCs may be impacted by cell type selection. Regarding availability, number, quality, multipotency, immunogenicity, and ethical considerations, each tissue-derived MSC has pros and cons. Comparing the traits and capabilities of various tissue-derived MSCs for chondrogenic differentiation and cartilage regeneration is part of the optimization of cell type selection [121]. The process of cultivating two or more cell types in vitro or in vivo is known as coculture system optimization. Through the synergistic effects of several cell types, coculture systems may improve MSC chondrogenic development. Coculture systems may also help heal and integrate cartilage abnormalities by simulating the natural cartilage microenvironment. MSCs including chondrocytes, endothelial cells, macrophages, platelets, and extracellular vesicles are a few types of coculture systems that are used in cartilage tissue engineering. In order to optimize coculture systems for chondrogenic differentiation and cartilage regeneration, the best cell type, ratio, arrangement, and interaction style must be chosen [120]. The six steps in this procedure serve as the main themes in the scaffold manufacturing symphony:

Fabrication Method: The symphonic overture starts with the electrospinning of complex polymer threads onto a revolving drum (I). The careful removal of cylinders from the electrospun mesh, which forms the scaffold's surface layer (II), is the next step in the ethereal dance of creation. The careful removal of cylinders from the electrospun mesh, which forms the scaffold's surface layer (II), is the next step in this ethereal dance of creation. The story then shifts to a little rectangle that is tastefully weaved into the GO-collagen hydrogel (III), revealing the harmonious integration of various components. Each thread is a crucial component of the scaffold's structure as the middle layer is delicately spun into life within an ethanol-water bath (IV). As the layers are flawlessly put together and bonded by a freezing technique (V), the story builds to a crescendo. A stunning three-layered electrospun scaffold is created as a result of this dramatic conclusion, demonstrating the painstaking artistry required. An anisotropic 3D electrospun structure is incorporated into the GO-collagen hydrogel as the process's culmination, directing the development of a dynamic and interconnected microporous network that envelops and supports the scaffold's complex architecture following freeze-drying.

Alternate Geometric Buildings: As the story progresses, more intricate symphonic compositions are revealed via alternate geometric buildings. Geometry 2, a combination of the superficial and middle levels that represents an early investigation in assembly (superficial + middle layers), appears as it moves from right to left. Then, Geometry 3 takes the stage, bringing the deep and superficial layers together in a way that indicates more intricate scaffold construction (deep + surface layers). Lastly, Geometry 4 shows how the middle and deep levels interact to create a beautiful tapestry of shape and function (middle + deep layers). The process of altering a cell's genome by adding, removing, or swapping out certain DNA sequences is known as gene editing optimization. By overexpressing or shutting down certain genes that control chondrogenesis and cartilage homeostasis, gene editing approaches may improve the chondrogenic differentiation of MSCs. Additionally, genetic disorders or flaws that impact cartilage function may be prevented or corrected via the use of gene editing methods. Base editors, zinc finger nucleases (ZFNs), transcription activator-like effector nucleases (TALENs), and clustered regularly interspaced short palindromic repeats (CRISPR)/CRISPR-associated protein 9 (Cas9) are a few gene editing methods that have been applied to cartilage tissue engineering. For chondrogenic differentiation and cartilage regeneration, gene editing approach optimization include choosing the best target gene, vector, delivery method, and gene editing efficiency [61, 122]. Three-dimensional (3D) porous scaffolds provide biochemical signals for cell adhesion, proliferation, differentiation, and matrix deposition in addition to mechanical support and spatial guidance. Additionally, bioactive compounds may be carried by scaffolds and progressively released into the cartilage defect. Scaffolds have a variety of qualities, including mechanical strength, porosity, biodegradability, and biocompatibility, and may be composed of natural or synthetic materials, or a mix of the two. To increase the biological activity of the cells and the scaffolds, scaffolds may also be functionalized with growth factors, medications, or genes. Scaffolds based on collagen, hyaluronic acid, chitosan, alginate, silk, polyethylene glycol, poly(lactic-co-glycolic acid), and polyurethane are a few examples of scaffolds that have been used for cartilage tissue engineering. In order to optimize scaffolds for chondrogenic differentiation and cartilage regeneration, the best possible material type, composition, structure, and functionalization must be chosen [122, 123]. In cartilage tissue engineering, these tactics may be used alone or in combination to promote MSC chondrogenic development. But these approaches also have drawbacks and restrictions that must be addressed, including expense, intricacy, scalability, repeatability, standardization, regulation, and translation. The performance and functionality of cartilage tissue engineering may be enhanced by new methods for boosting MSC chondrogenic differentiation by tackling these issues and taking use of these possibilities. The revolutionary potential of bioprinting has sparked a revolution in the constantly changing fields of tissue engineering and regenerative medicine [122].

8.10 The Pattern of Current Research

The emerging technologies of computer-aided design (CAD) and 3D printing play a revolutionary role in tissue engineering. These technologies mark the beginning of a new age in regenerative medicine by making it possible to fabricate biomimetic and custom-fit scaffolds. These scaffolds have the ability to smoothly combine bioactive compounds, cells, and biomaterials, coming together to enhance the healing of diseased or injured tissues, particularly cartilage. In joints, cartilage, an essential connective tissue, plays a crucial role in providing cushioning and lubrication. But the regret is that it is inherently incapable of significant self-healing. Irreparable harm may be caused by age, trauma, or conditions like osteoarthritis. As a result, restoring cartilage requires the coordinated delivery of the right signals and cells to the damaged area. In light of this, research is increasingly focusing on using CAD and 3D printing to create passive and stimuli-responsive drug delivery scaffolds for cartilage regeneration. This advanced kind of scaffold has the ability to deliver bioactive compounds in a controlled temporospatial manner, which may improve cellular reactions and reduce the risk of infection. The advancement also includes the simultaneous co-printing of biomaterials and cells, creating dynamic living structures that blend in perfectly with the host tissue. A thorough analysis explores drug delivery scaffolds for tissue regeneration and cell guidance that are shaped using computer-aided design and manufacturing. The effect of incorporating bioactive molecules into three-dimensional porous scaffolds on cellular fates and tissue formation is carefully shown by this investigation. Interestingly, it discusses how bioprinting may be used to create stimuli-responsive drug delivery scaffolds, which can be used in tissue engineering and the complex field of cancer precision medicine [124].

8.11 Perspective and Direction for the Future

The smooth integration of computer-aided design (CAD) and additive manufacturing (AM) methods is necessary to promote cartilage regeneration in tissue engineering. This offers a revolutionary platform for innovation and progress. The convergence of these methods, motivated by a multidisciplinary approach that combines engineering principles with biological insights, provides promise for future research paths in addressing the intricate problems associated with cartilage repair and regeneration. The future of cartilage tissue engineering depends on the creation of scaffold designs that accurately mimic the intricate structure and molecular cues of the natural extracellular matrix (ECM). Researchers are investigating innovative scaffold modification techniques with a focus on enhancing biocompatibility, structural integrity, and bioactivity utilizing advanced CAD tools. These cutting-edge scaffolds use biomimicry principles to enable successful tissue regeneration by creating an environment that is conducive to cell adhesion, proliferation, and differentiation. The creation of multifunctional platforms that can distribute bioactive substances, such growth factors, in a spatiotemporally controlled way is a key topic of study in scaffold design. There is great potential for regulating cellular activity and directing tissue formation via this intricate relationship between scaffold design and bioactive payload. Future studies will probably concentrate on improving encapsulation and release methods to maximize therapeutic effectiveness while preventing side effects. By adapting the delivery kinetics of bioactive molecules to the changing requirements of tissue regeneration, researchers aim to pave the way for future developments in precision medicine and regenerative therapies. A paradigm change in tissue engineering is brought about by the emergence of personalized medicine, which offers previously unheard-of opportunities for individualized treatment strategies. By adding patient-specific data, such as genetic information and medical imaging scans, to CAD platforms, researchers may customize scaffold design and manufacturing parameters to each patient's unique anatomical and physiological features. This patient-centered approach lays the foundation for truly customized regenerative treatments by improving treatment results and reducing the risk of immunological rejection and transplant failure. A major force behind the creation of new polymers, hydrogels, and composite materials with adjustable characteristics is biomaterial innovation in cartilage tissue engineering. To meet the many requirements of tissue regeneration, these advanced biomaterials are made to have desired qualities including mechanical strength, bioactivity, and biodegradability. Through careful material design and characterization, researchers hope to enhance scaffold performance and promote smooth integration with the host tissue, enabling tissue homeostasis and long-term functional restoration. The selection and optimization of cell sources for integration into the scaffold matrix is a crucial aspect of scaffold design. Because of their remarkable ability to regenerate, stem cells have become a promising option for enhancing tissue growth and repair. In order to enhance stem cell survival, proliferation, and differentiation inside the scaffold milieu, future research will focus on comprehending the intricate interactions among scaffold architecture, biomaterial characteristics, and cell activity. By understanding the complex processes behind stem cell fate determination, researchers hope to take advantage of its therapeutic potential for the regeneration of damaged cartilage tissues. Bridging the translational gap from bench to bedside is becoming more and more crucial as AM and CAD technologies develop. Strict attention to safety evaluations, quality control procedures, and regulatory standards is necessary for the effective clinical translation of novel scaffold designs. Before tissue-engineered structures are used in clinical settings, thorough preclinical research is necessary to confirm their safety and effectiveness. This includes both in vitro characterization and in vivo validation. Furthermore, navigating the complex world of medical device regulation and upholding strict quality standards need close cooperation between researchers, physicians, and regulatory agencies. The use of AM and CAD methods holds promise for treating more complicated tissue disorders such osteochondral lesions and osteoarthritis, in addition to isolated cartilage abnormalities. By creating integrated scaffolds that may promote cartilage and subchondral bone regeneration, researchers hope to provide holistic therapies for degenerative joint diseases. In addition to addressing the variety of musculoskeletal issues, this all-encompassing approach to tissue engineering emphasizes how regenerative medicine has the potential to transform orthopedic treatment. In conclusion, the integration of AM and CAD techniques represents a paradigm shift in cartilage tissue engineering, offering hitherto unheard-of opportunities for creativity and progress. Researchers aim to push the limits of scaffold design, biomaterial innovation, and regenerative therapies by using the synergistic potential of these technologies. Multidisciplinary collaboration and an unrelenting drive for excellence have enabled the search for functional cartilage regeneration to advance, offering patients worldwide improved clinical results and a greater standard of living.

8.12 Expert Viewpoint: Promoting Cartilage Tissue Engineering via Additive Manufacturing

The use of Additive Manufacturing (AM) in tissue engineering, particularly in cartilage regeneration, has the potential to revolutionize actual medical results. It is becoming more and more clear that AM technologies have the potential to completely transform medication delivery, therapeutic effectiveness, and patient care in general. Effective tissue regeneration depends on the construction of constructs that closely resemble natural cartilage, which is made possible by AM's precise control over scaffold design. In addition to offering patients better functional results, this capacity may hasten healing and lessen the need for follow-up treatments. Additionally, targeted, sustained drug administration is made easier by the incorporation of bioactive molecules into these scaffolds, which reduces systemic adverse effects and maximizes therapeutic benefit. For example, adding growth factors like TGF-β to scaffolds has shown potential in improving the formation and quality of cartilage tissue, which may result in better therapies for degenerative diseases and cartilage damage. Nevertheless, a number of obstacles need to be overcome before AM may be completely incorporated into clinical practice, even with these encouraging advancements. The problem of vascularization in synthetic tissues is one of the main obstacles. Although the surrounding bone and other tissues need a blood supply to operate, cartilage is avascular in contrast to many other tissues. The survival and integration of these synthetic tissues after implantation are restricted by the inability of current AM methods to form functioning circulatory networks inside massive tissue constructions. Larger structures are more likely to fail if there is inadequate vascularization since there will be less oxygen and nutrients available. For AM-engineered tissues to remain viable in clinical settings over the long term, this is a major obstacle that has to be removed. The standardization of bioprinting procedures is another significant obstacle. It is challenging to establish universal procedures for bioprinting because to the wide range of bioinks, patient-specific variances, and experimental circumstances. This delays the translation of research into practical applications and makes repeatability problematic. It will be crucial to standardize these procedures to guarantee the safe and dependable use of AM technology in medical practice. Another major obstacle to the broad use of AM in clinical settings is material restrictions. Even while biocompatible materials for AM procedures have advanced significantly, many of these materials still lack important properties including biocompatibility, mechanical strength, and degradation rates. For example, the mechanical strength required for load-bearing applications is lacking in hydrogels, which are often utilized to replicate the natural cartilage environment. However, although bioactive ceramics and other materials provide the requisite strength, their degradation rate is too slow to be beneficial for cartilage regeneration. The advancement of AM in tissue engineering will depend on the creation of novel materials or hybrid strategies that integrate the advantages of several materials. Furthermore, the implementation of AM procedures in clinical practice is severely hampered by their high cost and complexity. Many medical institutions may find AM prohibitive due to the high cost of the equipment required and the specialist knowledge required for the procedures. To make AM technologies more affordable and accessible, further research and development is required. This might include creating equipment that is easier to use or streamlining the design and production processes. In the next ten years, the field of AM in tissue engineering is probably going to undergo major change. Enhancing the efficacy of AM-engineered tissues is anticipated to be largely dependent on developments in biomimicry, such as the creation of stimuli-responsive materials that may release medications in response to certain triggers. Furthermore, combining AM with microfluidics and soft lithography may result in even more precise control over scaffold design and medication administration, enhancing the accuracy and efficacy of these therapies. The complexity and expense of tissue engineering processes might rise as a result of these developments, however, which would delay their use in clinical settings. Notwithstanding these difficulties, AM has the potential to improve patient outcomes, making it a worthwhile subject to pursue. In order to improve the quality of life for patients with cartilage-related disorders and lessen the need for intrusive procedures, the ultimate objective is to create fully functioning, patient-specific tissues that can replace damaged cartilage and restore joint functionality. In conclusion, there are a number of noteworthy obstacles in the area of AM in cartilage tissue engineering, but there are also a lot of potential advantages. The effective integration of AM technologies into conventional clinical practice will depend on overcoming the obstacles associated with vascularization, uniformity, material restrictions, and cost. Realizing AM's full potential in tissue engineering will need ongoing multidisciplinary research and innovation, which will eventually improve patient outcomes and therapies. It will be fascinating to see how these technologies advance and how they may be incorporated into clinical practice to enhance patient care as the area develops further.

8.13 Conclusion

An academic paper by Santos et al. in the journal Virtual and Physical Prototyping discusses additive manufacturing methods for scaffold-based cartilage tissue engineering. The limits of conventional approaches for producing cartilage tissue are examined in this chapter, along with the potential applications of additive manufacturing (AM) techniques to get around them. The authors provide a thorough rundown of many AM technologies, such as inkjet printing, extrusion-based procedures, powder bed fusion, and photo-polymerization, and explain how each may be used to produce scaffolds for cartilage tissue engineering. The benefits of employing AM for cartilage regeneration are highlighted in the chapter conclusion, which also indicates that this technique has a lot of potential for creating efficient therapies for cartilage damage. A potential area of regenerative medicine is cartilage tissue engineering, which attempts to produce tissues that behave similarly to cartilage in order to replace or repair damaged or deteriorated cartilage. Nevertheless, creating scaffolds that closely resemble the complex structure and molecular makeup of cartilage's natural extracellular matrix (ECM) is one of the field's major problems. When combined with computer-aided design (CAD) tools, additive manufacturing (AM) or 3D printing processes have become very effective instruments for creating intricate, highly customizable scaffolds with exact control over their characteristics. The benefits, drawbacks, and uses of AM and CAD methods for designing and manufacturing drug-delivery scaffolds for cartilage regeneration have all been covered in this chapter study. It has investigated several AM and CAD scaffold materials, such as hydrogels, composite materials, and natural and synthetic polymers. It emphasized their advantages for tissue engineering, such as mechanical strength, porosity, bioactivity, biodegradability, biocompatibility, and processability. Additionally, it talks about how bioactive compounds may be added to AM and CAD scaffolds to promote tissue regeneration and cell activity. A wide range of bioactive molecule types, including cytokines, growth factors, small molecules, anti-inflammatory, and chondroprotective compounds, have been evaluated. It described how they may alter cellular functions such matrix production, migration, differentiation, and proliferation. Additionally, it looked at ways to improve tissue formation by integrating various cell sources into AM and CAD scaffolds. It has examined the potential, accessibility, and suitability of many cell sources for cartilage tissue engineering, including stem cells, primary cells, and cell lines. Additionally, it has shed light on the latest uses of AM and CAD methods in cartilage tissue engineering, with a particular emphasis on three key areas: osteoarthritis treatment, intervertebral disc regeneration, and osteochondral defect repair. It has shown how AM and CAD scaffolds may help with these clinical issues by promoting chondrogenic differentiation, improving the production of extracellular matrix, and facilitating tissue integration. It has also covered issues like scalability, standardization, legislation, and ethics that provide potential and problems in bringing innovative methods from the bench to the bedside. Nonetheless, there are still a number of shortcomings and holes in this subject. More research is needed to improve scaffold design and manufacturing procedures, create accurate medication distribution plans, and combine patient-specific data with computer modeling. Furthermore, further testing in bigger animal models and clinical studies are required to assess the long-term safety and effectiveness of these 3D-printed scaffolds. Future studies should focus on resolving these problems and refining these methods for regenerative medicine and cartilage tissue engineering. In conclusion, there is a great deal of promise for developing the area of regenerative medicine via the use of AM and CAD methods in tissue engineering, especially for cartilage regeneration. New approaches to treating osteoarthritis and cartilage abnormalities are made possible by the development of biomimetic scaffolds with customized characteristics and drug delivery capabilities, which gives patients hope for better clinical results and an improved quality of life. Undoubtedly, further research and creativity will help create new treatments and approaches for regenerative medicine and cartilage tissue engineering.

References

1. B. Mikaeeli Kangarshahi, S.M. Naghib, N. Rabiee, 3D Printing and computer-aided design techniques for drug delivery scaffolds in tissue engineering. Expert Opin. Drug Deliv. **21**(11), 1615–1636 (2024)
2. T. Jafari, S.M. Naghib, M. Rahmanian, M.R. Mozafari, 3D Printing of self-healing materials for drug delivery applications: promises, advances and outlooks. Bioprinting **42**, e00353 (2024). https://doi.org/10.1016/j.bprint.2024.e00353

3. S.M. Naghib, M. Zarrineh, M.R. Mozafari, 3D Printing chitosan-based nanobiomaterials for biomedicine and drug delivery: recent advances on the promising bioactive agents and technologies. Curr. Org. Chem. **28**(7), 510–525 (2024)
4. K. Khodaverdi, S.M. Naghib, M. Mozafari, Chitosan/Bioglass nanocomposites for bone tissue engineering and regenerative medicine: an overview of promising biomaterials. Curr. Org. Chem. **28**(18), 1437–1451 (2024)
5. H.S. Ghazali, E. Askari, Z.S. Ghazali, S.M. Naghib, T. Braschler, Lithography-based 3D printed hydrogels: from bioresin designing to biomedical application. Colloid Interface Sci. Commun. **50**, 100667 (2022). https://doi.org/10.1016/j.colcom.2022.100667
6. R. Gholami, S.M. Naghib, Bone extracellular matrix, in *Nanobioceramics for Bone Tissue Engineering and Regenerative Biomedicine: Tunable Biological Characteristics* (Springer Nature, Singapore, 2025), pp. 79–105. https://doi.org/10.1007/978-981-96-0041-0_4
7. R. Gholami, S.M. Naghib, Bone cells, in *Nanobioceramics for Bone Tissue Engineering and Regenerative Biomedicine: Tunable Biological Characteristics* (Springer Nature, Singapore, 2025), pp. 43–77. https://doi.org/10.1007/978-981-96-0041-0_3
8. R. Gholami, S.M. Naghib, Bone ECM proteins—Part II, in *Nanobioceramics for Bone Tissue Engineering and Regenerative Biomedicine: Tunable Biological Characteristics* (Springer Nature, Singapore, 2025), pp. 149–175. https://doi.org/10.1007/978-981-96-0041-0_6
9. R. Gholami, S.M. Naghib, Bone ECM proteins—Part I, in *Nanobioceramics for Bone Tissue Engineering and Regenerative Biomedicine: Tunable Biological Characteristics* (Springer Nature, Singapore, 2025), pp. 107–148. https://doi.org/10.1007/978-981-96-0041-0_5
10. R. Gholami, S.M. Naghib, Bioactivity and osteogenic features, in *Nanobioceramics for Bone Tissue Engineering and Regenerative Biomedicine: Tunable Biological Characteristics*, ed. by R. Gholami, S.M. Naghib (Springer Nature, Singapore, 2025), pp. 177–215. https://doi.org/10.1007/978-981-96-0041-0_7
11. R. Gholami, S.M. Naghib, Nano-bioceramics, in *Nanobioceramics for Bone Tissue Engineering and Regenerative Biomedicine: Tunable Biological Characteristics* (Springer Nature, Singapore, 2025), pp. 217–249. https://doi.org/10.1007/978-981-96-0041-0_8
12. R. Gholami, S.M. Naghib, *Nanobioceramics for Bone Tissue Engineering and Regenerative Biomedicine*
13. Y. Zhu, 3D Printing biocompatible polymers for biomedical applications, Arizona State University (2024)
14. H.S. Ghazali, E. Askari, Z.S. Ghazali, S.M. Naghib, T. Braschler, Colloid and interface science communications
15. B. Mikaeeli Kangarshahi, S.M. Naghib, Multicomponent 3D-printed collagen-based scaffolds for cartilage regeneration: recent progress, developments, and emerging technologies. Curr. Org. Chem. **28**(8), 576–594 (2024). https://doi.org/10.2174/011385272829343724022 7065230
16. M. Aghajani, H.R. Garshasbi, S.M. Naghib, M. Mozafari, 3D Printing of hydrogel polysaccharides for biomedical applications: a review. Biomedicines **13**(3), 731 (2025)
17. H.S. Ghazali, E. Askari, A. Seyfoori, S.M. Naghib, A high-absorbance water-soluble photoinitiator nanoparticle for hydrogel 3D printing: synthesis, characterization and in vitro cytotoxicity study. Sci. Rep. **13**(1), 8577 (2023). https://doi.org/10.1038/s41598-023-35865-3
18. K. Mohammad-Jafari, S.M. Naghib, 3D Printing of microfluidic-assisted liposomes production for drug delivery and nanobiomedicine: a review. Curr. Med. Chem. (2024). https://doi.org/10.2174/0109298673285199231210170549
19. S.M. Naghib, S.N. Hosseini, A. Beigi, 3D 4D Printing of chitosan-based scaffolds for wound dressing applications. Carbohydr. Polym. Technol. Appl. **8**, 100594 (2024). https://doi.org/10.1016/j.carpta.2024.100594
20. S. Soleymani, S.M. Naghib, 3D and 4D printing hydroxyapatite-based scaffolds for bone tissue engineering and regeneration. Heliyon **9**(9) (2023). https://doi.org/10.1016/j.heliyon.2023.e19363
21. S.M. Naghib, S. Amiri, M. Mozafari, Stimuli-responsive chitosan-based nanocarriers for drug delivery in wound dressing applications: a review. Carbohydr. Polym. Technol. Appl. 100497 (2024)

22. Y. Cao, P. Cheng, S. Sang, C. Xiang, Y. An, X. Wei, Y. Yan, P. Li, 3D printed PCL/GelMA biphasic scaffold boosts cartilage regeneration using co-culture of mesenchymal stem cells and chondrocytes: in vivo study. Mater. Des. **210**, 110065 (2021). https://doi.org/10.1016/j.matdes.2021.110065
23. N. Shanehband, S.M. Naghib, Microfluidics-assisted tumor cell separation approaches for clinical applications: an overview on emerging devices. Comb. Chem. High Throughput Screening **28**(2), 202–225 (2025)
24. H.V. Singh, L. Das, R. Malayil, T. Singh, S. Singh, T. Goyal, A. Munshi, Comprehensive analysis of culture conditions governing differentiation of MSCs into articular chondrocytes. Regen. Med. **18**(7), 573–590 (2023)
25. D.G. O'Shea, C.M. Curtin, F.J. O'Brien, Articulation inspired by nature: a review of biomimetic and biologically active 3D printed scaffolds for cartilage tissue engineering. Biomater. Sci. **10**(10), 2462–2483 (2022)
26. A. Salerno, P.A. Netti, Review on computer-aided design and manufacturing of drug delivery scaffolds for cell guidance and tissue regeneration. Front. Bioeng. Biotechnol. **9**, 682133 (2021)
27. K. Khodaverdi, S.M. Naghib, M.R. Mozafari, M. Rahmanian, Chitosan/Hydroxyapatite hydrogels for localized drug delivery and tissue engineering: a review. Carbohydr. Polym. Technol. Appl. **9**, 100640 (2025). https://doi.org/10.1016/j.carpta.2024.100640
28. N. Top, İ Şahin, H. Gökçe, H. Gökçe, Computer-aided design and additive manufacturing of bone scaffolds for tissue engineering: state of the art. J. Mater. Res. **36**(19), 3725–3745 (2021)
29. R. Gholami, S.M. Naghib, Introduction, in *Nanobioceramics for Bone Tissue Engineering and Regenerative Biomedicine: Tunable Biological Characteristics* (Springer Nature, Singapore, 2025), pp. 1–24. https://doi.org/10.1007/978-981-96-0041-0_1
30. B. Pavan Kalyan, L. Kumar, 3D Printing: applications in tissue engineering, medical devices, and drug delivery. Aaps Pharmscitech **23**(4), 92 (2022)
31. W. Sun, A. Darling, B. Starly, J. Nam, Computer-aided tissue engineering: overview, scope and challenges. Biotechnol. Appl. Biochem. **39**(1), 29–47 (2004)
32. D. Lacroix, J.A. Planell, P.J. Prendergast, Computer-aided design and finite-element modelling of biomaterial scaffolds for bone tissue engineering. Philos. Trans. R. Soc. A Math. Phys. Eng. Sci. **367**(1895), 1993–2009 (2009)
33. P.J. Kondiah, P.P. Kondiah, Y.E. Choonara, T. Marimuthu, V. Pillay, A 3D bioprinted pseudo-bone drug delivery scaffold for bone tissue engineering. Pharmaceutics **12**(2), 166 (2020)
34. H. Garshasbi, S. Salehi, S.M. Naghib, S. Ghorbanzadeh, W. Zhang, Stimuli-responsive injectable chitosan-based hydrogels for controlled drug delivery systems. Front. Bioeng. Biotechnol. **10** (2023). https://doi.org/10.3389/fbioe.2022.1126774
35. H.R. Garshasbi, S.M. Naghib, Smart stimuli-responsive alginate nanogels for drug delivery systems and cancer therapy: a review. Curr. Pharm. Des. **29**(44), 3546–3562 (2023). https://doi.org/10.2174/0113816128283806231211073031
36. S. Gooneh-Farahani, M.R. Naimi-Jamal, S.M. Naghib, Stimuli-responsive graphene-incorporated multifunctional chitosan for drug delivery applications: a review. Expert Opin. Drug Deliv. **16**(1), 79–99 (2019). https://doi.org/10.1080/17425247.2019.1556257
37. Z. Mazidi, S. Javanmardi, S.M. Naghib, Z. Mohammadpour, Smart stimuli-responsive implantable drug delivery systems for programmed and on-demand cancer treatment: An overview on the emerging materials. Chem. Eng. J. **433**, 134569 (2022). https://doi.org/10.1016/j.cej.2022.134569
38. S.M. Naghib, B. Ahmadi, B. Mikaeeli Kangarshahi, M.R. Mozafari, Chitosan-based smart stimuli-responsive nanoparticles for gene delivery and gene therapy: recent progresses on cancer therapy. Int. J. Biol. Macromol. **278**, 134542 (2024). https://doi.org/10.1016/j.ijbiomac.2024.134542
39. S.M. Naghib, B. Ahmadi, M.R. Mozafari, Stimuli-sensitive chitosan-based nanosystems-immobilized nucleic acids for gene therapy in breast cancer and hepatocellular carcinoma. Curr. Top. Med. Chem. **24**(17), 1464–1489 (2024). https://doi.org/10.2174/0115680266293173240506054439

40. S.M. Naghib, H.R. Garshasbi, Smart stimuli-responsive alginate nanogels for drug delivery systems and cancer therapy: a review. Curr. Pharm. Des. (2023). https://doi.org/10.2174/0113816128283806231211073031
41. B. Singh, K. Kim, M.-H. Park, On-demand drug delivery systems using nanofibers. Nanomaterials **11**(12), 3411 (2021)
42. A. Sadraei, S.M. Naghib, 4D Printing of physical stimuli-responsive hydrogels for localized drug delivery and tissue engineering. Polym. Rev. **65**(1), 104–168 (2025). https://doi.org/10.1080/15583724.2024.2427184
43. A. Sadraei, N.S. Morteza, N. Rabiee, 4D Printing biological stimuli-responsive hydrogels for tissue engineering and localized drug delivery applications—Part 1. Expert Opin. Drug Deliv. **22**(4), 471–490 (2025). https://doi.org/10.1080/17425247.2025.2466772
44. A. Sadraei, N.S. Morteza, N. Rabiee, 4D Printing chemical stimuli-responsive hydrogels for tissue engineering and localized drug delivery applications—Part 2. Expert Opin. Drug Deliv. **22**(4), 491–510 (2025). https://doi.org/10.1080/17425247.2025.2466768
45. S. Salehi, S.M. Naghib, H.R. Garshasbi, S. Ghorbanzadeh, W. Zhang, Smart stimuli-responsive injectable gels and hydrogels for drug delivery and tissue engineering applications: a review. Front. Bioeng. Biotechnol. **11**, 1104126 (2023)
46. J.-H. Lee, K.-J. Chen, S.-H. Noh, M.A. Garcia, H. Wang, W.-Y. Lin, H. Jeong, B.J. Kong, D.B. Stout, J. Cheon, On-demand drug release system for in vivo cancer treatment via self assembled magnetic nanoparticles. Angew. Chem. (Int. Ed. Engl.) **52**(16), 4384 (2013)
47. J. Xu, M. Harasek, M. Gföhler, From soft lithography to 3D printing: current status and future of microfluidic device fabrication. Polymers **17**(4), 455 (2025)
48. O. Sartipzadeh, S.M. Naghib, F. Haghiralsadat, F. Shokati, M. Rahmanian, Microfluidic-assisted synthesis and modeling of stimuli-responsive monodispersed chitosan microgels for drug delivery applications. Sci. Rep. **12**(1), 8382 (2022). https://doi.org/10.1038/s41598-022-12031-9
49. O. Sartipzadeh, S.M. Naghib, A. Seyfoori, M. Rahmanian, F.S. Fateminia, Controllable size and form of droplets in microfluidic-assisted devices: effects of channel geometry and fluid velocity on droplet size. Mater. Sci. Eng. C **109** (2020). https://doi.org/10.1016/j.msec.2019.110606
50. N. Shanehband, S.M. Naghib, Recent advances in nano/microfluidics-based cell isolation techniques for cancer diagnosis and treatments. Biochimie **220**, 122–143 (2024). https://doi.org/10.1016/j.biochi.2024.01.001
51. H.S. Santana, M.S. Palma, M.G. Lopes, J. Souza, G.A. Lima, O.P. Taranto, J.L. Silva Jr., Microfluidic devices and 3D printing for synthesis and screening of drugs and tissue engineering. Ind. Eng. Chem. Res. **59**(9), 3794–3810 (2019)
52. Y. Zheng, Y. Huang, J. Luo, X. Peng, X. Gui, G. Liu, Y. Zhang, Supercritical fluid technology: a game-changer for biomacromolecular nanomedicine preparation and biomedical application. Chin. Chem. Lett. **35**(7), 109169 (2024)
53. S. Ahadian, J.A. Finbloom, M. Mofidfar, S.E. Diltemiz, F. Nasrollahi, E. Davoodi, V. Hosseini, I. Mylonaki, S. Sangabathuni, H. Montazerian, Micro and nanoscale technologies in oral drug delivery. Adv. Drug Deliv. Rev. **157**, 37–62 (2020)
54. C. Tomasina, T. Bodet, C. Mota, L. Moroni, S. Camarero-Espinosa, Bioprinting vasculature: materials, cells and emergent techniques. Materials **12**(17), 2701 (2019)
55. Y. Wang, K.F. Bruggeman, S. Franks, V. Gautam, S.I. Hodgetts, A.R. Harvey, R.J. Williams, D.R. Nisbet, Is viral vector gene delivery more effective using biomaterials? Adv. Healthcare Mater. **10**(1), 2001238 (2021)
56. D. Peixoto, I. Pereira, M. Pereira-Silva, F. Veiga, M.R. Hamblin, Y. Lvov, M. Liu, A.C. Paiva-Santos, Emerging role of nanoclays in cancer research, diagnosis, and therapy. Coord. Chem. Rev. **440**, 213956 (2021)
57. L. Hao, Z. Tianyuan, Y. Zhen, C. Fuyang, W. Jiang, Y. Zineng, D. Zhengang, L. Shuyun, H. Chunxiang, Y. Zhiguo, Biofabrication of cell-free dual drug-releasing biomimetic scaffolds for meniscal regeneration. Biofabrication **14**(1), 015001 (2021)

58. Q. Chen, Y. Pei, K. Tang, M.G. Albu-Kaya, Structure, extraction, processing, and applications of collagen as an ideal component for biomaterials-a review. Collagen Leather **5**(1), 20 (2023)
59. P. Gil-Cabrerizo, I. Scacchetti, E. Garbayo, M.J. Blanco-Prieto, Cardiac tissue engineering for myocardial infarction treatment. Eur. J. Pharm. Sci. **185**, 106439 (2023)
60. X. Zhao, D.A. Hu, D. Wu, F. He, H. Wang, L. Huang, D. Shi, Q. Liu, N. Ni, M. Pakvasa, Applications of biocompatible scaffold materials in stem cell-based cartilage tissue engineering. Front. Bioeng. Biotechnol. **9**, 603444 (2021)
61. J.J. Chung, H. Im, S.H. Kim, J.W. Park, Y. Jung, Toward biomimetic scaffolds for tissue engineering: 3D printing techniques in regenerative medicine. Front. Bioeng. Biotechnol. **8**, 586406 (2020)
62. L. Winnington, A mirror of flesh: an exploration of materiality in living bioengineered art, ResearchSpace@Auckland (2021)
63. P.D. Costa, D.C. Costa, T.R. Correia, V.M. Gaspar, J.F. Mano, Natural origin biomaterials for 4D bioprinting tissue-like constructs. Adv. Mater. Technol. **6**(10), 2100168 (2021)
64. X. Wang, D. Zhang, Y.P. Singh, M. Yeo, G. Deng, J. Lai, F. Chen, I.T. Ozbolat, Y. Yu, Progress in organ bioprinting for regenerative medicine. Engineering **42**, 121–142 (2024). https://doi.org/10.1016/j.eng.2024.04.023
65. S. Bitting, T. Derme, J. Lee, T. Van Mele, B. Dillenburger, P. Block, Challenges and opportunities in scaling up architectural applications of mycelium-based materials with digital fabrication. Biomimetics **7**(2), 44 (2022)
66. Q. Zhang, H.-P. Bei, M. Zhao, Z. Dong, X. Zhao, Shedding light on 3D printing: printing photo-crosslinkable constructs for tissue engineering. Biomaterials **286**, 121566 (2022)
67. X. Niu, N. Li, Z. Du, X. Li, Integrated gradient tissue-engineered osteochondral scaffolds: challenges, current efforts and future perspectives. Bioactive Mater. **20**, 574–597 (2023)
68. Y.S. Zhang, G. Haghiashtiani, T. Hübscher, D.J. Kelly, J.M. Lee, M. Lutolf, M.C. McAlpine, W.Y. Yeong, M. Zenobi-Wong, J. Malda, 3D Extrusion bioprinting. Nat. Rev. Methods Primers **1**(1), 75 (2021)
69. C. Dou, V. Perez, J. Qu, A. Tsin, B. Xu, J. Li, A state-of-the-art review of laser-assisted bioprinting and its future research trends. ChemBioEng Rev. **8**(5), 517–534 (2021)
70. B. Grigoryan, D.W. Sazer, A. Avila, J.L. Albritton, A. Padhye, A.H. Ta, P.T. Greenfield, D.L. Gibbons, J.S. Miller, Development, characterization, and applications of multi-material stereolithography bioprinting. Sci. Rep. **11**(1), 3171 (2021)
71. A. Zaszczyńska, M. Moczulska-Heljak, A. Gradys, P. Sajkiewicz, Advances in 3D printing for tissue engineering. Materials **14**(12), 3149 (2021)
72. Y. Ma, B. Deng, R. He, P. Huang, Advancements of 3D bioprinting in regenerative medicine: exploring cell sources for organ fabrication. Heliyon **10**(3), e24593 (2024). https://doi.org/10.1016/j.heliyon.2024.e24593
73. P. Zhu, L. Wang, Microfluidics-enabled soft manufacture of materials with tailorable wettability. Chem. Rev. **122**(7), 7010–7060 (2021)
74. C.E. Larson, E. Meng, A review for the peripheral nerve interface designer. J. Neurosci. Methods **332**, 108523 (2020)
75. E. Stocco, A. Porzionato, E. De Rose, S. Barbon, R. De Caro, V. Macchi, Meniscus regeneration by 3D printing technologies: Current advances and future perspectives. J. Tissue Eng. **13**, 20417314211065860 (2022)
76. V.H. Mouser, R. Levato, L.J. Bonassar, D.D. D'Lima, D.A. Grande, T.J. Klein, D.B. Saris, M. Zenobi-Wong, D. Gawlitta, J. Malda, Three-dimensional bioprinting and its potential in the field of articular cartilage regeneration. Cartilage **8**(4), 327–340 (2017)
77. X. Yang, S. Li, Y. Ren, L. Qiang, Y. Liu, J. Wang, K. Dai, 3D Printed hydrogel for articular cartilage regeneration. Compos. B Eng. **237**, 109863 (2022)
78. Q. Li, S. Xu, Q. Feng, Q. Dai, L. Yao, Y. Zhang, H. Gao, H. Dong, D. Chen, X. Cao, 3D Printed silk-gelatin hydrogel scaffold with different porous structure and cell seeding strategy for cartilage regeneration. Bioactive Mater. **6**(10), 3396–3410 (2021)
79. N.K. Singh, 4.32—Stem cell therapy and retinal regeneration, in *Comprehensive Hematology and Stem Cell Research*, 1st edn., ed. by N. Rezaei (Elsevier, Oxford, 2024), pp. 473–480. https://doi.org/10.1016/B978-0-443-15717-2.00077-9

80. Y. Cao, S. Sang, Y. An, C. Xiang, Y. Li, Y. Zhen, Progress of 3D printing techniques for nasal cartilage regeneration, Aesthetic Plastic Surgery (2022) 1–18.
81. Y. Wang, Y. You, H. Chen, J. Liu, Q. Wu, K. Dai, Y. Sun, 3D-bioprinted anti-senescence organoid scaffolds repair cartilage defect and prevent joint degeneration via miR-23b/ELOVL5-mediated metabolic rewiring. Chem. Eng. J. **491**, 152049 (2024)
82. J.G. Kim, Y.A. Rim, J.H. Ju, The role of transforming growth factor beta in joint homeostasis and cartilage regeneration. Tissue Eng. Part C Methods **28**(10), 570–587 (2022)
83. N. Mamidi, F. Ijadi, M.H. Norahan, Leveraging the recent advancements in GelMA scaffolds for bone tissue engineering: an assessment of challenges and opportunities. Biomacromol **25**(4), 2075–2113 (2023)
84. P.C. Shanto, S. Park, M. Park, B.-T. Lee, Physico-biological evaluation of 3D printed dECM/TOCN/alginate hydrogel based scaffolds for cartilage tissue regeneration. Biomater. Adv. **145**, 213239 (2023)
85. S. Shen, M. Chen, W. Guo, H. Li, X. Li, S. Huang, X. Luo, Z. Wang, Y. Wen, Z. Yuan, Three dimensional printing-based strategies for functional cartilage regeneration. Tissue Eng. Part B Rev. **25**(3), 187–201 (2019)
86. R. Gholami, S.M. Naghib, Composites for BTE, in *Nanobioceramics for Bone Tissue Engineering and Regenerative Biomedicine: Tunable Biological Characteristics*, ed. by R. Gholami, S.M. Naghib (Springer Nature, Singapore, 2025), pp. 251–282. https://doi.org/10.1007/978-981-96-0041-0_9
87. M.A. Khorasani, S.M. Naghib, A review of chitosan-based multifunctional nanocomposites for drug/gene/protein delivery and gene therapy in cancer treatments: promises, challenges and outlooks. Int. J. Biol. Macromol. **306**, 141394 (2025). https://doi.org/10.1016/j.ijbiomac.2025.141394
88. D. Primorac, V. Molnar, D. Tsoukas, I. Uzieliene, C. Tremolada, P. Brlek, E. Klarić, D. Vidović, M. Zekušić, J. Pachaleva, Tissue engineering and future directions in regenerative medicine for knee cartilage repair: a comprehensive review. Croat. Med. J. **65**(3), 268 (2024)
89. G. Liu, X. Wei, Y. Zhai, J. Zhang, J. Li, Z. Zhao, T. Guan, D. Zhao, 3D printed osteochondral scaffolds: design strategies, present applications and future perspectives. Front. Bioeng. Biotechnol. **12**, 1339916 (2024)
90. A. Szwed-Georgiou, P. Płociński, B. Kupikowska-Stobba, M.M. Urbaniak, P. Rusek-Wala, K. Szustakiewicz, P. Piszko, A. Krupa, M. Biernat, M. Gazińska, Bioactive materials for bone regeneration: biomolecules and delivery systems. ACS Biomater. Sci. Eng. **9**(9), 5222–5254 (2023)
91. Q. Zhang, J. Yang, N. Hu, J. Liu, H. Yu, H. Pan, D. Chen, C. Ruan, Small-molecule amines: a big role in the regulation of bone homeostasis. Bone Res. **11**(1), 40 (2023)
92. J. Dorogin, H.B. Hochstatter, S.O. Shepherd, J.E. Svendsen, M.A. Benz, A.C. Powers, K.M. Fear, J.M. Townsend, J.S. Prell, P. Hosseinzadeh, Moderate-affinity affibodies modulate the delivery and bioactivity of bone morphogenetic protein-2. Adv. Healthcare Mater. **12**(26), 2300793 (2023)
93. H. Liu, P. Liu, Kartogenin promotes the BMSCs chondrogenic differentiation in osteoarthritis by down-regulation of miR-145-5p targeting Smad4 pathway. Tissue Eng. Regenerative Med. **18**, 989–1000 (2021)
94. H. Zhou, Y. He, W. Xiong, S. Jing, X. Duan, Z. Huang, G.S. Nahal, Y. Peng, M. Li, Y. Zhu, MSC based gene delivery methods and strategies improve the therapeutic efficacy of neurological diseases. Bioactive Mater. **23**, 409–437 (2023)
95. T. Hodgkinson, D.C. Kelly, C.M. Curtin, F.J. O'Brien, Mechanosignalling in cartilage: an emerging target for the treatment of osteoarthritis. Nat. Rev. Rheumatol. **18**(2), 67–84 (2022)
96. P. Baei, H. Daemi, F. Aramesh, H. Baharvand, M.B. Eslaminejad, Advances in mechanically robust and biomimetic polysaccharide-based constructs for cartilage tissue engineering. Carbohyd. Polym. **308**, 120650 (2023)
97. A. Thakur, A. Kumar, S. Kaya, R. Marzouki, F. Zhang, L. Guo, Recent advancements in surface modification, characterization and functionalization for enhancing the biocompatibility and corrosion resistance of biomedical implants. Coatings **12**(10), 1459 (2022)

98. O. Craciunescu, M. Icriverzi, P.E. Florian, A. Roseanu, M. Trif, Mechanisms and pharmaceutical action of lipid nanoformulation of natural bioactive compounds as efficient delivery systems in the therapy of osteoarthritis. Pharmaceutics **13**(8), 1108 (2021)
99. S.R. Falsafi, A.C. Karaca, L. Deng, Y. Wang, H. Li, G. Askari, H. Rostamabadi, Insights into whey protein-based carriers for targeted delivery and controlled release of bioactive components. Food Hydrocolloids **133**, 108002 (2022)
100. S. Tavasoli, Q. Liu, S.M. Jafari, Development of Pickering emulsions stabilized by hybrid biopolymeric particles/nanoparticles for nutraceutical delivery. Food Hydrocolloids **124**, 107280 (2022)
101. P. Vosoughi, S.M. Naghib, M.R. Mozafari, Stimuli-responsive nano/biomaterials for smart drug delivery in cardiovascular diseases: promises, challenges and outlooks. Curr. Med. Chem. (2024). https://doi.org/10.2174/0109298673319981241021063524
102. A. Ebrahimiverkiani, S.M. Naghib, G. Shafiei Sabet, P. Arghavani, M. Edrisi, N. Rabiee, Recent advances in microfluidic technologies for stimuli-sensitive alginate particles: applications in biosensing and drug delivery. TrAC Trends Anal. Chem. **184**, 118124 (2025). https://doi.org/10.1016/j.trac.2024.118124
103. E. Behjat, M. Tajabadi, S.M. Naghib, H.R. Rezaie, Multifunctional stimuli-responsive self-healing polyvinyl alcohol/tannic acid/thyme-honey/aloe vera hydrogel as promising antibacterial drug delivery platforms: physicochemical and in vitro biological assays. Polymer **317**, 127897 (2025). https://doi.org/10.1016/j.polymer.2024.127897
104. S.M. Naghib, S. Amiri, M.R. Mozafari, Stimuli-responsive chitosan-based nanocarriers for drug delivery in wound dressing applications: a review. Carbohydr. Polym. Technol. Appl. **7**, 100497 (2024). https://doi.org/10.1016/j.carpta.2024.100497
105. A. Moheb Afzali, M.A. Kheradmand, S.M. Naghib, Bioreactor design-assisted bioprinting of stimuli-responsive materials for tissue engineering and drug delivery applications. Bioprinting **37**, e00325 (2024). https://doi.org/10.1016/j.bprint.2023.e00325
106. H.R. Garshasbi, S. Soleymani, S.M. Naghib, M.R. Mozafari, Multi-stimuli-responsive hydrogels for therapeutic systems: an overview on emerging materials, devices, and drugs. Curr. Pharm. Des. **30**(26), 2027–2046 (2024). https://doi.org/10.2174/0113816128304924240527113111
107. M. Yazdan, S.M. Naghib, Smart ultrasound-responsive polymers for drug delivery: an overview on advanced stimuli-sensitive materials and techniques. Curr. Drug Deliv. (2025)
108. S. Amirthalingam, A.K. Rajendran, Y.G. Moon, N.S. Hwang, Stimuli-responsive dynamic hydrogels: design, properties and tissue engineering applications. Mater. Horiz. **10**(9), 3325–3350 (2023)
109. C. Pitta Kruize, S. Panahkhahi, N.E. Putra, P. Diaz-Payno, G. Van Osch, A.A. Zadpoor, M.J. Mirzaali, Biomimetic approaches for the design and fabrication of bone-to-soft tissue interfaces. ACS Biomater. Sci. Eng. **9**(7), 3810–3831 (2021)
110. P.A. Everts, T. Mazzola, K. Mautner, P.S. Randelli, L. Podesta, Modifying orthobiological PRP therapies are imperative for the advancement of treatment outcomes in musculoskeletal pathologies. Biomedicines **10**(11), 2933 (2022)
111. T. Jafari, S.M. Naghib, M.R. Mozafari, Chitosan-based nano/biomaterials in bone tissue engineering and regenerative medicine: recent progress and advances. Curr. Org. Synth. **22**(4), 457–480 (2025)
112. P. Vosoughi, S.M. Naghib, T. Jafari, B.M. Kangarshahi, Chitosan-encapsulated lipid-based nanovesicles for therapeutic applications and tissue engineering: a review. Carbohydr. Polym. Technol. Appl. 100805 (2025)
113. A.S. Teymourlouei, S.M. Naghib, M.R. Mozafari, Stimuli-responsive graphene-polysaccharide nanocomposites for drug delivery and tissue engineering. Curr. Org. Synth. **22**(2), 211–233 (2025)
114. T. Jafari, S.M. Naghib, M.R. Mozafari, Bio-printing of materials for bone tissue engineering. Curr. Org. Chem. **29**(1), 19–40 (2025)
115. J. Liang, P. Liu, X. Yang, L. Liu, Y. Zhang, Q. Wang, H. Zhao, Biomaterial-based scaffolds in promotion of cartilage regeneration: recent advances and emerging applications. J. Orthop. Transl. **41**, 54–62 (2023)

116. G. Zhou, R. Xu, T. Groth, Y. Wang, X. Yuan, H. Ye, X. Dou, The combination of bioactive herbal compounds with biomaterials for regenerative medicine. Tissue Eng. Part B Rev. **30**(6), 607–630 (2024)
117. P. Bartolo, A. Malshe, E. Ferraris, B. Koc, 3D bioprinting: materials, processes, and applications. CIRP Ann. **71**(2), 577–597 (2022)
118. Z. Mao, X. Bi, C. Wu, Y. Zheng, X. Shu, S. Wu, J. Guan, R.O. Ritchie, A cell-free silk fibroin biomaterial strategy promotes in situ cartilage regeneration via programmed releases of bioactive molecules. Adv. Healthcare Mater. **12**(1), 2201588 (2023)
119. K. Zha, Z. Sun, Y. Yang, M. Chen, C. Gao, L. Fu, H. Li, X. Sui, Q. Guo, S. Liu, Recent developed strategies for enhancing chondrogenic differentiation of MSC: impact on MSC-based therapy for cartilage regeneration. Stem Cells Int. **2021**(1), 8830834 (2021)
120. P. De Kinderen, J. Meester, B. Loeys, S. Peeters, E. Gouze, S. Woods, G. Mortier, A. Verstraeten, Differentiation of induced pluripotent stem cells into chondrocytes: methods and applications for disease modeling and drug discovery. J. Bone Miner. Res. **37**(3), 397–410 (2020)
121. H. Yang, H. Yang, Q. Wang, H. Ji, T. Qian, Y. Qiao, J. Shi, M. Cong, Mesenchymal stem cells and their extracellular vesicles: new therapies for cartilage repair. Front. Bioeng. Biotechnol. **13**, 1591400 (2025)
122. A.F. Girao, A. Semitela, G. Ramalho, A. Completo, P.A. Marques, Mimicking nature: fabrication of 3D anisotropic electrospun polycaprolactone scaffolds for cartilage tissue engineering applications. Compos. B Eng. **154**, 99–107 (2018)
123. A. Lolli, F. Colella, C. De Bari, G.J. van Osch, Targeting anti-chondrogenic factors for the stimulation of chondrogenesis: a new paradigm in cartilage repair. J. Orthopaedic Res. **37**(1), 12–22 (2019)
124. P. Lakkala, S.R. Munnangi, S. Bandari, M. Repka, Additive manufacturing technologies with emphasis on stereolithography 3D printing in pharmaceutical and medical applications: a review. Int. J. Pharm. X **5**, 100159 (2023)

Chapter 9
Lipid-Based Micro/Nanovesicles as a Potential Strategy for Drug Delivery in Breast Cancer Treatment

Abstract Among the most prevalent and deadly types of cancer in the world, breast cancer is very critical. Gene therapy shows potential as a favourable treatment for breast cancer, despite a number of drawbacks including poor selectivity, immunological rejection and limited transport efficiency. Many cells spontaneously create exosomes, which are nanocarriers that have garnered a lot of interest due to their possible use in gene transfer. Exosomes have the ability to effectively transport nucleic acids, including DNA, mRNA, circRNA, siRNA and miRNA, to target cells and alter their biological functions. Because of their remarkable design flexibility, low immunogenicity and compatibility with live organisms, exosomes are ideal for gene therapy. The aim of this study is to provide a thorough evaluation of the current state of exosome-mediated gene delivery for breast cancer treatment, taking into account both its advantages and disadvantages. In this chapter, I provide a general information of exosome synthesis and properties, as well as their function in the development, spread and resistance to treatment of breast cancer. Here, I provide an overview of the existing approaches and tactics for encoding nucleic acids into exosomes, together with the variables influencing the precision and efficiency of their transportation. I also go over the many kinds of nucleic acid exosomes that are employed in breast cancer treatment, as well as their results and modes of action. We also discuss the latest advancements and uses of exosome-based delivery nanoplatforms, which improve the stability, targeting and functioning of exosomes by combining them with other nanomaterials or biomolecules. We conclude by outlining the existing state and difficulties of exosome-based clinical trials as well as the prospects for the future and possible medical uses based on exosomal gene nanoplatforms for drug delivery. I anticipate that this evaluation will provide insightful information and helpful recommendations for creating and refining exosome-mediated gene delivery in order to treat breast cancer.

Keywords Nucleic acids · Exosomes · Breast cancer · Gene delivery · Nanoplatforms · Gene therapy

S. M. Naghib, *Nano-Biosensors, Drug Delivery and Tissue Engineering*, Series in BioEngineering, https://doi.org/10.1007/978-981-95-1888-3_9

9.1 Introduction

The most common disease affecting women globally and one kind of cancer that starts in the breast tissue is called breast cancer [1–7]. Genetic abnormalities, hormone changes, reproductive history, lifestyle decisions, and exposure to the environment are risk factors for breast cancer [8, 9]. The stage, subtype, and molecular features of the tumor, together with the patient's preferences and general health, all affect the diagnosis and course of therapy for breast cancer. Immunotherapy, hormone treatment, targeted therapy, radiation therapy, chemotherapy, and surgery are the main methods of treating breast cancer (Fig. 9.1) [10–12]. By introducing, altering, or silencing genes in the target cells, gene therapy is a new method that attempts to cure or prevent illnesses. By focusing on the underlying cause of the illness at the genetic level, gene therapy may be able to address the drawbacks of traditional treatments, including toxicity, resistance, and recurrence. Gene editing and gene insertion are the two categories of gene therapy [13]. Delivering exogenous genes to express therapeutic proteins or antigens, or to replace or replenish missing or damaged genes, is known as gene addition [14]. Using tools such as zinc finger nucleases, effector nucleases that resemble transcription activators, or short palindromic repeats that are clustered regularly interspaced (CRISPR)/Cas9, gene editing involves precisely altering or repairing endogenous genes [15]. Gene therapy has had encouraging outcomes in preclinical and clinical studies for a number of cancer types, including breast cancer [16–19]. However, there are several challenges with gene therapy, especially when it comes to delivering the therapeutic genes to the target cells [20–23]. Vectors, which are carriers of the genes that may shield them while promoting their entry and expression in the target cells, are necessary for gene delivery. Vectors may be classified as either viral or non-viral. The pathogenicity of the viruses that produce viral vectors has been eliminated, and therapeutic genes have been added [24]. Viral vectors have a number of drawbacks, including as immunogenicity, toxicity, insertional mutagenesis, and limited cargo capacity, despite their high transfection efficiency and persistent expression [25]. Synthetic or natural materials that can combine with the therapeutic genes to create complexes and transfer them to the target cells are known as non-viral vectors [26]. Although non-viral vectors are simple to produce, have a high cargo capacity, low immunogenicity, and low toxicity, they also have drawbacks such poor stability, quick clearance, and low transfection effectiveness [27, 28].

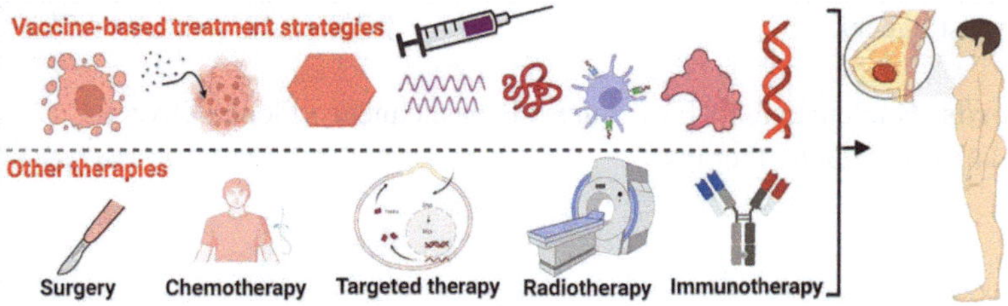

Fig. 9.1 Treatment methods for breast cancer [12]

Because cancer cells' mitochondria have far greater activity and transmembrane potential than those of healthy cells, they may be a target for the administration of chemotherapeutic medicines [29]. One effective strategy for treating tumors with gene rearrangement seems to be to target chromosomal rearrangement or gene fusion. Oncogenic NRG1 fusions are relatively infrequent drivers that are seldom seen in most tumor types. In malignant tumors with NRG1 fusions, a variety of therapies that target the ErbB signaling system, such as monoclonal antibodies and ErbB kinase inhibitors, have shown encouraging anticancer effectiveness. Breast tissue is made up of immune cells and mammary epithelial cells, including T and B cells and other cell types. Recent studies in the pathophysiology of breast cancer have shown that B lymphocytes may either stimulate or prevent the formation of tumors. Immune cell subtypes, including CD4+ T cells, CD8+ T cells, and CD68+ macrophages, seem to be associated with certain molecular subtypes of breast cancer and patient outcomes. Numerous gene and epigenetic changes, in addition to the impact of the local microenvironment, drive the differentiation and progression of breast cancer [30]. Particularly the impact of the local microenvironment, regarding breast cancer stem cells (BCSCs), a knowledge of stem cells is crucial to an understanding of breast cancer. Specialized cells called BCSCs have the remarkable ability to renew and change into many cell types. Increased migration, invasion, proliferation, and resistance to therapy are characteristics of a subgroup of BCSCs seen in breast malignancies [31]. The intricate anatomy of the human breast is made up of the epidermis, breast parenchyma (lobules and ducts), subcutaneous tissue, and a supporting network of lymphatic and fat veins. Among the many connections in this system are arteries, veins, ligaments, and nerves (Fig. 9.2) [32].

The development of breast cancer is influenced by many nerves. Adipocytes and fibroblasts are two examples of cells found in breast tissues that are known to significantly influence the onset and progression of breast cancer. When some fibroblasts in

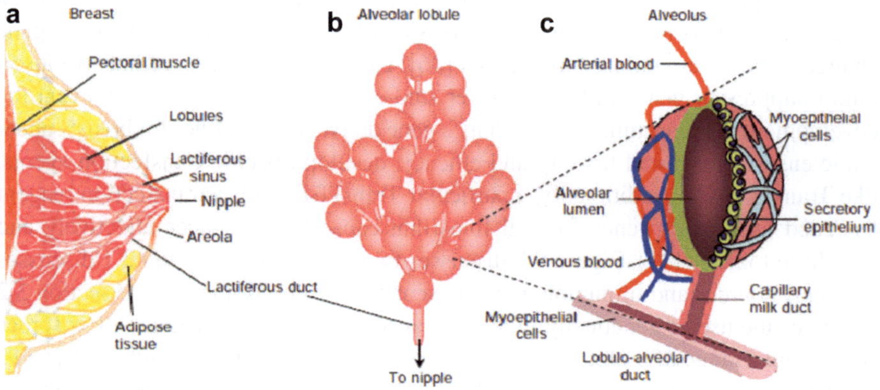

Fig. 9.2 **a** Breast schematic illustrating the connection between the duct system and alveolar lobules. **b** A mammary lobule's alveolar arrangement. **c** The mammary alveolus, which is made up of cells that produce milk and is encircled by a network of contractile myoepithelial cells [32]

breast cancer are activated, the tumor begins, extracellular matrix (ECM) remodeling occurs, and immune cell modulation may occur. These pathways lead on the aggressive features of breast cancer [20, 33]. Adipocytes are vital for the development of breast cancer. These neurological components influence several biological mechanisms, including angiogenesis, tumor microenvironment, inflammation, transition between epithelial and mesenchymal, cell cycle progression, and cancer in general spread and development. The connection between cancerous cells, especially those of the breast, and nerves has also been thoroughly examined and recorded. Breast cancer is linked to a range of immunological cell types, like B cells, neutrophils, macrophages, monocytes, NK cells, CD3+ cells (CD4+ and CD8+), and monocytes [28].

One of the elements that contributes to the aggressiveness about breast cancer stem cells' (BCSCs') ability to differentiate and self-renew. The differentiation of BCSCs is controlled by several signaling pathways. The action, carcinogenesis, and proliferation of BCSCs are governed by a variety of signaling mechanisms. Breast cancer cells may exhibit characteristics of stem cells, acquire chemotherapy resistance, and inhibit their immune systems when these pathways are out of balance [34]. The assisting cells, sometimes known as stromal cells, are essential for the development and maintenance of breast tissue. To fully understand breast tissue's complex biology, one must have a solid comprehension of differentiation. During embryonic development, the gland that produces breast milk develops through complex morphogenesis in ectodermal tissue to produce a fully functioning breast [9, 35]. Gene transfection is the procedure of bringing outside genetic material in the cells of breast tissue. Cells frequently receive the genetic component, which could be RNA, DNA, or more nucleic acids, through specific procedures known as vectors. Gene transfection is a widely used method for delivering foreign genetic material into cells, like RNA or DNA. Gene transfection is essential in breast cancer research to comprehend the intricate genetic processes underlying the disease and investigate novel therapeutic approaches. By inserting certain genes into cells that cause breast cancer, researchers may regulate the expression of genes, evaluate the impact of gene changes, and create tailored treatments [36]. Gene transfection is utilized to insert tumor suppressor genes in breast cancer studies, such p53 and p21, to prevent the development of breast tumors. To enhance gene delivery in cell lines for breast cancer while ensuring minimal toxicity and efficacy, develop Specific transfection agents, like TransIT-BrCa. Additionally, a particular TransIT-BrCa is a transfection reagent designed to transfer genes for cell lines that cause breast cancer effectively and with little toxicity [37]. Figure 9.3 discusses two main topics: section 3.1 discusses the many sources and immunological consequences of exosomes, while section 3.2 discusses the use of genetically modified exosomes in cancer treatment. Exosomes come from a range of biological sources, including cancer cells, immune cells, and a number of other cell types. It graphically illustrates how exosomes affect the immune system, particularly in relation to immunoregulation and intercellular communication, and how these factors connect to diseases and therapeutic approaches. As we move toward version 1.2, the figure offers a thorough analysis of strategies that use genetically altered exosomes to treat cancer. This section's panels each illustrate a different technique, including using exosomes loaded with MiR-29a-3p for gliomas, CRISPR-Cas9 for cervical cancer, and Si-VEGF-laden exosomes for breast cancer treatment, among others. The graphic provides a comprehensive visual depiction of the many origins and medicinal applications of exosomes in immune system regulation and cancer therapy [35, 38].

To address the drawbacks of existing vectors and increase the therapeutic efficacy of gene therapy for breast cancer, new and effective gene delivery methods are thus needed. One of the most recent options for gene delivery is exosomes. Different cell types release exosomes, which are tiny vesicles that contain proteins, lipids, and nucleic acids. Exosomes may act as natural gene delivery vehicles because of their efficiency in transferring nucleic acids to target cells and altering their biological functions. Exosomes' high designability, biocompatibility, and minimal immunogenicity make them ideal candidates for gene therapy. This chapter will provide a thorough overview of the most recent advancements and challenges in exosome-mediated gene delivery of the therapy of breast cancer [37].

Breast cancer therapy has advanced significantly, but these advancements have also presented a number of challenges [1–7]. These changes provide a number of issues, including. Cost and Accessibility, Ethics and Society Consequences, Combining and Implementing: Evidence-based decision-making, Patient education and informed consent, Health Inequalities [39].

Breast cancer is a fatal illness that often affects women and is the second most prevalent cause of cancer-related deaths. In 2020, there were almost 2.3 million new instances of breast cancer that had spread in women globally, and 685,000 people died from the illness. Compared to initial tumors, metastases are responsible for around 90% of cancer-related fatalities. When the cancer cells expand to distant sites outside the breast that are near nodes of auxiliary lymph, it is referred to as stage IV, advanced-stage, or metastatic breast cancer [40, 41]. Usually, it spreads to the liver, brain, lungs, and bones, among other organs. Sometimes, metastasized cancer cells are discovered prior to the main tumor being diagnosed. Along with the capacity of metastatic cells to invade different organs (Fig. 9.4), metastasis poses a significant challenge in the assessment and treatment of cancer because of the intricate tumor cell plasticity, variability, and unique TME (tumor microenvironment) that change the response to therapy [42]. Mesenchymal stem cells are a primitive cell type with unlimited proliferative potential and the ability to repopulate entire tumors. These cells interact and communicate with cancer cells by secreting cytokines, growth factors and by transferring mitochondria or microRNAs through specialized structures such as tunneling nanotubes or exosomes. This crosstalk facilitates tumor recurrence both at the primary tumor site and at distant locations, even after apparently successful treatment. For these reasons, cancer stem cells including mesenchymal stem cells, have gained significant attention as therapeutic targets within the tumor microenvironment. The ECM plays a critical role in maintaining structural and biochemical support to cells and acts as a key regulator of tumor invasion and progression, as depicted in Fig. 9.4. This dynamic interplay between mesenchymal stem cells, cancer stem cells and the ECM contributes to cancer persistence and metastasis, underscoring the importance of targeting these interactions for effective therapy.

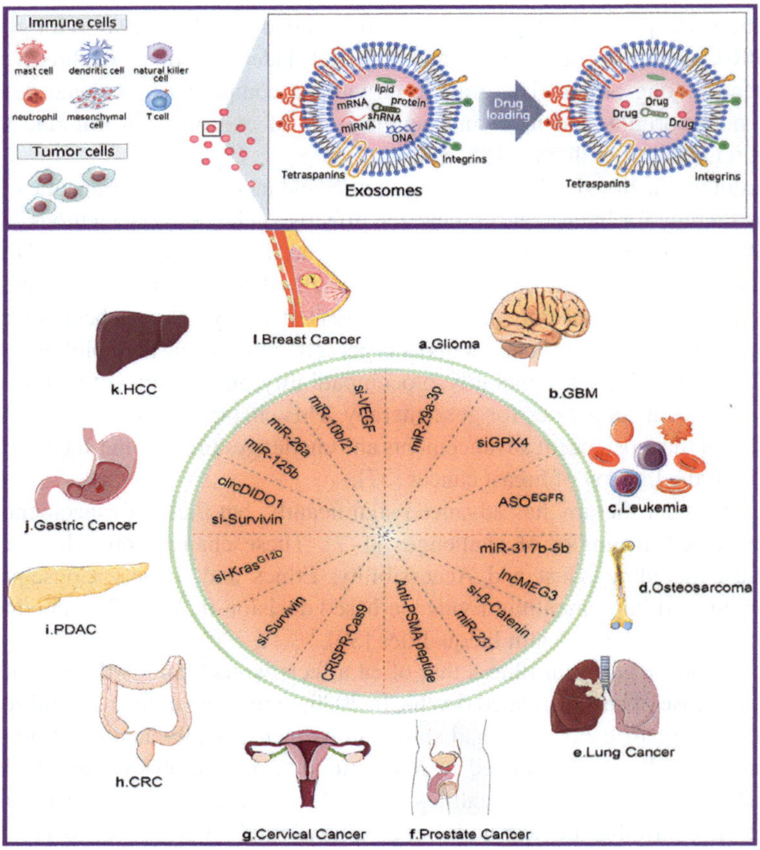

Fig. 9.3 3.1: Synopsis of exosomes' effects on the immune system from various sources. Many biological origins of exosome immune cells, cancer cells, and many other cell types are highlighted in this figure. Additionally, the image depicts several ways that exosomes impact the immune system, including immunoregulation, intercellular communication, and potential consequences for diseases and therapeutic modalities [38]. 3.2: Treating cancer with genetically altered exosomes. The use of genetically engineered exosomes in cancer therapy is shown in this figure. Each panel represents a different tactic: **a** glioma migration and vasculogenic mimicry are inhibited by exosomes that attach to ROBO1 and contain MiR-29a-3p. **b** Ferroptosis in glioblastomas is triggered by MNP@BQR@ANG-EXOsiGPX4. **c** The EGFR receptors seen in leukemia cells are the particular binding site for ASOEGFR. **d** The development of osteosarcoma is influenced by internalized cRGD-Exo-MEG3 and miR317b-5b. **e** In lung cancer, miRNA-231-Exo interferes with the PTEN/PI3K/AKT signaling pathway, whereas Si-β-Catenin inhibits cell proliferation. **f** Drugs containing anti-PSMA peptides are used to treat prostate cancer. **g** CRISPR-Cas9 is transported by exosomes to reduce cervical cancer. **h** Si-Survivin inhibits genes associated with colorectal cancer. Si-KrasG12D is used in conjunction with PDAC to stop the spread of pancreatic cancer. **j** By controlling SOSC2 levels, CircDIDO1 inhibits the growth of gastric cancer. **k** Hepatocellular carcinoma migration and progression are inhibited by exosomes containing MiR-26a. **l** Exosomes loaded with Si-VEGF show promising therapeutic promise in the management of breast cancer. These cancers include colorectal, prostate, pancreatic ductal, and prostate-specific membrane antigen (PSMA). Glutathione peroxidase 4, antisense oligonucleotide, exoASO-STAT6, epidermal growth factor receptor, modified c(RGDyK), exosomes containing MEG3, and maternally expressed genes that include long non-coding RNA (LncRNA) [35]

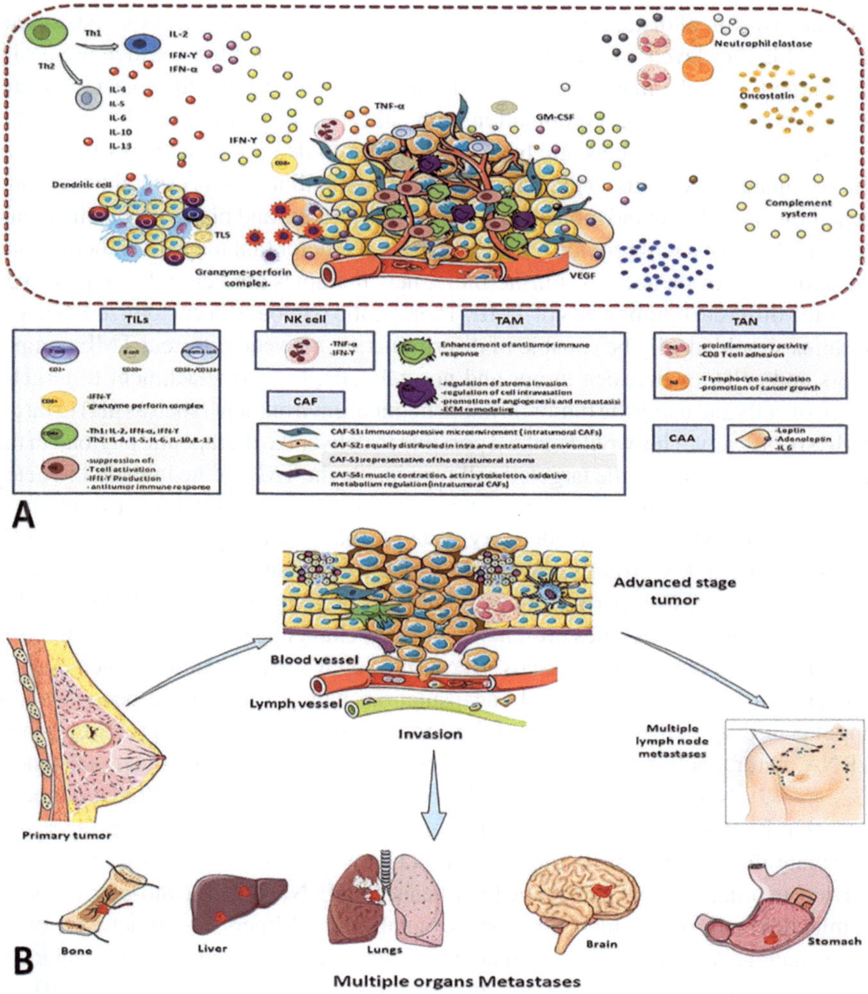

Fig. 9.4 Section 1: The diagram illustrates how the tumor microenvironment, or TME, contributes to the development of cancer. Section 2: Diagrammatic depiction of the spread of breast cancer from its original location to other organs [42]

Consequently, recent evidence indicates that TME may play a key part in the creation of and metastasis of metastatic cancers because of the changed epigenetic alteration that results in aberrant expression of genes and may be a predictor of better clinical outcomes. Because of our growing understanding of the components of TME, including lymph nodes, blood vessels, stromal cells, extracellular matrix (ECM), immune cells, secreted molecules, and cancer-associated fibroblasts (CAFs), as well as their roles in tumor growth, we have an exceptional chance to target TME as a cancer treatment [43]. These cells, all connected to components, produce the hypoxic setting, which mechanically encourages the development of tumor mass and the invasion of distant organs. Most patients having breast cancer that has spread also had higher levels of tumor-derived, going-around fibroblasts, which are easily discharged from the main tumor site into the circulation as one fibroblast, a cluster of fibroblasts, or even in combination with cancer cells. Conventional treatment is still unsuccessful for metastatic breast cancer due to its poor prognosis [44]. Given the recent advancements in the biology of metastatic illness, it is critical to comprehend how the cellular and molecular components interact and play a part in the breast cancer TME. This will provide a very important predictor that may aid in therapy and the avoidance of metastases. Furthermore, new therapies are being developed to go after the different components of TME. Therapeutic payloads in customized delivery systems should ideally be capable of discriminating between different TME components and cells, or between tumor and normal cells. Ligand attachment to target a range of cells and using the tumor physicochemical environment (such as pH) to target TME specifically. This would make it possible to deliver the therapeutic payload in the appropriate amounts to the targeted cell type inside the TME at the intended targeted tumoral region. To show mastery of the targeting skills and qualities essential for the clinical translation of nanodelivery systems, many challenges must undoubtedly be overcome first. Currently, several FDA-approved nanoparticles are being used to combat cancer. Comparing a huge number of pre-clinical tests with the few that might be employed in clinics presents an obvious challenge since only a tiny fraction of them participate in clinical trials. Because of their special qualities, such as their exceptional biocompatibility and encapsulation, nanoparticles have become more popular in drug delivery systems (DDS) for the administration of proteins, genes, siRNA, natural active ingredients, vaccines, anticancer drugs, and other substances [22, 45–49]. Nano DDS may be developed as multifunctional, intelligent nanocarriers that have better pharmacokinetics, are easier for cells to absorb, protect, and distribute the medications to the targeted place over time, and can cross biological barriers in order to target metastatic breast cancer ME. Numerous nano-carriers, such as micelles made of polymers, liposomes, metallic, dendrimers, biomimetic nanoparticles, nanogels, and inorganic, can convey a number of medications with enhanced anticancer activity and showed potential therapeutic benefits [6, 23, 50–58]. These nanoparticles' distinct physicochemical characteristics may be used to recognize and target circulating metastatic cells, therefore eradicating the cancerous cells and eventually improving the patient's prognosis. It is challenging to do this using just traditional means. HER2 has emerged as a distinct target and may be used as the agent for targeting if TME among individuals with breast cancer that has spread is well understood. For clinical translation for the management of cancer that has spread, only a few nanoparticles have received approval, such as Liposomal polymer nanoparticles of doxorubicin (DOX) (Doxil) and docetaxel (DTX) (BIND-014), and PTX-albumin nanoparticles (Abraxane) [59]. Tumor heterogeneity is another reason to be concerned since the different behaviors are shown by the same cancer cells in various organs. As a result, there are phenotypic differences in immunological physiognomies, levels of aggressiveness, rates of proliferation, and susceptibility to various treatments. The effectiveness of NPs in live systems that imitate the TME in humans has been evaluated using a variety of 3-D culture models, such as the extracellular matrix (ECM), vascularization of tumors, and assisting cells that may impact dispersion and penetration. Examining tumor spheroids might enable the NPs to spread and build up. To address the clinical problems and ascertain the safety in vivo and effectiveness of NPs, however, animal models are equally important [60].

9.2 Exosomal Micro/Nanovesicles

Exosomes are tiny, membrane-bound vesicles that are released by several cell types and are crucial for intercellular communication. Through endocytosis or fusion, exosomes may deliver a wide variety of materials, including as proteins, nucleic acids, and lipids, to certain cells. Numerous physiological and pathological processes, including cancer, immunological responses, tissue repair, and stem cell maintenance, may be impacted by exosomes [61, 62].

The endosomal system, namely the multivesicular bodies (MVBs) that house intraluminal vesicles (ILVs), is the source of exosomes. Through its ability to identify and seize ubiquitinated proteins inside intraluminal vesicles (ILVs), the endosomal sorting complex needed for transport (ESCRT) mechanism facilitates their production. Rather, ESCRT-independent processes such tetraspanins, ceramide, or lipid rafts may be involved in ILV synthesis. The multivesicular bodies (MVBs) may merge with lysosomes during breakdown, whereas they may fuse with the plasma membrane during exosome secretion. The diameter of exosomes ranges from 30 to 150 nm, and their density ranges from 1.13 to 1.19 g per ml [63]. Similar to the membrane of plasma, the lipid makeup of exosomes is mostly composed of cholesterol, phospholipids, and sphingolipids. Heat shock proteins (HSP70, HSP90), integrins, annexins, and extracellular trap components (ESCRT) are often found in exosomes. However, the kind of cell and its physiological state determine the precise makeup of exosomes. Exosomes' nucleic acid content consists of DNA and several RNA species, including circular, long non-coding, micro, and mRNA. The composition of exosomes depends on a number of parameters, including the cellular origin, the stimuli, the sorting processes, and the separation techniques [41, 64]. The enhanced treatment outcomes achieved with altered exosomes and modified RNA nanoparticles are shown in Fig. 9.5. Example of how RNA nanoparticles modify exosomes are shown in the picture. Furthermore, exosomes show promise as carriers, enabling the delivery of oxaliplatin (OXA), a chemotherapeutic agent, and gal-9 siRNA, a medication made of nucleic acids, to improve immunotherapy and modify the tumor microenvironment (TME) [65].

Drug transportation is hampered by biological barriers (BBs), despite the fact that they provide protection against illness and invasive diseases. Several BBs surround certain organs and tissue areas in the human body, including the placental barrier (PB), stromal barrier (SB), blood–air barrier (BAB), blood–retinal barrier (BRB), blood–labyrinth barrier (BLaB), blood–lymph barrier (BlyB), blood–cerebrospinal fluid barrier (BCSFB), blood–brain barrier (BBB), and many more (Fig. 9.6) [8, 45, 62, 66]. The intrinsic activity of BBs hinders the transport and absorption of

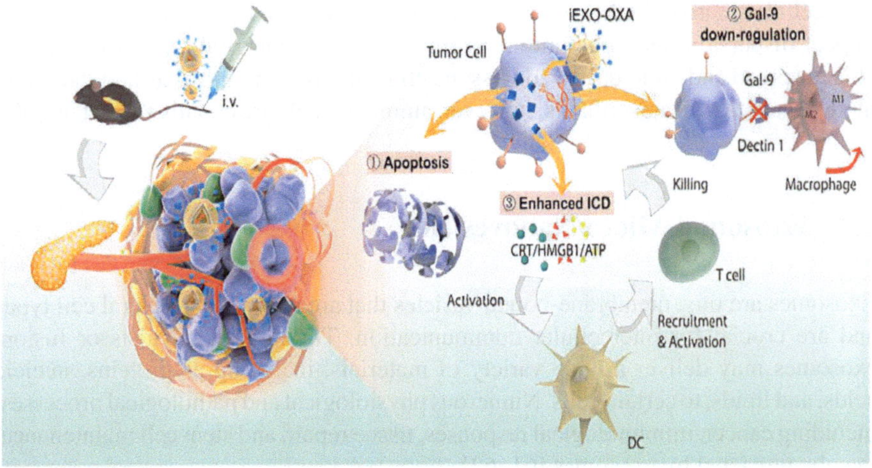

Fig. 9.5 Modified exosomes carrying oxaliplatin and Gal-9 siRNA trigger apoptosis and immunogenic cell death, reprogramming the tumor microenvironment to boost immune response and improve cancer therapy outcomes [65]

medications, preventing effective therapeutic interventions. Advances in a variety of fields, including as ophthalmic illnesses, genetic abnormalities, prenatal morbidities, auditory dysfunction, and cancer therapy, are nevertheless hampered by the different physical structures and environments found in different BBs. Inappropriate intervention methods, however, have the potential to impair the body's defenses, increasing the danger of infection or opening the door for pathogenicity [67, 68]. Therefore, it is crucial to carefully balance the protections offered and the restrictions imposed by BBs in order to provide availability of therapeutic treatments without endangering the physical structures' integrity. Apart from restricting the range of available therapies, BBs decrease the absorption of drugs in BB-protected regions, potentially making bacteria and other illnesses more resistant to therapy. Moreover, pharmaceutical treatments and BB crossing-specific nanoparticles may inhibit or dysregulate biochemical pathways, increasing the risk of negative consequences [69]. The primary emphasis of medication delivery techniques has been access to pharmaceuticals for tissues protected by BB. Dendrimers, hydrogels, liposomes, nanoparticles, and CPPs (cell-penetrating peptides) have all emerged as intriguing options. However, therapeutic access has been challenging due to non-specificity, immunogenicity, and insufficient absorption in vivo. To address these issues, changes to address these issues, gold-enhanced nanoparticles and hybrid nanoparticles have been used. Recently developed materials called nanogels (NGs) are designed to duplicate internal particles. Techniques that allow for controlled payload release in response to light, pH, and enzyme activity are still challenging, despite their application to enhance penetration into BBs. Enzyme saturation, tight junctions, receptor-mediated transport, molecule size for passive transport, and efflux pumps all contribute to the effective absorption and bioavailability of medications in the BB microenvironment, which is necessary for the creation of secure and efficient delivery strategies [66, 70]. Due to their innate ability to flow through BBs, EVs (extracellular vesicles) have recently been shown to give access that is much required. The endosome pathway's nanosized exosomes (30–150 nm) have attracted attention for their intrinsic capacity to go past BBs because of their easily accessible targeting abilities as well as naturally occurring, particular homing markers, despite the fact that how EV subcategories are defined, such as exosomes, apoptotic vesicles, and microvesicles, is still up for debate [71]. For example, studies reveal that exosomes include transferrin, insulin, and LDL receptors that are known to allow absorption across the blood–brain barrier. Exosomes were also swallowed by microglial cells, which are part of the neurovascular unit of the blood–brain barrier, according to a recent research on psychopathological diseases. To improve exosomes' inherent transport and targeting properties, their surfaces may also be modified. Although the exact transport properties of exosomes are unknown, their advantageous properties have been strategically used to produce a variety of inventions that may increase the efficacy of delivery and absorption by improving penetration across various BBs. Exosomes seem to be a potential tool for overcoming the many obstacles that are present both within and outside the BLyB. Because exosomes have the perfect size and makeup lipids instead than soluble they enable transit through and movement within the BLyB, including the RN. A systemic comprehension of the impact of BLyB in connecting many BBs may be possible using exosomes. Dendritic cell-derived exosomes, or DC-Exos, have been utilized to carry antigens, stimulate T-cell lymphocytes, and regulate the inflammatory response in arthritic conditions. Despite being crucial for effectively stopping the transmission of diseases, the BAB also inhibits the application of therapeutic treatments [72]. Pathogens and pharmaceutical drugs are blocked and eliminated by 3 main barriers: chemical, immunological, and mechanical barriers. The mucosa, made up of smooth muscle and the lamina propria, and epithelium, contains a variety of immune cells, such as neutrophils, B and T lymphocytes, chemokines, cytokines, defensins, and macrophages. The intricate "bronchial tree" network in the alveolated area shields the pulmonary air space's epithelial layer. The surfactant secretion and gas exchange occur in the alveolar epithelium. Pattern recognition receptors (PRRs) in the cytoplasm of lung epithelial cells, microbicidal effector chemicals, and tight junctions in epithelium that may contain targetable signaling molecules are new research areas that also safeguard and monitor the BAB. Reprogramming or using exosomes to control defense systems inside the BAB might provide the possibility of useful strategies for functioning within the BAB instead of attempting to circumvent it. The SB responds differentially to inflammation and may be related with malignancy refractory, and significant variation in the stroma's cancer-associated fibroblasts (CAFs) reduces the effectiveness of cancer therapies [73]. These as well as further studies have made SBs appealing candidates for cancer therapy, but conflicting findings underscore the need for more study. The recently proposed portion of the BLaB's obstacles consists of five distinct but interconnected membranous labyrinth barriers: the middle ear–labyrinth barrier, the perilymph–cerebrospinal fluid barrier, the blood–perilymph barrier, and the endolymph–perilymph barrier. Vincristine, kanamycin, adriamycin, and styrene are examples of ototoxic drugs that may affect

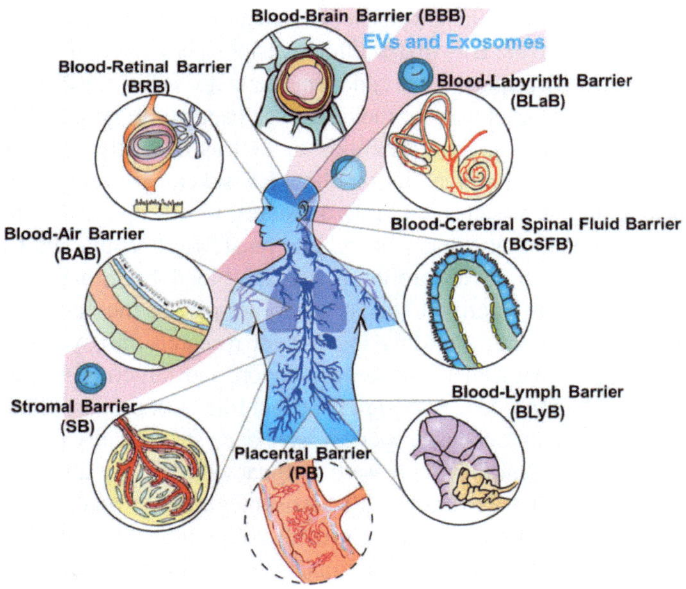

Fig. 9.6 Illustration of essential biological barriers penetrated by exosomes for cellular regulation and delivery [66]

the cochlear outer hair cells, endolymphatic function, and cochlear barrier function, respectively. The retinal vascular epithelium (RVE), which comprises blood vessels resembling both the BLaB and BBB, is one layer of the BRB. It is composed of tight junctions. The RPE (retinal pigment epithelium), which serves as the BRB's second layer, making it more challenging to reach the inner regions. The outer BRB and the blood–thymus barrier are comparable to the placental barrier (BPB), which is made up of cells called epithelium with tight intercellular connections, few pinocytotic vesicles, and increased transporter expression that enhance selectivity for molecule exchange [66].

Via invagination of the plasma membrane and endocytosis, extracellular and fluid substances like lipids, metabolites, proteins, ions, and small molecules may enter cells in addition to cell surface proteins. The creation of a plasma membrane bud as a consequence on the cell's luminal side exhibits orientation of the plasma membrane from the outside [62]. This budding mechanism produces ESEs, or the bud may fuse with ESEs made by the ER (endoplasmic reticulum), trans-Golgi network, and mitochondrial components. The ESEs merging with the TGN and ER may also explain how they are reached by the endocytic cargo. Therefore, certain ESEs may have luminal and membrane components from different sources. ESEs are the source of LSEs. The second invagination in the LSE creates ILVs, and by enabling cytoplasmic components to enter the newly produced ILV, this step may lead to further cargo modification for future exosomes. As ILVs form, proteins first found on the cell surface may be distributed differentially within the cell [74]. The technique may produce

ILVs with different contents and sizes based on the invagination volume. LSEs create MVBs with a certain set of ILVs, future exosomes. MVBs and autophagosomes may merge, and lysosomes may ultimately degrade the contents. The breakdown products may be recycled by the cells. MVBs have the ability to immediately combine with lysosomes for disintegration. MVBs that don't take this route may be transported to the plasma membrane via the cell's microtubule and cytoskeletal network, where they can interact with MVB-docking proteins on the membrane's luminal side. Exosomes having a lipid bilayer orientation similar to the plasma membrane are released during exocytosis [75]. Proteins involved in exosome formation include Rab GTPases, ESCRT proteins, and other proteins used as exosome markers (flotillin, CD63, CD81, CD9, TSG101, Alix, and ceramide). The surface of exosomes contains integrins, tetraspanins, immunomodulatory proteins, and other proteins. Exosomes include a wide range of DNA, RNA, amino acids, metabolites, cell surface proteins, and intracellular proteins [8, 61].

Because of a lipid bilayer that protects them from enzymatic degradation and pH changes, exosomes are stable in a variety of body fluids, including as saliva, urine, blood, and cerebrospinal fluid. Additionally, exosomes have cryo-resistance, which makes them suitable for transit and storage even after many cycles of freezing and thawing. However, extrinsic factors including temperature, pH, oxidation, and mechanical stress may affect how stable exosomes are. A number of mechanisms, such as direct fusion, receptor-mediated endocytosis, phagocytosis, macropinocytosis, or lipid raft-mediated internalization, may be used by recipient cells to internalize exosomes. The number of exosomes, the kind of cells involved, the surface chemicals present, and the physiological state are some of the factors that affect exosome internalization. The contents of exosomes may be transferred to the cytoplasm, nucleus, or endosomal compartments as a result of recipient cells absorbing them (Fig. 9.7) [67, 76–79].

9.3 Functions of Exosomes in Cancer

Numerous mechanisms in cancer biology, including tumor initiation, development, angiogenesis, invasion, drug resistance, and metastasis, are associated with exosomes. By influencing the actions of immune cells, stromal cells, cancer cells, and endothelial cells, exosomes may alter the tumor microenvironment. By linking primary tumors and distant organs, exosomes may promote the development of pre-metastatic settings and the colonization of metastatic cells. Exosomes have the ability to alter gene expression, signaling pathways, and properties in cancer cells by delivering proteins, lipids, and nucleic acids to recipient cells [62]. Oncogenes that control the proliferation and survival of cancer cells, such as EGFRvIII, KRAS, and c-Myc, as well as tumor suppressors like PTEN and p53, may be transferred by exosomes. In order to control the expression of target genes linked to the advancement of cancer, exosomes may also transport long non-coding RNAs (H19), MALAT1, miR-21, miR-155, and miR-10b. Genetic instability and heterogeneity may result from

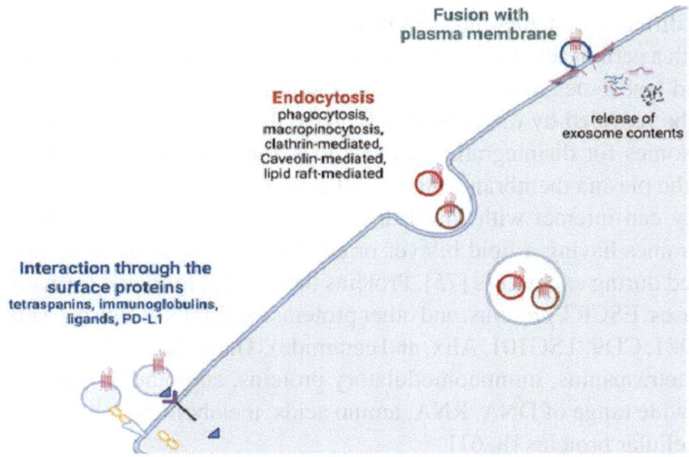

Fig. 9.7 Exosome uptake. Exosome uptake occurs via one of three pathways: (1) fusion with the plasma membrane to release exosome contents; (2) endocytosis; (3) direct interaction with surface receptors [79]

exosomes transferring extrachromosomal, genomic, or mitochondrial DNA to cancer cells [80]. Exosomes may also influence how cancer cells react to different treatments, including immunotherapy, radiation, chemotherapy, and targeted therapy. Through the sequestration or exportation of medicines, the activation of survival or detoxification pathways, the induction of stemness or epithelial-mesenchymal transition (EMT), or immune system modulation, exosomes may impart drug resistance. By modifying the DNA damage response, oxidative stress, the cell cycle, or the inflammatory response, exosomes may also affect how sensitive cancer cells are to radiation [81].

Exosomes have the potential to enhance cancer diagnosis, assessment, and therapy. Exosomes are helpful as cancer biomarkers because they represent the physiological condition and molecular makeup of the tumor cells. Because exosomes can be separated from body fluids such urine, saliva, blood, and cerebrospinal fluid, they are very accessible and minimally invasive. Additionally, exosomes may provide information on the effectiveness of therapy, the unique features of the environment around the tumor, and the diversity within the tumor [47]. Exosomes may also transport certain molecules to target cells in cancer, acting as therapeutic agents or carriers. To increase their effectiveness and specificity, exosomes may be modified to deliver genetic tools like siRNA, shRNA, or CRISPR/Cas9, or anti-cancer medications like doxorubicin, paclitaxel, or cisplatin. By displaying tumor antigens, stimulating dendritic cells, or releasing immunomodulatory chemicals, exosomes may also boost the immune system's ability to fight cancer. Small vesicles called exosomes are crucial for intercellular communication, particularly in cancer [62]. By delivering their cargo of proteins, and nucleic acids to recipient cells, exosomes may affect a number of cancer biology processes, including tumor development, angiogenesis, invasion, metastasis, and medication resistance. Because exosomes may transport certain chemicals to target cells and reflect the molecular profile and physiological condition of tumor cells, they can also be used as biomarkers or treatments in cancer. Consequently, new information and prospects for cancer diagnosis, prognosis, and treatment may arise from a knowledge of the biosynthesis, composition, stability, uptake, and activities of exosomes in cancer [1, 6, 82].

9.4 The Significance of Gene Delivery via Exosomes

Exogenous nucleic acids are introduced into certain cells as part of gene therapy in order to alter their function, phenotype, or gene expression. Numerous illnesses, including cancer, genetic problems, and infectious diseases, may benefit from this therapy [22, 47, 83]. Low transfection effectiveness, poor stability, immunogenicity, toxicity, and off-target effects of the gene carriers are some of the difficulties that gene delivery must overcome [23, 83, 84]. In order to overcome these challenges and accomplish targeted and efficient gene transfer, safe and efficient gene delivery technologies must be developed. Different cell types produce tiny extracellular vesicles called exosomes. By moving lipids, proteins, and nucleic acids to nearby cells, exosomes play a critical role in promoting intercellular communication. Because exosomes work better than conventional delivery methods, they have been known as a novel and appealing way to transfer genes [85]. The benefits include:

Biocompatibility: Because exosomes may elude the immune system and the reticuloendothelial system, they are obtained from natural biological sources and have minimal immunogenicity and toxicity [86].

Stability: The lipid bilayer membrane of exosomes shields their nucleic acid cargo from pH shifts in the extracellular milieu and the endosomal compartments, as well as nuclease destruction [87].

Targetability: To improve the specificity and effectiveness of gene delivery, exosomes may be modified to exhibit certain ligands or antibodies on their surface that can attach to the receptors or antigens on the target cells [88].

Versatility: Exosomes have the ability to transport and deliver a variety of nucleic acids, including DNA, RNA, and microRNA, to a wide range of cell types, including difficult-to-transfect cells like endothelial cells, neurons, and stem cells [89].

Overcoming biological obstacles including the immune system, extracellular matrix, and cellular membranes that restrict the effectiveness and specificity of gene transfer is one of the major problems of gene delivery. Foreign gene carriers may be identified and eliminated by the immune system, which can lead to immunological tolerance, allergic responses, or inflammation. Because of the existence of collagen fibers, glycosaminoglycans, and proteoglycans, the extracellular matrix may impede the diffusion and penetration of gene carriers, particularly those that are big and positively charged. Depending on their size, charge, and surface characteristics, the cellular membranes may block gene carriers' entrance and release

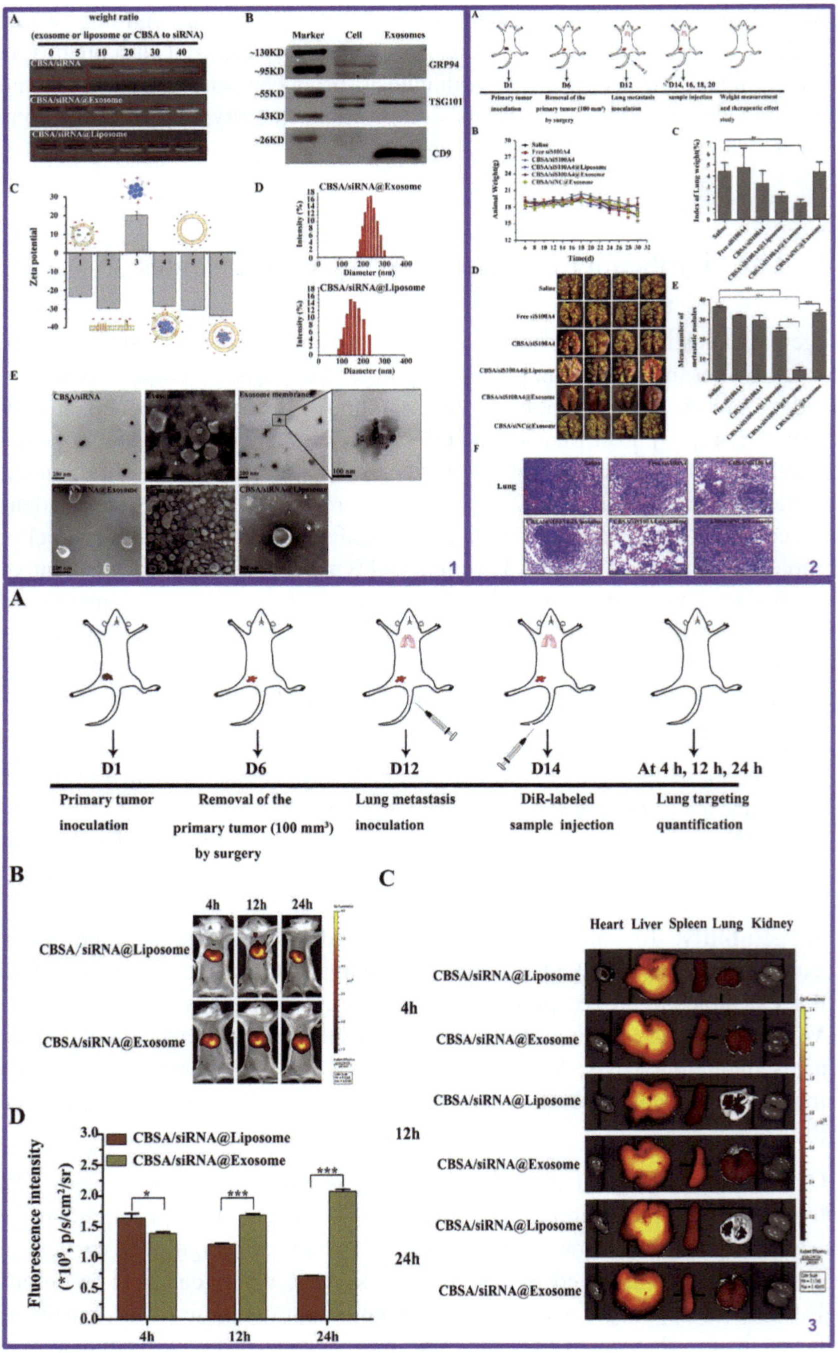
weight ratio
(exosome or liposome or CBSA to siRNA)
0 5 10 20 30 40
CBSA/siRNA
CBSA/siRNA@Exosome
CBSA/siRNA@Liposome
Marker Cell Exosomes
~130KD
~95KD
~55KD
~43KD
~26KD
GRP94
TSG101
CD9
Zeta potential
CBSA/siRNA@Exosome
CBSA/siRNA@Liposome
Intensity (%)
Diameter (nm)
Exosome membrane
Animal Weight(g)
Time(d)
Index of Lung weight(%)
Mean number of metastatic nodules
Lung
1
2
A
D1
D6
D12
D14
At 4 h, 12 h, 24 h
Primary tumor inoculation
Removal of the primary tumor (100 mm³) by surgery
Lung metastasis inoculation
DiR-labeled sample injection
Lung targeting quantification
B
4h 12h 24h
CBSA/siRNA@Liposome
CBSA/siRNA@Exosome
C
Heart Liver Spleen Lung Kidney
4h
12h
24h
CBSA/siRNA@Liposome
CBSA/siRNA@Exosome
D
Fluorescence intensity (*10⁹, p/s/cm²/sr)
3.0
2.5
2.0
1.5
1.0
0.5
0.0
*
***
***
4h 12h 24h
3

◄**Fig. 9.8** Provides a thorough examination of CBSA/siS100A4@exosome's physicochemical properties and therapeutic potential in avoiding lung metastases after surgery. A variety of weight ratios (1:0, 1:5, 1:10, 1:20, 1:30, and 1:40) of CBSA/siRNA, CBSA/siRNA@Exosome, and CBSA/siRNA@Liposome were assessed using agarose gel electrophoresis in the first panel (8.1). Information on the amounts of liposome lecithin, exosome protein, and CBSA protein related to siRNA was supplied by this evaluation. Using Western blot analysis, the expression levels of GRP94, TSG101, and CD9 in exosomes and breast cancer cells were determined. A number of compounds, including exosomes, membranes, liposomes, CBSA/siRNA, CBSA/siRNA@Exosome, and CBSA/siRNA@Liposome, were tested for zeta potentials using the NanoBrook 90Plus Zeta device. Dynamic light scattering (DLS) was used to evaluate the size distribution of CBSA/siRNA@Exosome and CBSA/siRNA@Liposome, and labeled transmission electron microscopy pictures showed CBSA/siRNA@Exosome and its predecessors. The ability of CBSA/siS100A4@exosome to prevent postoperative lung metastases is seen in the second panel (8.2). The figure shows models for lung metastases from surgery and medication. In the research, mice's body weight changes over certain time periods are shown, with a control group being treated with saline. The material includes pictures of lung tissues from different treatment groups as well as variations in the lung weight index after therapy. Importantly, the average number of visible lung metastases and the histological analysis using tissue sections stained with H&E provide further insight into the therapeutic outcomes. Exosome and CBSA/siS100A4@ precursor qualities are described in detail in the last panel (8.3). The best percentage ratios were found using agarose gel electrophoresis, and Western blot analysis was used to confirm that the proteins in exosomes and breast cancer cells were different. Using the NanoBrook 90Plus Zeta equipment, the samples' Zeta potentials were determined. CBSA/siRNA, CBSA/siRNA@Exosome, liposomes, exosomes, membranes, and CBSA/siRNA@Liposome were among the samples. Furthermore, both CBSA/siRNA@Exosome and CBSA/siRNA@Liposome size distributions were investigated by means of dynamic light scattering. The earliest forms of CBSA/siRNA@Exosome and their corresponding markers were seen in transmission electron microscope images, which further corroborated this [93]

[90]. By taking use of their inherent characteristics and modes of action, exosomes may get past these obstacles. Because they originate from the same or related cell types as the target cells and carry immunomodulatory molecules including CD47, CD59, and HLA-G that may prevent macrophage, natural killer, and T cell phagocytosis and activation, exosomes can avoid immune detection. Exosomes may penetrate interstitial gaps and penetrate the extracellular matrix to degrade its components because they are tiny and flexible enough. They create matrix metalloproteinases, including MMP-2 and MMP-9, to do this. Because of their similar lipid compositions and membrane proteins, exosomes and target cells may combine and enter cells via endocytic pathways like macropinocytosis, phagocytosis, clathrin-mediated endocytosis, and caveolin-mediated endocytosis [91]. The stromal barrier (SB), blood–retinal barrier (BRB), blood–labyrinth barrier (BLaB), blood–lymph barrier (BlyB), blood–air barrier (BAB), blood–brain barrier (BBB), blood–cerebrospinal fluid barrier (BCSFB), and placental barrier (PB) are among the difficult biological barriers that exosomes can pass through. The majority of chemicals can't get through the barriers, which are made up of basement membranes, tight junctions, and specialized transporters. Transferrin, lactoferrin, low-density lipoprotein, insulin, angiotensin-converting enzyme, or integrins are examples of receptors or ligands that exosomes may express to cross barriers. By appropriating the barrier cells' natural transport mechanisms, such as transcytosis, paracellular transport, or

receptor-mediated transport, exosomes may also get past these barriers [92]. The physical and chemical properties of CBSA/siS100A4@exosome are examined in Fig. 9.8, along with its potential as a treatment to stop lung metastases after surgery. In Fig. 9.8(1), the weight ratios of CBSA/siRNA, CBSA/siRNA@Exosome, and CBSA/siRNA@Liposome were evaluated using agarose gel electrophoresis. This investigation provides information on liposome lecithin, exosome protein, and CBSA protein levels in relation to siRNA. Western blot analysis was used to measure the expression levels of GRP94, TSG101, and CD9 in breast cancer cells and exosomes. The zeta potential of several substances, including liposomes, exosomes, membranes, CBSA/siRNA, CBSA/siRNA@Exosome, and CBSA/siRNA@Liposome, may be measured using the NanoBrook 90Plus Zeta apparatus. Using dynamic light scattering (DLS), the size distributions of CBSA/siRNA@Exosome and CBSA/siRNA@Liposome are examined. Transmission electron microscopy images labeled and showing CBSA/siRNA@Exosome and its progenitors provide evidence for the investigation. The effectiveness of CBSA/siS100A4@exosome in preventing the development of lung metastases after surgery is shown in Fig. 9.8(2). The research shows the variations in mouse body weight across certain time periods using models for surgical lung metastases and pharmaceutical treatment. Saline therapy was given to the control group. Complete understanding is achieved by comparing lung tissue pictures from different treatment groups and the changes in lung weight index after therapy. Additional information about the treatment outcomes is provided by histological evaluations utilizing tissue sections stained with H&E and the average number of visible lung metastases. The characteristics of exosomes and CBSA/siS100A4@ precursors are thoroughly reviewed in Fig. 9.8(3). Western blotting analysis is used to confirm the presence of certain proteins in exosomes and breast cancer cells, while agarose gel electrophoresis is used to identify the optimal percentage ratios. A variety of compounds, including liposomes, exosomes, membranes, CBSA/siRNA, CBSA/siRNA@Exosome, and CBSA/siRNA@Liposome, may have their zeta potential analyzed using the NanoBrook 90Plus Zeta device. The size distributions of CBSA/siRNA@Exosome and CBSA/siRNA@Liposome are investigated using the dynamic light scattering method. The earliest forms of CBSA/siRNA@Exosome and its associated markers are seen in transmission electron microscope pictures in this work [93].

9.5 Exosome Engineering and Targeted Delivery

Improving the capacity of gene carriers to target certain cells and tissues is another difficulty in gene delivery, which may increase therapeutic effectiveness and lessen negative effects of gene therapy. Gene carriers' surface characteristics, such as charge, hydrophobicity, and functional groups, influence their capacity to target cells and may have an impact on their pharmacokinetics, biodistribution, and interactions with the target cells. By altering their surface with certain ligands or antibodies that can attach to the receptors or antigens on the target cells, gene carriers may be made more targeted and increase the affinity and specificity of gene delivery. Because they may preferentially carry their payload to cells that share their parents' origin, phenotype, or tissue tropism, exosomes have an intrinsic ability to target cells. The brain, liver, kidney, and heart are among the damaged organs where mesenchymal stem cell-produced exosomes may selectively target and promote tissue regeneration. By specifically targeting antigen-presenting cells, such as B cells, dendritic cells, and macrophages, exosomes produced from immune cells may alter the immune response. Cancer cell-derived exosomes have the ability to interact and impact stromal cells, such as immune cells, fibroblasts, and endothelial cells, as well as malignancies. They may also increase resistance to treatment, encourage tumor growth, encourage the creation of new blood vessels, aid invasion, and help cancer cells move to other areas of the body [94]. For certain gene therapy applications, however, exosomes' inherent targeting capability may not be enough or precise enough, particularly if the target cells are uncommon, diverse, or inaccessible. Thus, by altering their surface with different molecules, including peptides, proteins, aptamers, or antibodies, that can detect the receptors or antigens on the target cell, exosome engineering has been created to improve the targeting ability of exosomes. To carry siRNA to the brain, exosomes may be modified to show the rabies virus glycoprotein (RVG) peptide, which can attach to the neuronal cells' nicotinic acetylcholine receptor. The epidermal growth factor receptor (EGFR) antibody, which may attach to the EGFR on cancer cells and transport doxorubicin to the tumor, can also be made to appear on exosomes [95]. Exosome engineering may enhance the loading capacity and release kinetics of exosomes for gene transport in addition to their targeting capabilities, which can impact the quantity and duration of gene expression in the target cells. Numerous variables, including the cellular origin, the stimuli, the sorting processes, and the separation techniques, may affect the biogenesis, composition, and interaction of exosomes with the target cells, which in turn affects their loading capacity and release kinetics. By altering the membrane or content of exosomes using techniques like membrane permeabilization, electroporation, sonication, or chemical conjugation, which can help incorporate or release nucleic acids into or from the exosomes, the loading capacity and release kinetics of exosomes can be improved. For instance, exosomes may be loaded with mRNA using electroporation, which uses an electric field to temporarily break the membrane, and permeabilized with saponin, which can enlarge the pore size of the exosomal membrane. Additionally, click chemistry may be used to conjugate with DNA, creating covalent connections between the DNA and the exosomal lipids, and ultrasound can be used to sonicate exosomes, producing cavitation bubbles that can rupture the exosomal membrane [96].

Because they can deliver their cargo of proteins, lipids, and nucleic acids to recipient cells, exosomes small extracellular vesicles released by a variety of cell types play crucial roles in intercellular communication. As possible biomarkers and treatments for a number of illnesses, including cancer, heart disease, neurological disorders, and infectious diseases, exosomes have also garnered a lot of attention. To fully use exosomes, however, effective and trustworthy techniques for their isolation and modification are needed. The current approaches of separating exosomes from different sources, including cell lines or body fluids, will be discussed in this chapter. Additionally, the strategies for efficiently altering exosomes to contain therapeutic genes will be highlighted [97].

Ultracentrifugation with Differentiation

The most popular and standardized technique for isolating exosomes based on their size is differential ultracentrifugation. In order to separate the exosomes from other components according to their sedimentation rate, the exosome-containing samples are subjected to a series of centrifugation stages with increasing speed and duration. Exosomes are found in the final pellet, while soluble proteins and other impurities are found in the supernatant. Washing and resuspending the exosome pellet in an appropriate buffer will help to purify it. The ability to separate exosomes from a variety of sources, including cell culture supernatants, plasma, urine, saliva, and cerebrospinal fluid, as well as the ability to produce a large number of exosomes with high purity and intact morphology, are two benefits of differential ultracentrifugation. Nevertheless, differential ultracentrifugation has the drawbacks of being labor-intensive, time-consuming, and requiring certain tools and knowledge. Additionally, exosomes may be mechanically damaged and aggregated by differential ultracentrifugation, which may also co-isolate other vesicles and pollutants of comparable size and density, including lipoproteins, apoptotic bodies, microvesicles, and protein aggregates [98, 99].

Another technique for isolating exosomes based on size is size-exclusion chromatography. In order to separate the exosomes from other components according to their molecular weight, the exosome-containing samples are passed down a column filled with porous beads. Smaller components, including soluble proteins and other pollutants, are retained in the beads and elute later, while the exosomes, which are more important than the pore size of the beads, are eluted in the void volume. Size-exclusion chromatography has the benefit of being able to separate exosomes from a variety of sources, including cell culture supernatants, plasma, urine, saliva, and cerebrospinal fluid, while still maintaining the integrity and biological activity of exosomes. Additionally, size-exclusion chromatography may readily be used with additional techniques like ultracentrifugation, precipitation, or immunoaffinity to further purify exosomes and can eliminate the majority of soluble proteins and other impurities. The drawbacks of size-exclusion chromatography include its lengthy processing time, high sample volume requirements, and potential to co-isolate contaminants and other vesicles of comparable sizes, including lipoproteins, apoptotic bodies, and microvesicles [100].

A technique for isolating exosomes based on the chemical property of charge is precipitation. By neutralizing their negative charge, a precipitation agent, such as polyethylene glycol, protamine sulfate, or organic solvents, may cause exosomes to aggregate and precipitate when added to samples that contain them. Low-speed centrifugation or filtering may be used to collect the exosome precipitate, which can then be resuspended in an appropriate buffer. Precipitation has the benefits of being easy, quick, and requiring no specific tools or expertise. Furthermore, a large number of exosomes with good recovery rates may be obtained by precipitation from a variety of sources, including cell culture supernatants, plasma, urine, saliva, and cerebrospinal fluid. Precipitation's drawbacks include the potential to change the composition and functionality of exosomes as well as co-precipitate other substances and impurities including soluble proteins, lipids, and nucleic acids [101].

A technique for isolating exosomes that relies on the biological characteristic of affinity is immunoaffinity. The idea behind this technique is to utilize antibodies that can identify certain exosome surface indicators, including integrins, heat shock proteins, or tetraspanins, in order to extract and concentrate exosomes from materials that contain them. To make exosome separation and elution easier, the antibodies may be immobilized on magnetic beads, microplates, or columns. The ability to separate exosomes with high specificity and purity and to distinguish between exosomes from various cell types or physiological situations using their surface markers are two benefits of immunoaffinity. Additionally, immunoaffinity can maintain the biological activity and integrity of exosomes and may be readily combined with other techniques for exosome purification, such as size-exclusion chromatography, ultracentrifugation, or precipitation. Immunoaffinity's drawbacks include its high cost, the need for certain antibodies and tools, and its limited ability to separate exosomes that express the target markers while excluding those that either do not express them at all or just express them weakly [102].

A technique for isolating exosomes that relies on the engineering characteristic of microfabrication is called microfluidics. The idea behind this technique is to separate exosomes from materials that contain them by using microscale machines that can work with tiny fluid volumes and carry out different processes such filtration, centrifugation, electrophoresis, or acoustic separation. To get the best possible performance and integration, the microfluidic devices may be developed and manufactured using a variety of materials, geometries, and functions. Microfluidics has the ability to manage tiny sample quantities, minimize sample loss and contamination, and separate exosomes with high speed, efficiency, and sensitivity [103]. Additionally, microfluidics may improve the automation and uniformity of exosome separation by combining many exosome isolation and analysis procedures into a single device, including sample preparation, exosome capture, exosome release, exosome characterisation, and exosome quantification. Microfluidics' drawbacks include its high cost, need for specific tools and expertise, and potential for technical difficulties with device construction, fluid management, exosome recovery, and compatibility [104]. A brief summary of the characteristics and functions of exosomes is given in Table 9.1. The size feature is discussed first, emphasizing the exosome's 30–150 nm range. It also highlights how important they are for enabling intercellular communication and cargo movement. The similarities to the plasma membrane, its function in preserving membrane integrity, and its capacity to facilitate fusion with recipient cells are highlighted in the following description of the lipid composition. The protein composition, which varies according to the particular cell type and physiological condition, is the subject of the following description. It draws attention to the protein's involvement in enzyme activity, immunological control, and receptor binding. The makeup of nucleic acids, including DNA, mRNA, microRNA, long non-coding RNA, and circular RNA, is well covered throughout the book.

The primary emphasis is on the role these chemicals play in regulating genes and producing proteins. The stability of exosomes in various body fluids as well as their capacity for transit and storage are the main topics. A number of processes, such as direct fusion, receptor-mediated endocytosis, phagocytosis, macropinocytosis, and lipid raft-mediated internalization, are thoroughly explained in the section that follows, which concentrates on absorption mechanisms. It becomes clear how important these mechanisms are for moving cargo to the endosomal, cytoplasmic, or nuclear compartments [104].

Exosome modification is a necessary step in their use as it may enhance their specificity and functionality. However, because of their complexity, variability, and tiny size, exosomes are difficult to change. Consequently, a number of techniques, including loading, labeling, and exosome engineering, have been devised to alter exosomes according to distinct tactics and goals. Here, we will emphasize the strategies for successfully changing exosomes to deliver therapeutic genes and provide a quick overview of the main exosome modification approaches [105].

By adding foreign molecules, such as proteins, medicines, or nucleic acids, to exosomes either before or after they are isolated, exosome loading is a technique for altering exosomes. By delivering the required chemicals to the target cells or tissues, exosome loading increases the therapeutic potential of exosomes. Numerous techniques, including electroporation, sonication, saponin treatment, freeze-thaw cycles, and incubation, may be used to load the exogenous molecules into the exosomes. These techniques can help the molecules enter the exosomes by rupturing or permeabilizing the exosomal membrane [106]. Exosome loading offers the benefits of increasing the loading capacity and delivery efficiency of exosomes as well as facilitating the delivery of different molecules, including proteins, peptides, antibodies, microRNA, siRNA, shRNA, CRISPR/Cas9, DNA, RNA, microRNA, siRNA, and

Table 9.1 Features and roles of exosomes

Feature	Function	Description
Size	Cellular communication, intercellular cargo transfer	30–150 nm
Lipid composition	Membrane stability, fusion with recipient cells	Reflects plasma membrane
Protein composition	Receptor binding, immune modulation, enzyme activity	Varies depending on cell type and physiological condition
Nucleic acid composition	Gene regulation, protein synthesis	mRNA, microRNA, long non-coding RNA, circular RNA, DNA
Stability	Potential for storage and transportation	Stable in various biological fluids
Uptake	Delivery of cargo to the cytoplasm, nucleus, or endosomal compartments	Direct fusion, receptor-mediated endocytosis, phagocytosis, macropinocytosis, lipid raft-mediated internalization

drugs, to modify the target cells' function, phenotype, or gene expression. Furthermore, exosome loading may be carried out on exosomes that have been separated from other sources, including cell lines or body fluids, and it can be used in conjunction with other techniques, like exosome labeling or engineering, to maximize the use of exosomes even more [107].

Small vesicles called exosomes, which are secreted by cells, have become important intercellular communication mediators and offer enormous potential for developing gene therapy for breast cancer. Because of their inherent capacity to carry a variety of chemicals, these amazing vesicles are a great choice for administering therapeutic drugs. By using the capacity of transporters to deliver nucleic acids, like DNA, mRNA, microRNA, and siRNA, researchers may modify and accurately regulate the expression of genes linked to cancer [108]. Additionally, exosomes can contain therapeutic proteins like peptides, antibodies, and enzymes that can be directed specifically at cancer cells to activate the immune system or exert their therapeutic effects, strengthening the body's defenses against malignant growths. Exosomes improve treatment efficacy and reduce side effects by enabling the exact delivery of chemotherapeutic drugs and targeted therapy to cancer cells. Finding certain genes and medicinal substances has shown encouraging outcomes in a pilot investigation. In breast cancer models, exosomes containing tumor suppressor genes including p53, BRCA1, and PTEN have shown promise in preventing tumor development and cancer cell metastasis [109]. Furthermore, it has been shown that exosomes containing miRNA or siRNA molecules that target oncogenes or important signaling molecules may effectively decrease tumor growth and induce death in breast cancer cells. Promising results from preclinical research examining exosomes with siRNA that selectively targets the HER2 gene highlight the potential of exosomes as a targeted therapy for breast cancer. Exosomes may be altered and loaded with immune checkpoint inhibitors, like PD-1 and CTLA-4 antibodies, which have shown promise in boosting the immune response against tumors and boosting the efficacy of immunotherapy in breast cancer models. By combining immunotherapeutic techniques with exosome-mediated gene therapy, this novel strategy offers potential for the synergistic treatment of breast cancer [110]. The potential of exosomes to carry enzymes that activate prodrugs is another fascinating use for them in gene therapy. These enzymes are unique in that they can only change inactive prodrugs into their active form within the tumor microenvironment, greatly increasing chemotherapy's effectiveness while causing the least amount of damage to healthy tissues. Chemotherapy treatments for breast cancer may be much more successful with this tailored activation approach [111]. When it comes to breast cancer, exosomes are a very promising and focused delivery system for gene therapy. Because of their ability to carry a variety of chemicals and their natural affinity to cancer cells, they are a good option for improving therapy outcomes. Exosome-based gene therapy has the potential to completely transform the way that cancer is treated, especially breast cancer and other cancers. As further research and developments in this area continue, this novel technique gives patients fresh hope and a higher quality of life [112]. The advantages of delivering genes via exosomes are listed in Table 9.2. First off, natural exosomes are biocompatible, which means they are not very harmful or immunogenic. This enables them to evade immune detection and reticuloendothelial system clearance. Second, the lipid bilayer membrane of exosomes gives them a stability advantage by preventing cargo degradation and preserving the integrity and functioning of nucleic acids. Additionally, exosomes may be modified to exhibit certain ligands or antibodies, which makes them targetable and enhances the efficiency and specificity of gene delivery. Finally, since exosomes are adaptable and may transport various nucleic acids, target cells' gene expression, function, or phenotype can be altered.

In Vitro Studies: Research on breast cancer has shown that gene transfer via exosomes is a viable approach, according to experiments carried out in a controlled laboratory setting. Exosomes have been shown in studies to efficiently transfer therapeutic genes to breast cancer cells, leading to a number of beneficial effects. The inhibition of tumor development and proliferation is a noteworthy outcome that has been noted. By delivering exosomes encoding tumor suppressor genes such p53, BRCA1, and PTEN, or by introducing siRNA molecules that selectively target oncogenes, it is possible to successfully block the development and proliferation of breast cancer cells in vitro. This discovery highlights the ability of exosomes to transport gene therapies that may directly disrupt cancer cell signaling pathways and encourage the suppression of tumor development. Furthermore, it has been shown that exosomes may cause breast cancer cells to undergo apoptosis, or programmed cell death. Exosomes containing pro-apoptotic genes or miRNA molecules have been shown in vitro to efficiently initiate cell death pathways and cause apoptosis. Researchers have identified a possible strategy to lower tumor cell viability and slow the spread of cancer by using exosomes' innate capacity to transfer pro-apoptotic proteins [113]. Exosomes may transport chemotherapeutic medications, enabling their targeted administration to certain cells, and they have been shown to improve the responsiveness of breast cancer cells to therapy. Tumor development is inhibited and cell death is increased by this focused administration. Chemotherapy for breast cancer may be more successful with this tailored drug delivery strategy, which might result in better treatment outcomes. Furthermore, by modifying exosomes and expressing immune checkpoint inhibitors or other immunomodulatory substances

Table 9.2 Benefits of gene delivery via exosomes

Advantage	Explanation	Significance
Biocompatibility	Derived from natural biological sources, low immunogenicity and toxicity	Evades immune recognition, avoids clearance by reticuloendothelial system
Stability	The lipid bilayer membrane protects cargo from degradation	Maintains integrity and functionality of nucleic acids
Targetability	It can be engineered to display specific ligands or antibodies	Enhanced specificity and efficiency of gene delivery
Versatility	Can carry different types of nucleic acids	Modulates gene expression, function, or phenotype of target cells

that might improve the immune system's capacity to identify and combat cancer cells, it is possible to control the immune response in breast cancer. Exosomes' ability to modulate immune systems offers a chance to combine immunotherapeutic and exosome-based gene therapy, which might have a synergistic effect on breast cancer treatment [114]. Overall, the outcomes of in vitro tests on exosome-mediated gene transfer in breast cancer research have been encouraging. Exosomes have the ability to alter the immune system's response, promote medication responsiveness, cause programmed cell death, and obstruct tumor development. These traits demonstrate their potential use as tools for targeted gene therapy in the treatment of breast cancer. The prognosis and quality of life for individuals with breast cancer may be improved by further research and development in this area, which may convert these discoveries into efficient therapy strategies [115].

In Vivo Studies: The effectiveness and safety of exosome-mediated gene transfer as a treatment for breast cancer have been evaluated in large part via animal studies. This study has shed important light on the possible advantages and disadvantages of this strategy. In vivo experiments using animal models of breast cancer have shown that exosome-mediated gene transfer is effective in causing tumor regression and preventing tumor development. When researchers load exosomes with therapeutic genes or siRNA molecules that precisely target genes linked to cancer, they have seen a reduction in tumor growth and an overall regression of tumors. According to the findings, exosomes may be useful delivery systems for therapeutic genes to be delivered to animal models of breast cancer cells. Exosomes have the ability to prevent breast cancer cells from spreading, according to research on animals [116]. Researchers using gene therapy based on exosomes have shown a reduction in the development of metastases in distant organs. This illustrates how exosomes may prevent the spread of cancer cells, impacting the course of the illness and the results for patients. Exosome-mediated gene transfer has also been shown to increase survival rates in breast cancer models in animal studies. In breast cancer models, mice's lifespans have been shown to be extended by exosome-mediated trafficking of therapeutic genes or siRNA molecules [117]. These results imply that utilizing exosomes for gene therapy might be a viable therapeutic option to enhance overall survival in breast cancer, but further study is needed to validate these findings and comprehend the underlying processes. In animal investigations, exosomes have shown a generally positive safety profile. Exosome-based gene therapy has been shown to have no toxicity or side effects. It is crucial to remember that further research is required to assess the long-term safety of this approach and any potential unforeseen consequences. Before using these discoveries in therapeutic settings, it is essential to comprehend the possible dangers and guarantee the security of exosome-mediated gene delivery [118]. In conclusion, studies conducted on animals to assess the effectiveness of exosome-mediated gene delivery in breast cancer have shown encouraging findings. Exosomes have been shown in animal models to improve survival, decrease metastasis, and impede tumor development. Even though the short-term safety record seems good, further research is required to evaluate the long-term safety and any side effects. These results provide the groundwork for more research

Table 9.3 Exosome gene delivery's potential for breast cancer treatment

Aspect	Key points	Implications
Preclinical studies	Exosome-based gene delivery has shown promising results in inhibiting tumor growth, metastasis, and drug resistance	Demonstrates the therapeutic potential of exosomes in breast cancer
Targeted delivery	Exosomes can be engineered to target breast cancer cells specifically	Reduces off-target effects, increases therapeutic efficacy
Combination therapies	Exosome-based gene delivery can be combined with other modalities, such as chemotherapy, radiotherapy, or immunotherapy	Synergistic anti-tumor effects overcome resistance mechanisms
Clinical applications	Clinical trials are needed to evaluate the safety and efficacy of exosome-based gene delivery in breast cancer patients	Assess long-term effects compared with standard treatments

and the possible use of exosome-mediated gene therapy in breast cancer clinical treatment [119]. Many aspects of employing exosomes to transfer genes in breast cancer investigations are included in Table 9.3. Preclinical research has shown that this approach effectively prevents tumor growth, cancer cell metastasis to other body areas, and medication resistance. This demonstrates how exosomes may be used to treat breast cancer. The focus on targeted delivery draws attention to exosomes' ability to be engineered to specifically target breast cancer cells, which may reduce side effects and increase the overall efficacy of treatment. The discussion of combination therapies also emphasizes how exosome-based gene delivery might be used in conjunction with therapies like chemotherapy, radiation, or immunotherapy to fight cancers in a way that reinforces each other. Resistance mechanisms are effectively overcome by this method. The need of conducting clinical trials to assess the efficacy and safety of delivering genes to patients with breast cancer via exosomes is emphasized in the section on clinical applications. Gaining insight into the long-term effects in contrast to traditional therapy is the goal.

9.6 Exosomal Gene Delivery Drawbacks and Restrictions in Breast Cancer Treatment

There are significant manufacturing and scaling challenges for exosome-based gene therapy. Exosome production and separation are complex and time-consuming processes that make it difficult to produce them in sufficient quantities for medical applications. It is still challenging to guarantee homogeneity and consistency in the synthesis and characterisation of exosomes since different batches may differ in terms of cargo quantity, purity, and content. For exosome-based therapies to be widely used, these challenges must be overcome [120]. The potential for off-target effects and a lack of selectivity are problems with exosome-mediated gene delivery. Unwanted side effects and unforeseen biological repercussions may arise from exosomes' ability to deliver their contents to cells that are not their intended targets. To decrease off-target delivery and boost treatment efficiency, exosomes must be precisely and efficiently targeted to breast cancer cells. Other considerations for exosome-based gene therapy include immunogenicity and immune response. Exosomes may trigger an immune response because they are made from external cells or contain exogenous substances. Depending on their origin and modifications, exosomes might have varying immunogenicities, which may lead to immunological rejection or adverse reactions. For exosomes to be employed effectively in medical therapies, it is essential to learn more about them and lessen the possibility of immunological reactions they may induce. Exosomes' ability to cross the blood–brain barrier (BBB) presents another challenge to their therapeutic use, namely in the treatment of cancers of the central nervous system or brain metastases [121]. In other cases, the effectiveness of molecules like exosomes is limited by the blood–brain barrier (BBB), which prevents them from entering the brain. Finding strategies to improve exosome distribution to the brain and their capacity to traverse the blood–brain barrier (BBB) is crucial [122]. In order to optimize the therapeutic advantages of exosomes in the treatment of brain cancer, this is essential. Exosomes have a limited ability to transport therapeutic cargo, including nucleic acids. Both the overall therapeutic efficacy and the efficiency of gene delivery may be impacted by this limitation [123]. Optimizing the processes for loading exosomes and using engineering approaches are required to increase the capacity of exosomes to transport cargo and improve the efficacy of gene delivery. The long-term safety and any negative consequences of exosome-based gene therapy need more investigation. The possible risks associated with long-term or frequent usage of exosomes, such as possible accumulation, immunological reactions, or adverse effects on healthy cells or tissues, need to be thoroughly examined. For exosome-based medicines to be successfully used in clinical settings, a thorough grasp of their long-term safety features is essential [124]. In conclusion, a number of challenges need to be addressed even if exosome-based gene therapy has great promise in the treatment of breast cancer. It is essential to thoroughly examine a number of important research topics in order to completely realize the promise of exosome-mediated gene therapy in breast cancer and other disorders. These include overcoming manufacturing and scalability issues, minimizing off-target effects and guaranteeing specificity, understanding immunogenicity and immune response, improving blood–brain barrier penetration, increasing cargo loading capacity, and assessing long-term safety and toxicity [125].

One practical way to address issues with exosome-based gene therapy is to develop exosomes for better distribution. By altering their surface molecules or ligands, exosomes may be engineered to have a greater affinity for certain receptors on breast cancer cells. This will improve the drug's distribution to the targeted area and lessen its adverse effects. This targeted approach maximizes the effectiveness of gene delivery while minimizing potential side effects. Optimizing exosome loading methods may also increase the effectiveness of gene delivery. Innovative loading methods like click chemistry or microfluidics can more successfully and efficiently load exosomes with therapeutic cargo. These advancements make it possible to deliver more therapeutic genes, increasing the efficacy of therapy. Combining exosome-mediated gene delivery with other therapeutic modalities like radiation, chemotherapy, or immunotherapy may result in synergistic anti-tumor effects. It is possible to overcome resistance mechanisms and enhance the overall treatment response by combining various medicines. Investigating novel exosome sources, such as immune or stem cells, presents exciting potential. These alternative sources' cargo, targeting, or immunomodulatory properties may be beneficial and enhance exosome-mediated gene delivery to breast cancer cells. Immunomodulation techniques are essential for lowering immunological rejection and improving therapy efficacy. By altering them to express immune-modulatory substances or by decreasing their immunogenicity using immunosuppressive medications, exosomes may be rendered safer and more efficient. To overcome this challenge, exosomes that are designed to produce peptides or ligands that cross the blood–brain barrier (BBB) may be more effectively carried across the barrier. This advancement may lead to a greater usage of exosomes in the treatment of brain metastases and cancers of the central nervous system. Clinical studies must be carried out to evaluate the efficacy and safety of exosome-based gene delivery in breast cancer patients. While phase I/II studies can identify the safety profiles and optimal dosage schedules, phase III trials can assess the effectiveness of exosome-based therapy in clinical settings by comparing it with traditional therapies. Translational research is crucial to bridging the knowledge gap between basic research and clinical applications. To aid in the clinical translation of exosome-based gene therapy for conditions like breast cancer, translational research focuses on exosome production, characterisation, and delivery mechanism improvement. Focusing on a few key areas is necessary to fully realize the potential of exosome-mediated gene delivery in the treatment of breast cancer. These areas include engineering exosomes for improved delivery, improving loading methods, looking into new exosome sources, creating immunomodulation plans, addressing BBB penetration, carrying out clinical trials, and advancing translational research.

9.7 Clinical Trials and Ongoing Research

The safety and effectiveness of exosome-based treatments for breast cancer are being assessed in a number of clinical studies. Researchers are looking at treating patients with locally advanced or metastatic breast cancer using exosomes made from mesenchymal stem cells loaded with doxorubicin in one Phase I/II study (NCT03644679). Exosomes carrying therapeutic payload are being investigated as a possible source from mesenchymal stem cells. The safety and viability of exosomes made from dendritic cells modified to express tumor-specific antigens in patients with HER2-positive breast cancer are being evaluated by another Phase I study (NCT04042616). By creating exosomes from dendritic cells, researchers hope to elicit an immunological response against HER2-positive breast cancer cells.

Exosomes produced from macrophages modified to express a chimeric antigen receptor (CAR) targeting HER2 are also being assessed in a Phase I/II study (NCT04532207). The purpose of the modified exosomes is to improve the targeting and destruction of HER2-positive breast cancer cells, since macrophages are essential for immune responses. Finally, in patients with HER2-positive breast cancer, a Phase I study (NCT04860333) examines the safety and viability of exosomes made from natural killer cells modified to produce a CAR targeting HER2. By using the special qualities of exosomes and their potential for targeted delivery and immune regulation, these studies together show the continuous attempts to create exosome-based treatments that may enhance treatment results for patients with breast cancer [126–128].

9.8 Conclusion

In the context of treating breast cancer, exosomes have shown great promise as a gene delivery vehicle. Their special qualities as naturally occurring nanocarriers made from different cell types such as tumor cells, dendritic cells, and mesenchymal stem cells (MSCs) offer clear benefits over conventional delivery methods. This last part will examine the state of exosome-based treatments for breast cancer, highlighting new developments, present obstacles, and potential avenues for further study and clinical implementation. Exosomes are perfect candidates for therapeutic uses because of their intrinsic benefits. The transfer of bioactive molecules like proteins, lipids, RNAs, and DNAs is made possible by their tiny size (30–150 nm), lipid bilayer membrane, and surface proteins, which all aid in effective absorption by recipient cells. Exosomes have been thoroughly investigated in the field of breast cancer because of their capacity to alter gene expression, rewire signaling networks, and affect phenotypic alterations in cancer cells. Exosome-mediated gene delivery has been shown in preclinical research to be effective in stopping tumor growth, halting metastasis, and overcoming treatment resistance mechanisms in models of breast cancer. For example, exosomes designed to deliver therapeutic payloads like miRNAs controlling important signaling pathways or siRNAs targeting oncogenes have shown encouraging outcomes in slowing the growth of tumors. Furthermore, exosomes' natural biocompatibility and capacity to avoid immune detection add to their allure as secure and efficient delivery systems. The therapeutic potential of these nanovesicles has increased due to recent developments in exosome engineering. Optimizing treatment results has mostly depended on methods to target certain cell types within the tumor microenvironment, increase exosome loading efficiency, and improve stability throughout circulation. To improve the selectivity and effectiveness of exosome-mediated treatments, methods including surface modification with ligands that target receptors unique to cancer cells or the addition of stimuli-responsive materials have been investigated. Additionally, there is potential for synergistic therapeutic benefits when exosome-based gene delivery is combined with other treatment modalities including immunotherapy, radiation therapy, and chemotherapy. By circumventing resistance mechanisms and improving the delivery of therapeutic drugs directly to cancer cells while reducing systemic toxicity, exosomes may increase the effectiveness of traditional treatments. The clinical translation of exosome-based treatments for breast cancer is hampered by a number of issues and restrictions, despite their great promise. Significant obstacles include things like standardization of therapeutic cargo loading protocols, scalability of manufacturing, and variation in isolation and purification techniques. Furthermore, careful preclinical testing and delivery protocol improvement are required due to worries about the immunogenicity of exosomes and possible off-target effects. Therapeutic targeting of exosomes is made more difficult by the intricacy of the tumor microenvironment, which is defined by dynamic interactions between immune cells, stromal cells, and cancer cells. A crucial topic of research is how to maximize exosomes' selectivity for tumors while reducing their interactions with non-target organs. There are a number of research directions that might advance exosome-based treatments for breast cancer in the future. Standardizing procedures for exosome separation, characterisation, and cargo loading should be the primary goal in order to guarantee repeatability and regulatory compliance. Exosomes with better pharmacokinetics and biophysical characteristics may be engineered for systemic distribution thanks to developments in nanotechnology and biomaterial sciences. Investigating new exosome sources, such as genetically modified cell lines or engineered stem cells, may provide exosomes with improved targeting specificity and customized therapeutic potential. Furthermore, new opportunities for treating metastatic breast cancer and other cancers affecting inaccessible locations are made possible by the creation of exosomes that can penetrate biological barriers like the blood–brain barrier. To prove the safety, effectiveness, and long-term advantages of exosome-based treatments for patients with breast cancer, clinical validation via well planned and sufficiently powered clinical studies is crucial. Critical information on the therapeutic usefulness and possible superiority of exosome-based methods will be provided by comparative studies comparing the results of exosome-mediated gene delivery with standard-of-care therapies. To sum up, exosome-based gene delivery is a revolutionary strategy that will completely change the field of breast cancer treatment. Researchers and medical professionals may create tailored and focused therapies that tackle the intricacies of breast cancer biology by using exosomes' inherent capabilities as nanocarriers. Overcoming present obstacles and achieving the full therapeutic potential of exosome-based medicines in clinical practice need sustained multidisciplinary cooperation and investment in translational research. The next generation of precision medicine approaches to treating breast cancer will be made possible by developments in exosome engineering and thorough clinical validation as the field develops. We may aim to enhance patient outcomes, reduce treatment-related toxicity, and eventually accomplish more efficient control and management of breast cancer by using exosomes' special ability to transport therapeutic payloads directly to cancer cells.

References

1. M. Sadeghi, S. Sadeghi, S.M. Naghib, H.R. Garshasbi, A comprehensive review on electrochemical nano biosensors for precise detection of blood-based oncomarkers in breast cancer. Biosensors **13**(4), 481 (2023)
2. R. Salahandish, A. Ghaffarinejad, S.M. Naghib, K. Majidzadeh-A, H. Zargartalebi, A. Sanati-Nezhad, Nano-biosensor for highly sensitive detection of HER2 positive breast cancer. Biosens. Bioelectron. **117**, 104–111 (2018). https://doi.org/10.1016/j.bios.2018.05.043
3. R. Salahandish, A. Ghaffarinejad, E. Omidinia, H. Zargartalebi, K. Majidzadeh-A, S.M. Naghib, A. Sanati-Nezhad, Label-free ultrasensitive detection of breast cancer miRNA-21 biomarker employing electrochemical nano-genosensor based on sandwiched AgNPs in PANI and N-doped graphene. Biosens. Bioelectron. **120**, 129–136 (2018). https://doi.org/10.1016/j.bios.2018.08.025
4. N. Shahbazi, R. Zare-Dorabei, S.M. Naghib, Design of a ratiometric plasmonic biosensor for herceptin detection in HER2-positive breast cancer. ACS Biomater. Sci. Eng. **8**(2), 871–879 (2022). https://doi.org/10.1021/acsbiomaterials.1c01369
5. P. Vosoughi, S.M. Naghib, M. Rahmanian, M.R. Mozafari, A comprehensive review of mRNA-based vaccines for COVID-19, a new era in pharmaceuticals: unspecified and unknown aspects, effects and challenges. Curr. Top. Med. Chem. (2025). https://doi.org/10.2174/0115680266325847241121034100
6. M. Yazdan, S.M. Naghib, M.R. Mozafari, Liposomal nano-based drug delivery systems for breast cancer therapy: recent advances and progresses. Anti-Cancer Agents Med. Chem. (Formerly Curr. Medi. Chem. Anti-Cancer Agents) **24**(12), 896–915 (2024). https://doi.org/10.2174/0118715206293653240322041047
7. A.-A. Zare, H. Naderi-Manesh, S.M. Naghib, M. Shamsipur, F. Molaabasi, Label-free electrochemical cancer cell detection leveraging hemoglobin-encapsulated silver nanoclusters and Cu-MOF nanohybrids on a graphene-assisted dual-modal probe. Sci. Rep. **13**(1), 21980 (2023). https://doi.org/10.1038/s41598-023-49418-1
8. S.M. Naghib, B. Ahmadi, M. Mozafari, Stimuli-sensitive chitosan-based nanosystems-immobilized nucleic acids for gene therapy in breast cancer and hepatocellular carcinoma. Curr. Top. Med. Chem. **24**(17), 1464–1489 (2024)
9. N. Shanehband, S.M. Naghib, Microfluidics-assisted tumor cell separation approaches for clinical applications: an overview on emerging devices. Comb. Chem. High Throughput Screening **28**(2), 202–225 (2025)
10. A. Behera, A.K. Nayak, R.K. Mohapatra, A.A. Rabaan, *Smart Micro- and Nanomaterials for Drug Delivery* (CRC Press, 2024)
11. T. Zafar, A.Q. Naik, M. Kumar, V.K. Shrivastava, Epidemiology and risk factors of breast cancer, in *Breast Cancer: From Bench to Personalized Medicine* (Springer, 2022), pp. 3–29
12. A.M. Karim, J. Eun Kwon, T. Ali, J. Jang, I. Ullah, Y.-G. Lee, D.W. Park, J. Park, J.W. Jeang, S.C. Kang, Triple-negative breast cancer: epidemiology, molecular mechanisms, and modern vaccine-based treatment strategies. Biochem. Pharmacol. **212**, 115545 (2023). https://doi.org/10.1016/j.bcp.2023.115545
13. Z. Mirza, S. Karim, Nanoparticles-based drug delivery and gene therapy for breast cancer: recent advancements and future challenges, in *Seminars in Cancer Biology* (Elsevier, 2021), pp. 226–237
14. A. Mimoun, M. Bou-Jaoudeh, S. Delignat, V. Daventure, A.R. Ruiz, M. Lecerf, A. Azam, R. Noe, I. Peyron, O.D. Christophe, Transplacental delivery of therapeutic proteins by engineered immunoglobulin G: a step toward perinatal replacement therapy. J. Thromb. Haemost. **21**(9), 2405–2417 (2023)
15. S.G. Yilmaz, Genome editing technologies: CRISPR, LEAPER, RESTORE, ARCUT, SATI, and RESCUE. EXCLI J. **20**, 19 (2021)
16. S.M. Naghib, B. Ahmadi, B. Mikaeeli Kangarshahi, M.R. Mozafari, Chitosan-based smart stimuli-responsive nanoparticles for gene delivery and gene therapy: recent progresses on

cancer therapy. Int. J. Biol. Macromol. **278**, 134542 (2024). https://doi.org/10.1016/j.ijbiomac.2024.134542
17. N.A. Abtahi, S.M. Naghib, F. Haghiralsadat, M. Akbari Edgahi, E. Askari, A comparative study on biopharmaceutical function of curcumin and miR-34a by multistimuli-responsive nanoniosome carrier: in-vitro and in-vivo. Front. Mol. Biosci. **9**, 1043277 (2022). https://doi.org/10.3389/fmolb.2022.1043277
18. N.A. Abtahi, S.M. Naghib, F. Haghiralsadat, M. Akbari Edgahi, Development of highly efficient niosomal systems for co-delivery of drugs and genes to treat breast cancer in vitro and in vivo. Cancer Nanotechnol. **13**(1), 1–18 (2022)
19. A. Beigi, S.M. Naghib, A. Matini, M. Tajabadi, M.R. Mozafari, Lipid-based nanocarriers for targeted gene delivery in lung cancer therapy: exploring a novel therapeutic paradigm. Curr. Gene Ther. (2024). https://doi.org/10.2174/0115665232292768240503050508
20. N.A. Alhakamy, D.T. Curiel, C.J. Berkland, The era of gene therapy: from preclinical development to clinical application. Drug Discov. Today **26**(7), 1602–1619 (2021)
21. B. Mikaeeli Kangarshahi, S.M. Naghib, N. Rabiee, DNA/RNA-based electrochemical nanobiosensors for early detection of cancers. Crit. Rev. Clin. Lab. Sci. 1–23. https://doi.org/10.1080/10408363.2024.2321202
22. G.K. Kashani, S.M. Naghib, S. Soleymani, M.R. Mozafari, A review of DNA nanoparticles-encapsulated drug/gene/protein for advanced controlled drug release: current status and future perspective over emerging therapy approaches. Int. J. Biol. Macromol. **268**, 131694 (2024). https://doi.org/10.1016/j.ijbiomac.2024.131694
23. P. Vosoughi, S.M. Naghib, B.M. Kangarshahi, M.R. Mozafari, A review of RNA nanoparticles for drug/gene/protein delivery in advanced therapies: current state and future prospects. Int. J. Biol. Macromol. **295**, 139532 (2025). https://doi.org/10.1016/j.ijbiomac.2025.139532
24. N. Sayed, P. Allawadhi, A. Khurana, V. Singh, U. Navik, S.K. Pasumarthi, I. Khurana, A.K. Banothu, R. Weiskirchen, K.K. Bharani, Gene therapy: comprehensive overview and therapeutic applications. Life Sci. **294**, 120375 (2022)
25. K. Entzian, A. Aigner, Drug delivery by ultrasound-responsive nanocarriers for cancer treatment. Pharmaceutics **13**(8), 1135 (2021)
26. S. Lin, Z. Yu, D. Chen, Z. Wang, J. Miao, Q. Li, D. Zhang, J. Song, D. Cui, Progress in microfluidics-based exosome separation and detection technologies for diagnostic applications. Small **16**(9), 1903916 (2020)
27. S. Gandhi, I. Roy, Lipid-based inhalable micro-and nanocarriers of active agents for treating non-small-cell lung cancer. Pharmaceutics **15**(5), 1457 (2023)
28. C. Wang, C. Pan, H. Yong, F. Wang, T. Bo, Y. Zhao, B. Ma, W. He, M. Li, Emerging non-viral vectors for gene delivery. J. Nanobiotechnol. **21**(1), 272 (2023)
29. F.S. Basingab, O.A. Alshahrani, I.H. Alansari, N.A. Almarghalani, N.H. Alshelali, A.H. Alsaiary, N. Alharbi, K.A. Zaher, From pioneering discoveries to innovative therapies: a journey through the history and advancements of nanoparticles in breast cancer treatment. Breast Cancer: Targets Therapy 27–51 (2025)
30. M.R. Cring, V.C. Sheffield, Gene therapy and gene correction: targets, progress, and challenges for treating human diseases. Gene Ther. **29**(1), 3–12 (2022)
31. K. Cheng, H. Di, C. Wu, H. Zhao, W. Li, Mapping knowledge landscapes and themes trends of breast cancer stem cells: a comprehensive data-mining-based study. Ann. Surg. Oncol. 1–19 (2025)
32. H.M. Goodman, Chapter 14—Hormonal control of pregnancy and lactation, in *Basic Medical Endocrinology*, 4th edn., ed. by H.M. Goodman (Academic Press, San Diego, 2009), pp. 277–301. https://doi.org/10.1016/B978-0-12-373975-9.00014-8
33. S. Kabiri, R. Rahimi, M. Mozafari, S.M. Naghib, Porphyrin-based metal–organic frameworks in drug delivery for cancer therapy: promises, advances and prospects. Discover Appli. Sci. **7**(4), 313 (2025)
34. R. Gholami, S.M. Naghib, *Bone Cells, Nanobioceramics for Bone Tissue Engineering and Regenerative Biomedicine: Tunable Biological Characteristics* (Springer, 2025), pp. 43–77

35. M. Zhang, S. Hu, L. Liu, P. Dang, Y. Liu, Z. Sun, B. Qiao, C. Wang, Engineered exosomes from different sources for cancer-targeted therapy. Signal Transduct. Target. Ther. **8**(1), 124 (2023)
36. R. Mohammadinejad, A. Dehshahri, V.S. Madamsetty, M. Zahmatkeshan, S. Tavakol, P. Makvandi, D. Khorsandi, A. Pardakhty, M. Ashrafizadeh, E.G. Afshar, In vivo gene delivery mediated by non-viral vectors for cancer therapy. J. Control. Release **325**, 249–275 (2020)
37. Z. Ma, Y. Wu, G. Li, J. Liu, Z. Geng, J. Su, Extracellular vesicles-loaded DNA hydrogels: a promising candidate for cartilage organoids engineering. Chem. Eng. J. **477**, 147146 (2023)
38. B. Moon, S. Chang, Exosome as a delivery vehicle for cancer therapy. Cells **11**(3), 316 (2022)
39. A. Matini, S.M. Naghib, M. Mozafari, Quantum dots in cancer theranostics: a thorough review of recent advancements in bioimaging, tracking, and therapy across various cancer types. Curr. Pharm. Biotechnol. (2025)
40. Y. Li, F. Liu, Q. Cai, L. Deng, Q. Ouyang, X.H.-F. Zhang, J. Zheng, Invasion and metastasis in cancer: molecular insights and therapeutic targets. Signal Transduct. Target. Ther. **10**(1), 57 (2025)
41. Z. Han, H. Chen, Z. Guo, J. Shen, W. Luo, F. Xie, Y. Wan, S. Wang, J. Li, J. He, Circular RNAs and their role in exosomes. Front. Oncol. **12**, 848341 (2022)
42. K. Niveria, M. Yadav, K. Dangi, A.K. Verma, Overcoming challenges to enable targeting of metastatic breast cancer tumour microenvironment with nano-therapeutics: current status and future perspectives. OpenNano **8**, 100083 (2022). https://doi.org/10.1016/j.onano.2022.100083
43. H. Aheget, L. Mazini, F. Martin, B. Belqat, J.A. Marchal, K. Benabdellah, Exosomes: their role in pathogenesis, diagnosis and treatment of diseases. Cancers **13**(1), 84 (2020)
44. Y. Zhang, Q. Liu, X. Zhang, H. Huang, S. Tang, Y. Chai, Z. Xu, M. Li, X. Chen, J. Liu, Recent advances in exosome-mediated nucleic acid delivery for cancer therapy. J. Nanobiotechnol. **20**(1), 279 (2022)
45. H. Wang, L. Ding, F. Xu, L. He, L. Ye, L. Huang, L. Zhang, B. Luo, Construction of novel amphiphilic chitosan-polylactide graft copolymer nanodroplets for contrast enhanced ultrasound tumor imaging. J. Biomater. Appl. **36**(4), 613–625 (2021). https://doi.org/10.1177/08853282211011766
46. N.A. Abtahi, S.M. Naghib, S.J. Ghalekohneh, Z. Mohammadpour, H. Nazari, S.M. Mosavi, S.M. Gheibihayat, F. Haghiralsadat, J.Z. Reza, B.Z. Doulabi, Multifunctional stimuli-responsive niosomal nanoparticles for co-delivery and co-administration of gene and bioactive compound: In vitro and in vivo studies. Chem. Eng. J. **429**, 132090 (2022). https://doi.org/10.1016/j.cej.2021.132090
47. A. Beigi, S.M. Naghib, A. Matini, M. Tajabadi, M.R. Mozafari, Lipid-based nanocarriers for targeted gene delivery in lung cancer therapy: exploring a novel therapeutic paradigm. Curr. Gene Therapy **25**(2), 92–112 (2025). https://doi.org/10.2174/0115665232292768240503050508
48. H.R. Garshasbi, S.M. Naghib, Smart stimuli-responsive alginate nanogels for drug delivery systems and cancer therapy: a review. Curr. Pharm. Des. **29**(44), 3546–3562 (2023). https://doi.org/10.2174/0113816128283806231211073031
49. S. Gooneh-Farahani, S.M. Naghib, M.R. Naimi-Jamal, A. Seyfoori, A pH-sensitive nanocarrier based on BSA-stabilized graphene-chitosan nanocomposite for sustained and prolonged release of anticancer agents. Sci. Rep. **11**(1), 17404 (2021). https://doi.org/10.1038/s41598-021-97081-1
50. Y. Zhang, J.P. Thomas, T. Korcsmaros, L. Gul, Integrating multi-omics to unravel host-microbiome interactions in inflammatory bowel disease. Cell Rep. Med. **5**(9) (2024)
51. Z. Gharehdaghi, S.M. Naghib, R. Rahimi, A. Bakhshi, A. Kefayat, A. Shamaeizadeh, F. Molaabasi, Highly improved pH-responsive anticancer drug delivery and T2-Weighted MRI imaging by magnetic MOF CuBTC-based nano/microcomposite. Front. Mol. Biosci. **10**, 1071376 (2023)
52. Z. Gharehdaghi, R. Rahimi, S.M. Naghib, F. Molaabasi, Fabrication and application of copper metal–organic frameworks as nanocarriers for pH-responsive anticancer drug delivery. J. Iran. Chem. Soc. **19**(7), 2727–2737 (2022)

53. S.N. Hosseini, S.M. Naghib, G.K. Kashani, M.R. Mozafari, Chitosan-grafted graphene materials for drug delivery in wound healing. Curr. Pharm. Des. **31**(9), 691–715 (2025). https://doi.org/10.2174/0113816128333493241014134711
54. M.A. Khorasani, S.M. Naghib, A review of chitosan-based multifunctional nanocomposites for drug/gene/protein delivery and gene therapy in cancer treatments: promises, challenges and outlooks. Int. J. Biol. Macromol. **306**, 141394 (2025). https://doi.org/10.1016/j.ijbiomac.2025.141394
55. S.M. Naghib, S. Amiri, M.R. Mozafari, Stimuli-responsive chitosan-based nanocarriers for drug delivery in wound dressing applications: a review. Carbohydr. Polym. Technol. Appl. **7**, 100497 (2024). https://doi.org/10.1016/j.carpta.2024.100497
56. S.M. Naghib, K. Mohammad-Jafari, Microfluidics-mediated liposomal nanoparticles for cancer therapy: recent developments on advanced devices and technologies. Curr. Top. Med. Chem. **24**(14), 1185–1211 (2024). https://doi.org/10.2174/0115680266286460240220073334
57. S.M. Naghib, M. Zarrineh, M.R. Mozafari, 3D Printing chitosan-based nanobiomaterials for biomedicine and drug delivery: recent advances on the promising bioactive agents and technologies. Curr. Org. Chem. **28**(7), 510–525 (2024). https://doi.org/10.2174/0113852728298168240222114449
58. P. Vosoughi, S.M. Naghib, T. Jafari, B.M. Kangarshahi, Chitosan-encapsulated lipid-based nanovesicles for therapeutic applications and tissue engineering: a review, Carbohydr. Polym. Technol. Appl. 100805 (2025)
59. M. Heidarzadeh, Y. Gürsoy-Özdemir, M. Kaya, A. Eslami Abriz, A. Zarebkohan, R. Rahbarghazi, E. Sokullu, Exosomal delivery of therapeutic modulators through the blood–brain barrier; promise and pitfalls. Cell Biosci. **11**, 1–28 (2021)
60. H. Zhu, A. Guan, J. Liu, L. Peng, Z. Zhang, S. Wang, Noteworthy perspectives on microglia in neuropsychiatric disorders. J. Neuroinflamm. **20**(1), 223 (2023)
61. S. Soleymani, S.M. Naghib, M. Mozafari, Circulating tumor cells in cancer diagnosis, therapy, and theranostics applications: an overview of emerging materials and technologies. Curr. Pharm. Des. (2024)
62. Z.S. Hashemi, M. Ghavami, S. Khalili, S.M. Naghib, The emerging role of exosome nanoparticles in regenerative medicine, in *Nanopharmaceuticals in Regenerative Medicine* (CRC Press, 2022), pp. 67–93
63. F.A. Al-shubaily, M.H. Al-Zahrani, Characterization and fine structure of exosomes, in *Role of Exosomes in Biological Communication Systems* (2021), pp. 27–75
64. G.N. Alzhrani, S.T. Alanazi, S.Y. Alsharif, A.M. Albalawi, A.A. Alsharif, M.S. Abdel-Maksoud, N. Elsherbiny, Exosomes: isolation, characterization, and biomedical applications. Cell Biol. Int. **45**(9), 1807–1831 (2021)
65. W. Zhou, Y. Zhou, X. Chen, T. Ning, H. Chen, Q. Guo, Y. Zhang, P. Liu, Y. Zhang, C. Li, Y. Chu, T. Sun, C. Jiang, Pancreatic cancer-targeting exosomes for enhancing immunotherapy and reprogramming tumor microenvironment. Biomaterials **268**, 120546 (2021). https://doi.org/10.1016/j.biomaterials.2020.120546
66. R.O. Elliott, M. He, Unlocking the power of exosomes for crossing biological barriers in drug delivery. Pharmaceutics **13**(1), 122 (2021)
67. M. Yazdan, S.M. Naghib, Smart ultrasound-responsive polymers for drug delivery: an overview on advanced stimuli-sensitive materials and techniques. Curr. Drug Deliv. (2025)
68. A. Bakhshi, S.M. Naghib, N. Rabiee, Antibacterial and antiviral nanofibrous membranes, in *Antibacterial and Antiviral Functional Materials*, vol. 2 (ACS Publications, 2024), pp. 47–88
69. W. Gong, T. Zhang, M. Che, Y. Wang, C. He, L. Liu, Z. Lv, C. Xiao, H. Wang, S. Zhang, Recent advances in nanomaterials for the treatment of spinal cord injury. Mater. Today Bio **18**, 100524 (2023)
70. S. Kaya, Cellular structure and function, in *Pathological Basis of Oral and Maxillofacial Diseases* (2025), pp. 51–63

71. F. Soltanmohammadi, A.M. Gharehbaba, A.R. Zangi, K. Adibkia, Y. Javadzadeh, Current knowledge of hybrid nanoplatforms composed of exosomes and organic/inorganic nanoparticles for disease treatment and cell/tissue imaging. Biomed. Pharmacother. **178**, 117248 (2024)
72. R.I. Teleanu, M.D. Preda, A.-G. Niculescu, O. Vladâcenco, C.I. Radu, A.M. Grumezescu, D.M. Teleanu, Current strategies to enhance delivery of drugs across the blood–brain barrier. Pharmaceutics **14**(5), 987 (2022)
73. P.J. Asif, C. Longobardi, M. Hahne, J.P. Medema, The role of cancer-associated fibroblasts in cancer invasion and metastasis. Cancers **13**(18), 4720 (2021)
74. J. Ye, D. Li, Y. Jie, H. Luo, W. Zhang, C. Qiu, Exosome-based nanoparticles and cancer immunotherapy. Biomed. Pharmacother. **179**, 117296 (2024)
75. Z. Chen, X. Wang, The role and application of exosomes and their cargos in reproductive diseases: a systematic review. Vet. Sci. **9**(12), 706 (2022)
76. A.S. Teymourlouei, S.M. Naghib, M. Mozafari, Stimuli-responsive graphene-polysaccharide nanocomposites for drug delivery and tissue engineering. Curr. Org. Synth. **22**(2), 211–233 (2025)
77. A. Kyriakoudi, E. Spanidi, I. Mourtzinos, K. Gardikis, Innovative delivery systems loaded with plant bioactive ingredients: formulation approaches and applications. Plants **10**(6), 1238 (2021)
78. R. Tenchov, J.M. Sasso, X. Wang, W.-S. Liaw, C.-A. Chen, Q.A. Zhou, Exosomes–nature's lipid nanoparticles, a rising star in drug delivery and diagnostics. ACS Nano **16**(11), 17802–17846 (2022)
79. S.V. Krylova, D. Feng, The machinery of exosomes: biogenesis, release, and uptake. Int. J. Mol. Sci. **24**(2), 1337 (2023)
80. O. Abazari, P. Dayati, M. Shahidi, J. ZavarReza, M. Rahmanian, S.M. Naghib, Exploring a desirable quadr-mRNAs panel for non-invasive and ultrasensitive bladder cancer diagnosis: in-silico and clinical studies. Curr. Med. Chem. (2025)
81. S.M. Hosseinikhah, F. Gheybi, S.A. Moosavian, M.-A. Shahbazi, M.R. Jaafari, M. Sillanpää, P. Kesharwani, S.H. Alavizadeh, A. Sahebkar, Role of exosomes in tumour growth, chemoresistance and immunity: state-of-the-art. J. Drug Target. **31**(1), 32–50 (2023)
82. B.M. Kangarshahi, S.M. Naghib, Nanogenosensors based on aptamers and peptides for bioelectrochemical cancer detection: an overview of recent advances in emerging materials and technologies. Discover Appl. Sci. **6**(2), 47 (2024). https://doi.org/10.1007/s42452-024-05681-z
83. S.M. Naghib, B. Ahmadi, M.R. Mozafari, Stimuli-sensitive chitosan-based nanosystems-immobilized nucleic acids for gene therapy in breast cancer and hepatocellular carcinoma. Curr. Top. Med. Chem. **24**(17), 1464–1489 (2024). https://doi.org/10.2174/0115680266293173240506054439
84. S.M. Naghib, B. Ahmadi, M.R. Mozafari, Smart physicochemical-triggered chitosan-based nanogels for siRNA delivery and gene therapy: a focus on emerging strategies and paradigms for cancer therapy. Curr. Med. Chem. (2024). https://doi.org/10.2174/0109298673286052240426044945
85. H. Chen, L. Wang, X. Zeng, H. Schwarz, H.S. Nanda, X. Peng, Y. Zhou, Exosomes, a new star for targeted delivery. Front. Cell Dev. Biol. **9**, 751079 (2021)
86. J. He, W. Ren, W. Wang, W. Han, L. Jiang, D. Zhang, M. Guo, Exosomal targeting and its potential clinical application. Drug Deliv. Transl. Res. **12**(10), 2385–2402 (2022)
87. L. Jiang, P. Vader, R. Schiffelers, Extracellular vesicles for nucleic acid delivery: progress and prospects for safe RNA-based gene therapy. Gene Ther. **24**(3), 157–166 (2017)
88. F. Malekian, A. Shamsian, S.P. Kodam, M. Ullah, Exosome engineering for efficient and targeted drug delivery: current status and future perspective. J. Physiol. **601**(22), 4853–4872 (2023)
89. L. Duan, L. Xu, X. Xu, Z. Qin, X. Zhou, Y. Xiao, Y. Liang, J. Xia, Exosome-mediated delivery of gene vectors for gene therapy. Nanoscale **13**(3), 1387–1397 (2021)

90. Z. Belhadj, Y. Qie, R.P. Carney, Y. Li, G. Nie, Current advances in non-viral gene delivery systems: liposomes versus extracellular vesicles. BMEMat **1**(2), e12018 (2023)
91. Q. Xie, Y. Hao, N. Li, H. Song, X. Chen, Z. Zhou, J. Wang, Y. Zhang, H. Li, P. Han, Cellular uptake of engineered extracellular vesicles: biomechanisms, engineered strategies, and disease treatment. Adv. Healthcare Mater. **13**(2), 2302280 (2024)
92. J. Kim, Y. Eygeris, R.C. Ryals, A. Jozić, G. Sahay, Strategies for non-viral vectors targeting organs beyond the liver. Nat. Nanotechnol. **19**(4), 428–447 (2024)
93. L. Zhao, C. Gu, Y. Gan, L. Shao, H. Chen, H. Zhu, Exosome-mediated siRNA delivery to suppress postoperative breast cancer metastasis. J. Control. Release **318**, 1–15 (2020)
94. Z. Chen, M. Xiong, J. Tian, D. Song, S. Duan, L. Zhang, Encapsulation and assessment of therapeutic cargo in engineered exosomes: a systematic review. J. Nanobiotechnol. **22**(1), 18 (2024)
95. K. Haroon, H. Zheng, S. Wu, Z. Liu, Y. Tang, G.-Y. Yang, Y. Liu, Z. Zhang, Engineered exosomes mediated targeted delivery of neuroprotective peptide NR2B9c for the treatment of traumatic brain injury. Int. J. Pharm. **649**, 123656 (2024)
96. G. Patel, T.G. Agnihotri, M. Gitte, T. Shinde, S.S. Gomte, R. Goswami, A. Jain, Exosomes: a potential diagnostic and treatment modality in the quest for counteracting cancer. Cell. Oncol. **46**(5), 1159–1179 (2023)
97. J. Gao, A. Li, J. Hu, L. Feng, L. Liu, Z. Shen, Recent developments in isolating methods for exosomes. Front. Bioeng. Biotechnol. **10**, 1100892 (2023)
98. K. Seo, J.H. Yoo, J. Kim, S.J. Min, D.N. Heo, I.K. Kwon, H.-J. Moon, Ginseng-derived exosome-like nanovesicles extracted by sucrose gradient ultracentrifugation to inhibit osteoclast differentiation. Nanoscale **15**(12), 5798–5808 (2023)
99. T.S. Martins, M. Vaz, A.G. Henriques, A review on comparative studies addressing exosome isolation methods from body fluids. Anal. Bioanal. Chem. **415**(7), 1239–1263 (2023)
100. H. Contreras, P. Alarcón-Zapata, E. Nova-Lamperti, V. Ormazabal, M. Varas-Godoy, C. Salomon, F.A. Zuniga, Comparative study of size exclusion chromatography for isolation of small extracellular vesicle from cell-conditioned media, plasma, urine, and saliva. Front. Nanotechnol. **5**, 1146772 (2023)
101. Y. Zheng, M. Li, B. Weng, H. Mao, J. Zhao, Exosome-based delivery nanoplatforms: next-generation theranostic platforms for breast cancer. Biomater. Sci. **10**(7), 1607–1625 (2022)
102. G. Ströhle, J. Gan, H. Li, Affinity-based isolation of extracellular vesicles and the effects on downstream molecular analysis. Anal. Bioanal. Chem. **414**(24), 7051–7067 (2022)
103. J. Wang, P. Ma, D.H. Kim, B.-F. Liu, U. Demirci, Towards microfluidic-based exosome isolation and detection for tumor therapy. Nano Today **37**, 101066 (2021)
104. B. Lin, Y. Lei, J. Wang, L. Zhu, Y. Wu, H. Zhang, L. Wu, P. Zhang, C. Yang, Microfluidic-based exosome analysis for liquid biopsy. Small Methods **5**(3), 2001131 (2021)
105. J. Donoso-Quezada, S. Ayala-Mar, J. González-Valdez, State-of-the-art exosome loading and functionalization techniques for enhanced therapeutics: a review. Crit. Rev. Biotechnol. **40**(6), 804–820 (2020)
106. M. Xu, Q. Yang, X. Sun, Y. Wang, Recent advancements in the loading and modification of therapeutic exosomes. Front. Bioeng. Biotechnol. **8**, 586130 (2020)
107. F. Mehryab, S. Rabbani, S. Shahhosseini, F. Shekari, Y. Fatahi, H. Baharvand, A. Haeri, Exosomes as a next-generation drug delivery system: an update on drug loading approaches, characterization, and clinical application challenges. Acta Biomater. **113**, 42–62 (2020)
108. S. Reda El Sayed, J. Cristante, L. Guyon, J. Denis, O. Chabre, N. Cherradi, MicroRNA therapeutics in cancer: current advances and challenges. Cancers **13**(11), 2680 (2021)
109. N. Dastmalchi, R. Safaralizadeh, B. Baradaran, M. Hosseinpourfeizi, A. Baghbanzadeh, An update review of deregulated tumor suppressive microRNAs and their contribution in various molecular subtypes of breast cancer. Gene **729**, 144301 (2020)
110. M.Z. Wojtukiewicz, M.M. Rek, K. Karpowicz, M. Górska, B. Polityńska, A.M. Wojtukiewicz, M. Moniuszko, P. Radziwon, S.C. Tucker, K.V. Honn, Inhibitors of immune checkpoints—PD-1, PD-L1, CTLA-4—new opportunities for cancer patients and a new challenge for internists and general practitioners. Cancer Metastasis Rev. **40**, 949–982 (2021)

111. K. Goutsouliak, J. Veeraraghavan, V. Sethunath, C. De Angelis, C.K. Osborne, M.F. Rimawi, R. Schiff, Towards personalized treatment for early stage HER2-positive breast cancer. Nat. Rev. Clin. Oncol. **17**(4), 233–250 (2020)
112. G. Srinivasan, Cervical cancer: Novel treatment strategies offer renewed optimism. Pathol. Res. Pract. **254**, 155136 (2024)
113. X. Li, Z. Jing, X. Li, L. Liu, X. Xiao, Y. Zhong, Z. Ren, The role of exosomes in cancer-related programmed cell death. Immunol. Rev. **321**(1), 169–180 (2024)
114. X. Dai, Y. Ye, F. He, Emerging innovations on exosome-based onco-ther. Front. Immunol. **13**, 865245 (2022)
115. J.J. Mao, G.G. Pillai, C.J. Andrade, J.A. Ligibel, P. Basu, L. Cohen, I.A. Khan, K.M. Mustian, R. Puthiyedath, K.S. Dhiman, Integrative oncology: addressing the global challenges of cancer prevention and treatment. CA: Cancer J. Clin. **72**(2), 144–164 (2022)
116. Z. Xu, S. Zeng, Z. Gong, Y. Yan, Exosome-based immunotherapy: a promising approach for cancer treatment. Mol. Cancer **19**, 1–16 (2020)
117. H. Kim, H. Jang, H. Cho, J. Choi, K.Y. Hwang, Y. Choi, S.H. Kim, Y. Yang, Recent advances in exosome-based drug delivery for cancer therapy. Cancers **13**(17), 4435 (2021)
118. D.H. Ha, S.-D. Kim, J. Lee, H.H. Kwon, G.-H. Park, S.H. Yang, J.Y. Jung, J.H. Lee, S.R. Park, J. Youn, Toxicological evaluation of exosomes derived from human adipose tissue-derived mesenchymal stem/stromal cells. Regul. Toxicol. Pharmacol. **115**, 104686 (2020)
119. J.J. Klemeš, Y. Van Fan, R.R. Tan, P. Jiang, Minimising the present and future plastic waste, energy and environmental footprints related to COVID-19. Renew. Sustain. Energy Rev. **127**, 109883 (2020)
120. S.C. Thomas, J.-W. Kim, G.M. Pauletti, D.J. Hassett, N. Kotagiri, Exosomes: biological pharmaceutical nanovectors for theranostics. Front. Bioeng. Biotechnol. **9**, 808614 (2022)
121. J.M. Essola, M. Zhang, H. Yang, F. Li, B. Xia, J.F. Mavoungou, A. Hussain, Y. Huang, Exosome regulation of immune response mechanism: pros and cons in immunotherapy. Bioactive Mater. **32**, 124–146 (2024)
122. G.C. Terstappen, A.H. Meyer, R.D. Bell, W. Zhang, Strategies for delivering therapeutics across the blood–brain barrier. Nat. Rev. Drug Discov. **20**(5), 362–383 (2021)
123. R. Kar, R. Dhar, S. Mukherjee, S. Nag, S. Gorai, N. Mukerjee, D. Mukherjee, R. Vatsa, M. Chandrakanth Jadhav, A. Ghosh, Exosome-based smart drug delivery tool for cancer theranostics. ACS Biomater. Sci. Eng. **9**(2), 577–594 (2023)
124. M. Deng, S. Wu, P. Huang, Y. Liu, C. Li, J. Zheng, Engineered exosomes-based theranostic strategy for tumor metastasis and recurrence. Asian J. Pharm. Sci. **18**(6), 100870 (2023)
125. A.K. Dubey, E. Mostafavi, Biomaterials-mediated CRISPR/Cas9 delivery: recent challenges and opportunities in gene therapy. Front. Chem. **11**, 1259435 (2023)
126. R. Li, R. Chibbar, J. Xiang, Novel EXO-T vaccine using polyclonal CD4+ T cells armed with HER2-specific exosomes for HER2-positive breast cancer. OncoTargets Therapy 7089–7093 (2018)
127. H.S. Budi, F.N. Ahmad, H. Achmad, M.J. Ansari, M.V. Mikhailova, W. Suksatan, S. Chupradit, N. Shomali, F. Marofi, Human epidermal growth factor receptor 2 (HER2)-specific chimeric antigen receptor (CAR) for tumor immunotherapy; recent progress. Stem Cell Res. Ther. **13**(1), 40 (2022)
128. Q. Liu, X. Zhang, J. Zhang, Exosome-based nanoplatforms: the emerging tools for breast cancer therapy. Front. Oncol. **12**, 898605 (2022)

Chapter 10
Polysaccharides Based on Smart Chitosan Nanoparticles for Gene Delivery and Gene Therapy

Abstract Chitosan (Ch)-based nanoparticles have shown significant promise for targeted gene delivery in recent cancer treatment studies. Because of its remarkable qualities, chitosan is a biocompatible and biodegradable polymer that is perfect for delivering therapeutic genes. The exact transport and controlled release of genes are made possible by these nanoparticles' ability to react to certain stimuli, including pH, temperature and enzymes. These nanoparticles have shown efficacy in cancer treatment by delivering genes to tumor cells, hence decreasing the development of tumors. Using targeting ligands, encasing protective agents and modifying the nanoparticle surface have all increased the effectiveness of gene delivery. By boosting targeted delivery, decreasing off-target effects and selectively releasing genes in response to tumor circumstances, chitosan-based smart nanoparticles have shown potential in improving outcomes. Targeting ligands on the nanoparticles' surface also improves their efficacy and absorption. These findings open the door to novel gene-focused cancer treatments, even if further research is necessary to refine the composition and structure of these particles in nanoparticle form and evaluate their long-term safety.

Keyword Cancer treatment · Gene delivery · Chitosan nanoparticles · Sensitivity to stimuli

10.1 Introduction

Cancer is an extremely harmful disease that poses a serious threat to world health and claims many lives. It is estimated that by 2030, this illness would affect about 12 million people. Even while cancer therapies like chemotherapy, laser therapy, biotherapy, surgery, treatments via hyperthermia, radiation, and hormone therapy have advanced significantly, several constraints still limit their efficacy [1–4]. The demand for new targeted therapies that may increase therapeutic efficacy while minimizing adverse effects is increasing. Gene therapy is one approach to cancer treatment. In contrast to conventional protein-based therapies, gene therapy provides a

S. M. Naghib, *Nano-Biosensors, Drug Delivery and Tissue Engineering*, Series in BioEngineering, https://doi.org/10.1007/978-981-95-1888-3_10

number of benefits by introducing new genetic material into certain bodily cells [5–8]. Genomic expression may be efficiently regulated and controlled by introducing foreign nucleic acids, like RNA or DNA, into certain cells. With this method, a certain gene may be directly inserted into a person cells, causing those particular body cells to produce proteins [9].

Advantages of gene therapy are its accuracy, possibility for cure, and long-lasting effects, so gene therapy has several benefits in the treatment of cancer. One of the biggest advantages is its timeless and enduring effects, which include the possibility of a permanent cure by the introduction of changed genes with lifelong consequences. Typically, gene therapy uses a single dosage, lowering the therapeutic load on patients in contrast to traditional therapies that call for recurring doses. Furthermore, by carefully targeting cancer cells or certain tissues without endangering healthy cells, gene therapy offers tailored treatment. It does this by changing faulty genes, which interferes with vital functions including the development and survival of cancer cells. This method treats the underlying cause of cancer by repairing genetic defects, which may lead to a cure, particularly when other therapies are ineffective. Additionally, gene therapy improves patients' quality of life by controlling or even curing certain malignancies that were formerly thought to be incurable, which results in better health and fewer symptoms. Customization and personalization of medicines is another important benefit; gene therapies may be made to match a person's genetic composition, increasing effectiveness and reducing negative effects. This optimistic environment also includes cutting-edge strategies similar to CAR T-cell treatment, which alters immune cells that go after cancer selectively, oncolytic viruses, CAR NK-cell treatment, and therapeutic cancer vaccines. In conclusion, gene therapy is a promising new avenue for treating cancer that might greatly improve patient outcomes by providing accuracy, long-lasting benefits, and the possibility of a cure. These cells essentially function as tiny bioreactors that produce the targeted proteins for medical applications [10, 11]. Targeting certain cells, preventing early breakdown of the nucleotide sequences outside the cell, and making sure enough genetic material is delivered to create a therapeutic impact are all necessary for gene therapy to be successful [12, 13]. The primary obstacle in gene therapy is coming up with a safe and efficient delivery method that doesn't cause an immunological reaction. In an effort to reduce toxicity and immunological responses, researchers have developed both viral and non-viral delivery systems for therapeutic genetic material. Successful gene therapy requires overcoming physicochemical and biomolecular obstacles after introduction into cells, as shown in Fig. 10.1. Nucleic acids administered intravenously, intramuscularly, or via mucosal channels degrade quickly due to the presence of serum nucleases in the extracellular environment [14, 15].

The efficacy and patient outcomes of conventional cancer therapies are impacted by a number of limitations. For instance, photodynamic therapy (PDT) has restricted penetration of tissue because it uses activation of light that can only penetrate shallow levels (roughly 1 cm). As a result, it is best suited for tumors that are located just below the epidermis or that line inside organs. Furthermore, PDT can cause sensitivity to light, necessitating precautions from patients, and it is still difficult to target tumors precisely. Chemotherapy has a reputation for being systemically toxic, affecting

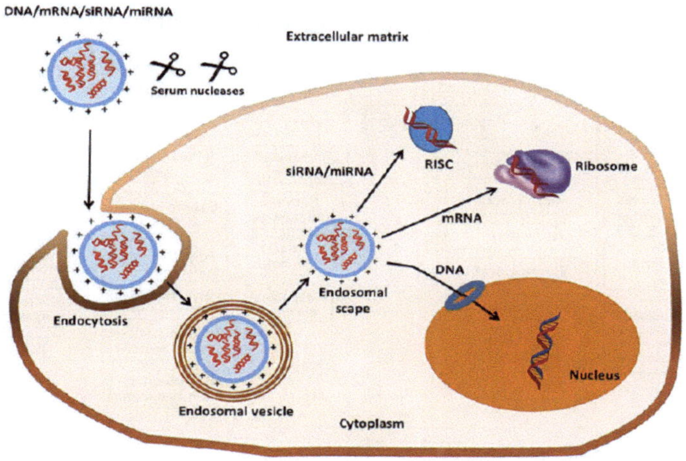

Fig. 10.1 Obstacles to effective nucleic acid transport in living organisms [14]

both healthy and cancerous cells and producing adverse effects like fatigue, nausea, and hair loss [16]. Furthermore, limited tumor penetration brought on by diffusion issues within solid tumors can impact drug distribution, and some cancers eventually become resistant to chemotherapy medications. Among the drawbacks of immunotherapy are response variability, which occurs when patients react differently; the possibility of autoimmune reactions brought on by inhibitors of immune checkpoints; and the existence of a tumor microenvironment that suppresses the immune system, which can reduce the effectiveness of treatment. Surgery has drawbacks of its own, including the incapacity to adequately treat metastatic disease, accessibility concerns for specific tumors, and invasiveness and related risks. Because hormone therapy only works for tumors that are susceptible to hormones and can have negative adverse consequences like fatigue, Osteoporosis and hot flushes, its use is limited. Hormonal treatments may also eventually cause tumors to become resistant. Because of its lack of accuracy, radiation therapy can harm healthy tissues next to the tumor, and prolonged exposure can cause secondary cancers. Because they mainly heat superficial tissues, hyperthermia treatments have depth limitations, and combining hyperthermia with other treatments can be challenging. For patients to tolerate, the ideal temperature must be maintained without damaging healthy tissues. All of these drawbacks of traditional cancer treatments highlight the necessity of less invasive and more focused treatment alternatives to enhance patient care and results. The glycoproteins on the plasma membrane and the strong negative charge of the nucleotide clusters make it hard for the cell surface to interact with DNA or RNA [17–21]. To improve the efficiency of gene transport over these obstacles and enable the transfer of foreign genes to specific cells, DNA/RNA may be incorporated into specialized gene transfer carriers. These carriers fall into two primary categories: synthetic and virus-derived vectors. Figure 10.2 provides a summary of the many kinds of carriers used in gene transfer. Although viral vectors are essential

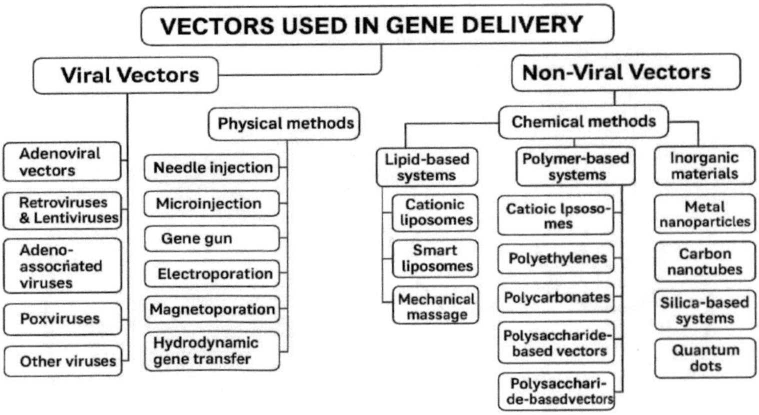

Fig. 10.2 Kinds of gene transfer vectors

for gene expression and transfer, they have serious disadvantages. These include the potential for significant systemic inflammation, toxic consequences, immunological responses, cargo capacity limitations, and the danger of inciting immune-related adverse reactions [22, 23]. Non-viral vehicles, on the other hand, are thought to be safer due to the fact that they elicit a low immune response. They also have the benefit of being simple to make in big quantities. Furthermore, these vectors are capable of carrying genetic materials of any size without any restrictions and can store a greater quantity of biomaterials [24].

Physical and chemical processes are the two primary categories of chemical vectors used to transmit genetic material. Biolistic delivery, needle injection, magnetoporation, microinjection, sonoporation, and electroporation are examples of physical techniques, while carriers derived from inorganic materials, lipids, and polymers are used in chemical techniques [25]. The non-vector gene transfer mechanism is shown in Fig. 10.3 [26].

Mostly positively charged, chemical vectors are made up of different molecules such as lipids, polypeptides, and inorganic vectors. By overcoming barriers like endocytosis and forming strong compounds with nucleic acids, these molecules may facilitate gene transfer, guaranteeing continuous transcriptional activity and defense against environmental harm [27]. By altering their size, shape, and porosity, inorganic nanoparticles (NPs) such as QDs, Au, and CNTs may evade the body's defenses. When genes are transferred, lipids with a hydrophilic head that is positively charged and a hydrophobic tail joined by a connecting agent create condensed complexes known as lipoplexes. Both organic and manmade cationic polymers may form stable nanoscale complexes called polyplexes by joining with nucleic acids. Examples include a variety of biopolymers, including proteins, peptides, and carbohydrates. However, synthetic polymers such as CDs and PEI are also noteworthy [28]. Numerous positively charged polymer bases that carry genetic information may bind natural nucleic acids that are negatively charged via electricity exchanges. Positively

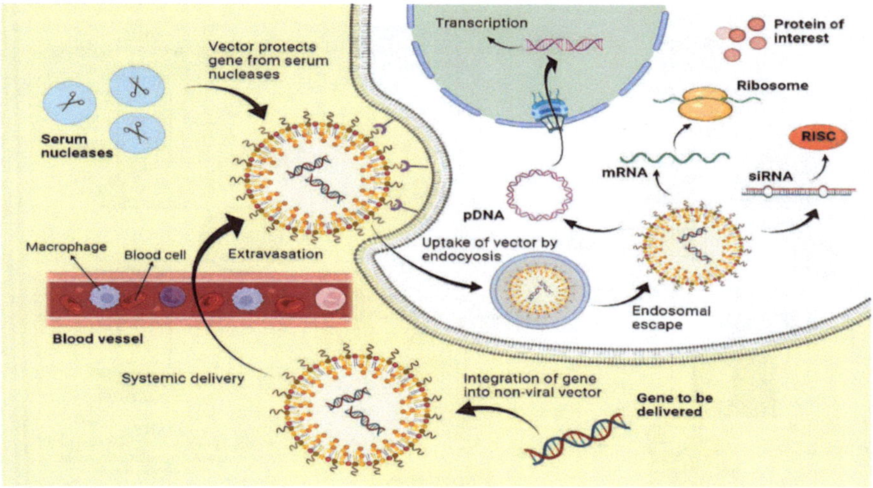

Fig. 10.3 Gene delivery by a nonvector process [26]

charged polyplexes, also known as NPs (polymer-nucleic acids), can thus effectively pass through the plasma membrane and move their contained components within the cell [29, 30]. Chitin, a polysaccharide composed of units of N-acetyl glucosamine arranged in a certain pattern (β-(1,4)), is the natural source of the polymer Ch. Chitin is converted to Ch via a process known as deacetylation, which may be accomplished biochemically using deacetylases or chemically with a strong NaOH solution at high temperatures (Fig. 10.4) [31]. Ch is primarily produced from chitin through a four-step process: (1) deproteinization, using a mild sodium hydroxide solution to remove proteins; (2) demineralization, involving hydrochloric acid to eliminate mineral content; (3) discoloration by acetone and ethanol to purify chitin; and (4) deacetylation, where concentrated sodium hydroxide at elevated temperatures, removes acetyl groups to convert chitin into ch (Fig. 10.4). This sequential treatment yields ch with desired properties for various applications [31].

Chitosan's degree of deacetylation (DD) has a major effect on its physicochemical characteristics, especially its viscosity and solubility. A greater DD denotes less acetylation. DD is the proportion of acetyl groups extracted from chitin to create chitosan. Greater solubility is shown by chitosan with a higher DD. Because of protonation, which creates ammonium ions that improve solubility, it dissolves easily in situations that are acidic. However, chitosan oligomers are still soluble at normal pH values (~ 7.4). On the other hand, even at greater DD levels, chitosan that has less DD needs increased acidity to be soluble. The DD of chitosan is also important in relation to viscosity. Greater viscosity is often associated with chitosan that has a greater DD (lower acetylation). This is because more accessible amino groups lead to longer chains, which in turn encourage more interactions between molecules. Its behavior in solutions and gels is influenced by its increased viscosity, which has an impact on applications like wound healing and medication administration. Conversely, chitosan

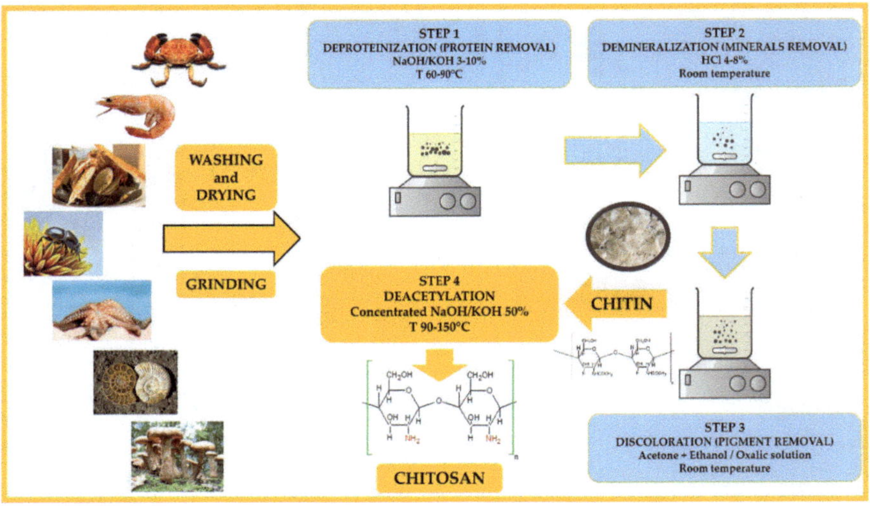

Fig. 10.4 An overview of the process of deacetylation that turns chitin into Ch [31]

with more acetylation (lower DD) has less viscosity; because there are fewer amino groups in shorter chains, there are fewer interactions between molecules, which lowers viscosity. This attribute affects its processing ease and flow properties in a range of industrial and biomedical applications. As a result, chitosan's degree of deacetylation plays a critical role in defining its solubility and viscosity, two essential aspects of its functional characteristics and uses [32]. By protonating its -NH groups and creating hydrogen bonds with water, Ch has the capacity to improve osmosis. Interactions with positively charged ions, particularly hydrogen ions, enhance this protonation, increasing the Ch molecule's total positive charge. A Donnan layer forms close to the solution's surface as a result of the positive charge of Ch drawing in molecules or ions that are negatively charged. Larger solutes are blocked by this layer, which functions as a barrier that only permits solvent molecules, such as water, to flow through [33]. Ch also has a lot of –OH and –NH groups that may attach to water molecules to create H-bonds. By lowering the concentration of water outside the Ch structure and generating a differential between the surrounding solution and the Ch surface, the osmotic pressure from these linkages helps keep water within the Ch structure. By encouraging water's pouring into the Ch structure, this pressure differential improves osmosis. The efficiency of Ch in boosting osmosis depends on several variables, including Ch's molecular weight, the level of deacetylation, and the ambient pH. Because of their exceptional physical, chemical, and biological characteristics, ChNPs may be used in cutting-edge drug delivery applications. These nanosystems remarkable properties may be enhanced by their capacity to modify protein incorporation and control many aspects of manufacturing. ChNPs may also be made in a handy powdered form for simpler handling, and they exhibit improved durability and efficient protein entrapment [34]. For Ch-based polyplexes to remain stable during nanoparticle fabrication, the pH of the solution must be kept between

5.6 and 6.5. While high pH circumstances diminish the positive charge in Ch and may lead to the instability of the complexes of polymers and nucleic acids, acidic pH increases the binding capacity of NH_2 groups via protonating Ch with nucleic acid. Complexation may also be impacted by other interactions, like hydrophobic interactions and hydrogen bonding. Although polyplexes generated in an acidic environment are stable, physiological circumstances may affect their stability and the effectiveness of transfection in vivo. When exposed to external stimuli or internal stimuli, drug/gene transfer systems that are responsive to stimuli can release drugs or genes. Intelligent polymeric systems are stimuli-sensitive drug/gene transfer devices, like Ch, poly(methacrylic acid)s, poly(vinylimidazole)s, cellulose, etc., that show a quick change in their physical and chemical properties upon minor changes in cues from the surroundings, resulting in a significant drug or gene release. Because of their unique characteristics, a lot of attention has been paid to examining the remarkable features of Ch among natural polymers. This natural polymer has very low levels of toxicity and immunogenic responses, yet it is biodegradable and mucoadhesive [29, 35]. The construction of a Ch-derived nano vector that can release medication and gene molecules in a regulated way in response to certain stimuli is shown in Fig. 10.5. This targeted delivery method offers a novel approach in the realm of nanocarriers by enabling cargo molecules' release as a result of individual or combination stimuli [36].

There is a lot of interest in how the pH levels in biological surroundings fluctuate, especially in solid tumors. It has been shown that aberrant metabolism of glucose in solid tumor cancer cells results in the synthesis of lactic acids. Following their release from the cells, these lactic acids infiltrate the tumor microenvironment extracellular fluid and cause an acidic change in the surrounding area. Knowledge solid tumor biology requires a knowledge of this phenomena. When in contrast to healthy tissues and cells, the intracellular spaces called lysosomes and endosomes in tumor cells have shown abnormally elevated acidity levels. Because of their special quality, they are a good target for nanovectors that respond to pH, which may deliver therapeutic drugs into the microenvironment of the tumor and the tumor cells' systems of vesicular trafficking. A targeted and regulated method of gene delivery is made possible by these nanocarriers' ability to release genes precisely in response to pH level changes [37–40].

By taking advantage of the variations in the redox process between the interior and exterior surroundings of cells, researchers have made notable progress in creating Ch-based nanovectors with controlled-release capabilities. It is crucial to highlight that these nanocarriers may be specifically engineered to react to the unique redox conditions present in malignant cells. Researchers have effectively developed a novel nanocarrier, CSO-ss-SA, in recent research. It resembles a glycolipid, but it possesses special characteristics that make it very susceptible to cancerous cells. The dispersion of this nanocarrier in vital organs such as the liver and spleen, as well as the actual tumor, was shown by the researchers. Interestingly, scientists discovered that the nanocarrier only broke down quickly and released the encapsulated medicine when it came into contact with intracellular glutathione (GSH) prevalent in cancer

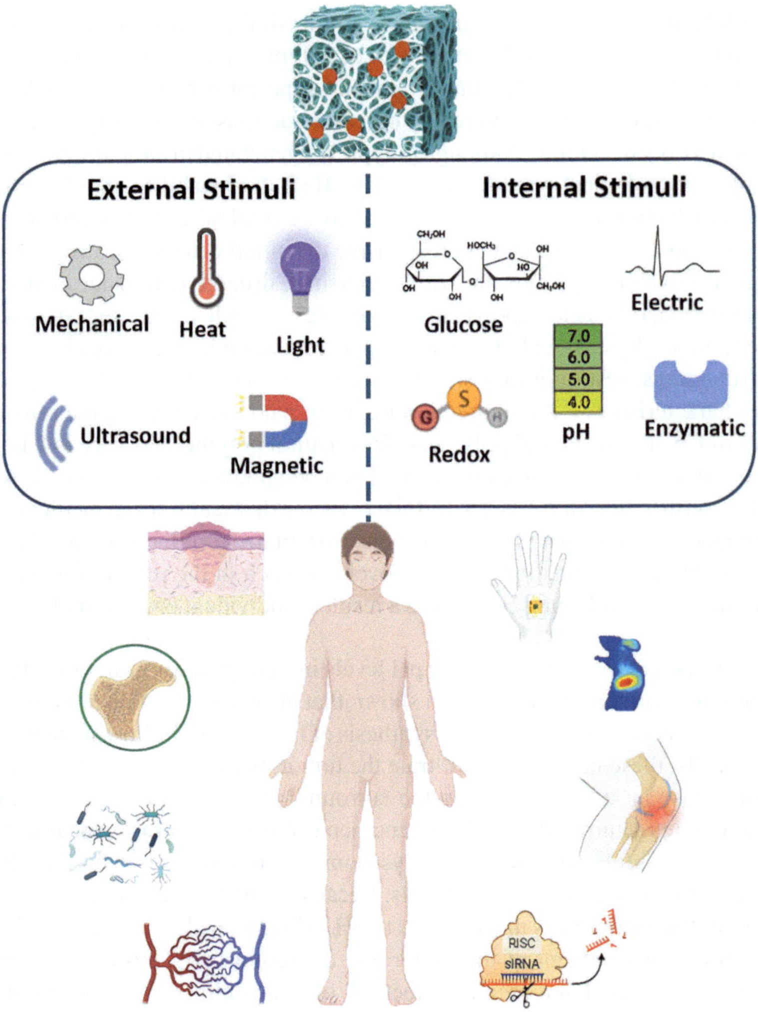

Fig. 10.5 The illustration displays nanovectors based on Ch that are stimuli-responsive. An internal or exterior stimulation, such as temperature, light, ultrasonography, redox gradients, magnetic fields, or pH changes, may trigger the carrier's capacity to release drug molecules [36]

cells. Significant tumor growth inhibition resulted from this ground-breaking finding, indicating enormous promise for future developments in cancer therapy [40, 41].

The use of nanodroplets with ultrasound has enormous promise for enabling the precise and regulated delivery of genes. As a wave of mechanical motion with controlled and accurate transmission, ultrasound provides a powerful delivery method. The use of externally positioned movable guiding for transducers and imaging allows for easy targeting of tumors. Moreover, nanodroplets have the potential to serve as genetic material carriers, facilitating the systemic dissemination of genes. A potential method for the spatiotemporally regulated delivery of genes is the combination of ultrasonography and nanodroplets. A process known as sonoporation occurs when bubbles exposed to certain ultrasonic radiation create brief, tiny holes in adjacent cell membranes. Increased permeability of cellular and vascular structures results from this, which is crucial for the medical profession. Genetic material is then released locally as a result of the bubbles being destroyed. This procedure, called ultrasound-targeted microbubble destruction (UTMD), provides a regulated way to distribute genes, increasing the efficacy of therapeutic therapies. This novel approach has enormous potential for attaining the best possible treatment results in a regulated way, both in terms of time and space [42, 43].

Researchers have used temperature-responsive materials, such as polymers, to manufacture temperature-responsive methods for gene transfer. Examples of N-isopropylacrylamide and its derivatives are well known and often used in creating temperature-sensitive methods for gene transfer. Multiple researchers have investigated the potential of Ch vectors with thermoresponsive elements for a range of applications. These polymers' temperature-dependent responsiveness depends on changes to their hydrophobic properties close to a certain temperature called the LCST (low critical solution temperature). These polymers are hydrophilic and water-soluble below the LCST, which causes them to swell primarily as a result of hydrogen bonds with water molecules. Nevertheless, the hydrogen bonding degrades at temperatures higher than the LCST, resulting in the production of hydrophobic insoluble polymers that shrink. When triggered, this unusual activity has the capacity to release genes and medication molecules in a precise and regulated manner [40, 44].

In light-sensitive drug delivery systems, utilizing light irradiation demonstrates significant promise in initiating drug and gene release. Numerous processes, including photo-isomerization and photo-induced heat impact inside the structure of nano vectors, are involved in the mechanisms causing this release. There are several advantages to using light-responsive approaches in many different domains, especially when it comes to delivering stimuli-triggered gene/drug release, releasing genes/drugs simultaneously with cancer imaging, and successfully stopping tumor development. The production and evaluation of Ch-IR820 conjugates was the subject of a recent study by researchers. According to their results, these conjugates have enormous promise as a multipurpose imaging and hyperthermia system in disease detection and treatment. In this work, scientists found that when subjected to an 808 nm laser, the pairing of Ch and IR820 may produce heat. Through a mechanism known as hyperthermia, it has been discovered that this heat generation efficiently inhibits the development of tumor cells. The researchers suggest using Ch-IR820 conjugates to treat hyperthermia in a localized and image-guided manner. Furthermore, it has been noted that the conjugates are removed from the hepatobiliary system of the body, which results in their finding in the feces [40, 45].

Magnetic minerals like magnetite, which may be found as magneto liposomes, porous metallic nanocapsules, or NPs, are often used in magnetically-sensitive gene transfer systems. When exposed to external magnetic fields, these systems are made to alter size or form reversibly. This characteristic enables precise control of gene release by allowing the gel loaded with genes to alternate between ON and OFF states. Usually produced in an organic phase, magnetic nanoparticles need to be modified to work in aqueous conditions. Amphiphilic polymers are essential for surface modification in order to accomplish this. These polymers have a longer half-life in the circulation, higher biocompatibility, and increased stability in terms of their chemical and physical characteristics. Adding different functional entities, such as targeting factors, imaging factors, and stimuli-sensitive moieties, to polymer chains is one possible way to change them and improve their performance. Accordingly, the derivatives of Ch have become well-liked materials for coating magnetic nanoparticles that are widely used in this industry [46].

10.2 Chitosan Nanoparticles for Specific Nucleic Acid Delivery

Since nucleic acid chemistry has advanced significantly, new therapeutic tools and targeted drug delivery systems have been developed. DNA and RNA are perfect for pharmaceutical applications because of their biostability, distinct spatial arrangement, and slow dissociation rates. Though its potential is frequently limited by issues with stability, translocation, and off-target effects that call for effective delivery systems, small interfering RNA (siRNA) is particularly noteworthy in this context because it plays a crucial role in gene regulation by targeting specific mRNA sequences to induce degradation and silencing of gene expression. Researchers have used this mechanism for therapeutic purposes, such as targeting oncogenes in cancer therapy. Chitosan nanoparticles, which are made from chitin, a naturally occurring polymer found in crustacean shells, offer a promising solution [47, 48]. They are known for their biocompatibility, biodegradability, positive charge for interacting with negatively charged nucleic acids, and capacity to encapsulate siRNA, preventing its degradation and enabling targeted delivery to particular cells or tissues. When combined with chitosan's mucoadhesiveness and pH responsiveness for site-specific drug release, these chitosan-siRNA nanoparticles are promising candidates for use in the treatment of infectious diseases, cancer, and genetic disorders. To move closer to customized medicine and more potent treatments, additional study is needed to improve tissue-specific targeting, stabilize siRNA, and optimize chitosan-based drug delivery systems [49].

Preventing the growth of cancer and boosting sensitivity against cancer therapies require altering gene expression, and RNA interference (RNAi) is a useful strategy for achieving this goal. A class of oligonucleotides with two strands with 20–25 pairs of nucleotides known as siRNA selectively breaks down targeted mRNA and induces gene repression. SiRNA has to be integrated into the RISC system in order to affect target genes. Potential benefits of using siRNA include low cytotoxicity, efficacy, and specificity [50].

Given our comparatively thorough knowledge of the underlying causes of carcinoma malignancy, a great deal of consideration has been given to the use of siRNA-mediated cancer therapy. When treating cancer, we may target these factors using siRNA. The development and immune evasion of cancer cells are influenced by PD-L1 (programmed death-ligand 1). PD-L1-siRNA is beneficial because it may enhance cell death brought on by T cells and reduce oncogenesis by blocking mammalian target of rapamycin (mTOR) activity. Oncogene expression may be decreased to increase the anti-neoplastic effect of chemotherapeutic agents. Co-administration of RRM2-siRNA and cisplatin suppresses tumor development and increases the risk of PCD [51]. In stomach cancer cells, elevated expression of c-Met is linked to a poor prognosis and treatment resistance. These cells' growth and metastasis may be inhibited, chemosensitivity can be increased, and apoptosis can be induced by using c-Met gene silencing RNA. Likewise, PYROXD1-siRNA may cause PCD and suppress cell division, making a patient more susceptible to chemotherapy. It is noteworthy that research on animals has confirmed siRNA's efficacy; siRNA-mediated reduction of Annexin A7 in transplantable models of gastric cancer significantly increases the number of cells in the G1/G0 stages while lowering the number in the S phases; consequently, this stops the growth of stomach neoplasm cells. Although these findings demonstrate the promise of siRNA in the treatment of cancer and in the silence of genes, siRNA use has a number of disadvantages. As an example, siRNA may be broken down by RNase enzymes, has trouble passing across the plasma membrane, and tends to impact undesired sites [52, 53].

A crucial component of cancer therapy is NPs-mediated gene silencing, and studies have been done to assess how well different nanostructures can carry siRNA. The role of NPs in the siRNA transition will be briefly covered in this section. For the therapy of siRNA-mediated cancer, many kinds of nanoparticles have been created, including metal nanoparticles, carbon nanotubes, polymers, and lipid-based nanomaterials. The high loading efficiency of nanoparticles, reaching up to 92.3% for liposomal siRNA, is one of the most desired features for delivering siRNA in cancer therapy [54]. Furthermore, liposomes may achieve a general efficiency of over 90% and a virtually full siRNA encapsulation efficiency of almost 99%, according to another research. When compared to isolated siRNA, nanoparticles significantly enhance the absorption of siRNA into cells, increasing its effectiveness in inducing cellular death in cancer. One possible strategy to enhance nanostructures' identification and absorption by source cells is to alter them via cell membrane coating. A recent study used DOX and siRNA to suppress PD-L1 with NPs coated with PLGA that had undergone neoplastic cell membrane modification. The intracellular transport of these nanocomplexes was aided by the transfer integrins, which enhanced the tumor-suppressive effectiveness of doxorubicin and siRNA-blocking PD-L1. The hematoencephalic barrier may make it difficult for anticancer agents to enter brain tumors during treatment. Through vesicle-related or vesicle-related methods, nanoscale particles may penetrate the hematoencephalic barrier and distribute siRNA systemically to treat brain cancer. The development of pH-responsive smart nanocarriers for siRNA delivery is a promising avenue. Poly(l-histidine)-based multifunctional nanoparticles have been developed recently for the delivery of siRNA in

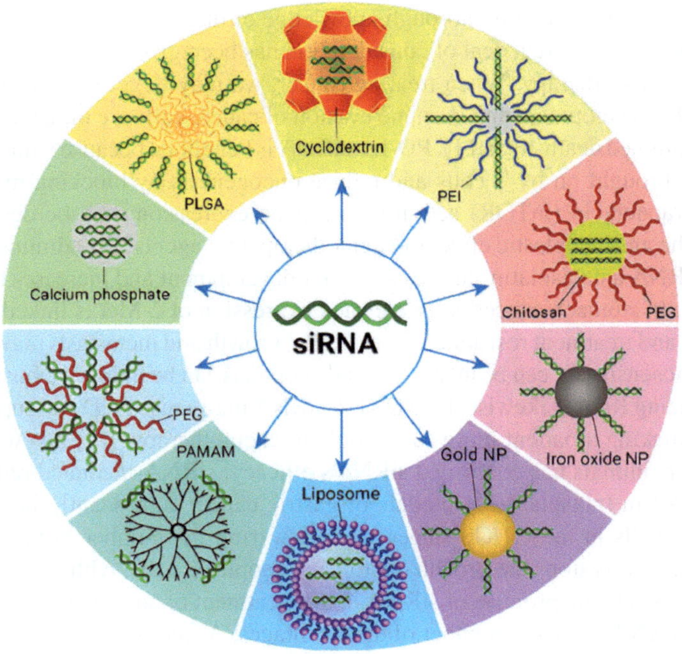

Fig. 10.6 A variety of nanoparticles, including ChNPs, have been used to carry siRNA. PEI (polyethylene imines), PEG (polyethylene glycol), NPs, PAMAM, and PLGA are further instances of nanoparticles that have been used for siRNA distribution [56]

the treatment of pulmonary neoplasia. SiRNA is released when there is a small tumor acidosis, and intracellular release is facilitated by polyhistidine copolymer [55]. Overall, as shown in Fig. 10.6, nanoparticles provide a flexible and potential method for cancer siRNA delivery treatment; Ch is a biopolymer that may improve the biocompatibility and biodegradability of nanoparticles. nanostructures based on Ch and their possible role in effective control of gene therapy will be covered in the next sections [56].

10.3 Chitosan Vector Gene Transfer

For an siRNA/DNA transfer method to be successful, it must remain steady in the biological setting. Nevertheless, the medication may thin down rapidly when administered intravenously due to components of blood, jeopardizing the transfer mechanism. Polyanions in the blood may compete with positively charged Ch vectors, causing NAs to be released prematurely before they reach their target location. Problems like increased toxicity and embolism may arise from the interaction of cationic Ch formulations with anionic serum proteins. To test vector consistency, vectors are usually incubated within a biological setting while their size is tracked with time. Strongly positively charged Ch/NAs polyplexes have been shown in one investigation to prevent aggregation and maintain stability for at least seven days. A research study assessed absorbance at 350 nm after incubating Ch/DNA polyplexes with varying doses of BSA. Particle instability was suggested by the results, which showed enhanced variability in optical density in pDNA/Ch nanocomplexes treated with 2 mg/ml BSA after an hour. When BSA is present, ChNPs' capacity to silence RNA with stability is shown; in complexes of Ch-siRNA and Ch-NPsTPP-siRNA, experiments indicated that ChNPs did not alter when exposed to less than 10% BSA. According to a research, siRNA isolated remained from ChNPs-TPP viable for twenty-four hours (it was fully damaged after 72 h), siRNA that was unbound was eliminated in FBS within 30 min. Additionally, siRNA from ChNPs-siRNA remained unaltered for up to 7 h (totally destroyed after 48 h) after incubation, but free siRNA rapidly dissolved when exposed to a 50% serum concentration. According to the research, Ch/siRNA formulations that are not PEGylated may degrade in plasma from humans, rendering them inappropriate for siRNA delivery in living beings. Some Ch groups' deprotonation, which reduces their ability to bind to siRNA, is thought to be the cause of this breakdown. Disintegration may also result from plasma proteins that are negatively charged interfering with the binding of siRNA. PEGylation of Ch increases stability for siRNA administration, as shown by the fact that PEGylated NPs were stable in human plasma. These NPs' remarkable stability was shown by their positive surface charge, which ranged from + 35 to + 41 mV. According to the research, Ch/siRNA formulations that are not PEGylated may degrade in human plasma, rendering them inappropriate for siRNA delivery in living beings. The deprotonation of certain Ch groups, which reduces their ability to bind to siRNA, is thought to be the cause of this breakdown. Disintegration may also result from negatively charged plasma proteins interfering with siRNA binding. PEGylation of Ch increases stability for siRNA administration, as shown by the fact that PEGylated NPs were stable in human plasma. These NPs' remarkable stability was shown by their positive surface charge, which ranged from + 35 to + 41 mV. The characteristics of Ch in gene therapy for cancer may change depending on how we alter PEG; this is because the addition of PEG alters Ch's chemical and physical properties. These alterations may impact the interaction, absorption by cells, and transport of Ch and nucleic acids throughout the body. Mechanisms for delivering genes based on Ch may be made more stable and more successful using PEGylation of Ch to improve pharmacodynamics and pharmacokinetics. The resilience, soluble nature, immunogenicity, and residence duration of gene-based Ch vectors may all be enhanced by PEGylating Ch, which will eventually increase the efficacy of treatment. It is crucial to remember that further research and development is needed in this field to improve and optimize PEGylated ChNP formulation and design, guaranteeing its effectiveness in uses of gene therapy [9].

Effectively targeting and entering specified cells is crucial when delivering siRNA utilizing Ch vectors to avoid premature siRNA release from potentially harmful agents. Whereas targets of siRNA are found inside the cytoplasm, gene therapy targets are found in the nucleus of cells. Vectors that lack target cell-specific ligands are not cell-selective; instead, they primarily use their exterior characteristics to go into cells via adsorptive endocytosis without specificity. Effective internalization may be facilitated by vectors with surfaces that are positively charged, such as those derived from formulations of Ch, which may improve electrostatic exterior attaching to cellular membranes that are negatively charged have shown that the size and form of NPs may significantly affect how well they penetrate cells [57]. When nanoparticles are internalized, depending on their size are unaffected by the presence of serum. Serum opsonin proteins may change the shape of nanoparticles, resulting in density or sedimentation. Confocal microscopy is used to detect fluorescently tagged particles after they have been treated with cells for 4–6 h in order to investigate cellular internalization. The fraction of cells that contain the particles is measured using flow cytometry. Depending on the protein receptors of the target cell, Ch carriers' surface modification with ligands enhances cellular entrance. Through receptor-mediated endocytosis, Ch carriers with a positive charge attach to bilayers of negatively charged lipids to enable cellular entrance. To improve cellular penetration, ligands such as mannose, FA, and others are transformed into formulas of Ch. When providing ligands to carriers, a number of factors are crucial, including the ligand's quality, replacement method, combination processes, spacer size in polymer ligands, and the potency of the binding between the ligand and receiver. Peptides that penetrate cells have also been used by researchers to improve Ch-carriers' capacity for cell penetration. For neurodegenerative therapies, researchers developed TAT peptide-modified ChNPs/PEGylated ChNPs to transport siRNA. Other research revealed that adding ChNPs and folic acid enhanced delivery of genes in cancer cells with excessive folate receptor expression. Chemically, folate was linked to Ch to target FA receptor-containing tumor cells, allowing unique transfer of DNA. The particular gene was effectively transmitted via folacin-Ch/pDNA nanocomplexes by the use of folate-mediated endocytosis. This improvement implies that the efficacy of gene transport may be increased by mixing folic acid with ChNPs. The findings suggest that supplying cancer cells with genes that elevate expression of the folate receptor may be possible with this focused strategy [58].

Nucleic acids with functions must be published in the targeted cells after siRNA vectors have entered them. After cell internalization, the messengers are encased in main organelles that sort cells within cells. Using a proton ATPase transporter, these main endosomes turn into late endosomes when their pH falls to 5–6; Lysosomes and late endosomes merge to create an acidic environment with a pH of 4–5. Vectors must avoid endosomes to prevent combining with lysosomes; this may be accomplished by Ch formulations using the proton sponge mechanism, in which the vectors are kept within endosomes by the slow $-NH_2$ groups of Ch being protonated, which has about 6.5 pKa and reacts to the endosomal acid milieu. Water and chloride ions enter endosomes to offset the larger favorable fields caused by increased positive charges. This causes the endosomes to enlarge osmotically and eventually burst, releasing Ch enters the cytoplasm via vectors. Fusogenic peptides and pH-sensitive neutral lipids are two methods to improve vector release from endosomes. At Lower pH causes fusogenic peptides to shift structurally from coil to amphiphilic α-helical form, which enhances their capacity to pass through endosomal membranes [59].

Depending on the target and design of the vector, the introduction of DNA carriers may result in transfection rates that are either transient or persistent. Effective DNA transport vectors must effectively reach the nucleus for successful transfection, while resilient formulations may permanently connect DNA with the desired genes. Likewise, siRNA-transporting vectors are anticipated to release and transmit the siRNA in the cytoplasm of the cell to stop the creation of certain proteins. The siRNA delivery techniques used now are not particularly good in permanently silencing genes. Many delivery techniques have only been successful in briefly silencing genes due to the difficulties in delivering siRNA into target cells and keeping it active. To improve this, carrier designs must be changed to effectively carry siRNA into cells and release it for a longer amount of time inside the cytoplasm of the cells, which will result in more effective gene silencing. For instance, in order to stop the loss of alveolar bone in periodontal disorders, RNAi was used to suppress RANK signaling [60].

10.4 ChNPs and Gene Therapy Nanodrugs for the Treatment of Various Cancer Types

Several aspects of Ch must be carefully considered and adjusted to achieve robust and effective gene suppression. These include its molecular weight, specific features, level of acidity, how much N-acetyl groups are removed, and the N:P ratio [61]. Along with genetic changes, recent research suggests N-acetyl group removal, particularly the control of ncRNA such as microRNAs, may contribute to the development of cancer. Making use of mRNA JAM-A as a sign of effective delivery, researchers investigated the utilization of nanoparticles of miRNA-2 as a possible vehicle of miRNA-2 nanoparticles in breast cancerous cells. The effect of nanocomplexes of Ch-miRNA on MCF-7 breast cancerous cells was investigated inside the lab by Santos-Carballal et al. Ch is used with varying degrees of deacetylation and MV; they produced conjugates of self-organized miRNA-Ch with variable densities of charges. According to the research, MCF-7 cells target mRNA levels were effectively suppressed by Ch-miRNA complexes with acetylation levels of 12 and 29%. This implies suggesting miRNA-ChNPs may be a viable gene treatment approach for the treatment of breast cancers [62]. Mirzaie and colleagues investigated the effects of these NPs on the genes for BCL2 and BAX in MCF-7 cells after experimenting with attaching DTX to Ch using HA. Although the BAX and BCL-2 levels were marginally greater compared to those using just DTX, the research demonstrated that the particles decreased control of genes in comparison to regular cell activity. ChNPs covered in PEG to increase steadiness were used to deliver SVN-siRNA in breast cancer research; when compared to ChNPs alone, the PEG-ChNPs were more stable and smaller. Breast tumor development was prevented and SVN transcription was dramatically reduced when siRNA-SVN was loaded onto ChNPs-PEG. Because of Ch's 6.6 acidity constant, NH_4 ions may form with a pH of acid values, which facilitates interactions between anions and materials such as siRNA. Ch's minimal potential for electrostatics encourages the possible dissociation of NPs and ensures cell safety; TPP, acting as a cross-linking agent, enhances sustainability [63]. PEG is incorporated into ChNPs to enhance their chemical and physical properties, creating PCL NPs. These NPs have a positive surface charge of around + 11 mV, a uniform size of about 100 nm, and minimal dispersion (< 0.3). PCL NPs are affixed with silencing RNA that targets A2AR in order to improve immune-oncological effectiveness. By efficiently entering T cells and inhibiting A2AR expression, these RNA/PCL NPs prevent cell division, cause PCD, and cause the release of CKs, all of which impede the development of tumors (Fig. 10.7) [64].

Since tumor cells often resist PCD, it is an essential treatment method for HCC. In order to cause PCD and inhibit tumor development, researchers use $C_6H_6N_2O_2$-conjugated Ch nanoparticles to transfer PCD-related genes, such as p53. By raising p53 and caspase-3 transcription levels, administering these nanoparticles in conjunction with DOX therapy improves the efficacy of tumor treatment and causes PCD in HCC cells. Combining an LHRH hormone with lower MW Ch may improve Ch for gene transfer in the treatment of HCC. This alteration produces a stable nano-sized carrier with a diameter of around 85 nm, a charge of + 30 mV, and a good DNA packing capacity. For the treatment of HCC, this strategy shows promise. Using shRNA and siRNA as essential agents to target neoplastic genes such as VEGF, RNAi is a potential therapy for cancer. HCC progression may be postponed by using UDCA

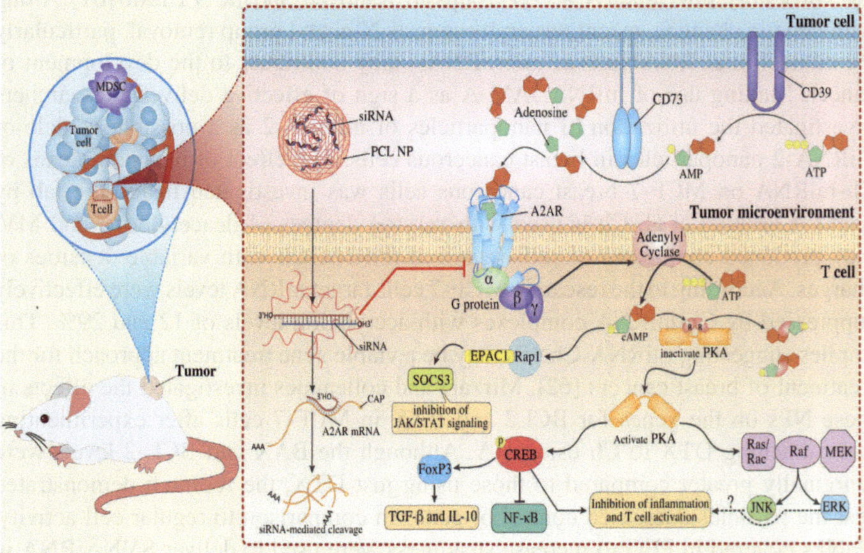

Fig. 10.7 By activating the CREB/PKA signaling pathway, A2AR signaling reduces the immune system's anti-cancer potential. Due of its association with a Gs protein, A2AR causes adenylyl cyclase to produce more cAMP. This in turn promotes the activation of CREB and protein kinase A (PKA) in a cAMP-dependent manner. CREB may increase the expression of FoxP3 and inhibitory cytokines while suppressing NF-kB. Furthermore, A2AR signaling influences additional pathways, such as MAPK and EPAC-1/COCS3 [64]

to down-regulate VEGF. In HCC, the lengthy PAARH/ncRNA promotes HOTTIP transcription, which activates the VEGF signal to promote the development of new blood vessels; therefore, VEGF regulation is essential for the therapy of HCC [65]. Connecting VEGF-shRNA to low MW Ch is a novel liver cancer therapeutic strategy. When this substance is administered to mice with tumors, the quantity of shRNA in the malignant tissues rises, which lowers the transcription of VEGF. In liver cancer, this complex has shown promise in slowing the proliferation of cancerous cells and preventing the creation of blood vessels. To cure HCC, oxide of ferric nanoparticles was developed to provide interference of RNA treatment. These NPs are made up of a GPC3-specific antibody attached to a ferric oxide core that has been altered with Ch-PEG. In animal trials, they have shown efficacy in delivering siRNA and halting the development of HCC. By modulating signaling mechanisms and gene transcription in the tumor microenvironment, miRNAs small molecules composed of 22–19 nucleic acids play a crucial role in limiting the development of HCC and have a major influence on tumor growth [66]. To treat HCC, researchers are delivering miR-122 using Ch-5-FU NPs, which improves the stability of miR-122 in the body and promotes tumor suppression and cell death. By lowering the expression of certain proteins (Bcl-2 and ADAM17), miR122 delivered by ChNPs impairs the viability of HCC. This strategy strikes a balance between systemic toxicity and microRNA transfer by efficiently suppressing tumors in live beings, causing cell death in HepG2 cells, and preventing the formation of HCC. Because VEGF plays a role in tumor development, ChNPs are employed to administer VEGF and SO at the same time in the therapy of HCC. By lowering VEGF levels and administering SO, this method increases the anti-tumor actions and causes PCD. Researchers discovered that controlling NEGF2 may affect the development of HCC cells in a recent study. Furthermore, IGF-1R has the ability to trigger STAT3 signaling, which raises NEGF2 expression and works in concert with STAT3 to promote the formation of HCC. In order to transport NEGF2-siRNA, researchers created a hybrid nanostructure utilizing Ch. The nanoparticles may effectively bundle RNA silencing at a mass ratio larger than three by altering Ch by amino-dehalogenation. These structures had a + 3 mV surface charge and were spherical in form, measuring between 100 and 200 nm. Along with the silencing RNA, the ChNPs effectively lowered NEGF2 expression and halted the growth of HCC cells [67]. The delivery of genes and medications to treat HCC by NPs containing Ch has been studied, with an emphasis on a tumor suppressor called nutlin-3a that blocks MDM2. By rearranging the cytoskeleton on the tumor suppressor nutlin-3a, inducing the death of cells and inhibiting the proliferation of cells of osteosarcoma. Researchers co-delivered the p53 and nutlin-3a genes using core/shell nanoparticles to increase their impact on HCC. When compared to normal Ch, these nanoparticles showed five times as many cells entering. They also disrupted the membrane of the cell and interfered with interactions between p53 and MDM2, which hindered the development of HCC. RASSF1A is essential in liver cancer because it increases sensitivity to chemotherapy, promotes cell death, and inhibits the development of the disease. In one work, ferric oxide and CMC were used to generate nanostructures; Ch-NH_2 groups served as cross-linkers, while ligands for galactose were added via a decrease in the ammoniation process. With a size of

40.1 nm and a positive surface charge of + 6.5 mV, the resultant nanoparticles efficiently compressed DNA while maintaining good biocompatibility. When a ferric oxide field is applied externally, these NPs increase the levels of caspase-3 in tumors, causing PCD. By boosting its transcription via LINC00460, PAK1 plays a critical function in encouraging the formation of HCC, preventing miR-485-5p and hastening the advancement of liver cancer. By decreasing PAK1 transcription, myricetin causes liver cancer cells to die; HCC growth may be hampered by IPA-3-induced PAK1 inhibition. Scientists have developed NPs covered with GA and LA to improve endocytosis in liver cancer cells and use Ch to transport PAK1-silencing RNA. By inhibiting PAK1 and blocking the route of MEK/ERK, these NPs increase RNA absorption, preventing tumor development and migration and causing PCD [68]. Like PAK1, PLK1 is an important target for the treatment of HCC. Liver cancer cells develop as a result of STK39's interaction with PLK1 to increase ERK transcription. Increased FBXO45 transcription levels increase PLK1 expression, which promotes the development of cells that cause liver cancer even more. GCP NPs have the ability to encapsulate siRNA-PLK1, which prevents the G2/M phase of cell division and prevents cell duplication. Furthermore, by downregulating Bcl-2 and upregulating BAX, p21, and p53, GCP nanoparticles with siRNA induce PCD. To increase the effectiveness of ChNPs in delivering RNA/DNA for the treatment of HCC, smart nanocomplexes that are susceptible to redox conditions and pH have been developed [69]. Combining DCs with cancer cells to make a vaccine is a potential tumor therapy approach. Early clinical studies of this strategy, however, have shown limited effectiveness, perhaps as a result of an immunosuppressive milieu. As chemical vectors, researchers used folacin/ChNPs in a study that attach to cancerous cells with a strong amount of transcription of folacin receptors. To boost gene activity in living organisms, a plasmid for the mouse mIP-1 gene was sent into cells using these folacin-coated ChNPs. MIP-10 is a potent CTL attractor. In one research, a DC/cancer cell fusion vaccination combined with folacin-Ch/mIP-10 effectively reduced the progression of liver cancer and increased mouse survival rates. The medication enhanced anti-tumor immunity via IFN-γ and dramatically decreased MDSC levels in the bone marrow, spleen and local cancer of mice. It increased cell death and suppressed the progression of cancer, indicating that mIP-10 delivery counteracts immune suppression at the tumor site and increases the efficacy of a DC/tumor cell vaccine [70]. Immunologic treatment for liver cancer used plasmid for the transcription of the mouse mIP-1 gene. To aim at cancer cells, ChNPs and a folacin-containing IP-10 plasmid were given to H22 mice with transplanted tumors; the success of the treatment was monitored using optical imaging in living subjects. The tumor and spleen were examined for MDSC and CD44+CD25+FoxP3+ Treg levels, respectively. ELISPOT was used to measure the number of IFN-γ-positive cells, and immunohistochemistry examination showed that treated mice had higher survival rates and less cancer development. The treatment increased IP-10 expression in the tumors and prevented neovascularization. When folacin and the IP-10 plasmid were administered together to cancer-stricken mice, the number of splenic Tregs decreased and cell death occurred, while the number of MDSCs and cells that produce IFN-γ that infiltrated the malignancy increased. This indicates potential for HCC immunotherapy utilizing this gene transfer platform.

When human liver cancer cells are present in naked mice, the combined application of CD8+ CTLs of folacin-Ch-hIP-10 and pMAGE-A1278-286 enhanced IFN-γ production, reduced cancer development, and enhanced the survival rate. This novel therapeutic strategy reduces cell division, promotes cancer cell death, and stops the formation of tumor blood vessels. The findings show that improving tumor-specific CTLs may be an effective way to cure cells of liver cancer in humans [71]. These results show that by increasing their concentration at the tumor site and prolonging their blood circulation duration, ChNPs may increase gene efficacy in liver cancer treatment. ChNPs may better bind to cure cells of liver cancer in humans by adding immunoglobulins and binding molecules to their outermost layer. The potential of ChNPs to treat liver cancer has been shown in animal experiments, underscoring the advantages of employing Ch delivery vehicles for the treatment of genes. Their capacity to be employed for two genes and medication transfer combined treatment for carcinoma of the liver is a significant benefit. One important chemokine that draws CTLs to cancer cells and increases their capacity to combat cancer is IP-10. It has been discovered that boosting IP-10 expression in tumors improves adoption. To preferentially target liver cells that are cancerous and increase transcription of IP-10, one research study sought to develop ChNPs decorated with folacin and coated with the HIP-10 gene. The goal of this strategy was to strengthen the immune system's defenses against MAGE-A12, which might increase the effectiveness of therapy [72].

Polymers with functional properties may be added to formulations of Ch to increase gene transfer and Ch efficacy [5, 73]. With several $-NH_2$ groups that energize it and give it a sponge-like quality, PEI is a well-researched polymer that is a strong choice for the transfer of genes. Although PEI's strong favorable charge helps transfer DNA and RNA, it may also damage healthy cells. The toxicity may be decreased by adding PEI to ChNPs as a carrier or by chemically altering them. Compared to employing native PEI-NPs, this method increases the intracellular transport ChNP's efficiency, leading to improved effectiveness of transfection and decreased cytotoxicity [74]. Jiang and colleagues developed ChNPs-g-PEI coupled with LND to deliver Bcl-2 siRNA with the goal of targeting mitochondria and tumor cells. In Fig. 10.8, the targeting method is shown: following internalization by target cells by folacin mediation, the release of siRNA into the cytoplasm to lower Bcl-2 expression, and TPP transports LND to mitochondria, where it causes cell death; together with LND, this mitochondrial signaling-based strategy has strong anti-tumor effects. When HeLa cells were treated with these NPs, their rate of proliferation decreased to 11.9% after 72 h, which was much lower than when LND or siRNA alone was used [75].

Particularly in lung cancers, Akt1 stimulates the growth of tumor cells, PEI-g-Ch was created to deliver Akt1-siRNA to lung tumors, which down-regulates Akt1 and prevents A549 cells from proliferating and surviving. Because of its higher transfection efficiency, the PEI-g-Ch combination enhanced the efficacy of siRNA-Akt1 in gene silencing. HPV is the main cause of cervical cancers, and the E6 and E7 genes are essential because they produce oncoproteins that control the cell cycle. In cervical cancers, ChNPs may efficiently deliver silencing RNA that targets HPVE7. The synthesis of a compound containing ChNPs is facilitated by the positive charge of Ch, which results in effective tumor cell uptake. Cervical tumor PCD is increased

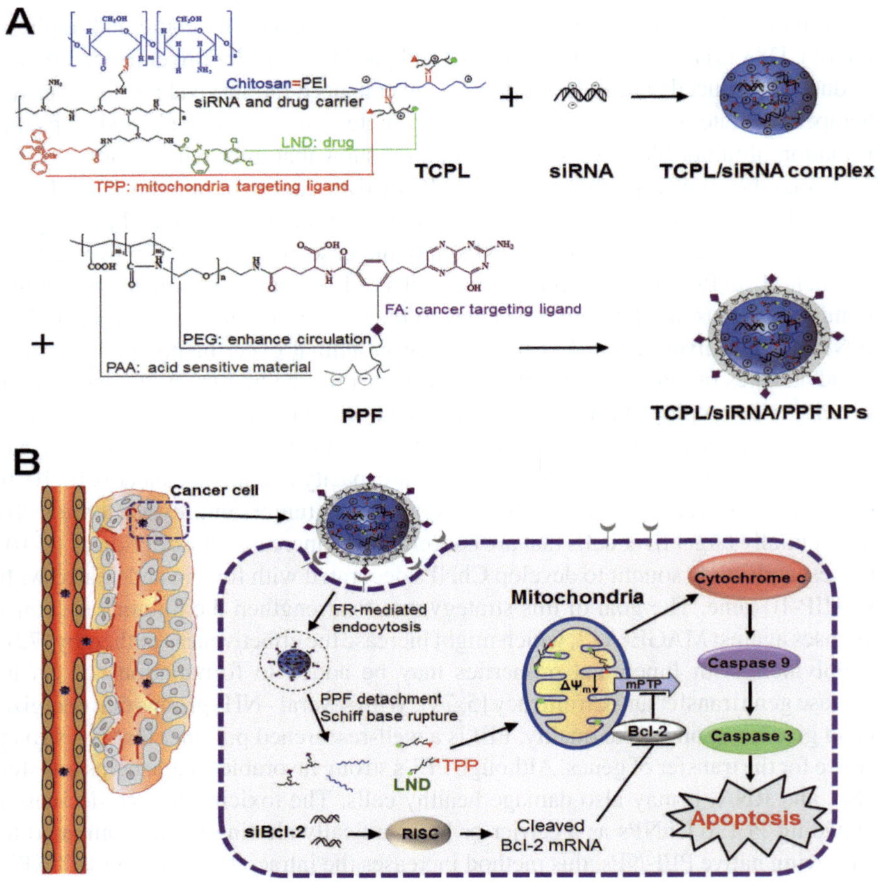

Fig. 10.8 Shows the composition of PEI-ChNPs that co-deliver lonidamine (**a**) and siRNA-Bcl-2. The process of using siRNA-Bcl-2 and lonidamine (**b**) to induce the mitochondrial PCD pathway and the hierarchically targeted approach. RNA-induced silencing complex (RISC), lonidamine (LND), poly (acrylic acid) (PPA), PAA-PEG-FA (PPF), and mitochondrial permeability transition pore (mPTP) [75]

by this HPVE7 silencing technique. When employed for nanoscale delivery, ChNPs exhibit superior qualities such as extended release and high-capacity loading. Additionally, adding certain binding molecules to ChNP surfaces enhances their absorption and targeting capabilities. For instance, it has been shown that ChNPs treated with folacin improve tumor endocytosis. Tumor development may be inhibited by BV6, an increased PCD inhibitor transcription. Cancer treatment could get better by targeting IL-6 along with other factors; to deliver BV6 and IL-6-silencing RNA for treating breast and colon cancers, researchers used PEG-ChNPs-lactic acid. This co-delivery strategy works together to prevent tumor growth and duplication; it appears that CHIMI may assist the nanoparticles in escaping from the endosomes. Making use of TMC, which is created by trimethylation to maintain its cationic charge, whereas rendering it soluble in water and able to stick to surfaces of mucosa, is an additional strategy. The capacity of CHIMI and TMC to successfully cross the gastrointestinal barrier is crucial to provide silencing RNA and the treatment of digestive system cancers. Ch has useful characteristics for delivering siRNA and guaranteeing long-term drug release, but chemical changes may improve its features further more. Trimethylation is one efficient technique that increases Ch's water solubility. Conjugating CMD to Ch is a further strategy that increases its solubility while inhibiting aggregation. This breakthrough led to the creation of a nanostructure that combines CMD and TMC. It is intended to administer BV6 (which causes PCD in tumors) and STAT3 (which targets genes that promote cancer) inhibitors simultaneously for the treatment of breast, colorectal, and melanoma malignancies [76]. Using NIK, a BV6 mediator, in conjunction with siRNA-STAT3. The capacity of CMD/TMCNPs to slow the growth of tumors has been further enhanced by the integration of NIK/siRNA/STAT3 [77].

Similar to multiplication, ChNPs carrying silencing RNA may also affect tumor cell emigration. Finding the factors that cause tumor emigration is essential to achieving this objective. One illustration of a substance that causes tumor cells to mutate and promotes their malignant development is BRAF, a person with the gene type RAF. The BRAF protein, which is essential for controlling cell division and proliferation, is produced by the BRAF gene. Colorectal cancer, melanoma, and other malignancies may result from mutations in this gene. BRAF mutations increase the invasiveness and propensity for metastasis of cancer cells in the setting of tumor metastasis. A frequent mutation that causes hyperactive BRAF kinase and the unchecked cell growth and division characteristic of cancer is the v600e mutation. Furthermore, by its exchanges with several mechanisms for cell signaling, such as MMPs, VEGF, and EGFR in cancer cells that have spread, IGF-1R contributes significantly to the development of tumor malignancy [78]. ChNPs containing DTX and IGF-1R-siRNA have been developed by researchers to stop the spread of breast tumors. By altering the outer layer of nanovectors, aptamers artificial molecules that bind preferentially to ligands have been used to improve breast targeting carcinoma cells. Aptamers provide a number of benefits, including cost-effectiveness, minimal immunogenicity, excellent specificity, and compact size. Targeting MUC1 with MUC1-aptamers is essential for the selective therapy of breast cancers since MUC1 is extensively expressed on tumor cell surfaces. Researchers created NPs that efficiently transport siRNA by altering ChNPs with anti-MUC1 aptamers. These altered ChNPs suppressed chemicals linked to metastasis, including MMP-9, STAT3, and VEGF, by lowering 1R-IGF transcription in breast carcinoma cells [79]. To be efficient vectors for siRNA transfer, ChNPs must overcome some drawbacks, such as their poor cell membrane penetration and limited buffering capacity. The remarkable capacity of ChNPs to load substances is one of their primary features; research has shown β-catenin-silencing RNA loading efficiencies of up to 97%, a part of the route of Wnt signaling connected to tumor progression. It has been shown that PEGylated ChNPs efficiently transport β-catenin-siRNA, which prevents tumor metastases. Furthermore, by lowering epithelial markers and raising mesenchymal markers, the EMT process contributes to tumor spread. The remarkable capacity of ChNPs to load substances is one of their primary features; research has shown β-catenin-silencing RNA loading efficiencies of up to 97%, a part of the route of Wnt signaling connected to tumor progression. It has been shown that PEGylated ChNPs efficiently transport β-catenin-siRNA, which prevents tumor metastases [80].

Drugs used in chemotherapy, like cisplatin, first show promise in the treatment of cancer. On the other hand, prolonged usage may result in medication resistance, which may reduce the treatment's efficacy. Breast cancers have significant expression of the ligand for CXCR4. The tumor cells' capacity to migrate and proliferate is enhanced by this elevated expression of the ligand for CXCR4. In addition to encouraging metastasis, which might heighten resistance to CP, CXCR4 also plays a role in chemotherapy resistance [81]. Researchers have developed a unique delivery method using nanoparticles to increase the efficacy of the chemotherapeutic medication cisplatin in the treatment of malignancies. In particular, they have created ChNPs that can carry and encapsulate siRNA molecules that target the CXCR4 gene. The resultant CXCR4-siRNA-loaded ChNPs have certain physicochemical characteristics, such as a diameter of 80–110 nm and a surface charge range of 20–50 mV. It has been shown that ChNPs loaded with siRNA that targets CXCR4 effectively transport the siRNA cargo into breast cancer cells. PCD induction, metastasis, and proliferation are among the important tumor processes that have been suppressed as a result of this CXCR4-siRNA administration. Additionally, it has been shown that using these CXCR4 siRNA-loaded ChNPs significantly reduces the CXCR4 gene's expression in breast cancer cells. Consequently, these tumor cells are now more susceptible to the chemotherapeutic actions of CP. In cancer treatment, nanoparticles based on chitosan are being used as efficient tools for delivering medications straight to tumors. These nanoparticles are great choices for enhancing therapeutic efficacy and minimizing side effects because of their unique characteristics and compatibility with the body. Nanoparticles may include a variety of medicinal materials, including immunomodulators, siRNA, and chemotherapeutic medications. They may interact with negatively charged cell membranes because of their positive charge, which facilitates the delivery of the materials straight to tumor locations. By increasing the concentration of the medications at the tumor and decreasing their influence on non-targeted locations, this targeted delivery minimizes exposure to the body's other parts. Furthermore, chitosan's mucoadhesive properties help tumor cells absorb it more efficiently, while nanoparticles' small size allows them to enter tumor tissues and improve intracellular medication release. Consequently, this enhances medication bioavailability, cellular absorption, and therapeutic effectiveness. Furthermore, the pH-dependent properties of chitosan nanoparticles allow them to release their cargo in the tumor's acidic microenvironment. This guarantees tailored drug action, increasing efficacy and reducing damage to healthy tissues. To improve treatment outcomes by leveraging the combined effects, these nanoparticles may also be utilized in combination with other therapeutic approaches, like the concurrent delivery of chemotherapeutic medications and siRNA that targets certain genes. By promoting presenting antigens and stimulating dendritic cells and triggering cytokine production, chitosan-based nanoparticles have immunomodulatory qualities. This promotes better immune recognition of tumor cells and aids in tumor regression. Compared to unbound medications, the biocompatibility of chitosan reduces side impacts and toxicity on the system, which lowers side effects and improves the general quality of life of patients while undergoing treatment. Nevertheless, challenges remain to improve the effectiveness of delivery methods based on chitosan. Improving surface alterations, cargo loading, and customized techniques are some of these challenges. Furthermore, careful evaluation and validation are required for the clinical use of these systems. In conclusion, chitosan-based nanoparticles have a lot of promise for treating malignancies since they may enhance medication delivery, lower toxicity, and complement existing therapies. This presents a huge chance to create individualized, efficient cancer therapies that put patients' health first [82]. A transporter protein called MDR-1/P-gp may reduce the effectiveness of chemotherapy in cells. To target MDR-1 and perhaps the general quality of life of patients, NPs have been used. To improve stability and condense the nanostructures, researchers altered GCh NPs. Glutathione within cells breaks down GCNPs, which have an average diameter of 269.8 nm and a charge on the surface of 7.7 mV; this allows GCNPs to exit endosomes. They have been shown to be successful in overcoming resistance to chemotherapy (Fig. 10.9) [83].

LDL receptors are more often seen as a charge on the surface, and these receptors may be altered to enhance the absorption of using nanoparticles to deliver medications and genes. However, MDR, in which medications are actively transported out of tumor cells, may result in chemotherapy resistance. NPs, including ChNPs, may be

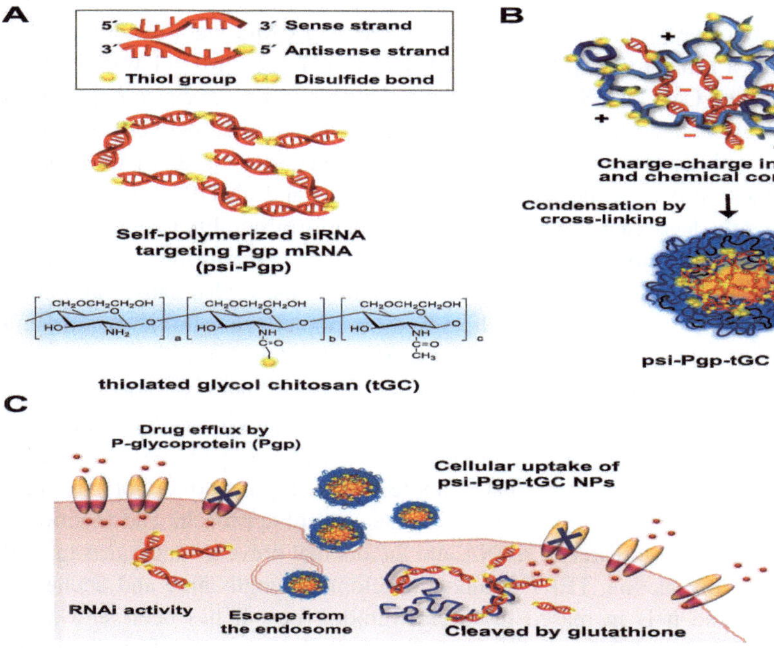

Fig. 10.9 Target gene silencing by the intracellular transport of psi-P-gp-tGC NPs [83]

delivered to tumors because of aberrant blood arteries. The EPR effect may be used to increase the efficacy of cancer medications by altering ChNPs with PEG. To more precisely target tumor cells, other tactics include the use of polysaccharides, small compounds, proteins, and antibodies. As the tumor develops and expands, cancer cells produce more of the EGFR protein. The capacity of nanoparticles to concentrate inside tumor cells may be enhanced by designing them to precisely identify and bind to EGFR. Cell division may be stopped by the Mad2 gene, which functions as a mitotic checkpoint. Across several tumor sample types, elevated Mad2 levels are often seen [84]. As a cancer therapy, NPs loaded with siRNA directed against the gene Mad2 and engineered to target tumor cells' EGFR protein have been created; the tumor tissue may be effectively penetrated by these specific NPs. When EGFR-targeted nanoparticles are used instead of non-targeted NPs, the Mad2-silencing siRNA's targeted delivery is six times greater. As a consequence, the Mad2 protein is more strongly downregulated, making the tumor cells more susceptible to the CP [85]. In a different work, polylactic acid nanoparticles covered in Ch were synthesized to co-transfer cisplatin and pDNA/siRNA. Ionic grafting was used to attach the siRNA to Ch. Globular ChNPs with particle diameters between 280 and 350 nm were able to encapsulate siRNA with a high efficiency of 78.5% and cisplatin with 82%. Benefits of using ChNPs for drug transfer include great serum durability and prevention of siRNA degradation. Treatment of tumors with siRNA/ChNPs enhanced susceptibility to cisplatin chemotherapy and dramatically decreased tumor cell proliferation [86]. To administer combining the therapy of DOX with MDR1 gene-siRNA, Hui Peng and her associates developed a versatile nano-platform employing CMC. They used a redox-responsive prodrug model, joining DOX and CMC via a disulfide link at the core. As a protective layer that is sensitive to pH, the siMDR1 was drawn to the structure of CMC-DOX and encapsulated inside a substance known as OEI. AHA/GHA was added to the nanoplatform to enable targeted delivery to tumors. It enhances cellular absorption and accumulation in cancer tissues by releasing its payload under redox or acidic conditions. P-gp activity and siMDR1 gene expression are significantly decreased as a result of the nanoplatform's ability to facilitate cargo molecules that leak into the cytoplasm from lysosomes or endosomes. When MCF-7/ADR cells have low exposure to DOX concentrations, it also causes apoptosis. Furthermore, when evaluated on live beings, the nanoplatform exhibits encouraging therapeutic outcomes in terms of preventing tumor development [87].

Although several anti-cancer factors are used in cancer therapy, tumors are often not eradicated by chemotherapy alone, and various strategies have been investigated to increase the efficacy of chemotherapy [88–92]. Utilizing gene-based materials, such as siRNA, to suppress genes that promote the development of tumors and make them more sensitive to chemotherapy is one successful approach. The function of ChNPs during the co-transfer of siRNA and factors that prevent cancer therapy will be covered in this section. TERT is linked to telomere lengthening and unchecked cell division, and it is increased in several tumor types. It has been shown that siRNA targeting the TERT gene may effectively slow the growth of malignancies. The possibility of employing HTCC NPs as a method to co-deliver TERT-siRNA and the chemotherapeutic medication PTX for the treatment of cancer was examined in research. Both TERT-siRNA and the chemotherapeutic medication PTX may be effectively encapsulated by HTCC NPs. The HTCC NPs' positive charge and porous nature made this possible. Furthermore, the HNPs improved the encapsulated siRNA's cellular absorption and offered enzyme inhibition capabilities. This platform for NP produced a synergistic decrease in the growth of malignant cells when utilized as a combined delivery method. This was accomplished by inhibiting the growth of malignant cells, and PTX's intracellular accumulation is growing inside the tumor cells [93]. Because of its cationic charge, which allows it to use a negative charge to silence RNA, Ch has drawn interest as a material for the synthesis of NPs. Furthermore, Ch's biodegradability and biocompatibility make it a suitable choice for experiments carried out within living things; one study used ChNPs/CMD to co-entrap DOX and Snail-silencing RNA in the treatment of colorectal cancers. EMT is known to be induced by snail, and greater tumor cell spread is linked to its overexpression [94]. Vimentin and MMP-9 synthesis were significantly decreased, and cellular absorption was enhanced as a consequence of CMD/ChNPs' successful trapping of Snail-silencing RNA and DOX. Additionally, it has been shown that E-cadherin is upregulated, which inhibits colorectal tumor cell motility. By stabilizing HMGA2, an oncogene promotes tumor growth and metastasis. Moreover, chemoresistance has been shown to result from HMGA2 overexpression. ChNPs were developed to treat lung cancers by delivering DOX and HMGA2-siRNA at the same time. ChNPs effectively encapsulated both drugs and increased cellular absorption, demonstrating a synergistic effect. When compared to utilizing each drug individually, the nanocomplex's improved encapsulation and absorption demonstrated greater efficacy in inhibiting lung cancers. To treat tumors, GCNPs were created to DOX and Bcl-2-siRNA co-transfer, and GC-CA interlacings that are amphiphilic and hydrophobic characteristics were created by chemically combining the GCh with the water-repellent CA. While downregulating Bcl-2 by silencing RNA promotes DOX-mediated PCD, overexpressing Bcl-2 prevents cell death in malignancies. Anticancer medications and siRNA, which can block the genes that promote development and survival of malignant cells, may be co-delivered by ChNPs, increasing the tumors' vulnerability to anti-tumor actions. According to studies, breast cancers that have elevated amounts of IL-17B and IL-17RB, its receptor, are more likely to survive and have a worse prognosis. This implies that breast cancer may be promoted by this signaling system [95]. By building compounds with chemotherapeutic medicines and siRNA, ChNPs may enhance cellular absorption and prevent siRNA degradation. It has been shown that IGF-1R overexpression promotes tumor development by triggering signaling pathways that encourage lung tumor growth, metastasis, and treatment resistance. ChNPs have been developed to deliver together the IGF-1R-siRNA and DOX to overcome this issue, increasing the lung tumor cells' susceptibility to DOX treatment. With an average diameter of 176 nm and a charge on the surface of 11 mv, the produced ChNPs carrying dox and siRNA demonstrated favorable characteristics. Due to decreased production of MMP-9, STAT3, and VEGF, lung tumor cells suffered from greater dox-induced cell death when siRNA downregulated IGF-1R. Since siRNA has a negative charge and is unable to pass through cell membranes, ChNPs are crucial for increasing its cellular absorption. GONPs may be harmful due to their drawbacks, which include limited water solubility, big size, and poor biocompatibility. GO and Ch together, however, may enhance biocompatibility and shield blood cells from harm. In order to create NPs for tumor therapy, a research interlaced CGO with HA and TMC. While carboxylation decreased toxicity, TMC improved the NPs' solubility, biodegradability, and biocompatibility. Due to decreased production of MMP-9, STAT3, and VEGF, lung tumor cells suffered from greater dox-induced cell death when siRNA down-regulated IGF-1R. Utilizing these nano-vectors enhanced cellular absorption, prevented siRNA degradation, and raised dinaciclib's bioavailability. Since hif-1α-silencing RNA inhibited the production of TGF-β, VEGF, MMP-9, and MMP-2, it decreased tumor cell viability, metastasis, and neovascularization. Dinaciclib, a CDK2 and CDK1 suppressor, inhibited tumor growth and interfered with hif-1α activation [96].

10.5 Gene Delivery from Stimuli-sensitive Chitosan Nanoparticles in Cancer

ChNPs responsive to stimuli, have plenty of possibilities for targeted distribution of medications for cancer therapy [17, 97–100]. Chitin-derived Ch, a polymer that is both biodegradable and biocompatible, offers an appropriate matrix for creating nanoparticles that may contain medicinal substances [101–103]. These nanoparticles have the unusual capacity to let go of the laden medications only at the location of the tumor in response to certain external or internal stimulation, like variations in redox, pH, and ultrasound. This exceptional characteristic makes it possible to regulate medication release, minimizing side impacts and enhancing medicinal effectiveness while lowering systemic poisoning. Additionally, using Ch nanoparticles provides higher cellular absorption, extended circulation duration, and increased tumor penetration, all of which contribute to better therapy results [104].

The TME has a major impact on how cancer develops and how well a therapy works, and pH changes are one important factor [39, 105]. While internal pH fluctuates across compartments, tumors have an acidic extracellular pH because of increased glycolysis and lactate generation, which is exacerbated by inadequate perfusion and metabolic changes. Because they react to these pH dynamics, pH-sensitive carriers are essential to cancer treatment because they enable tailored medication delivery. These carriers, which include membrane-camouflaged nanoparticles and pH-responsive polymers like PEG and PLGA, are designed to change their characteristics or release medications in reaction to acidic environments. They take advantage of the acidity of the TME to increase specificity and efficiency. They preserve healthy tissues, enhance medication accumulation in tumors, and may be combined with additional mechanisms that respond to stimuli to provide treatment outcomes that work in concert. Despite encouraging preclinical findings, the heterogeneity of the tumor microenvironment (TME) and the need for thorough clinical confirmation present significant challenges, highlighting the potential of pH-sensitive carriers to transform individualized cancer therapy approaches. Numerous studies have shown that the introduction of siRNA and medications targeting using nano-vectors for MDR genes successfully suppresses P-gp expression in drug-resistant cell lines. As a result, this method facilitates the effective encapsulation of the delivered medications inside the nano-vectors, which in turn increases their intracellular accumulation. The hydrophilic character of siRNA and the hydrophobic nature of the majority of anticancer medications make constructing nano-vectors for siRNA and drug co-transfer very difficult. A significant barrier to this potential strategy is the resulting reduction in loading efficiency [106]. In another work, a team of researchers under the direction of Yongmao synthesized charge-switching Agm-Ch bioconjugates. Ch was modified using DMA, and it was then subjected to a nucleophilic reaction with primary –NH AGM and tosylated Ch. The goal of this study was to investigate the creation of altered AGM-Ch structures with controlled charge change capabilities. It has been shown that the aforementioned bioconjugates, in particular AGM-DMA-Ch, efficiently condense siRNA into nano composites. Even at healthy pH values, these nanocomplexes remain stable when serum is present. Changes in pH inside the environment of the weak acid tumor may change the charge on complexes from (−) to (+). Tumor cells that are intended to be resistant to serum adsorption may effectively internalize thanks to this modification. In both lab and live animal scenarios, the previously reported intelligent biogenic system demonstrated remarkable effectiveness in reducing gene expression and causing substantial PCD in cancer cells. These nanocomplexes successfully compacted RNA-VEGF into NPs for silencing. The NPs' surface charge changed from (−) to (+) at an acidic pH, which helped the tumor endocytose the NPs. This led to PCD induction and a significant VEGF mRNA knockdown (almost 79.7%) [107]. A study was carried out to build a pH-sensitive DNA/polymer nanocomposite covered with biomaterial that incorporates an HMGB1 to establish an efficient non-viral gene transfer mechanism. A class of naturally occurring compounds containing NSL is called HMGB1; using PEG as a linker, researchers were able to successfully create carboxylated Ch modified by folacin (CCTS). CCTS-PEG-folacin, the modified chemical, was produced and evaluated since its buffering ability using the titration of base-acid, a pH-sensitive core-shell architecture called FPCPHDs, was created and examined. Electrophoresis was used to show that FPCPHDs were resistant to DNase I degradation and heparin substitution; HepG2 and KB cell lines showed less cytotoxicity from FPCPHDs. The results of the RFP fluorescence intensity measurement and the luciferase activity test indicate that FPCPHDs improved the effectiveness of expression and transfection of genes in KB cells. Furthermore, Free FA is present in KB cells, inhibiting transfection and expression of genes. It was shown that FPCPHDs underwent intracellular trafficking throughout their migration inside KB cells, quickly leaving endo-lysosomes. At the time point three hours after transfection, it was discovered that these FPCPHDs exclusively localized inside the nucleus. Furthermore, it was shown that FPCPHDs increased Expression of RFP at the location of the tumor in the S180 naked xenograft mice model. Together, our Findings indicate that FPCPHDs are a viable strategy to enhance transfection and increase expression effectiveness in cancer cell populations with elevated FA receptor levels. With consequences for enhancing targeted therapy in opposition to FR-positive malignancies, this technique has a lot of promise for oncology research [108].

In a different study, a team of researchers created a technique known as MLNs to precisely distribute medications like VEGF and DOX. Using MSN, which had been altered using TAT peptides to act as the central hub for DOX loading, these MLNs were constructed via electrostatic self-assembly. The structure of the MLNs was composed of an exterior layer made up of a cationic compound termed GTC, which was particularly created to encapsulate silencing RNA, and an innermost layer made up of PAH-Cit, which had anionic characteristics. This strategy aimed to develop an intelligent delivery method for these drugs, enabling improved efficacy and control. Silencing RNA is greatly protected by the strong GTC outer layers' stability at pH values of 7.4 and 6.5, which stops it from degrading in the circulation and the tumor microenvironment. Furthermore, via a mechanism known as receptor-mediated endocytosis, Galactose ligands are present on the outer layers of GTC, promoting the effective uptake of MLNs. When MLNs eventually get in the endosomes or lysosomes, the acidic pH 5.0 surroundings cause a reversal in PAH-Cit's charge, which causes MLNs to disassemble. MLNs may then flee into the cytosol for further activity as a result of this disintegration. Cytoplasmic glutathione helped release silencing RNA by breaking the disulfide bonds in the layers of GTC. The efficiency of gene silencing was greatly enhanced by this procedure; the DOX-containing cores were simultaneously revealed and moved into the nucleus using the aid of the TAT peptide; the DOX was then shown to discharge continuously and under control. Consequently, MLNs demonstrated potent tumor-suppressive properties without any discernible harm when administered intravenously at a modest dose. The MLNs' combination of chemo-drugs and therapeutic genes maximized their synergistic effects and proved to be a safe and effective strategy (Fig. 10.10) [109]. Section 2: In another work, researchers used a redox-responsive disulfide-thiol link to create a co-delivery system that mixes two different components into a single nanovehicle. Chemotherapy and RNA silencing drugs may be transferred simultaneously thanks to this novel technology. In particular, the system is made up of a nanovector made of HA-SH, and a complicatedly created Dox is loaded onto N, O-CMC (NOCC). A dopamine-conjugated SH-HA-dopamine-coated RNA nanovector with calcium phosphate silencing is also included. With their two-in-one structure, the chimeric NPs showed exceptional stability in the circulation and during storage. Both Dox molecules and silencing RNA may be delivered to the cytosol with precision using this clever platform. Through receiver-mediated endocytosis, the system of co-delivery, which includes the component HA that targets cancer cells, effectively penetrates the cells and demonstrates remarkable targeting specificity. Essentially, these clever two-in-one chimeric nanoparticles have potential uses in functional co-delivery and cancer therapy since they can react to both pH and redox environments (Fig. 10.10) [110].

Effective treatment results may be obtained when anti-cancer drugs are administered in situations when the P-gp level is decreased. In one research, DOX was used in conjunction with siMDR1, a silencing RNA that down-regulates P-gp, to target

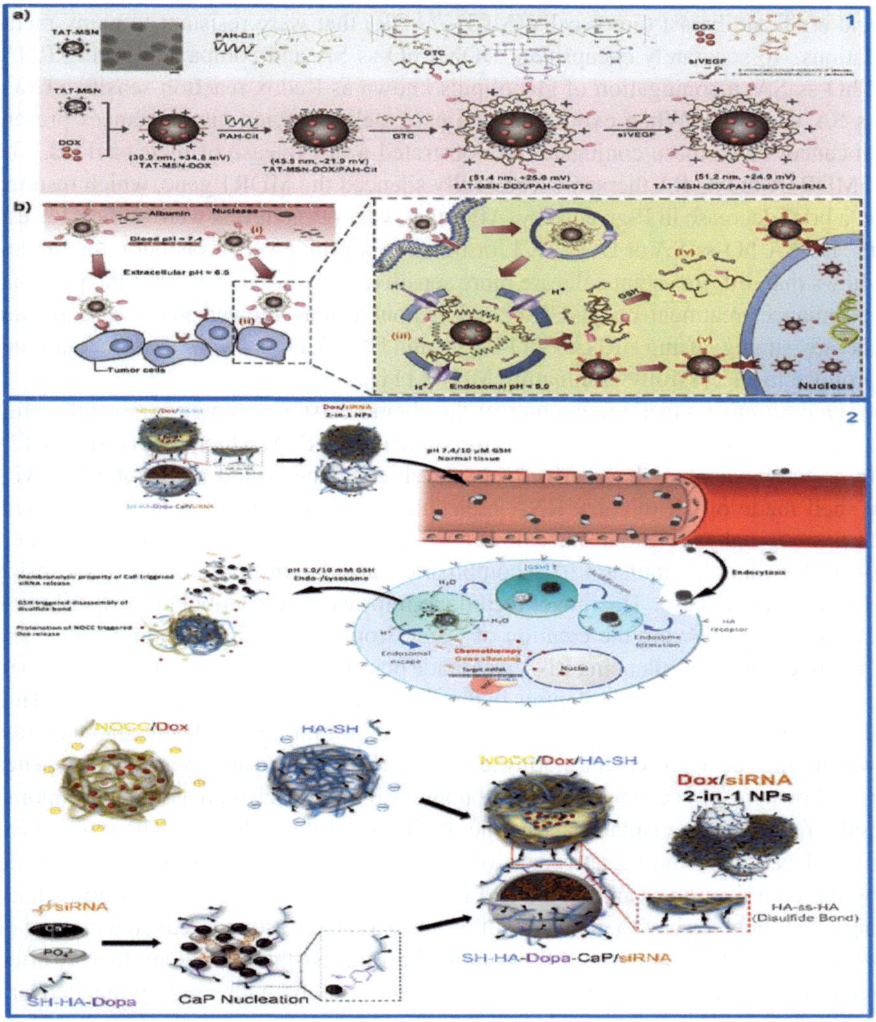

Fig. 10.10 The dynamics of MLN production and delivery are shown in the first half of the figure. According to the TEM picture, MLNs are created by covering the TAT-MSN core with GTC and PAH-Cit using a method of layer-by-layer self-assembly. With distinct processes at every level, the MLNs act as a method of siVEGF and DOX delivery. MLNs preserve their structural integrity in the surroundings of the tumor and blood (i). MLNs enter by receptor-mediated endocytosis (ii), which targets cancer cells' galactose receptors. When MLNs internalize, they react to the acidic intracellular pH by undergoing structural disintegration and escaping from endosomes (iii). GSH causes disulfide breakage, which releases siVEGF into the cytoplasm (iv). Moreover, MLNs use TAT-mediated targeting to help transport DOX to the nucleus (v) [109]. Section 2: A mechanism of delivery that is composite consisting of 2 parts is shown in the figure: SH-HA-Dopamine-coated CaP/siRNA nanoparticles and joined by A connection of disulfide bonds (HA-ss-HA) and distinct nanocarriers made from NOCC/HA-SH Dox-loaded. Together, these elements provide an integrated vehicle that functions as a combination and effective medication and gene delivery method [110]

and eradicate breast tumor cells (MCF-7/ADR) that were resistant to many medications. To separately encapsulate DOX/ChO-ss-SA and compact silencing RNA/ChO-ss-SA, a conjugation of glucolipids known as Redox reaction-sensitive ChO-ss-SA was used. When exposed to the extremely reducing circumstances present in cancer cells, these conjugates demonstrated a fast release of their payload. The siMDR1/ChO-ss-SA therapy successfully silenced the MDR1 gene, which resulted in a brief decrease in P-gp activity. After that, when P-gp levels were at their lowest, the DOX/ChO-ss-SA or DOX·HCl formulation was given. This encouraging finding shows that drug-resistant cells are more sensitive to cytotoxins and has the potential to enhance treatment results. The study's conclusions showed that the sequential therapy strategy using siRNA/ChO-ss-SA and DOX/ChO-ss-SA shows encouraging promise in successfully eliminating MDR [111].

One promising tactic to counteract the chemoresistance shown in several malignant tumors, including melanoma, is the silencing of RNA, which suppresses (Nrf2). In a study, scientists have effectively developed a new kind of nanobubble with a shell made of Ch to carry RNA molecules being silenced that target Nrf2 with ultrasonography. Scientists have developed a novel method using a W/O/W (water-oil-water) double-emulsion to accomplish encapsulation of silencing RNA inside the nanobubble core. This novel technique improves the nanobubbles' distribution efficiency by allowing the creation of water droplets carrying silencing RNA. The effective loading of silencing RNA was made possible by the successful fabrication of compact and stable nanostructures in the shape of tiny nanobubbles, around 100 nm in size, with a surface that is positively charged. An impressive 90% efficiency rate was attained using the encapsulating technique. The targeted downregulation of genes was shown in M14 cells by using nanobubbles-siNrf2, which increased the melanoma cells' resistance to cisplatin and vulnerability to cisplatin therapy. The internalization of cells and the effectiveness of transfection were both positively impacted by the combination of ultrasonic and nanobubbles. According to the study's findings, nanobubbles have a great deal of promise as an ultrasound-activated method of providing RNA that silences. Furthermore, the drug resistance problems that are often present in melanoma tumor cells have been successfully addressed by this delivery method [112]. The persistent challenge of developing transfer methods that are safe to use, able to selectively target, and possess high efficacy impedes the successful integration of silencing RNA treatment into clinical practice. With the capacity to react to ultrasonic stimulation to initiate the release of medicinal drugs, Fluorescein hydrochloride (CHC) and fluorescein isothiocyanate (FITC) NPs were created in research to act as medication delivery systems for colon tumors. The findings show that anti-β-catenin silencing RNA may be efficiently encapsulated by (FITC-CHC)-CMC nanoparticles via ionic gelation self-formation to increase the durability of RNA silencing. The FIC-CHC crosslinked-CMC NPs showed a poor cumulative silencing RNA release ratio in a pH 2.2 solution, totaling just 11.08% during a 120-min period. The pH-responsive transfer system based on FITC-CHC-CMC has shown that it can respond to an outside stimulus, in this case, ultrasonography, in a pH-appropriate environment to produce a controlled release of silencing RNA. The introduction of fluorescent-labeled silencing RNA into HT-29 cells was effectively shown via fluorescence microscopy. The β-catenin protein's expression was significantly reduced, reaching around 40.10%, after 48 h of exposure to RNA that silences anti-β-catenin in colon carcinoma cells. This decrease suggests that the protein that is essential for promoting colon tumor growth has been successfully inhibited. The findings of this study show that the use of RNAi treatment in impacted cells may benefit greatly from the silencing RNA-(FITC-CHC)-CMC transfer method. Because of their exceptional capacity to generate sonoporation, ultrasound nanodroplets (NDs) have become a very attractive nanovector for the effective transport of silencing RNA. The majority of methods used today to silence RNA transport using NDs are based on electrostatic interactions. However, this approach is severely hampered by the existence of barriers such cationic toxicity. Researchers recently utilized bacteriophage Phi29 DNA along with negatively charged O-CMS packing motor nanotechnology (3WJ-pRNA) to demonstrate a novel approach for delivering RNA, a type of genetic material that is sensitive to ultrasound [113]. By using complementary base pairing with an extended sequence, a 3WJ-pRNA scaffold was coupled with an A10-3.2 aptamer with 39 nucleotides that is specifically engineered to attach to a 21-nucleotide silencing RNA and the PSMA that targets Transporter 1 of cationic amino acids (SiCAT-1). Attaching a molecule of cholesterol to a single branch is essential for anchoring the 3WJ-pRNA nanoparticles onto nanodisks. By using RNA-aptamer modification and suppressing RNA-loading, the researchers were able to create the necessary O-CMS NDs, namely the A10-3.2/siCAT-1/3WJ-NDs. These NPs had core-shell structures and were uniformly spherical in shape. With a PDI of 0.3, the NPs' diameter was determined to be 198 nm. One common shell component, O-CMC, had strong anti-cancer effects. According to preclinical studies, siCAT-1/A10-3.2/3WJ-NDs exhibited good biocompatibility, buffering capacities, and efficient contrast-enhanced ultrasound imaging. Under ultrasonic irradiation, these nanostructures effectively transported siCAT-1 to PSMA-overexpressing prostate carcinoma cells. We can successfully stop 22RV1 cell migration and proliferation by suppressing CAT-1 expression. These findings point to a possible strategy for creating negatively charged, ultrasonic-sensitive NDs that can target tumors with RNA silencing. All things considered, these outcomes provide a possible path forward for the advancement of RNA transfer methods that silence malignancy. To transfer the gene DKK-2 in a lab setting in a geographically and temporally regulated manner, research was carried out to produce Ch nanodroplets that are echogenic combined Using nanodroplets of perfluorohexane (CNDs). The resultant CNDs had advantageous (+) charges on their surfaces, which allowed them to create nanocomplexes by efficiently attaching to genetic material that carries (−) electrical charges. Agarose gel electrophoresis was used to confirm that the CNDs and plasmid DNA had successfully bound. When CNDs loaded with DKK-2 plasmid DNA were utilized in conjunction with ultrasound, the DKK-2 plasmid DNA was disrupted and subsequently released. This finally resulted in a significant decrease in LNCaP cell reproduction by allowing LNCaP cells to enter via tiny holes in their cellular membranes. To achieve precise and controlled ultrasound-assisted gene transfer in specific tissues at the right time and place, stable and safe CNDs could be a good option [114].

10.6 Gene Release Mechanism from Polymeric Composite

The process by which a therapeutic agent is transferred from the matrix's inside to its outside surface and then released into the surrounding media is often referred to as "gene release" in polymeric systems. The medicinal agent may be delivered to its target environment via this approach. Through a variety of methods, devices with regulated release may facilitate the discharge of medicinal substances or genes. Diffusion is one of these processes, in which genes are released via holes that are filled with water. Apart from diffusion, the genes may also be liberated by the polymeric matrix eroding. Following an initial delay period regulated via dissemination, the holes created by this erosion become evident. In the polymer matrix, the main methods for controlled release are osmotic pumping, gene diffusion, erosion, and the movement of genes through water-filled pores [115–117]. Gene molecules disperse inside degradable polymeric materials due to random mobility caused by a variation in chemical potential. The diffusion mechanism via a network of pores that are linked regulates the gene's release rate when these systems break down. In contrast to gene release, polymeric implants absorb water quickly. Gradually, holes containing water are created when water is continuously taken up by the polymeric matrix. These voids progressively enlarge until they reach apertures big enough to allow the gene to be released [118].

Under such circumstances, the genes just diffuse from the matrix of polymers. Diffusion is the primary method of gene release in depots or reservoirs with regulated releases that are not biodegradable. The concentration gradient has no effect on the rate of release; instead, it is determined by the polymeric membrane's thickness and permeability [118]. Another way that genes are transferred via water-filled apertures is by convection. Osmotic pressure is produced when water enters a system without expansion. Thus, osmotic pumping is the term used to describe gene delivery driven by this force. Water entering the polymer bulk causes bulk erosion, which causes the matrix to deteriorate uniformly. The materials inside the bulk of the polymers hydrolyze because the water penetration rate into the bulk exceeds the rate of erosion. Bulk erosion is an ineffective method for precise gene/drug delivery because it is less dependable than erosion on the surface and cannot shield genes and drugs from outside influences [119]. As soon as the rate of erosion exceeds the speed with which water percolates through the substance made of polymers, surface erosion takes place. The process starts at the matrix surface and progressively shrinks the inner layer's matrices from the outer layer. Because the erosion kinetics may be efficiently controlled and reproduced, surface erosion is especially beneficial for applications requiring precise and controlled release [120]. To predict and explain polymeric gene transfer devices' gene release, a number of Tests have been conducted on mathematical models and developed. These models may be empirical/semi-empirical or mechanical, depending on the situation. Mechanistic models provide a fuller comprehension of gene dispersion, deterioration, and erosion by including biophysicochemical factors.

The idea of diffusivity, Fick's law serves as its foundation, is included in mechanistic models. Assuming a steady-state context, this basic equation connects the spread of materials with their gradients of concentration. Substances have a tendency to go from regions of greater focus to certain regions of lower concentration, per Fick's law. Fick's second rule, which explains how diffusion affects concentration fluctuation over time, is another crucial idea. Solute diffusion is often divided into two types in the context of gene transfer systems: Fickian and non-Fickian. When the solvent's diffusion time (td) and the polymer's relaxation time (tr) are about identical, non-Fickian diffusion takes place. Fick's conventional laws are insufficient to describe this kind of diffusion as it deviates from them. Designing efficient pharmacological formulations and maximizing therapeutic effects need an understanding of the sort of diffusion taking place [121].

10.7 Concluding Remarks

In the realm of gene therapy, Smart stimuli-sensitive nanoparticles based on Ch have attracted a lot of interest because they provide several advantages for precisely treating cancer. The natural polymer chitin is the source of Ch, a biocompatible and biodegradable material with remarkable properties including stability, non-toxicity, and ease of cellular barrier penetration. These nanoparticles have enormous promise for improving gene delivery, protecting genetic material from deterioration, and reducing any unwanted side effects. In terms of effectively delivering genes, ChNPs have a notable advantage. They have the rare capacity to get past biological obstacles that would often make gene delivery difficult. These nanoparticles can stop the extracellular environment's nuclease enzymes from breaking down the genetic material by forming a barrier. They may also pass through complex biological barriers including the Tumor microenvironment, endothelial cell layer, and extracellular matrix because of their tiny size. Targeted distribution to cancer cells is made much easier by this navigation capability, improving the likelihood of a favorable outcome. The capacity of these nanoparticles to sense and react to a variety of environmental stimuli, like pH levels, temperature variations, enzymes, or redox potential, is still another fantastic advantage. This implies that they may be made to respond to certain environmental cues. Our ability to precisely regulate and guide gene release at the tumor location is made possible by this exceptional sensitivity. Consequently, the detrimental effect on healthy tissues is reduced. The accuracy and efficacy of cancer therapy are greatly increased by this focused strategy. Aside from all of that, ChNPs have the significant benefit of being further improved by targeting ligands or antibodies. This indicates that they have the ability to identify and attach to cancer cells with specificity. The therapeutic genes or medications may more efficiently concentrate in the tumor when this targeted delivery method is used, increasing therapy efficacy while lowering potential harm to the body as a whole. When it comes to ChNPs for gene delivery and cancer treatment, there are a few obstacles that must be overcome. Effectively encasing the genetic material in these nanoparticles while maintaining its stability and security throughout transit is one specific problem. To achieve the ideal balance between condensing the genetic material with Ch and maintaining its integrity and bioactivity, this process may be quite complex and may need to be optimized. Furthermore, using nanoparticles by systemic administration presents a number of difficulties. The reticuloendothelial system's (RES) ability to remove these particles and the possibility of cytotoxic consequences are two significant challenges. The nanoparticles' quick clearance and restricted accumulation at the targeted tumor location might be caused by the interaction between the RES and the nanoparticles, which can hinder their circulation inside the body. Researchers are concentrating on altering the surface of nanoparticles to reduce immune detection, increase their concentration in tumors, and lengthen their circulation period in order to solve this problem. The complex tumor microenvironment presents researchers with yet another challenge in the fight against cancer. Depending on the kind of cancer, this environment might vary significantly. The efficacy of ChNPs may be impacted by elements such as the thick extracellular matrix, oxygen deprivation, acidic pH, and variations in enzyme activity. Research is still being done to figure out how to make these nanoparticles function properly in response to the particular circumstances seen in tumors. The potential of chitosan-based smart stimuli-sensitive nanoparticles for gene delivery and cancer treatment is quite promising, despite the difficulties we now face. The design, synthesis, and optimization of these nanoparticles are anticipated to improve as a result of ongoing research efforts, allowing us to get beyond current challenges. Significant developments in genetic engineering, material science, and nanotechnology are anticipated, all of which will help create more effective gene delivery methods with better therapeutic results. Additionally, there is a great deal of promise for synergistic treatment methods and better patient outcomes when ChNPs are used with other therapeutic modalities like immunotherapy, chemotherapy, or photodynamic therapy. We want to improve cancer therapy's efficacy while simultaneously addressing problems like drug resistance and minimizing adverse effects by combining different therapeutic approaches. This all-encompassing strategy marks a new era in cancer therapy, where many medicines coexist peacefully to provide patients the best care possible. The use of methods like 3D, 4D, and 5D printing technologies as well as artificial intelligence technology are some of the promising possibilities for the future. The potential of ChNPs to transport therapeutic substances, like as genes or medications, is amazing. Even more astounding is their ability to unleash these agents precisely where we want them to be in our bodies. This is where 3D printing is useful since it provides an excellent chance to create complex scaffolds or frameworks that are ideal for these nanoparticles. This implies that we can precisely administer drugs to cancer cells without endangering healthy organs. A revolutionary development made possible by 4D and 5D printing processes is the creation of dynamic systems that can react to certain stimuli or triggers. Anything from changes in temperature or pH levels to exposure to light may serve as one of these triggers. Because of this amazing technology, it is possible to create things that are not only static but also have the ability to change and adapt to their surroundings. ChNPs may be added to these systems to allow us to tailor medication release to the requirements of specific patients. This implies that by adjusting medication release kinetics and doses to each patient's unique needs, we may maximize the effectiveness of therapy. Furthermore, 3D printing has shown itself to be a useful technique for creating organoids and tissue constructions for drug screening and in vitro testing. We may be able to help with tissue regeneration or repair damaged tissues by combining 3D printing with ChNPs that contain therapeutic genes. The future of medicine might be bright thanks to this. The versatility of ChNPs is one of its many wonderful features. A variety of compounds, including targeted ligands, imaging agents, and immune-modulating agents, may be used to functionalize them. This enables us to further improve their capabilities. We may integrate many treatment approaches onto a single platform by integrating these functionalized nanoparticles into 3D-printed structures. This implies that gene therapy, immunotherapy, and targeted treatment may all be combined into a single application. All things considered, combining ChNPs with 4D and 5D printing methods opens up a world of fascinating possibilities for customized treatment. In addition to enabling tissue regeneration and integrating various treatment approaches, it enables us to design systems that react to certain stimuli and personalize medication release. In several facets of cancer gene therapy, artificial intelligence (AI) is essential. Helping to identify particular biological targets and provide individualized gene treatments for each patient is one of its main accomplishments. AI systems are able to identify distinct molecular signals in cancer cells that might be the focus of treatment by examining vast amounts of genomic and proteomic data. By specifically treating the genetic defects that cause cancer to develop and spread, this individualized approach has the potential to significantly improve treatment results. Additionally, AI may increase the effectiveness of ChNP-based gene delivery. These nanoparticles provide a number of benefits for gene transfer, including ease of manipulation, biocompatibility, and biodegradability. Regarding their capacity to encapsulate and transport therapeutic genes, there is still opportunity for improvement. AI can help with the creation of nanoparticles with the best physicochemical characteristics in this situation. AI may assist in maximizing the potential of ChNPs in efficiently delivering genes by improving elements including cellular absorption, toxicity reduction, and gene transfection efficiency. In order to create efficient nanoparticle designs for delivering therapeutic genes, researchers rely heavily on AI algorithms. By suggesting potential designs with a better chance of success, these algorithms may help researchers save time and money. AI's capacity to evaluate and comprehend enormous volumes of data from preclinical and clinical research is another useful feature. AI computers may find hidden patterns and connections that human researchers might not see right away by looking at gene expression profiles, medication response data, and patient outcomes. Finding possible biomarkers, forecasting treatment outcomes, and refining therapy approaches may all be substantially aided by this. In summary, AI helps us achieve important medical discoveries by bringing a priceless human-like knowledge to the intricate field of biomedical research. In order to guarantee the success of using AI in cancer gene therapy, particularly via the use of ChNPs, a number of obstacles must be addressed. These difficulties include a number of topics, including securing trustworthy and accurate data, encouraging cooperation between medical researchers from diverse specialties and AI specialists, and managing the ethical and legal ramifications that come up when using AI to treatment decision-making. In conclusion, it is impossible to ignore the amazing promise of integrating AI with ChNPs in cancer gene therapy. Researchers can improve gene delivery techniques, customize therapy regimens for each patient, and eventually improve treatment outcomes thanks to AI's analytical skills. Cancer therapy guided by AI and ChNPs may soon become a standard approach to cancer treatment as more research and development are conducted. In conclusion, there has been a lot of promise recently for the creation of Ch-based smart stimuli-sensitive nanoparticles for gene delivery, especially in cancer treatment. This novel method makes use of Ch's special qualities, which include low toxicity, biocompatibility, and biodegradability, which make it the perfect option for targeted gene delivery. These nanoparticles are unique in that they may react to certain stimuli and release genetic material in a regulated way. This reduces any unforeseen adverse effects and guarantees that the treatment is more successful. Additionally, new developments in ChNP-based cancer treatment have shown promising outcomes. Notable results include less systemic toxicity, enhanced transfection efficiency, and better tumor targeting. By providing tailored and targeted medicines, this innovative technology has the potential to completely transform the way cancer is treated. Nonetheless, further study is necessary to maximize these nanoparticles' functioning and design. It's also critical to get over present obstacles and convert these developments into practical uses. All things considered, the development of Ch-based smart stimuli-sensitive nanoparticles for gene delivery offers a potential avenue for next cancer treatments. It raises the prospect of more accurate and successful therapies.

References

1. S.M. Naghib, B. Ahmadi, B.M. Kangarshahi, M. Mozafari, Chitosan-based smart stimuli-responsive nanoparticles for gene delivery and gene therapy: recent progresses on cancer therapy. Int. J. Biol. Macromol. 134542 (2024)
2. O. Abazari, P. Dayati, M. Shahidi, J. ZavarReza, M. Rahmanian, S.M. Naghib, Exploring a desirable quadr-mRNAs panel for non-invasive and ultrasensitive bladder cancer diagnosis: in-silico and clinical studies. Curr. Med. Chem. (2024). https://doi.org/10.2174/0109298673302140240627053904
3. A. Matini, S.M. Naghib, M.R. Mozafari, Quantum dots in cancer theranostics: a thorough review of recent advancements in bioimaging, tracking, and therapy across various cancer types. Curr. Pharm. Biotechnol. (2024). https://doi.org/10.2174/0113892010294163240407153842
4. M. Sadeghi, S. Kashanian, S.M. Naghib, E. Arkan, A high-performance electrochemical aptasensor based on graphene-decorated rhodium nanoparticles to detect HER2-ECD oncomarker in liquid biopsy. Sci. Rep. **12**(1), 3299 (2022). https://doi.org/10.1038/s41598-022-07230-3
5. S.M. Naghib, B. Ahmadi, M.R. Mozafari, Smart physicochemical-triggered chitosan-based nanogels for siRNA delivery and gene therapy: a focus on emerging strategies and paradigms for cancer therapy. Curr. Med. Chem. (2024). https://doi.org/10.2174/0109298673286052240426044945

6. F. Yaghoubi, S.M. Naghib, N.S.H. Motlagh, F. Haghiralsadat, H.Z. Jaliani, D. Tofighi, A. Moradi, Multiresponsive carboxylated graphene oxide-grafted aptamer as a multifunctional nanocarrier for targeted delivery of chemotherapeutics and bioactive compounds in cancer therapy. Nanotechnol. Rev. **10**(1), 1838–1852 (2021)
7. M. Yazdan, S.M. Naghib, M.R. Mozafari, Liposomal nano-based drug delivery systems for breast cancer therapy: recent advances and progresses. Anti-Cancer Agents Med. Chem. (Formerly Curr. Med. Chem. Anti-Cancer Agents) **24**(12), 896–915 (2024). https://doi.org/10.2174/0118715206293653240322041047
8. A.-A. Zare, H. Naderi-Manesh, S.M. Naghib, M. Shamsipur, F. Molaabasi, Label-free electrochemical cancer cell detection leveraging hemoglobin-encapsulated silver nanoclusters and Cu-MOF nanohybrids on a graphene-assisted dual-modal probe. Sci. Rep. **13**(1), 21980 (2023). https://doi.org/10.1038/s41598-023-49418-1
9. S.M. Naghib, B. Ahmadi, M. Mozafari, Stimuli-sensitive chitosan-based nanosystems-immobilized nucleic acids for gene therapy in breast cancer and hepatocellular carcinoma. Curr. Top. Med. Chem. **24**(17), 1464–1489 (2024)
10. X.M. Anguela, K.A. High, Entering the modern era of gene therapy. Annu. Rev. Med. **70**(1), 273–288 (2019)
11. T. Jafari, S.M. Naghib, M. Mozafari, Bio-printing of materials for bone tissue engineering. Curr. Org. Chem. **29**(1), 19–40 (2025)
12. M.S. Huh, E.J. Lee, H. Koo, J.Y. Yhee, K.S. Oh, S. Son, S. Lee, S.H. Kim, I.C. Kwon, K. Kim, Polysaccharide-based nanoparticles for gene delivery, in *Polymeric Gene Delivery Systems* (2018), pp. 65–83
13. S.A. Bhawani, Z. Karim, M. Jawaid, *Polysaccharide-Based Nanocomposites for Gene Delivery and Tissue Engineering* (Woodhead Publishing, 2021)
14. B. Santos-Carballal, E. Fernández Fernández, F.M. Goycoolea, Chitosan in non-viral gene delivery: role of structure, characterization methods, and insights in cancer and rare diseases therapies. Polymers **10**(4), 444 (2018)
15. S.D. Yang, W.J. Zhu, Q.L. Zhu, W.L. Chen, Z.X. Ren, F. Li, Z.Q. Yuan, J.Z. Li, Y. Liu, X.F. Zhou, Binary-copolymer system base on low-density lipoprotein-coupled N-succinyl chitosan lipoic acid micelles for co-delivery MDR1 siRNA and paclitaxel, enhances antitumor effects via reducing drug. J. Biomed. Mater. Res. B Appl. Biomater. **105**(5), 1114–1125 (2017)
16. Y. Jiang, C. Yan, M. Li, S. Chen, Z. Chen, L. Yang, K. Luo, Delivery of natural products via polysaccharide-based nanocarriers for cancer therapy: a review on recent advances and future challenges. Int. J. Biol. Macromol. 135072 (2024)
17. S.M. Naghib, M. Zarrineh, M.R. Mozafari, 3D Printing chitosan-based nanobiomaterials for biomedicine and drug delivery: recent advances on the promising bioactive agents and technologies. Curr. Org. Chem. **28**(7), 510–525 (2024)
18. G.K. Kashani, S.M. Naghib, S. Soleymani, M.R. Mozafari, A review of DNA nanoparticles-encapsulated drug/gene/protein for advanced controlled drug release: current status and future perspective over emerging therapy approaches. Int. J. Biol. Macromol. **268**, 131694 (2024). https://doi.org/10.1016/j.ijbiomac.2024.131694
19. B. Mikaeeli Kangarshahi, N.S. Morteza, N. and Rabiee, DNA/RNA-based electrochemical nanobiosensors for early detection of cancers. Crit. Rev. Clin. Lab. Sci. **61**(6), 473–495 (2024). https://doi.org/10.1080/10408363.2024.2321202
20. P. Vosoughi, S.M. Naghib, M. Rahmanian, M.R. Mozafari, A comprehensive review of mRNA-based vaccines for COVID-19, a new era in pharmaceuticals: unspecified and unknown aspects, effects and challenges. Curr. Top. Med. Chem. (2025). https://doi.org/10.2174/0115680266325847241121034100
21. P. Vosoughi, S.M. Naghib, B.M. Kangarshahi, M.R. Mozafari, A review of RNA nanoparticles for drug/gene/protein delivery in advanced therapies: current state and future prospects. Int. J. Biol Macromol. **295**, 139532 (2025). https://doi.org/10.1016/j.ijbiomac.2025.139532
22. C. Hidai, H. Kitano, Nonviral gene therapy for cancer: a review. Diseases **6**(3), 57 (2018)
23. D. Pan, H. Büning, C. Ling, Rational design of gene therapy vectors. Mol. Ther. Methods Clin. Dev. **12**, 246–247 (2019)

24. M.Z. Ahmad, M. Rizwanullah, J. Ahmad, M.Y. Alasmary, M.H. Akhter, B.A. Abdel-Wahab, M.H. Warsi, A. Haque, Progress in nanomedicine-based drug delivery in designing of chitosan nanoparticles for cancer therapy. Int. J. Polym. Mater. Polym. Biomater. **71**(8), 602–623 (2022)
25. U. Piotrowska, K. Orzechowska, Advances in chitosan-based smart hydrogels for colorectal cancer treatment. Pharmaceuticals **17**(10), 1260 (2024)
26. M.H. Butt, M. Zaman, A. Ahmad, R. Khan, T.H. Mallhi, M.M. Hasan, Y.H. Khan, S. Hafeez, E.E.S. Massoud, M.H. Rahman, Appraisal for the potential of viral and nonviral vectors in gene therapy: a review. Genes **13**(8), 1370 (2022)
27. B. Sachdeva, P. Sachdeva, A. Negi, S. Ghosh, S. Han, S. Dewanjee, S.K. Jha, R. Bhaskar, J.K. Sinha, A.C. Paiva-Santos, Chitosan nanoparticles-based cancer drug delivery: application and challenges. Mar. Drugs **21**(4), 211 (2023)
28. A. Moharana, A. Tiwari, S. Perada, S. Yadav, O.P. Kumar, Polysaccharide-based biomaterials for gene delivery, in *Integration of Biomaterials for Gene Therapy* (2023), pp. 47–75
29. M. Dattilo, F. Patitucci, S. Prete, O.I. Parisi, F. Puoci, Polysaccharide-based hydrogels and their application as drug delivery systems in cancer treatment: a review. J. Funct. Biomater. **14**(2), 55 (2023)
30. P. Vosoughi, S.M. Naghib, T. Jafari, B.M. Kangarshahi, Chitosan-encapsulated lipid-based nanovesicles for therapeutic applications and tissue engineering: a review. Carbohydr. Polym. Technol. Appl. 100805 (2025)
31. D. Chicea, A. Nicolae-Maranciuc, A review of chitosan-based materials for biomedical, food, and water treatment applications. Materials **17**(23), 5770 (2024)
32. T. Jafari, S.M. Naghib, M. Mozafari, Chitosan-based nano/biomaterials in bone tissue engineering and regenerative medicine: recent progress and advances. Curr. Org. Synth. **22**(4), 457–480 (2025)
33. M. Faucher, V. Perreault, S. Gaaloul, L. Bazinet, Defatting of sweet whey by electrodialysis with bipolar membranes: effect of protein concentration factor. Sep. Purif. Technol. **251**, 117248 (2020)
34. Q. Li, E. Dunn, E. Grandmaison, M.F. Goosen, *Applications and properties of chitosan* (CRC Press, Applications of Chitan and Chitosan, 2020), pp.3–29
35. H. Choukaife, S. Seyam, B. Alallam, A.A. Doolaanea, M. Alfatama, Current advances in chitosan nanoparticles based oral drug delivery for colorectal cancer treatment. Int. J. Nanomed. 3933–3966 (2022)
36. A. Valentino, S. Yazdanpanah, R. Conte, A. Calarco, G. Peluso, Smart nanocomposite hydrogels as next-generation therapeutic and diagnostic solutions. Gels **10**(11), 689 (2024)
37. M. Ashrafizadeh, K. Hushmandi, S. Mirzaei, S. Bokaie, A. Bigham, P. Makvandi, N. Rabiee, V.K. Thakur, A.P. Kumar, E. Sharifi, Chitosan-based nanoscale systems for doxorubicin delivery: exploring biomedical application in cancer therapy. Bioeng. Transl. Med. **8**(1), e10325 (2023)
38. Z. Gharehdaghi, S.M. Naghib, R. Rahimi, A. Bakhshi, A. Kefayat, A. Shamaeizadeh, F. Molaabasi, Highly improved pH-responsive anticancer drug delivery and T2-weighted MRI imaging by magnetic MOF CuBTC-based nano/microcomposite. Front. Mol. Biosci. **10**, 1071376 (2023)
39. S. Gooneh-Farahani, S.M. Naghib, M.R. Naimi-Jamal, A. Seyfoori, A pH-sensitive nanocarrier based on BSA-stabilized graphene-chitosan nanocomposite for sustained and prolonged release of anticancer agents. Sci. Rep. **11**(1), 17404 (2021). https://doi.org/10.1038/s41598-021-97081-1
40. S. Soleymani, S.M. Naghib, M.R. Mozafari, PH/Temperature/redox and light-responsive polymersome structure and application in cancer therapy: smart drug delivery and targeted drug release. Curr. Med. Chem. (2024). https://doi.org/10.2174/0109298673303351240507113603
41. Q. Wu, Y. Hu, B. Yu, H. Hu, F.-J. Xu, Polysaccharide-based tumor microenvironment-responsive drug delivery systems for cancer therapy. J. Control. Release **362**, 19–43 (2023)
42. A.K. Lunawat, S. Thakur, B.D. Kurmi, G.D. Gupta, P. Patel, S. Raikwar, Revolutionizing cancer treatment: the role of chitosan nanoparticles in therapeutic advancements. J. Drug Deliv. Sci. Technol. 105661 (2024)

43. M. Yazdan, S.M. Naghib, Smart ultrasound-responsive polymers for drug delivery: an overview on advanced stimuli-sensitive materials and techniques. Curr. Drug Deliv. (2025)
44. M. Mu, X. Li, A. Tong, G. Guo, Multi-functional chitosan-based smart hydrogels mediated biomedical application. Expert Opin. Drug Deliv. **16**(3), 239–250 (2019)
45. F. Damiri, A. Fatimi, A.C.P. Santos, R.S. Varma, M. Berrada, Smart stimuli-responsive polysaccharide nanohydrogels for drug delivery: a review. J. Mater. Chem. B **11**(44), 10538–10565 (2023)
46. K. Osouli-Bostanabad, T. Masalehdan, R.M. Kapsa, A. Quigley, A. Lalatsa, K.F. Bruggeman, S.J. Franks, R.J. Williams, D.R. Nisbet, Traction of 3D and 4D printing in the healthcare industry: from drug delivery and analysis to regenerative medicine. ACS Biomater. Sci. Eng. **8**(7), 2764–2797 (2022)
47. S. Gooneh-Farahani, M.R. Naimi-Jamal, S.M. Naghib, Stimuli-responsive graphene-incorporated multifunctional chitosan for drug delivery applications: a review. Expert Opin. Drug Deliv. **16**(1), 79–99 (2019). https://doi.org/10.1080/17425247.2019.1556257
48. S.N. Hosseini, S.M. Naghib, G.K. Kashani, M.R. Mozafari, Chitosan-grafted graphene materials for drug delivery in wound healing. Curr. Pharm. Des. **31**(9), 691–715 (2025). https://doi.org/10.2174/0113816128333493241014134711
49. M. Shahidi, O. Abazari, P. Dayati, A. Bakhshi, A. Rasti, F. Haghiralsadat, S.M. Naghib, D. Tofighi, Aptamer-functionalized chitosan-coated gold nanoparticle complex as a suitable targeted drug carrier for improved breast cancer treatment. Nanotechnol. Rev. **11**(1), 2875–2890 (2022)
50. S. Yonezawa, H. Koide, T. Asai, Recent advances in siRNA delivery mediated by lipid-based nanoparticles. Adv. Drug Deliv. Rev. **154**, 64–78 (2020)
51. C. Li, X. Han, Melanoma cancer immunotherapy using PD-L1 siRNA and imatinib promotes cancer-immunity cycle. Pharm. Res. **37**, 1–10 (2020)
52. Q. Zhang, H. Zhang, T. Ning, D. Liu, T. Deng, R. Liu, M. Bai, K. Zhu, J. Li, Q. Fan, Exosome-delivered c-Met siRNA could reverse chemoresistance to cisplatin in gastric cancer. Int. J. Nanomed. 2323–2335 (2020)
53. H.-F. Yuan, Y. Li, B.-B. Tan, Q. Zhao, L.-Q. Fan, Z.-J. An, Inhibitory effect of siRNA-Annexin A7 on growth, migration, and invasion in BGC823 cells and gastric cancer xenograftsin nude mice. Int. J. Clin. Exp. Pathol. **13**(2), 122 (2020)
54. A. Beigi, S.M. Naghib, A. Matini, M. Tajabadi, M.R. Mozafari, Lipid-based nanocarriers for targeted gene delivery in lung cancer therapy: exploring a novel therapeutic paradigm. Curr. Gene Ther. **25**(2), 92–112 (2025). https://doi.org/10.2174/0115665232292768240503050508
55. M. Chen, M. Chen, J. He, Cancer cell membrane cloaking nanoparticles for targeted co-delivery of doxorubicin and PD-L1 siRNA. Artif. Cells Nanomed. Biotechnol. **47**(1), 1635–1641 (2019)
56. M. Ashrafizadeh, M. Delfi, F. Hashemi, A. Zabolian, H. Saleki, M. Bagherian, N. Azami, M.V. Farahani, S.O. Sharifzadeh, S. Hamzehlou, Biomedical application of chitosan-based nanoscale delivery systems: potential usefulness in siRNA delivery for cancer therapy. Carbohyd. Polym. **260**, 117809 (2021)
57. A. Petkova, Gene therapy in combination with an approach targeting the tumour microenvironment in glioblastoma, University College London (UCL) (2020)
58. H.O. Othman, E.T. Anwer, D.S. Ali, R.O. Hassan, E.E. Mahmood, R.A. Ahmed, R.F. Muhammad, S. Smaoui, Recent advances in carbon quantum dots for gene delivery: a comprehensive review. J. Cell. Physiol. **239**(11), e31236 (2024)
59. D. Sharma, S. Arora, B. dos Santos Rodrigues, S. Lakkadwala, A. Banerjee, J. Singh, Chitosan-based systems for gene delivery, in *Functional Chitosan: Drug Delivery and Biomedical Applications*, pp. 229–267 (2019)
60. A.K. Abosalha, W. Ahmad, J. Boyajian, P. Islam, M. Ghebretatios, S. Schaly, R. Thareja, K. Arora, S. Prakash, A comprehensive update of siRNA delivery design strategies for targeted and effective gene silencing in gene therapy and other applications. Expert Opin. Drug Discov. **18**(2), 149–161 (2023)

61. S.M. Asiri, F.A. Khan, A. Bozkurt, Synthesis of chitosan nanoparticles, chitosan-bulk, chitosan nanoparticles conjugated with glutaraldehyde with strong anti-cancer proliferative capabilities. Artif. Cells Nanomed Biotechnol. **46**(sup3), 1152–1161 (2018)
62. B. Santos-Carballal, L. Aaldering, M. Ritzefeld, S. Pereira, N. Sewald, B. Moerschbacher, M. Götte, F.M. Goycoolea, Physicochemical and biological characterization of chitosan-microRNA nanocomplexes for gene delivery to MCF-7 breast cancer cells. Sci. Rep. **5**(1), 13567 (2015)
63. Z.H. Mirzaie, S. Irani, R. Mirfakhraie, S.M. Atyabi, M. Dinarvand, R. Dinarvand, R. Varshochian, F. Atyabi, Docetaxel–chitosan nanoparticles for breast cancer treatment: cell viability and gene expression study. Chem. Biol. Drug Des. **88**(6), 850–858 (2016)
64. A. Masjedi, H. Hassannia, F. Atyabi, A. Rastegari, M. Hojjat-Farsangi, A. Namdar, H. Soleimanpour, G. Azizi, A. Nikkhoo, G. Ghalamfarsa, A. Mirshafiey, F. Jadidi-Niaragh, Downregulation of A2AR by siRNA loaded PEG-chitosan-lactate nanoparticles restores the T cell mediated anti-tumor responses through blockage of PKA/CREB signaling pathway. Int. J. Biol. Macromol. **133**, 436–445 (2019). https://doi.org/10.1016/j.ijbiomac.2019.03.223
65. Q. Hu, Y. Li, H. Chen, H. Liao, Y. He, Q. Zheng, Retracted: CCDC88A post-transcriptionally regulates VEGF via miR-101 and subsequently regulates hepatocellular carcinoma. Front. Immunol. **13**, 859331 (2022)
66. N. Liu, C.W. Chang, C.J. Steer, X.W. Wang, G. Song, MicroRNA-15a/16-1 prevents hepatocellular carcinoma by disrupting the communication between Kupffer cells and regulatory T cells. Gastroenterology **162**(2), 575–589 (2022)
67. C. Bie, Y. Chen, H. Tang, Q. Li, L. Zhong, X. Peng, Y. Shi, J. Lin, J. Lai, S. Wu, Insulin-like growth factor 1 receptor drives hepatocellular carcinoma growth and invasion by activating Stat3-midkine-Stat3 loop. Dig. Dis. Sci. 1–16 (2022)
68. M. Cheng, Q. Li, T. Wan, X. Hong, H. Chen, B. He, Z. Cheng, H. Xu, T. Ye, B. Zha, Synthesis and efficient hepatocyte targeting of galactosylated chitosan as a gene carrier in vitro and in vivo. J. Biomed. Mater. Res. B Appl. Biomater. **99**(1), 70–80 (2011)
69. K. Karimi, S. Mojtabavi, P.M. Tehrany, M.M. Nejad, A. Rezaee, S. Mohtashamian, E. Hamedi, F. Yousefi, F. Salmani, M.A. Zandieh, Chitosan-based nanoscale delivery systems in hepatocellular carcinoma: versatile bio-platform with theranostic application. Int. J. Biol. Macromol. **242**, 124935 (2023)
70. Z. Hu, J. Chen, S. Zhou, N. Yang, S. Duan, Z. Zhang, J. Su, J. He, Z. Zhang, X. Lu, Mouse IP-10 gene delivered by folate-modified chitosan nanoparticles and dendritic/tumor cells fusion vaccine effectively inhibit the growth of hepatocellular carcinoma in mice. Theranostics **7**(7), 1942 (2017)
71. S.C. Reghupaty, D. Sarkar, Current status of gene therapy in hepatocellular carcinoma. Cancers **11**(9), 1265 (2019)
72. S. Duan, M. Song, J. He, N. Zhou, S. Zhou, J. Zhao, Y. Fang, Y. Peng, X. Huang, G. Luo, Folate-modified chitosan nanoparticles coated interferon-inducible protein-10 gene enhance cytotoxic T lymphocytes' responses to hepatocellular carcinoma. J. Biomed. Nanotechnol. **12**(4), 700–709 (2016)
73. S.M. Naghib, B. Ahmadi, M.R. Mozafari, Stimuli-sensitive chitosan-based nanosystems-immobilized nucleic acids for gene therapy in breast cancer and hepatocellular carcinoma. Curr. Top. Med. Chem. **24**(17), 1464–1489 (2024). https://doi.org/10.2174/0115680266293173240506054439
74. O.-J. Kwon, E. Kang, J.-W. Choi, S.W. Kim, C.-O. Yun, Therapeutic targeting of chitosan–PEG–folate-complexed oncolytic adenovirus for active and systemic cancer gene therapy. J. Control. Release **169**(3), 257–265 (2013)
75. B.-F. Zhang, L. Xing, P.-F. Cui, F.-Z. Wang, R.-L. Xie, J.-L. Zhang, M. Zhang, Y.-J. He, J.-Y. Lyu, J.-B. Qiao, Mitochondria apoptosis pathway synergistically activated by hierarchical targeted nanoparticles co-delivering siRNA and lonidamine. Biomaterials **61**, 178–189 (2015)
76. M. Garg, M.K. Shanmugam, V. Bhardwaj, A. Goel, R. Gupta, A. Sharma, P. Baligar, A.P. Kumar, B.C. Goh, L. Wang, The pleiotropic role of transcription factor STAT3 in oncogenesis and its targeting through natural products for cancer prevention and therapy. Med. Res. Rev. **41**(3), 1291–1336 (2021)

77. A. Nikkhoo, N. Rostami, S. Farhadi, M. Esmaily, S.M. Ardebili, F. Atyabi, M. Baghaei, N. Haghnavaz, M. Yousefi, M.R. Aliparasti, Codelivery of STAT3 siRNA and BV6 by carboxymethyl dextran trimethyl chitosan nanoparticles suppresses cancer cell progression. Int. J. Pharm. **581**, 119236 (2020)
78. L. Croce, F. Coperchini, F. Magri, L. Chiovato, M. Rotondi, The multifaceted anti-cancer effects of BRAF-inhibitors. Oncotarget **10**(61), 6623 (2019)
79. S.M. Taghdisi, N.M. Danesh, M.A. Nameghi, A. Bahreyni, M. Ramezani, M. Alibolandi, A.S. Emrani, K. Abnous, Co-delivery of doxorubicin and α-PCNA aptamer using AS1411-modified pH-responsive nanoparticles for cancer synergistic therapy. J. Drug Deliv. Sci. Technol. **58**, 101816 (2020)
80. W. Chen, Z. Liu, W. Mai, Y. Xiao, X. You, L. Qin, FZD8 indicates a poor prognosis and promotes gastric cancer invasion and metastasis via B-catenin signaling pathway. Ann. Clin. Lab. Sci. **50**(1), 13–23 (2020)
81. V. Antoniou, E.A. Mourelatou, E. Galatou, K. Avgoustakis, S. Hatziantoniou, Gene therapy with chitosan nanoparticles: modern formulation strategies for enhancing cancer cell transfection. Pharmaceutics **16**(7), 868 (2024)
82. S. Yu, Y. Chen, X. Li, Z. Gao, G. Liu, Chitosan nanoparticle-delivered siRNA reduces CXCR4 expression and sensitizes breast cancer cells to cisplatin. Biosci. Rep. **37**(3), BSR20170122 (2017)
83. J.Y. Yhee, S. Song, S.J. Lee, S.-G. Park, K.-S. Kim, M.G. Kim, S. Son, H. Koo, I.C. Kwon, J.H. Jeong, S.Y. Jeong, S.H. Kim, K. Kim, Cancer-targeted MDR-1 siRNA delivery using self-cross-linked glycol chitosan nanoparticles to overcome drug resistance. J. Control. Release **198**, 1–9 (2015). https://doi.org/10.1016/j.jconrel.2014.11.019
84. T. Wang, K. Rahimizadeh, R.N. Veedu, Development of a novel DNA oligonucleotide targeting low-density lipoprotein receptor. Mol. Ther. Nucleic Acids **19**, 190–198 (2020)
85. A.V. Nascimento, F. Gattacceca, A. Singh, H. Bousbaa, D. Ferreira, B. Sarmento, M.M. Amiji, Biodistribution and pharmacokinetics of Mad2 siRNA-loaded EGFR-targeted chitosan nanoparticles in cisplatin sensitive and resistant lung cancer models. Nanomedicine **11**(7), 767–781 (2016)
86. A. Babu, Q. Wang, R. Muralidharan, M. Shanker, A. Munshi, R. Ramesh, Chitosan coated polylactic acid nanoparticle-mediated combinatorial delivery of cisplatin and siRNA/Plasmid DNA chemosensitizes cisplatin-resistant human ovarian cancer cells. Mol. Pharm. **11**(8), 2720–2733 (2014)
87. H. Peng, L. Qiao, G. Shan, M. Gao, R. Zhang, X. Yi, X. He, Stepwise responsive carboxymethyl chitosan-based nanoplatform for effective drug-resistant breast cancer suppression. Carbohyd. Polym. **291**, 119554 (2022)
88. N.A. Abtahi, S.M. Naghib, F. Haghiralsadat, M. Akbari Edgahi, Development of highly efficient niosomal systems for co-delivery of drugs and genes to treat breast cancer in vitro and in vivo. Cancer Nanotechnol. **13**(1), 28 (2022). https://doi.org/10.1186/s12645-022-00135-w
89. N.A. Abtahi, S.M. Naghib, F. Haghiralsadat, M. Akbari Edgahi, E. Askari, A comparative study on biopharmaceutical function of curcumin and miR-34a by multistimuli-responsive nanoniosome carrier: in-vitro and in-vivo. Front. Mol. Biosci. **9** (2022). https://doi.org/10.3389/fmolb.2022.1043277
90. N.A. Abtahi, S.M. Naghib, F. Haghiralsadat, J.Z. Reza, F. Hakimian, F. Yazdian, D. Tofighi, Smart stimuli-responsive biofunctionalized niosomal nanocarriers for programmed release of bioactive compounds into cancer cells in vitro and in vivo. Nanotechnol. Rev. **10**(1), 1895–1911 (2021)
91. N.A. Abtahi, S. Salehi, S.M. Naghib, F. Haghiralsadat, M.A. Edgahi, S. Ghorbanzadeh, W. Zhang, Multi-sensitive functionalized niosomal nanocarriers for controllable gene delivery in vitro and in vivo. Cancer Nanotechnol. **14**(1), 22 (2023). https://doi.org/10.1186/s12645-023-00175-w
92. H.R. Garshasbi, S.M. Naghib, Smart stimuli-responsive alginate nanogels for drug delivery systems and cancer therapy: a review. Curr. Pharm. Des. **29**(44), 3546–3562 (2023). https://doi.org/10.2174/0113816128283806231211073031

93. B. Wu, Y. Yuan, X. Han, Q. Wang, H. Shang, X. Liang, H. Jing, W. Cheng, Structure of LINC00511-siRNA-conjugated nanobubbles and improvement of cisplatin sensitivity on triple negative breast cancer. FASEB J. **34**(7), 9713–9726 (2020)
94. Y. Zhang, Y. Yuan, Y. Zhang, L. Cheng, X. Zhou, K. Chen, SNHG7 accelerates cell migration and invasion through regulating miR-34a-Snail-EMT axis in gastric cancer. Cell Cycle **19**(1), 142–152 (2020)
95. U. Krafft, S. Tschirdewahn, J. Hess, N.N. Harke, B. Hadaschik, C. Olah, S. Krege, P. Nyirády, A. Szendröi, M. Szücs, Validation of survivin and HMGA2 as biomarkers for cisplatin resistance in bladder cancer. Urol. Oncol. Semin. Original Invest. 810.e7–810.e15 (2019)
96. S. Izadi, A. Moslehi, H. Kheiry, F. Karoon Kiani, A. Ahmadi, A. Masjedi, S. Ghani, B. Rafiee, V. Karpisheh, F. Hajizadeh, Codelivery of HIF-1α siRNA and dinaciclib by carboxylated graphene oxide-trimethyl chitosan-hyaluronate nanoparticles significantly suppresses cancer cell progression. Pharm. Res. **37**, 1–20 (2020)
97. S.M. Naghib, S. Amiri, M.R. Mozafari, Stimuli-responsive chitosan-based nanocarriers for drug delivery in wound dressing applications: a review. Carbohydr. Polym. Technol. Appl. **7**, 100497 (2024). https://doi.org/10.1016/j.carpta.2024.100497
98. S.M. Naghib, S.N. Hosseini, A. Beigi, 3D 4D printing of chitosan-based scaffolds for wound dressing applications. Carbohydr. Polym. Technol. Appl. **8**, 100594 (2024). https://doi.org/10.1016/j.carpta.2024.100594
99. J. Saberi, M. Ansari, B. Ebrahimi Hoseinzadeh, S.S. Kordestani, S.M. Naghib, Chitosan-polyacrylic acid hybrid nanoparticles as novel tissue adhesive: synthesis and characterization. Fibers Polym. **19**(12), 2458–2464 (2018). https://doi.org/10.1007/s12221-018-8762-2
100. O. Sartipzadeh, S.M. Naghib, F. Haghiralsadat, F. Shokati, M. Rahmanian, Microfluidic-assisted synthesis and modeling of stimuli-responsive monodispersed chitosan microgels for drug delivery applications. Sci. Rep. **12**(1), 8382 (2022). https://doi.org/10.1038/s41598-022-12031-9
101. K. Khodaverdi, S.M. Naghib, M.R. Mozafari, M. Rahmanian, Chitosan/hydroxyapatite hydrogels for localized drug delivery and tissue engineering: a review. Carbohydr. Polym. Technol. Appl. **9**, 100640 (2025). https://doi.org/10.1016/j.carpta.2024.100640
102. M.A. Khorasani, S.M. Naghib, A review of chitosan-based multifunctional nanocomposites for drug/gene/protein delivery and gene therapy in cancer treatments: promises, challenges and outlooks. Int. J. Biol. Macromol. **306**, 141394 (2025). https://doi.org/10.1016/j.ijbiomac.2025.141394
103. N.F. Vahid, M.R. Marvi, M.R. Naimi-Jamal, S.M. Naghib, A. Ghaffarinejad, X-Fe_2O_4-buckypaper-chitosan nanocomposites for nonenzymatic electrochemical glucose biosensing. Anal. Bioanal. Electrochem **11**(2019), 930–942 (2019)
104. S. Fu, S. Liang, D. Jiang, R. Yang, Z. Zhang, L. Chang, X. Zhang, Y. Liu, N. Zhang, Gas-blasting nanocapsules to accelerate carboplatin lysosome release and nucleus delivery for prostate cancer treatment. Asian J. Pharm. Sci. **16**(2), 192–202 (2021)
105. S. Gooneh-Farahani, S.M. Naghib, M.R. Naimi-Jamal, A novel and inexpensive method based on modified ionic gelation for pH-responsive controlled drug release of homogeneously distributed chitosan nanoparticles with a high encapsulation efficiency. Fibers Polym. **21**(9), 1917–1926 (2020). https://doi.org/10.1007/s12221-020-1095-y
106. M. Saraswathy, S. Gong, Recent developments in the co-delivery of siRNA and small molecule anticancer drugs for cancer treatment. Mater. Today **17**(6), 298–306 (2014)
107. Y. Li, J. Yang, B. Xu, F. Gao, W. Wang, W. Liu, Enhanced therapeutic siRNA to tumor cells by a pH-sensitive agmatine–chitosan bioconjugate. ACS Appl. Mater. Interfaces **7**(15), 8114–8124 (2015)
108. M. Wang, H. Hu, Y. Sun, L. Qiu, J. Zhang, G. Guan, X. Zhao, M. Qiao, L. Cheng, L. Cheng, A pH-sensitive gene delivery system based on folic acid-PEG-chitosan–PAMAM-plasmid DNA complexes for cancer cell targeting. Biomaterials **34**(38), 10120–10132 (2013)
109. L. Han, C. Tang, C. Yin, Dual-targeting and pH/redox-responsive multi-layered nanocomplexes for smart co-delivery of doxorubicin and siRNA. Biomaterials **60**, 42–52 (2015)

110. H.-C. Wu, W.-T. Kuo, Redox/pH-responsive 2-in-1 chimeric nanoparticles for the co-delivery of doxorubicin and siRNA. Polymers **13**(24), 4362 (2021)
111. T. Meng, B. Lu, S. Shao, M. Yuan, X. Liu, H. Yuan, X. Huang, F. Hu, Sequential therapy with redox-responsive glucolipid nanocarrier separately delivering siRNA and doxorubicin to overcome multidrug resistance. Int. J. Pharm. **534**(1–2), 368–377 (2017)
112. M. Argenziano, F. Bessone, C. Dianzani, M.A. Cucci, M. Grattarola, S. Pizzimenti, R. Cavalli, Ultrasound-responsive Nrf2-targeting siRNA-loaded nanobubbles for enhancing the treatment of melanoma. Pharmaceutics **14**(2), 341 (2022)
113. L. Guo, D. Shi, M. Shang, X. Sun, D. Meng, X. Liu, X. Zhou, J. Li, Utilizing RNA nanotechnology to construct negatively charged and ultrasound-responsive nanodroplets for targeted delivery of siRNA. Drug Deliv. **29**(1), 316–327 (2022)
114. X. Liu, D. Shi, L. Guo, X. Zhou, M. Shang, X. Sun, D. Meng, Y. Zhao, J. Li, Echogenic, ultrasound-sensitive chitosan nanodroplets for spatiotemporally controlled DKK-2 gene delivery to prostate cancer cells. Int. J. Nanomed. 421–432 (2021)
115. S. Fredenberg, M. Wahlgren, M. Reslow, A. Axelsson, The mechanisms of drug release in poly (lactic-co-glycolic acid)-based drug delivery systems—a review. Int. J. Pharm. **415**(1–2), 34–52 (2011)
116. Y. Yang, H. Zhao, Y. Jia, Q. Guo, Y. Qu, J. Su, X. Lu, Y. Zhao, Z. Qian, A novel gene delivery composite system based on biodegradable folate-poly (ester amine) polymer and thermosensitive hydrogel for sustained gene release. Sci. Rep. **6**(1), 21402 (2016)
117. Z. Mazidi, S. Javanmardi, S.M. Naghib, Z. Mohammadpour, Smart stimuli-responsive implantable drug delivery systems for programmed and on-demand cancer treatment: An overview on the emerging materials. Chem. Eng. J. **433**, 134569 (2022). https://doi.org/10.1016/j.cej.2022.134569
118. D. Steinhauff, H. Ghandehari, Matrix mediated viral gene delivery: a review. Bioconjug. Chem. **30**(2), 384–399 (2019)
119. R.A. Keraliya, C. Patel, P. Patel, V. Keraliya, T.G. Soni, R.C. Patel, M. Patel, Osmotic drug delivery system as a part of modified release dosage form. Int. Sch. Res. Not. **2012**(1), 528079 (2012)
120. N. Kamaly, B. Yameen, J. Wu, O.C. Farokhzad, Degradable controlled-release polymers and polymeric nanoparticles: mechanisms of controlling drug release. Chem. Rev. **116**(4), 2602–2663 (2016)
121. W. Ahmad, V.S.S.C.S. Veeraghantla, A. Byrne, Advancing sustainable concrete using biochar: experimental and modelling study for mechanical strength evaluation. Sustainability **17**(6), 2516 (2025)

MIX
Papier aus verantwortungsvollen Quellen
Paper from responsible sources
FSC® C105338

If you have any concerns about our products,
you can contact us on
ProductSafety@springernature.com

In case Publisher is established outside the EU,
the EU authorized representative is:
Springer Nature Customer Service Center GmbH
Europaplatz 3, 69115 Heidelberg, Germany

Printed by Libri Plureos GmbH
in Hamburg, Germany